Space Flight

Space Flight

History, Technology, and Operations

LANCE K. ERICKSON

GOVERNMENT INSTITUTES
An imprint of
THE SCARECROW PRESS, INC.
Lanham, Maryland • Toronto • Plymouth, UK
2010

Published in the United States of America
by Government Institutes, an imprint of The Scarecrow Press, Inc.
A wholly owned subsidiary of
The Rowman & Littlefield Publishing Group, Inc.
4501 Forbes Boulevard, Suite 200
Lanham, Maryland 20706
http://www.govinstpress.com/

Estover Road
Plymouth PL6 7PY
United Kingdom

British Library Cataloguing in Publication Information Available

Library of Congress Cataloging-in-Publication Data

Erickson, Lance K., 1946-
  Space flight : history, technology, and operations / Lance K. Erickson.
     p. cm.
   Includes index.
     ISBN 978-0-86587-419-0 (cloth : alk. paper) – ISBN 978-1-60590-685-0 (pbk. : alk. paper) – ISBN 978-1-60590-684-3 (electronic)
     1. Astronautics–History. I. Title.
   TL788.5.E68 2010
   629.4'1–dc22   2010008262

Contents

Preface

The rigors of space flight are easy to spot in any undergraduate text book on space-craft engineering and space flight dynamics. At the other extreme are descriptive works that cover space flight applications and exploration operations, but rarely touch on the basic mathematics and operational principles of spacecraft systems and instruments. Nearly all of these books and those in between fail to discuss the development process and the historical foundation of even the most important space exploration programs. My objective in writing this book was to provide a general review using a historical format to weave together the spacecraft, the exploration programs, and the science gleaned from those programs. Several chapters also furnish sketches of our current knowledge on the solar system and the universe to underscore the importance of space exploration in understanding the world around us, and the influence these programs have on virtually every aspect of our lives.

One of the unfortunate omissions in most books on space flight and space technology is the influence that military weapons projects and military space programs had on the early space race. Spurred by the Cold War, the development of long-range military launchers and surveillance satellites were intimately linked to the first space launchers and the first manned missions. These early military and civil space projects that are still close relatives changed the direction of science, and together opened the door to human exploration of space, which had long been the dream of early space pioneers. A chapter on military space hardware and programs is included because of their historical importance, but also to emphasize the connection between the civil and military projects from the beginning. It is not by accident that the same designers responsible for the military long-range missiles and space projects for both superpowers during the early Cold War were also responsible for putting the first satellites, the first animals, and the first humans in space.

Another neglected subject in most books on space flight is the record of Russia and the Soviet Union, which makes up nearly half of space flight history logged since the space race began in 1957. Ignoring the details of the Soviet successes and failures leaves significant gaps in the complete picture of space exploration. I added a lengthy chapter on the Russian space flight history to help fill the voids, created in part by the secretive Stalinist Soviet Union, and because the difficulty in reaching the original materials favors published experts like Asif Siddiqi and James Oberg and their sublime summaries. Without insight into the competitive and antagonistic influence of the Soviet Union, the American race into space appears to be a mysterious blend of impulse and intuitive success, when, in fact, it was a battle of technology and political will over five decades that has recently culminated in the greatest cooperative effort in history between two Cold War enemies.

I have emphasized three major American space programs because of their colossal successes and even greater program management achievements, and because of their far-reaching influence on later space projects. The Mercury program instituted many of the operational procedures and hardware standards that are still used a half century later. In two years, the Mercury project was taken from its infancy to a new level of institutional, industrial, and military cooperation unrealized even in the two preceding World Wars. Astronaut selection and training were constructed from a loose framework of military pilot experience and psycho-social profiling that became the foundation used for selecting and training today's astronauts. Like Mercury, the Gemini project evolved from primitive hardware into advanced space flight systems and refined manufacturing skills that paved the way to the Moon for the Apollo astronauts. Although Gemini missions never reached beyond Earth orbit, the program engineers, scientists, and managers forged the hardware and drafted the operational details to reach the surface of the Moon with a guaranteed safe return. NASA's Apollo project was one of if not the most ambitious and spectacular program on record. The space shuttle and the International Space Station have rivaled Apollo's Moon landings in scale, even though the science returned, the challenges in perfecting entirely new hardware, and the collaboration between the federal agencies, the military branches, the aerospace industry, and the educational institutions involved with the Apollo program remain unmatched.

Space exploration has undergone many changes in many ways since Apollo, and it is no longer centered on grand programs or spectacular feats with small crews. Today, the space agencies around the world, including NASA, the Russian Space Agency (RSA), the European Space Agency (ESA), the Japan Aerospace Exploration Agency (JAXA), the Canadian Space Agency (CSA), and a host of other nations are engaged in even greater collaboration in space science and engineering, continually advancing our understanding of the physical world on virtually very scale. Apollo was perhaps the greatest achievement in space led by a single nation, but the journey continues in many new directions, and with many new partners. Join the author in a study of the remarkable space exploration programs that bring us closer to our origins, and offer a closer look at the universe that surrounds us.

CHAPTER 1

The Early History of Space Flight

Man's fascination with flight was first portrayed in cave drawings from prehistoric times and continued in more lavish detail from early mythology through the birth of science fiction a century and a half ago. But the first manned flight would have to wait until the 1950s when theory and technology would catch up with the fictional mastery of space flight.

Rockets used for centuries as weapons were transformed into modern rockets by Robert Goddard and others in the 1930s. Those liquid-fuel rockets would be converted from World War II guided missiles into satellite launchers in just a decade and set the stage for the first human space flight in the Mercury, Gemini, and Apollo programs. More recent space shuttle launches are just as spectacular, but are now rivaled by deep space missions and orbiting telescopes that provide breathtaking views of our planetary neighbors, unveiling beautiful spires, clouds, and clusters of stars in their birth inside our galaxy and others. So too have these probes given us dazzling views of the complex, bewildering, and elegant galaxies in every corner of the universe. Scores of robotic spacecraft have helped unlock many mysteries about the formation of our solar system, of planets, and of other stars and have helped solve some of the most puzzling details of the creation of the universe as well as expanded our fascination with space flight. Two thousand years of evolving science and mathematics have turned the imagination and curiosity of the human mind into exciting and powerful tools of exploration. And in just five decades since the launch of the first explorer, the tools of space exploration have uncovered stunning splendor and provided tremendous insight into what lies beyond our Earth. Space flight and the exploration of space began in the imagination of our ancestors and continue to inspire us with the beauty and excitement of the unknown.

To help understand the development of space technology and its growing importance in our everyday lives, we need to look at the development of science and its core tenet, the scientific method. It is also useful to trace out the role of science and technology in the development of space flight, as well as examine the technology advances needed to build the first modern rockets.

Technology's path is easy to follow through the history records, from the first liquid-fuel rockets, to the orbital spacecraft and their launchers—a process that took less than a half-century. But the record also shows that the progress in science and technology that pushed the industrial world into the space age came at a price. Along with the convenience and efficiency that sprung from the new technology came the development of the destructive weapons of the Cold War. The duality in the utility of space technology is still with us, examples being the Atlas and the Delta space launchers that were originally developed as nuclear-tipped long-range ballistic missiles.

The history record takes much longer to trace out the progress of science and technology from its roots in classical Greece to the first modern rockets launched by Robert Goddard in the 1930s. Even though the orbital mechanics and physics of propulsion were worked out centuries before the first modern rocket, the technology needed to reach space would have to be developed and refined before the vehicles could be built and flown. The rocket technologies included high-energy chemical propellants and combustion chambers, advanced metallurgy for constructing the engines and turbine pumps, electromechanical inertial guidance components, high-pressure fluid dynamic controls, supersonic aerodynamic controls, and many more. Yet as closely as science and technology are linked, the two have several important distinctions. Technology advances that span more than two centuries have improved our lives considerably, but debate rages over the role it plays in mechanizing society and in expanding the destructive weapons of war. Science, too, evolves, but its role has changed very little over the centuries. Science has brought neither peace nor war, but supplies the tools to let us succeed or fail in our survival on this fragile planet.

Early Astronomy and the Evolution of Science

CLASSICAL GREECE

The origins of science and mathematics that are often attributed to the classical Greek scholars actually began with the Egyptian, Babylonian, and Ionian cultures that made the first methodical observations of the sky and logged the first coherent descriptions of the physical world. One of the most profound applications of early science was the prediction of solar eclipses based on accurate celestial observations made by Thales of Miletus (624–546 BC) in the sixth century BC. Thales, who is generally recognized as the father of science, is also credited with creating the foundation of mathematics and philosophy. His curious belief that the Earth was a disk floating in a heavenly ocean surrounded by a sphere of stars was an early interpretation of the physical world influencing meaningful events such as earthquakes and volcanoes, rather than invoking mythological gods. By the third century BC, the universe was characterized more rationally, but incorrectly, as an Earth-centered (geocentric) world by Aristotle (384–322 BC), who described the motion of the heavens by rotating crystalline spheres. Although the geocentric "universe" is considered a childish view of the solar system today, it was favored over the Sun-centered (heliocentric) universe proposed by Aristarchus of Samos (310–230 BC) since the geocentric view furnished a plausible explanation of gravity by assuming the Earth was the center of the celestial world and had the ability to attract all matter to its center. While the geocentric theory is unsupported by common knowledge and even simple logic, a surprising number of the world's educated population unwittingly subscribe to the geocentric theory in that it serves as the cornerstone of astrology.

THE UNIVERSE ACCORDING TO ARISTOTLE

In addition to his dissertations on philosophy, biology, and medicine, Aristotle developed an elaborate model of the universe that was compelling enough to be embraced by the Western world for fifteen hundred years. Aristotle turned away from the early principles of science toward mysticism to describe the physical universe. He defined the forces of nature as directly related to the five basic elements representing matter in the physical world—air, earth, fire, water, and ether. He characterized the makeup and motion of the celestial bodies beyond Earth by the ether, with celestial motion contained in specific realms. Those realms were:

- Earth
  - Located at the center of the universe and did not move
  - Provided a simple explanation for the motion of everything in the sky
- Five visible planets, the Sun and the Moon
  - Located on individual solid crystalline spheres in the sky
  - Provided for the irregular motion of the planets and the moon
- Stars
  - Located on the outermost sphere
  - Provided for the predictable, periodic motion of the stars in the night sky
- Sublunar (imperfect) world
  - Consisted of everything under the Moon, with a natural motion toward Earth's center
  - Provided a form, however incorrect, for the force of gravity
- Superlunar (perfect) world
  - Consisted of everything beyond the Moon with circular motion around the Earth
  - Provided a separate natural force that kept the celestial world in motion without falling to Earth

In the second century BC, the mathematician and astronomer Hipparchus (190–120 BC) used simple technology, some of which he invented, to make the first reasonably accurate tabulation of the positions of the stars and planets, and the first accurate star catalog. Hipparchus's accurate observations and his newly developed tools of trigonometry allowed three-dimensional geometric position calculations that, for the first time, could accurately predict solar and lunar eclipses. Considered the father of astronomy and perhaps the greatest astronomer of ancient times, Hipparchus's measurements and discoveries ranged from accurate calculations of the Sun and Moon's distances from Earth and their diameters, to the rotation of the Earth's spin axis, called the precession of the equinoxes. And although Hipparchus rejected the heliocentric model of the solar system, his work has been recognized through the centuries as important in the development of the first realistic models of celestial motion—important enough to have an astronomical satellite named after him. The European Space Agency's Hipparcos (High Precision Parallax Collecting Satellite) orbital telescope in operation from 1989 to 1993 was, as you would expect, designed to make extremely precise measurements of star positions.

One of the most important compendiums of scientific works was written in the post-Hellenistic period by Ptolemy (Claudius Ptolemaeus, AD 90–168) and entitled *Almagest*. His works were compiled in the late Roman period, a period of decline of the sciences in the Western world some three centuries after the important Greek astronomical works were written. *Almagest*, also known as the *Great Syntaxis of Astronomy*, was a rich source of early astronomical data that included a star catalog and constellation identities from the Greek period derived centuries earlier from the Babylonians, Phoenicians, and Greeks. As important as the *Almagest* was, it was not handed down through Western cultures directly. Because of the rising power of the Christian religion that followed the relocation of the Roman seat of rule to Constantinople in the fourth century AD, interest in and the development of the sciences came to a standstill in the Western world. As a result, many of the Greek works of science were passed through Arabic scholars who copied and studied many of the original manuscripts. It was in this period that the Western world embraced the Ptolemaic (Earth-centered) view of the universe; a period suitably named the Dark Ages.

ARABIC RENAISSANCE

The science, mathematics, and philosophy of the Greek period flourished in the Arabic world for more than 800 years after the fall of Rome. Arab scholars advanced astronomy with their observations, stellar catalogs, and star classifications that can still be found today. Many of the familiar bright star names were named by the Arab astronomers a thousand years ago. Names like Algol (the Gouhl), Altair (Flying Eagle), Betelgeuse (Hand of Orion), Fomalhaut (Mouth of the Fish), Rigel (the Foot), Vega (Stooping Eagle), and Aldeberan (Follower of the Pleiades) refer to prominent stars positioned in the constellations named by the Greeks and renamed by the Romans, but originating in Babylonian and Phoenician cultures. Western convention for naming the brightest stars within each of the eighty-eight constellations follows in order of the star's brightness and the Greek alphabet. For example, Alpha Ursae Minoris, also known as Polaris, is the brightest star in the Small Bear constellation (Ursae Minoris) found at the North Celestial Pole. But because there are so many visible stars, and an overwhelming number visible even in a small telescope, current convention for naming stars incorporates the position of the star in celestial coordinates. Nevertheless, Arabic influence still lingers in the brightest stars in the sky.

By the thirteenth century, the Arabic cultures turned from the sciences to Islamic religion, and the advancement of knowledge would again pause until the Copernican Revolution three hundred years later. Yet, in that thousand years that science and mathematics progressed from classical Greece to the Arabic renaissance, the question of the true motion of the Sun and planets was not settled. The Sun-centered universe first proposed by Aristarchus of Samos in the third century BC was not generally accepted by Greek and Roman scholars. The popular view that the Earth-centered model remained entrenched because the Roman Catholic Church embraced mankind's central role in the universe and persecuted those that did not believe the same is only partly correct. The Greek and Roman era came well before gravity and the other forces were understood, which made a world with the Earth at the center a more convincing model of what kept us from floating off into space. Ptolemy's universe portrayed by Aristotle's perfect and imperfect world is considered foolish today, but was embraced for fifteen hundred years because it was convenient in representing the unknown physical forces that held the universe together. And for much the same reason, when the heliocentric solar system was reintroduced during the Copernican revolution, opposition to the idea was initially led by scientists, not by priests.

COPERNICAN REVOLUTION

One of the most pivotal publications in astronomy and science was *On the Revolutions of the Celestial Orbs* (*De revolutionibus orbium coelestium*), written in 1543 by Nicolaus Copernicus (1473–1543). In his work, which was dedicated to Pope Paul III and published with the help of a Catholic Bishop, Copernicus established that the Sun was the true center of the solar system with the Earth just one of the planets orbiting the Sun. Copernicus was reluctant to publish his text because of the growing controversy, not with the Church, but with the science community reluctant to accept the new concept. Moreover, his model of the solar system used circular orbits to describe planetary motion, which could not account for the unusual orbital motion of the planets such as the retrograde motion of Mars, while the artifice of epicyclic motion in Ptolemy's geocentric system could.

Copernicus delayed his publication until he was on his deathbed because of its controversial nature, yet he succeeded in igniting a revolution with his new and controversial ideas. Debate over the true motion of the planets was spirited and led to the rebirth of interest in the sciences—the scientific revolution. Controversy over geocentrism versus heliocentrism also sparked reaction from the Roman Catholic Church leadership several decades later with the pronouncement that Copernicus's work was heretical. Similar judgment by the Church resulted in the house arrest of Galileo Galilei nearly a century later for his refusal to recant his belief in Copernicus's heliocentric solar system.

TYCHO BRAHE

Tycho Brahe (1570–1601) made a significant contribution to early astronomy and the correct model of the solar system with his careful, accurate measurements of the stars and planets that were later used by Kepler to derive the first accurate heliocentric theory. Although Brahe was precise in his observations, he also believed in the Ptolemaic geocentric solar system. And because the model could not be reconciled with his own observations, Tycho chose to construct his own theory of celestial motion using a modified Ptolemaic model in which all of the planets except for the Earth and its moon orbited the Sun, while the remaining planets and the Sun orbited the Earth.

JOHANNES KEPLER

Because Brahe's observations were precise, his data were embraced by a number of astronomers, including his most celebrated assistant, named Johannes Kepler (1571–1630). Kepler was responsible for deriving elliptical planetary motion, which finally differentiated the less accurate and incorrect geocentric solar system from the correct heliocentric solar system model. Kepler's elliptical planetary orbit model solved the puzzling retrograde motion of Mars without using the contrived epicyclic motion of Ptolemy. Kepler's many contributions to astronomy were only a part of his life's work, which was also entangled with astrology, a pseudoscience that was soon after abandoned by the scientific community.

Kepler's first publication, which included the well-known first law of planetary motion, demonstrated quantitatively that the planets orbit the Sun, with the Sun positioned at one of the focal points of the ellipse. A later publication that became a widely used astronomy text entitled *Epitome Astronomia Copernicanae* (*Epitome of Copernican Astronomy*) included his two other laws of solar-celestial orbits, which defined planetary orbits sweeping out equal areas in equal times, and the period-distance law of the planetary orbits.

Kepler's Laws

1. The planets orbit the Sun in elliptical orbits, with the Sun at one focus of the ellipse.
2. Equal areas are swept out in equal times by the orbiting planets.
3. The period of orbit for a planet squared is equal to its semi-major axis cubed.

GALILEO GALILEI

In spite of the accurate orbital predictions that could be made using Kepler's laws, there was no corresponding physical law that could account for the attraction between celestial objects or that could explain their resulting motion. Observations and publications by Galileo Galilei (1564–1642), who corresponded with Kepler and other astronomers of the time, provided convincing proof of satellites around other planets, including Jupiter, and that Venus showed phases like the Moon. Galileo's telescope observations and theories published in his many works also supported the heliocentric solar system, further provoking the increasingly intolerant rulers of the Roman Catholic Church. Galileo refused to denounce his support for scientific proof of the Sun-centered universe, and was imprisoned in his home for the latter part of his life. The Church convicted Galileo of heresy and refused to recognize his extensive contribution

Galileo's Major Astronomical Discoveries

Galileo Galilei was a physicist, scientist, mathematician, astronomer, inventor, professor, philosopher, and prolific author responsible for many of the advances in science during the Renaissance-era scientific revolution. Several of Galileo's significant astronomical discoveries included:

- The inner planets have orbital phases similar to the Moon (proving the heliocentric theory).
- Jupiter has four large moons (the four Galilean moons) which implied that the Earth-Moon system was not unique.
- The Milky Way is composed of many stars that appear to the naked eye as a diffuse cloud.
- Seen from a telescope, the mountains, valleys, and seas on the Moon look similar to those on Earth.

to the many fields of science. The Church also refused to allow him to be buried with his family. It was not until 1992 that Galileo was recognized by the Catholic Church for his significant scientific discoveries, though not yet forgiven for advocating what the Church has now accepted.

Galileo, considered by many to be the father of physics and modern science, played an important part in the scientific revolution sparked by Copernicus just decades earlier. Galileo developed a mathematical description of gravity and acceleration that, for the first time, equated gravity with acceleration, both a function of elapsed time squared. But the mathematical formulation of gravity would have to wait until Sir Isaac Newton's derivations at the turn of the eighteenth century.

ISAAC NEWTON

Isaac Newton (1642–1726) was one of the most important and influential physicists and mathematicians in history, but also a prominent theologian who spent most of his later life writing on Christianity and practicing alchemy. In order to derive many of his theories relating to mechanics and motion, gravity, celestial orbits, optics, and many other phenomena, Newton developed one of the most powerful mathematical tools available—calculus. Credit is shared with Gottfried Leibniz for creating the powerful methods of integral and differential calculus, since both men developed the mathematical tools independently, and roughly at the same time.

Newton's scientific and mathematical advances are too numerous to describe in this brief review, but several of his monumental contributions to orbital mechanics and space flight are found in three of his basic laws of motion.

Newton was also responsible for developing analytical tools and numerical approximation methods used to solve two-body and n-body orbit calculations for which there are no analytic (closed equation) solutions.

Newton's Three Laws of Motion

1. Newton's first law: the law of inertia
   - An object at rest will stay at rest and an object in motion will stay in motion unless acted upon by an external force.
2. Newton's second law: force and acceleration
   - An applied force equals the rate of change (derivative) of the momentum of an object.
   - $F = dp/dt$, where p is momentum and the d/dt signifies a derivative with respect to time.
   - If mass is constant, this becomes the familiar $F = ma$, or force equals mass times acceleration.
3. Newton's third law: the principle of action-reaction
   - For every action there is an equal and opposite reaction. This is the basic principle describing rocket propulsion, and virtually every other propulsion method.

JOSEPH LAGRANGE

Joseph Lagrange (1736–1813) was a mathematician intrigued with planetary and orbital motion. In addition to developing a variety of useful mathematical tools for general physics and astronomy applications, Lagrange derived a powerful technique for approximating the motion of small objects orbiting in the solar system. Lagrange's method employs simplified orbits for three bodies, each with significantly different mass. His work led to the useful characterization of orbital motion, and the identification of orbit stability zones that apply to the largest planets as well as the smallest particles. The Lagrange stability points and the restricted 3-body analysis have become a foundation for most numerical models used in orbital operations, beginning with the Gemini spacecraft rendezvous operations and continuing today with the International Space Station, the space shuttle, and Russian Soyuz flights. The Lagrange stability approximations are also used for positioning spacecraft in orbits that provide continual solar exposure, such as Solar Heliospheric Observatory (SOHO), and in orbits that can protect spacecraft from the Sun's heat, such as the Wilkinson Microwave Anisotropy Probe (WMAP), and the European Space Agency's Planck satellite.

Lagrange's three-body dynamics and stability analysis have also been used for computing low-energy trajectories for spacecraft travel throughout the solar system, in effect, reducing propellant mass and improving the efficiency of mission operations. Lagrange's analysis can be, and has been, used to lower the propulsion requirements for reaching the Moon and the other planets, and for reducing propellant mass needed to transfer cargo beyond geosynchronous orbits.

MODERN PHYSICS AND QUANTUM MECHANICS

The industrial revolution that began in the eighteenth century was the first large-scale application of science that would benefit society as a whole. Simple technology that had evolved slowly over several thousand years suddenly vaulted from simple metal alloys and primitive chemical compounds to powered transportation, improved medical and scientific instruments, specialized chemicals, stabilized foods, and untold other advances. With these advances, industry and technology brought wealth and increased trade and expanded the world's largest cities. Industrialization also introduced the perils of pollution and resource depletion, ironically, with solutions derived primarily from advances in technology.

Advances in the pure sciences accelerated with the expansion of industrialization as well. New theories that established electric and magnetic properties of matter as fields were followed by the discovery of unstable particles, particle radiation, and the foundation of physics on the smallest scale. Arguably, the most profound changes in physics and astronomy took place in the beginning of the twentieth century with the development of quantum mechanics. Revolutionary concepts that enabled researchers and theorists to transform particles and to manipulate electromagnetic radiation on the smallest scale were introduced in the theoretical works of Werner Heisenberg, Max Planck, Louis de Broglie, Albert Einstein, Niels Bohr, Erwin Schrödinger, Max Born, John von Neumann, Paul Dirac, Wolfgang Pauli, and others. With the introduction of quantum mechanics came the answers to many of the puzzles of the "classical" physical world, revealing unknown aspects of optics, radioactive decay, discrete spectra and colors, even heat and energy transfer. In spite of the profound powers of quantum physics, the reconciliation of what we see on the largest scale and what we see on the smallest scale is not complete. Inconsistencies between Einstein's general theory of relativity, which represents the physical world on the large scale, and quantum physics, which represents particles and forces on the subatomic scale, are hidden within the character of mass and the force of gravity. More recent discoveries of dark matter and dark energy have compounded the problems of deriving comprehensive mathematical models that define everything measurable in the physical universe, known informally as the "theory of everything."

The equation describing all forces and particles that was first pursued by Einstein continues to be a monumental challenge for physicists and mathematicians. The ultimate solution is being approached from many directions, since a number of important pieces are missing in the complex physics of particles, fields, and forces. Little is understood about gravity at the quantum level, and less about dark matter that makes up some 90% of the mass of larger galaxies. Even less is known about dark energy that appears to be locked into the expanding empty space of our universe, and makes up three-fourths of its total mass-energy content. Several of NASA's newest space exploration initiatives have been targeted specifically to study dark matter and dark energy because of their importance in understanding the makeup and evolution of the universe.

Advent of Space Flight

Space flight is a visual and engineering marvel, but it requires much more than just design skills for success. The first rockets capable of reaching space relied on three fundamentals for their development and, somewhat ironically, were designed independently by three different engineers in three different countries. These three fundamental components also had to be in place to take basic rocket propulsion and guidance through the development and testing stages for creating the first modern rockets. Those three fundamentals were:

1. Theory
2. Technology
3. Advocates

These were not three equal ingredients that made space flight a reality, but were the necessary components for advancing rocket propulsion and guidance to a level that could reach space. Theory and technology can be easily argued as the two most logical elements in developing advanced rockets, while the concept of advocacy being a necessary component for modern rocket development needs some explanation.

A quick review of the history of rockets shows that the first primitive devices were created more than three thousand years ago for ceremonial displays and for celebrations, like our fireworks today. Military weapons were the first serious application of the rocket, with improvements that were slow and incremental, taking centuries to evolve into guided missiles. Limitations on the size and accuracy of the rocket and warheads restricted their effectiveness until the introduction of the liquid-fuel rocket. Then, in less than two decades, rockets went from solid to liquid propellants, which increased their range dramatically. Significant improvements were also made with the development of active guidance systems. From the early 1920s to the late 1930s, Robert Goddard in America, Wernher von Braun and Hermann Oberth in Germany, and Fredrik Tsander and Sergei Korolev in Russia built and launched the first liquid-fuel rockets that within another decade reached above the atmosphere. The three pioneers of advanced rocketry were determined to reach space, and succeeded in three decades to do just that. The one exception was Robert Goddard who died in 1945.

World War II brought the rocket interests and military funding together to build the first long-range modern missile—unfortunately, by the Nazis. Fortunately, Germany lost the war, and the chief designer of the V-2, Wernher von Braun, surrendered to the American forces in hopes of furthering his ambition to reach space. It was then that the two proponents of space exploration, Wernher von Braun, working for the United States, and Sergei Korolev of the Soviet Union, set out to launch the first satellites into orbit and explore the solar system. Without these pioneers of space, the history record would not be quite the same and the space race may instead have been a stroll.

1. THEORY

The fundamental theories of space flight are a broad collection of physics and mathematics principles that include basic mechanics; celestial mechanics; orbital dynamics and kinematics; propulsion; subsonic, supersonic, and hypersonic aerodynamics; guidance and control dynamics; and a host of other disciplines. The basic theories were introduced by Newton and Lagrange, and others, who were responsible for deriving the underlying theory three centuries ago. However, the basics of rocket propulsion, rocket performance, and aerodynamics would have to wait until the beginning of the twentieth century when the American Robert Goddard, Konstantin Tsiolkovsky of the Soviet Union, and Hermann Oberth of Germany published their works on the principles of rocket propulsion and on orbiting satellites. Primarily because of the secretive Soviet Union, little was shared between the scientists, although Oberth requested and received a copy of Goddard's *A Method of Reaching Extreme Altitudes* in 1922 before publishing his own text *The Rocket into Planetary Space* in 1923. However unlikely Oberth's plagiarism of Goddard's ideas may have been, accusations and denials dogged Oberth for much of his career in Germany, and later in the United States.

Konstantin Tsiolkovsky of the USSR introduced important derivations of rocket propulsion and propulsion concepts that included rocket staging, and discussed high-energy propellants. Without the exchange of scientific ideas from the Soviet Union, many of the same derivations and fundamentals had to be developed in the Western world before the launch of the first liquid-fuel rockets.

The V-2 and the longer-range missiles developed after World War II needed sophisticated test equipment and advanced theory to develop the supersonic aerodynamic controls that would allow their rockets to fly beyond Mach 1, the speed of sound. The two-stage experimental rockets built around the V-2s captured by the Americans and the Soviets extended rocket capabilities to hypersonic flight—flight beyond Mach 5—and improved guidance control to allow more accurate and near-vertical flight to reach higher altitudes. Soon after, von Braun and Korolev began developing the first missiles that they knew would also be able to launch the first satellites.

2. TECHNOLOGY

Those involved with the early theory and practice of space flight were well aware that solid-fuel rockets were impractical in reaching space, even if staged. Early pioneers, including Goddard, von Braun, and Tsiolkovsky also knew that liquid hydrogen and liquid oxygen were ideal propellants, although liquid hydrogen was not a reasonable choice of fuel for the early rocket engines because of its very low storage temperature and its handling difficulties. As a result, the first rocket prototypes that could, in time, reach space were the first liquid-fuel rockets with modest propellant performance. Other basic design features of the early liquid-fuel engines were high-pressure, supersonic exhaust, fed by either pressurized tanks or small turbopumps. However, the metal alloys, fuels, and other materials needed for these engines, their turbopumps, and the propellant tanks, were not readily available until the early 1900s. Not long after the technology appeared, the first liquid-fuel rockets were being built on three continents. The first to succeed was Robert Goddard in 1926.

Goddard's first liquid oxygen and gasoline–propelled rocket went from bench tests to a successful inaugural launch on March 16, 1926. Germany's VfR rocket club began its first launches in 1933, while the Russian rocket clubs led by Tsander and Nikolai Tikhomirov launched their first liquid rocket in 1936. All three were developed independently by engineer-scientists determined to reach space. As rockets grew larger and more complex, technology advances were needed, often fostered by those designing the rockets. Turbopumps that force propellants into the combustion chamber of the liquid rocket motors were major design hurdles in the early rocket programs, and still are. Guidance systems that were based on inertial gyroscopic platforms were a challenge to develop with enough precision and reliability to successfully guide the earliest modern rockets. Guidance systems required even greater sophistication with the progression to multistage rockets that could place payloads into orbit, and even more so for landing on the Moon, and for reaching distant planets.

Early guidance systems developed by Goddard and copied by the Germans for their V-2 missile consisted of external fins for aerodynamic stability, and control vanes placed in the exhaust stream for thrust vectoring. In the low-speed flight regime at liftoff speeds, the rocket needed augmented stability since the airflow over the fins was ineffective. Goddard's exhaust vane configuration furnished the low-speed flight stability. This technique also proved useful for stable flight control of the V-2 at supersonic speeds, since the aerodynamic fins were unusable in supersonic flight.

One of the greatest challenges in building larger, higher-performance rocket engines is overcoming the combustion instability in large thrust chambers. Scaling up the size of liquid rocket engines was not, and still is not, a simple process. Without computers to help solve the acoustic wave and thermodynamic problems in turbulent combustion, trial and error played a significant role in building larger engines. By the time the huge F-1 engine was completed for the Saturn V, digital computers had become an indispensable tool in rocket engine design. Because of Russia's lagging technology in electronics, their rocket engines were scaled up by building multiple thruster units instead of larger, single-chamber engines. Even though the technique was successful in reaching the required thrust, the engine efficiency suffered. It was not until the 1980s that the USSR acquired the skills and technology to produce high-thrust engines comparable to the F-1.

Technology limitations also paced the development of reentry vehicles because of the extreme heating during the hypersonic return through the atmosphere. Reentry temperatures can reach 1,090°C (2,000°F) on the surface of the reentry vehicle for suborbital missions, and more than twice that for reentry from eccentric orbits, with shock temperatures exceeding 5,540°C (10,000°F) in front of the vehicle. The problem first arose in the ballistic missile programs because of the need to protect warheads from the extreme temperatures, pressures, and even high g-loading during reentry. The intuitive solution of a pointed nose cone for the leading surface is unworkable since heat is concentrated on the point of the cone at hypersonic speeds, even if needle-shaped. The answer came in two parts; the first was the use of a broad, curved face to dissipate most of the kinetic energy as lateral shock waves instead of dissipating the kinetic energy as heat energy. This method was first used by NASA for the Mercury capsule design, followed by the Gemini and Apollo capsules, and now, for the Orion capsule. The second part of the solution was to reduce the extreme temperatures encountered during reentry by incorporating a volatile insulation material for the reentry heat shield. The surface of the material slowly vaporizes from the extreme reentry heat, creating a cool boundary layer of gas between the very hot shock front ahead of the surface and the heat shield. Gas is released continuously from the surface of the ablation shield, so named since it partially vaporizes, or ablates, during heating.

The Apollo program introduced an impressive number of new technologies and materials because of the unique and demanding missions to the Moon. Two of the greatest technological advancements created for the Apollo vehicles were the digital electronics and the first miniaturized digital flight computer. Since the Apollo vehicles and astronauts would be far from Earth and sometimes blocked from Earth view by the Moon, autonomous navigation and vehicle operations had to be included in the Apollo flight hardware. The small digital control systems used for rendezvous and docking on Gemini were replaced with the first programmable, digital computers using integrated circuits. Apollo's flight computer, designed by the MIT Instrumentation Lab, and the digital microcircuitry, built specifically for Apollo guidance computers, started the microcomputer revolution that has grown into one of the largest industries in the world and has entered almost every corner of our world of communications, computation, automation, and control.

3. ADVOCATES

Space flight was little more than fantasy before the first liquid-fuel rockets were launched in the 1920s and 1930s. Yet, a handful of men in three different countries were responsible for taking space flight theory to the first stages of modern rocket development in just a few years. By the 1960s, many of the same designers helped put the first satellite in Earth orbit, prepared the first robotic space exploration missions, and at the same time, began planning the first manned missions. Independently, these pioneers of space flight and modern rocketry developed the boosters and infrastructure that powered the space race between the Soviets and the Americans well before the race began.

Robert Goddard

The first step in creating the modern rocket was taken by Robert Goddard with his historic launch of the first liquid-fuel rocket in 1926. In less than a decade, Goddard designed and launched the first self-guided, liquid-fuel modern rocket and continued to advance engine design, thrust control, turbopumps, guidance systems, injectors, even structures, until his death from tuberculosis in 1945. His work on rocket design spawned a long list of innovations and patents in propulsion systems and rocket design, including ion engines and control electronics.

Dr. Goddard, who is considered the father of American rocketry, began patenting liquid- and solid-fuel rocket designs as well as multistage rockets as early as 1913. His research and design work continued with additional support from a variety of civil and military sources before being interrupted by World War I. Immediately after, Goddard began working on small solid-fuel rockets and launchers, one of which would be the precursor to the bazooka, a recoilless rocket launcher. At the end of the war in 1918, Dr. Goddard returned to Clark University to teach, but continued to pursue rocket design through his later years. Research grants allowed him to concentrate on his theoretical work and practical design of liquid fuel rockets, although he still held a post at Clark University (Clary 2003). At the onset of World War II, Goddard again joined the war effort by designing rocket-assisted boosters for U.S. Navy amphibian aircraft.

In spite of his extensive work in rocket design, Goddard had little influence on American missile and rocket development after World War II because of the slow pace of rocket development even after the captured German V-2 long-range missiles proved their utility as a high-altitude research booster. Many of Goddard's extensive contributions to rocket design were used in America's first long-range missiles and launchers, but without acknowledgment until a formal recognition of his efforts and patents was made by NASA and the Department of Defense (DoD) in 1960. More from a legal ruling than from magnanimous inspiration, NASA and the DoD awarded Goddard's widow $1 million for the use of his patents over the years. NASA also named its primary space flight control and operations center in Greenbelt, Maryland, the Goddard Space Flight Center.

Goddard's work did, however, interest others. Jet Propulsion Lab rocket designers contacted Goddard in an attempt to work with him on liquid propulsion designs, but with little success because of Goddard's penchant for secrecy about the details of his rockets. The press was also eager to publish articles on Goddard's rocket designs and exploits, and contrary to popular myth, Goddard often used media publicity to further his efforts and to improve his chances of finding funding.

Another popular myth of German weapons planners spying on Goddard and his work in Roswell, New Mexico, is mostly exaggeration of Germany's interest in his liquid-rocket designs. Although Nazi Germany was interested in developing missiles for the looming war, the difficult technology development process for advanced rockets was equally challenging for the American, Russian, and German designers, and few technologies were actually shared (Clary 2003).

Wernher von Braun

Pioneers of Germany's modern rockets emerged from an amateur rocket club called the Society for Space Flight (Verein fur Raumschiffahrt, or VfR) in the early 1930s, much like the Soviet rocket pioneers that were first brought together in the GIRD and GDL rocket clubs. The VfR members were enthusiastic about future space flight, but were derailed in their efforts to build larger liquid-fuel rockets by the German Gestapo. The club was disbanded and the rocket launches halted to help conceal the Nazi's ongoing missile development. A few of the promising engineers were invited to participate in the Germany Army's missile efforts, including Wernher von Braun, who led the Nazi A-4 (the prototype used for the V-2 missile design) development effort under General Dornberger. Von Braun was just one of a number of distinguished members of the VfR in its short history that included Willy Ley, Hermann Oberth, Hermann Noordung, and Eugen Sänger (Reuter).

Wernher von Braun's early career as a missile designer for the Nazis allowed him to develop engineering and management skills that would prove invaluable after the war, but would also brand him as a member of the German war machine. His true ambition to build rocket launchers and spacecraft for space exploration was advanced by his work developing the long-range V-2 missile. Yet, these were only part of von Braun's conflicted principles which included membership in the SS and his arrest by the Gestapo for a perceived lack of commitment to the war effort. Although debate continues, it was his knowledge of the use of concentration camp workers for constructing the V-2 and its facilities that blemished von Braun's stellar NASA career.

As the German war effort was collapsing in 1945, several hundred members of the V-2 rocket team led by von Braun escaped to Bavaria to surrender to the American forces. After several weeks of interrogation, more than 100 of von Braun's team were sent to the United States under Operation Paperclip. A number of the German rocket scientists, including von Braun, were assigned to begin work on reconstructing and launching captured V-2s, and helping build advanced missiles. His assignment to the U.S. Army Ordnance Research and Development at Ft. Bliss, Texas, and the White Sands missile range in New Mexico was followed by transfer to the new Redstone Arsenal in Alabama. Reorganization of the Army Ordnance research and development branch as the Ordnance Guided Missile Center, then later to the Army Ballistic Missile Agency (ABMA), placed von Braun at the head of the Army's Redstone ballistic missile development team. Von Braun stayed in Huntsville for twenty years, first as an Army research and development director, then from 1960 to 1972 as a NASA program director.

Wernher von Braun was firmly committed to space exploration even as a missile designer for the ABMA, and well before joining NASA in 1960. He promoted space exploration at nearly every level possible, including print and television media. Von Braun was responsible for a series of illustrated articles in the popular *Collier's* magazine between 1952 and 1954 that portrayed spectacular manned space flight projects, beginning with orbital stations, then manned lunar flights, lunar outposts and bases, and finally, manned Mars missions and outposts. He also helped popularize space flight by joining Walt Disney in the *Tomorrowland* television series on space flight. Many of the hardware concepts in these articles came from scaled-up versions of von Braun's work on German multistage missiles. Nevertheless, his popular articles offered a view of space exploration far more realistic and more detailed than anything previously published, while captivating America with the wonders of flying in space. Von Braun promoted manned and unmanned space exploration in the halls of Congress, in the Armed Services, in the aerospace industry, and to anyone of influence willing to listen.

The creation of NASA in 1958 and the transfer of manned programs and the ABMA to the new agency in 1960 included von Braun and his Saturn boosters, and the preliminary plans for the Apollo program. Following the launch of the first man in space by the Soviets, Apollo was chosen as the program in which America would surpass the Soviets in the race to the Moon. Apollo, NASA's most ambitious program, was spearheaded by von Braun, cheered on by the American public, and supported by nearly everyone, from politicians to the aerospace industry. Von Braun's promotion of space exploration also extended into areas he felt important that were overlooked by NASA at the time. The National Space Society that was started by von Braun is still active today, as is the Space Camp at the U.S. Space and Rocket Center in Huntsville, Alabama.

Wernher von Braun served in a number of managerial positions at NASA, including director of the Marshall Space Flight Center, until 1972. Precipitated by the termination of the Apollo project, von Braun left NASA to join Fairchild Industries as vice president for engineering and development. He remained with Fairchild until stricken with cancer in 1976 and passed away in 1977.

Sergei Korolev

After joining the United States and allies to help defeat the Germans in World War II, Joseph Stalin split the Soviet Union from the Western allies and began a massive offensive and defensive weapons buildup. Stalin was interested primarily in gaining parity in nuclear weapons and building long-range missiles, both of which set the stage for the Cold War between the USSR and the Western allies. Stalin began his missile program with the captured German V-2 rockets, like the United States, but more aggressively. By coincidence, the same V-2 projects would advance both countries toward competing space programs and would later pit Wernher von Braun's American missile design teams against the Soviet design teams led by their chief missile designer, Sergei Korolev.

Sergei Korolev, recognized as the father of the Russian space program, was a pilot and flight enthusiast, although more interested in space flight than in aviation. In 1931, Korolev joined the GIRD rocket club while working for the famous aircraft designer Andrei Tupolev, and was soon appointed director. The Russian GIRD rocket club was similar in function to the German VfR club, with interests in building advanced rockets, ultimately to reach space. Although not one of the founders, Sergei Korolev later joined GIRD, and became its director before it was merged under military direction with the GDL rocket propulsion club. Personality conflicts kept Korolev from the director's position of the newly combined research institute, which was fortunate for Korolev, since the director was executed in the 1937–1938 Stalin purges. Korolev and the other rocket designers did not escape persecution, however, as most were imprisoned and tortured or executed. Few engineers remained in the rocket research institutions, and those who survived did little to advance rocketry until the end of WW II. After release from prison, Korolev headed a group sent to Germany to capture the V-2 scientists and return their missile technology for building advanced weapons.

Because of his earlier expertise in rocket design and his participation in the 1946 V-2 recovery effort, Korolev was appointed head of the new Soviet rocket and missile design bureau named OKB-1. His initial directives were to build a heavy-lift, long-range missile for the Soviet's new atomic weapon, beginning with constructing copies of the V-2 using only Russian components. The R-1 all-Russian copy of the V-2 provided much of the necessary technology and manufacturing capability for follow-up missile programs, including their first intercontinental ballistic missile (ICBM), the R-7. Korolev was well aware that the R-7 could also place a satellite in orbit, and he had successfully pushed the project through the bureaucracy for approval. Following the world's first ICBM launch in August 1957, Korolev adapted the booster for the world's first satellite launch, which was timed for the 1957–1958 International

Geophysical Year. The satellite launch offered a unique opportunity for Korolev to demonstrate his engineering and management prowess, and for the Soviet scientific community to prove their level of theory and technology. But more important, it offered Premier Khrushchev and the Politburo an opportunity to display their leadership in advanced military weaponry. Soviet propaganda gained from the Sputnik 1 launch on October 4, 1957, helped put Korolev in a position unparalleled in the Soviet political structure. Following the successful launch of the second and third Sputnik satellites, and the first flights to the Moon with his Luna spacecraft launched with his R-7 boosters, Korolev was able to command most of the military missile development programs while directing the development of most of the Soviet spacecraft and exploration missions. His skill and success as a designer were important in his ability to lead the Soviet's first space exploration programs, and instrumental in his gaining support for creating military hardware that could also support the exploration of space. But it was Korolev's skillful influence over bureaucrats, whether military or political, that contributed most to his success in the space race with the United States.

In spite of the bureaucratic headwinds, Korolev succeeded in finding approval and funding to develop his boosters, scientific payloads, and exploration programs, even though nearly all were under competition from other design bureaus, and often opposed by powerful Soviet military leaders. As his space exploration project successes grew in number and importance, Korolev extended the scale of his exploration programs. Korolev's desire to put cosmonauts in orbit, then on the Moon before America, was supported by much of the Soviet leadership, but opposed by many others. One of Korolev's most ambitious projects was the Soviet version of the Saturn V named the N-1. Much of the challenge of the N-1 booster and the lunar vehicles came from the design and funding competition, due to conflicting politics and competing political alliances. Competitive programs were encouraged by the top bureaucrats, who often divided project efforts and funds. Internal forces that included bureaucratic indecisiveness, military interference, personality conflicts, changing political alliances, and the lack of a coordinating organization like NASA, doomed the Soviet's manned lunar project. Yet all of those obstacles did not compare to the death of Korolev in 1966 in condemning the program to failure. It was Korolev's political skills, determination, and forceful style that took the Soviet Union from Stalinist-era nuclear weapons delivery programs after World War II to a leadership position in space exploration and space technology during the race to the Moon. His death did not signal the end of the space programs in the Soviet Union, but it did signal the end of one man's unmatched ability to direct enormous space programs within the largest bureaucratic maze in history.

References

Clary, David. 2003. *Rocket Men: Robert Goddard and the Birth of the Space Age.* New York: Hyperion.

Reuter, Claus. 2000. *The V-2 and the Russian and American Rocket Program.* New York: S. R. Research and Publishing.

CHAPTER 2

Rockets and Propulsion

Rocket propulsion is the simplest and one of the most dramatic applications of Newton's third law—the principle of action-reaction. A rocket's propulsive force is derived from the expanding gas escaping from the combustion chamber, which generates an opposing force and pushes the rocket forward. In much the same way, the recoil of a rifle from the bullet forced forward by expanding gas is the impulse equivalent of rocket propulsion. The difference in the two is that the continual gas expansion released from the rocket motor would represent a smoothed average of a sequence of rifle shots, provided that both have an equivalent exhaust (bullet) mass and velocity over the same time period.

Slightly less intuitive but just as useful in describing rocket propulsion is the concept of momentum. Like the propulsion force of a rocket, momentum has equal and opposite components, but describes the two components of thrust more clearly. Exhaust momentum, which is the exhaust mass times the exhaust velocity, or $m_{exhaust}v_{exhaust}$, is exactly equal and opposite to the forward momentum of the vehicle. Conservation of momentum between the vehicle and the exhaust implies that the low-mass exhaust gas must have a very high velocity in order to accelerate the much heavier rocket. High-velocity rocket exhaust is essential for efficient propulsion. The higher the exhaust velocity, the higher the propulsion efficiency in converting propellant into thrust.

The second half of the action-reaction principle of rocket propulsion includes the exhaust mass flow. This also represents the mass variable (m) in momentum (mv). Albeit obvious, it is also necessary to point out that high thrust can only be generated by high exhaust gas mass flow. Just as obvious, high-thrust rockets require large engines that have a high propellant flow rate. Combustion efficiency, which is equivalent to exhaust velocity, and the exhaust gas mass flow rate together describe the fundamental performance of a rocket motor. Simply put, the size of a rocket engine determines its lift capacity, or thrust, while its exhaust velocity determines how much thrust is produced from a given propellant mass.

Basic Propulsion

To demonstrate simple rocket propulsion and simple flight through the atmosphere for yourself, inflate a balloon, then release it and notice carefully what happens. Several basic concepts of rocket propulsion can be seen in this simple exercise. As you know already, the escaping air is what produces the balloon's propulsive force. That gas is forced out by the pressure on the gas squeezed by the elastic balloon. The expelled gas will propel the balloon forward as it escapes rearward in rough proportion to the momentum of the exhaust gas. Momentum is not something that you can accurately gauge by the demonstration, but it is important. Momentum represents the propulsive thrust for any rocket propulsion system, from chemical rocket motors to electric ion engines. Forward thrust of the rocket is directly related to the escaping gas, or more precisely, the change in its momentum.

What about the toy balloon's limited speed and its changing direction? Any toy balloon that is inflated and released will trace out an unpredictable path with little straight-line motion. As you've guessed already, the balloon's chaotic motion is related to the changing forces from the balloon's flapping nozzle, and the air's opposing aerodynamic drag. The aerodynamic effects are also common to rocket flight in the atmosphere, while the fluttering nozzle is not. Flutter of the balloon nozzle contributes to the continually changing direction of flight from its somewhat random influence on the direction of the escaping gas. Because of that, accurate directional control of the balloon is not possible, but can be improved slightly. A similar, modest directional control was used by William Congreve to stabilize solid rocket weapons in the early 1800s, and can be found today on bottle rocket fireworks that have a narrow trailing stick. Similar stabilization can be made by taping a long ribbon or a straw to the nozzle end of the balloon and repeating the exercise.

In addition to basic rocket propulsion, the balloon is useful for demonstrating some of the principles of rocket flight through the air. The aerodynamic drag accentuates the nearly random influence of the flexible nozzle on the balloon's path. The opposing force of aerodynamic drag limits the balloon's speed through the air. Without the atmosphere, the balloon would fly closer to a straight-line trajectory, although in a vacuum, it still could not follow a direct path because of the flopping, fluttering nozzle, and because the thrust is behind the mass and pressure center of the balloon. Instability in the balloon flight comes from the two opposing forces—the thrust forward and the air resistance rearward. This is an inherent instability in any rocket, due to the thrust located behind both the center of pressure and the center of mass of the rocket (balloon) as it is pushes through the air.

A simple analog of the flight stability of a rocket or a balloon can be made using a pencil. The forward thrust on a rocket or a balloon is opposed by aerodynamic drag in the atmosphere. Similarly, the force of gravity on a pencil is opposed by the table it sits on, or by the pressure of your hand. If you hold the pencil by its top, or balance it at its center, it is relatively stable. If you try to balance it on its bottom by your finger it is not. The difference is the relative position of the two forces. Pushing against the pencil from the bottom is as difficult as balancing a pen or pencil on a table. Holding the pencil by its top is stable because the lifting force is above the average pull of gravity at the pencil center. But, holding the pencil at its top seems like cheating, since you could also grip the pencil by its bottom and stabilize it. So, for the skeptics, drill a half dozen holes in a thin strip of wood, or, even easier, punch a half dozen holes in a thin strip of cardboard or heavy paper. Hold the wood/cardboard/paper strip up, then insert a pin or stretched-out paper-clip in the top hole and release the strip. It's stable since the strip doesn't flip over. Place the pin in holes farther and farther down and notice where the strip becomes unstable. It's below the center of mass, a position easily found on the strip by balancing it on your finger.

Stability in a rocket is generally of two types: static and dynamic, if we ignore more complex motion. Both are used to describe the condition of remaining upright, or in the direction of flight. Another simple but effective analog of stability is the bowl and marble. If the bowl is upright (gravity required), the marble will roll from side to side if moved to one side, and gradually come to a stop in the center. This is a condition of positive stability. If, on the other hand, the bowl is inverted, the marble will roll off of the bowl regardless of where it's placed. This is negative stability. Positive and negative stability are both conditions in static and dynamic motion. For the more curious, neutral stability is a condition analogous to a marble on a flat, level plane. The marble will be displaced as far as the force is applied, and it is best visualized with a bit of friction on the rolling marble.

STATIC STABILITY

Static stability is the stable motion of a vehicle resulting from an applied force on an object at rest, while dynamic, or flight, stability is the resulting motion of a vehicle in motion due to applied forces, including aerodynamic forces. Other forces involved with dynamic stability, such as perturbations and rotational forces, will be omitted from this discussion for simplicity.

Static stability of the rocket, the pencil, or the balloon, is going to represent the balance of two opposing forces. Thrust on a rocket from the bottom is like pushing upward on a pencil tip with your finger, or a toy balloon being pushed from behind by escaping gas. Each of these represents static instability. If the rocket or the pencil or the balloon were pulled from the top, each would exhibit static stability. Since it is impractical to position the thrust of a rocket above its center of mass, the typical rocket, whether large or small, requires some method of static, low-speed, stabilization. The simplest solution for small rockets is to launch with a device to guide it until reaching a speed where aerodynamic forces can stabilize the rocket. This is usually accomplished with fins located at the bottom of the rocket that would then place it in a condition of dynamic stability. A more complex solution that is used on larger rockets employs thrust vectoring, or a set of independent guidance thrusters used to stabilize the vehicle throughout its powered flight.

DYNAMIC STABILITY

Flight stability, also called dynamic stability, generally refers to the stability of an aircraft or rocket in its flight through the atmosphere. Dynamic stability also involves more complex motion, including flight outside of the atmosphere, but will be simplified for this discussion by ignoring torques, rotations, and other perturbations.

There are similarities between the opposing forces in static stability and the opposing forces in dynamic stability. The rocket pushed from its aft or rear section by the motor requires a compensating force to stabilize the vehicle's trajectory. After the rocket motor thrust is cut off, the rocket will continue in stable flight if the vehicle's aerodynamic center of pressure is located behind the center of mass. In coasting flight, the rocket requires a stabilizing force toward the rear of the rocket. This is where you find stabilizing fins on small rockets. They are not intended to create aerodynamic drag, but to push in the direction opposite to the rocket's rotation from uneven air flow. The fins create the stabilizing force in the same way fins stabilize an arrow in its flight. You can imagine what would happen if the fins were placed on the front of the arrow, or if stabilizing fins were placed on top of the rocket (and not actively controlled as canards would be). One of the earliest applications of dynamic stability was implemented by Mongol warriors, who attached arrows to small rockets to improve the rocket's accuracy. Other advantages of the rocket-powered arrows were an increase in the arrow's range, and the psychological impact of long-range flaming arrows. And yes, the rockets were placed at the front of the arrow.

Rocket Propulsion Performance

The three primary measures of propulsion system performance are thrust, thrust duration, and thrust efficiency.

1. Thrust—the force available for launch or flight
2. Thrust duration—the time available for acceleration
3. Thrust efficiency—measured by the thrust produced by a specific mass of fuel, or specific impulse = I_{sp}

1. THRUST

Rocket thrust is the forward force created by the expulsion of exhaust gas rearward. For the chemical rocket motor, three factors dominate the magnitude of thrust produced, which are:

- Fuel flow rate
- Pressure difference between internal nozzle pressure and external (ambient) pressure
- Exhaust velocity

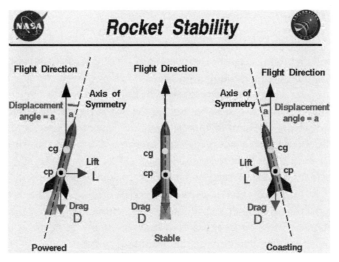

Figure 2.1. Rocket stability is portrayed in this diagram for two phases of flight: powered and coasting. Note first that the center of mass (shown as cg, or center of gravity) is ahead of the center of pressure (cp) for flight stability. Any movement of the rocket from stable flight (center) will push on the fins, creating lift in the opposite direction, and returning the rocket to stable flight. The cg ahead of cp provides positive dynamic stability for both powered and unpowered flight. Source: NASA Glenn Research Center.

A. Fuel Flow Rate

The magnitude of a rocket's forward thrust is proportional to the gas forced from the rocket's exhaust. Therefore, a high-thrust launch vehicle, for example, the Saturn V first stage, requires an enormous mass of gas coming from the engines at high velocity. Consequently, an enormous mass of fuel and oxidizer must be forced into the combustion chamber to provide the high-velocity, high-thrust exhaust. The mass fed into the engine is the same as the mass ejected, since the two must be conserved, but the volumes and speeds are much different. For example, each of the five first-stage F-1 engines on the Saturn V generated 1.5 million pounds force (lb_f) for a total thrust of 7.5 million lb_f at liftoff. That's not to say that 7.5 million pounds of fuel and oxidizer are fed through the engines each second to produce the equivalent thrust. The thrust generated in pounds (lb_p), or kilograms (kg_p), or Newtons (N) is equal to the propellant mass flow rate in pounds per second (kilograms per second) times the exhaust velocity in feet per second (meters per second).

Thrust = T = fuel mass flow rate × $V_{exhaust}$ = dm/dt × $V_{exhaust}$ where dm/dt is the fuel mass flow rate

B. Pressure Difference between Internal Nozzle Pressure and External (Ambient) Pressure

High-combustion chamber pressure is important in forcing the exhaust gas out of the nozzle. However, the exhaust gas flow is opposed by any atmospheric or gas pressure outside of the nozzle. A vacuum offers no flow resistance, hence, the maximum flow rate and the highest exhaust gas velocity through the exhaust nozzle are attained in the vacuum of space. The two largest factors in generating maximum pressure difference in a rocket motor are the combustion chamber pressure and the combustion chamber temperature that creates the pressure.

i. Combustion Chamber Pressure Chemical rockets use either solid or liquid propellants that burn rapidly, almost explosively. This produces high-pressure gas in the thrust (combustion) chamber that flows rapidly through the nozzle. The nozzle is designed to increase the exhaust speed even faster. Actually, as much as one-half of the chemical rocket's thrust comes from the optimized nozzle design. As combustion pressure increases, the expanding gas exhausts at higher speed, which produces greater exhaust gas momentum and greater resulting thrust. Is there a limit to chamber pressure? Yes, the maximum chamber pressure is limited by the strength of the combustion chamber, but is generally well below the maximum design pressure of the nozzle and combustion chamber, since exhaust velocity increases more slowly than nozzle pressure. As you might expect, the internal gas pressure is a function of the reaction rate, the reaction energy, the resulting maximum temperature of the reaction, and the nozzle's constriction and shape.

ii. Combustion Chamber Temperature To maximize the exhaust velocity coming from a combustion chamber, the combustion temperature must be maximized within design limits of the chamber and nozzle. Maximum combustion temperature is limited by the fuel and oxidizer's specific heat. There are a number of other factors that affect and/or limit the combustion temperature, but the maximum combustion chamber temperature cannot exceed the integrity of the chamber, nozzle, and injectors, all of which include a safety margin. High-performance rockets can have combustion temperatures that reach 3,320 to 3,870°C (6,000 to 7,000°F), which is well above the melting temperatures of any useful material. However, a cooler laminar exhaust flow keeps the nozzle surfaces lower than the gas temperatures. Often the rocket motor design employs some method to transfer heat away from the combustion chamber and nozzle surfaces by conduction. For liquid-fueled engines, the fuel or oxidizer is circulated in a chamber surrounding the nozzle and thrust chamber before injection, which removes much of the heat. Thermally conductive materials are commonly applied to the inner surface of combustion chambers and nozzles to draw heat away from the hottest regions in the engine, which are the inner surfaces. For large solid-fuel rocket motors, an ablative material can be used in the nozzle throat to protect the nozzle structure from being damaged by the high-temperature exhaust gas.

C. Exhaust Velocity

Thrust efficiency, which is most accurately represented by specific impulse, or I_{sp}, is proportional to exhaust velocity. There can be variations with nozzle geometries and atmospheric conditions; hence, an equivalent exhaust velocity must be used. Regardless, high exhaust velocity is the primary design objective, since that represents higher fuel efficiency and forward thrust. Factors that influence chemical rocket exhaust velocity include:

- **Exhaust gas molecular weight**: Lower-mass gas flows at a higher velocity, therefore the most efficient propellants for chemical rockets (not true for ion engines) have the lowest molecular weight. Hydrogen and oxygen, which, when combined, produce water, are currently the most efficient fuels available and have no adverse impact on the environment. Lower is better for exhaust gas molecular weight.
- **Combustion temperature**: Greater combustion chamber temperature produces greater combustion chamber pressure, which produces a greater expulsion force, which increases exhaust velocity. Although higher combustion chamber temperatures are better at generating thrust, limitations are imposed by the combustion chamber and nozzle materials. Higher combustion temperature is better.
- **Combustion chamber pressure**: Greater combustion chamber pressure produces a greater pressure difference between the combustion chamber and the outside, which in turn increases the pressure force on the exhaust gas's outward flow. Combustion temperature, exhaust flow, and nozzle constriction are the primary factors regulating combustion chamber pressure for the chemical rocket motor. Higher combustion pressure is better.
- **Specific heat ratio**: The chemical energy available to convert fuel into hot exhaust gas is fixed by the reaction chemistry of the fuel and oxidizer chosen. Some fuels are better than others.
- **Exhaust nozzle geometry**: Energy of the expanding gas in the combustion chamber is not just kinetic (velocity), but also potential. As the expanding gas flows through the exhaust nozzle, its character is different between subsonic and supersonic flow speeds. Optimal design for the nozzle is a convergent section for the gas exiting the combustion chamber, and a divergent section for the gas exiting the nozzle. Lots of design details are involved here.

2. THRUST DURATION

An important, although obvious, point is that two rockets with the same engines, but with different sized propellant tanks, would have the same thrust (lift), but a different thrust duration, or payload-distance capacity. A rocket's propellant mass which is proportional to the size of its tanks is equivalent to thrust duration because it represents the fuel burn time. Fuel tank size is also a measure of the energy available in the rocket's propulsion system, since it represents the available combustion energy that can be released during the entire propellant burn. We should also be able to see a direct link between the thrust duration of the rocket and the propulsion energy of the rocket, but there is a wrinkle. Thrust, which is equivalent to force times thrust duration, which is time, has dimensions of impulse, not energy. Force times distance has the correct dimensions of energy. Nevertheless, thrust duration represents the lift capacity of a rocket, which can be equated to an orbital or trajectory distance. Ignoring a few rather important details in launching payloads into space, the size of a launch vehicle approximates its propulsion energy. Another way to look at this is the intuitive approach—larger launch vehicles are required for larger payloads, and larger launch vehicles are needed for more distant destinations.

A useful representation of the thrust and thrust duration is the product of the two, or impulse, defined as force times time. Total impulse, which is the thrust for the entire burn duration, is often listed for specific orbits, or for interplanetary destinations. This is comparable to ΔV, which is the velocity change required to get to the same destination, but using different units.

3. THRUST EFFICIENCY

Thrust efficiency has been traditionally defined as specific impulse, or I_{sp}, which is the thrust produced for a specified amount of fuel. The expression of I_{sp} in relation to thrust fuel consumed is simply the ratio of thrust produced divided by the fuel weight flow rate. Fuel flow rate is used instead of fuel mass or weight since thrust produced is an instantaneous measure. Note that thrust is equivalent in dimensions to both force and weight, which cancels weight in both numerator and denominator. This gives I_{sp} the dimensions of seconds.

I_{sp} can be used as a measure of the theoretical efficiency of a propellant in generating thrust, or a measure of how efficient an engine is in converting fuel into thrust. Another way to look at specific impulse is that the I_{sp} value is the burn time of one kilogram (or lb) fuel with one kilogram (or lb) of thrust. A fuel with an I_{sp} that is two times another would burn twice as long with the same thrust. Because specific impulse represents the efficiency of a propellant or propulsion system, bigger is better. A brief list of I_{sp} values and ranges is given in the tables below.

The advantage of a larger I_{sp} is that a smaller fuel mass is required for a specified payload-distance, meaning that a smaller booster could be used for the same payload, or a larger payload could be carried for the same amount of fuel. Since smaller boosters are generally less costly than larger boosters, a higher I_{sp} propulsion system can offer a lower-cost launcher, although there are tradeoffs because of the more expensive technology typically needed for higher-efficiency engines.

Table 2.1. Approximate I_{sp} Ranges

Very high	1,000–10,000 sec	Ion and plasma engines
High	350–500 sec	Liquid bipropellant (liquid fuel + liquid oxidizer)
Moderate	200–350 sec	Solid fuel or liquid monopropellant (liquid fuel combined with oxidizer)
Low	0–200 sec	Cold (compressed) gas

Table 2.2. I_{sp} Examples

Engine	I_{sp}	Thrust
Space shuttle main engine (SSME)	453 s (vac) 363 s (sea level)	233,295 kg$_f$ (513,250 lb$_f$, 2.3 MN) (vac)
Space shuttle solid rocket boosters (SRB)	269 s (vac) 237 s (sea level)	1,500,000 kg$_f$ (3,300,000 lb$_f$, 14.8 MN) (sea level)
Saturn V F-1 first-stage engine	260 s (sea level)	681,180 kg$_f$ (1,500,000 lb$_f$, 6.7 MN) (sea level)

Very high-efficiency propulsion engines are available with plasma/ion (charged particle) engines that have I_{sp} values ranging from hundreds to thousands of seconds. These engines are extremely efficient, but are limited to very low thrust—typically 10^{-6} to 10^{-3} N (10^{-7} to 10^{-4} lb). Higher-thrust ion prototype engines have also been developed with nuclear reactor cores heating hydrogen gas to very high temperatures and pressures, with some incorporating electrical ion acceleration, although none have been flown.

Propellants

Propellant selection for chemical rockets is important, not just for maximizing combustion energy and exhaust gas velocity, but also for gaining advantage in density, cooling, storage, handling, cost, and more. In general, the easiest rocket fuels to make and handle are solids composed of a combined fuel and oxidizer stabilized in a fast-burning mixture. Solid-fuel motors are fairly simple to scale in size from very small to very large, and are normally cast in a combustion chamber as a unit. Liquid propellants can also have a combined fuel and oxidizer, a combination called monopropellant, and are generally simpler and less costly to store and handle than cryogenic bipropellants. The primary disadvantage of the liquid monopropellants and solid fuels are their lower I_{sp} ratings.

Liquid bipropellants consist of a separate fuel and oxidizer and have the highest I_{sp} of the chemical rocket propellants, although bipropellants can be found that have lower performance. A further consideration for a liquid propellant choice is the specific gravity, or density. Low-density propellants require a high-volume tank compared to higher-density fuels, making the rocket structure larger and potentially more costly. If powering a first stage, the aerodynamic drag and flight loads can be a problem for lower-density fuels, especially liquid hydrogen. Hydrogen makes an ideal propellant, but has the lowest density of any fuel, and also requires the lowest storage temperature of any of the cryogenic propellants.

A balance between fuel performance and density is seen in two examples—the Saturn V and the Space Transportation System (STS), the space shuttle. For first-stage boosters, aerodynamic drag and loading are important, hence, higher-density fuels are used. The Saturn V used RP-1 (refined kerosene) for the first stage, but higher-I_{sp} liquid oxygen and liquid hydrogen on the second and third stages. Tradeoffs in propellant selection for the space shuttle were more involved because of the high-thrust, solid-fuel boosters capable of making up for the drag induced by the large external liquid hydrogen fuel tank. Nonetheless, aerodynamic loading remains a concern for the STS because of the bulky solid-fuel rocket boosters and the large external tank.

As mentioned earlier, propellant choice is based on qualities that go well beyond performance and density, with two of the more important being safety and cost. A list of the more important considerations in the selection of rocket propellants includes:

- Specific impulse
- Cost
- Toxicity and health hazards
- Explosion and fire hazard
- Corrosion characteristics
- Handling safety
- Propulsion system computability
- Freezing/boiling-point temperatures
- Stability
- Heat-transfer properties
- Ignition, flame, and combustion properties

A nearly ideal propellant combination that furnishes the highest I_{sp}, and is both noncorrosive and nontoxic, is hydrogen and oxygen. The drawback to liquid oxygen (LOX) and liquid hydrogen (LH_2) is that they must be stored cryogenically, which makes the propellants expensive to transport, store, pump, and handle. For large boosters, this combination provides the best performance and one of the lowest-weight propellant choices. These performance parameters were recognized for nearly a century before being implemented for the first time in the Saturn V's second and third stages. Liquid oxygen and liquid hydrogen are the propellants of choice that have been used to get astronauts and cosmonauts to orbit, to space stations, and to the Moon and back.

SOLID PROPELLANTS

Solid rocket propellants contain a variety of chemicals in addition to the basic fuel and oxidizer, partly for stabilization and partly for performance. The stable, solid form of propellant has the advantages of not needing cryogenic or cold temperatures for storage, ease of handling, and long-term stability at low, moderate, and even elevated temperatures. Solid fuels used for larger rockets are composite mixtures containing separate granulated or powdered fuel and oxidizer, with a chemical binder, a stabilizer, and often an accelerant or catalyst added for improved performance and stability. The final mixture of solid fuel used today is a dense, rubber-like material that is cast into the combustion chamber. The cast fuel has a central cavity to allow burning through-

**Table 2.3. Common Liquid Oxidizer Propellants
(adapted from Sutton and Biblarz (2001))**

Oxidizer	Characteristics
Liquid oxygen (LOX)	• Not hypergolic but can combust spontaneously with many materials at elevated pressures • Most commonly used propellant oxidizer • Nontoxic and noncorrosive
Hydrogen peroxide (H_2O_2)	• Hydrogen peroxide decomposes into water, oxygen, and heat • Spontaneous decomposition on a catalyst such as platinum or iron oxide • Used for generating gas to drive turbopumps in the V-2, X-1 and X-15 rocket engines
Nitric acid (HNO_3)	• Highly corrosive • Hypergolic when combined with hydrazine and amines • Used for oxidizer with gasoline, amines, and hydrazine fuels • Red fuming nitric acid is nitric acid with the addition of 5–20% nitrogen dioxide • More stable, less corrosive than pure nitric acid • Inhibited red fuming nitric oxide has <1% fluorine ion (HF) added • Reduces corrosion
Nitrogen tetroxide (N_2O_4, NTO)	• Mildly corrosive unless mixed with water • Spontaneous combustion occurs when exposed to many materials • NTO is hypergolic when combined with most fuels • High vapor pressure requires relatively heavy tank • Used in numerous Russian rockets, the Titan booster series, and the space shuttle attitude control system • Highly toxic • Exposure limit < 5 ppm
Fluorine	• Highest theoretical Isp oxidizer • Fluorine and fluorides have been proposed in various fuel combinations that are highly corrosive, difficult to handle, and toxic • No fluorine oxidizers have been used for production rocket engines

out the length of the rocket motor, in shapes that can be anything from a simple cylinder to a star.

Solid rocket fuel is typically identified by the type of chemical binder used—either HTPB or PBAN. Hydroxyl-terminated polybutadiene, or HTPB, is a rubberlike binder that is stronger, more flexible, and faster-curing than PBAN, but suffers from a slightly lower I_{sp}, and uses fast-curing, toxic isocyanates. Polybutadiene acrylic acid acrylonitrile (PBAN), on the other hand, has a slightly higher I_{sp}, is less costly, and less toxic, which makes it popular for amateur rocket-makers. PBAN is also used in the large boosters, including the Titan III, the space shuttle's solid rocket boosters (SRBs),

**Table 2.4. Common Liquid Fuel Propellants
(adapted from Sutton and Biblarz (2001))**

Fuel	Characteristics
RP-1 (Highly refined kerosene)	• Developed as a fuel that could also be used for cooling high-temperature nozzles and combustion chambers • Sulfur, aromatics, and unwanted isomers removed to permit use at high temperatures • Greater stability, lower toxicity, less residue, higher performance than other hydrocarbons • High flash point 336 K • Safer, less explosive than many hydrocarbon fuels including gasoline • Used in early U.S. launchers, and Russian R-7 booster and its derivatives: Soviet N-1, Atlas, Thor, Delta I-III, Titan I, Saturn I, IB, V (1st stage)
Liquid hydrogen (H_2, LH_2)	• High specific impulse • Highly flammable when mixed with air • Increased density possible with supercooled solid or slush hydrogen (not yet used) • Nontoxic (breathable gas; can replace nitrogen in an artificial atmosphere) • Nontoxic exhaust gas when reacted with oxygen • Hydrogen makes most metals brittle, making turbopump design more challenging than with other fuels
Methane (CH_4)	• Hydrocarbon fuel • Stored cryogenically • Low cost; freely available from gas wells, biomass decomposition • Potentially useful for Mars return missions • Under research but not used in production liquid fuel engines • Possible fuel for arc-jet or resistojet thrusters
Hydrazine (N_2H_4)	• Used as both a monopropellant and a bipropellant fuel • Hypergolic • Can be used as a monopropellant fuel when exposed to a catalyst • Bipropellant with spontaneous combustion when mixed with nitrogen tetroxide or nitric acid • Also combustible if exposed as a liquid to air, or in contact with many materials • Used in the production of stainless steel, nickel, and some aluminum alloys • Not used in iron, copper, or some aluminum alloys • Very long storage life • Highly toxic • Exposure limit < 0.1 ppm • Known carcinogen • Has been used as a monopropellant for gas generators (e.g., space shuttle hydraulic system) • Often used for spacecraft attitude control

Monomethylhydrazine (CH_3NHNH_2, MMH)	• Better liquid temperature range than hydrazine • Lower reaction threshold to shock waves than hydrazine • Slightly lower I_{sp} than hydrazine • Highly toxic • Exposure limit < 0.2 ppm • Suspected carcinogen
Unsymmetrical dimethylhydrazine (($CH_3)_2NNH_2$, UDMH)	• Generally more stable than hydrazine, with increased high-temperature stability • Often used as mixture with 30% hydrazine • Highly toxic • Exposure limit < 0.5 ppm • Known carcinogen

Table 2.5. Characteristics of Common Liquid Bipropellant Combinations (adapted from Sutton and Biblarz (2001))

Oxidizer	Fuel	I_{sp} (theoretical, vacuum)	I_{sp} (theoretical, 1 atm)
Liquid oxygen (LOX)	Liquid hydrogen (LH_2)	477 s	390 s
LOX	Kerosene (RP-1)	370 s	300 s
LOX	Monomethyl hydrazine	365 s	301 s
LOX	Methane (CH_4)	368 s	296 s
Liquid ozone (O_3)	Hydrogen	580 s	
Nitrogen tetroxide (N_2O_4)	Hydrazine (N_2H_4)	334 s	292 s
Red fuming nitric acid	RP-1		269 s
Hydrogen peroxide (H_2O_2)	Monopropellant	154 s (90% H_2O_2)	
H_2O_2	RP-1		279 s
Fluorine (Fl)	Lithium (powdered)	542 s	
Fl	Hydrogen	580 s	410 s

and NASA's new Constellation Ares I and Ares V launchers. HTPB is or has been used in the Delta II, Delta III, Delta IV, Titan IVB, and Ariane launchers.

Arguably, the most well-known solid rockets are the space shuttle's SRBs, powered by a PBAN-based ammonium perchlorate composite propellant (APCP). The dual SRBs used to launch the space shuttle develop a specific impulse of 242 seconds at sea level, and 268 seconds in a vacuum. The SRB composition consists of the following by weight:

- Ammonium perchlorate oxidizer (69.6%)
- Powdered aluminum fuel (16%)
- Iron oxide catalyst (0.4%)
- PBAN polymer binder that is also a secondary fuel (12.0%)
- Epoxy curing agent (2.0%)

Aluminum is used as the fuel for the SRB and a variety of other solid rocket motors because it has a reasonable specific energy density, and a high volumetric energy density. Aluminum powder is also difficult to ignite accidentally. Ammonium perchlorate is an attractive oxidizer because of its high oxygen content, its stable crystal form, and its relatively fast-burning character.

Large solid rocket motors are also used for auxiliary boost on the first stage of a variety of expendable launchers. These include the Delta rocket's graphite epoxy motors (GEMs), the earlier Titan III and IV solid rocket motors, the Arianne V's solid rocket boosters, the Chinese Long March strap-on boosters, and the Russian Proton strap-on solid boosters.

Stand-alone solid rocket launchers are available as suborbital boosters and ICBMs, although several are able to place small payloads in low Earth orbit, including the Scout and Pegasus. In general, the earlier liquid-fuel ICBMs have been replaced by the solid-fuel varieties because of their low cost and long storage life. Suborbital solid rocket launchers include the Taurus, the Scout, and the Minotaur, while the solid-fuel ICBMs include the U.S. Minuteman (land-based), the Trident (submarine-launched), the Russian RT-23, the Chinese DF-31, and a host of others.

Smaller solid rocket boosters are used as upper stages that can take payloads to geostationary orbit, to the Moon, or even on interplanetary trajectories. Two of the most common upper-stage solid boosters are the single-stage Payload Assist Module (PAM) and the Boeing Inertial Upper Stage (IUS). The PAM includes the Star-48 motor that was developed for satellites launched from the space shuttle, and is currently used on the Delta launcher. The dual-stage IUS booster was used on larger launchers, including the Titan, Atlas, and the space shuttle. IUS was employed primarily as a booster for military DSP surveillance satellites and NASA's TDRSS communication satellites, but later used for several large interplanetary spacecraft. Because of the *Challenger* accident, the space shuttle orbiters could not use the more powerful and more efficient Centaur liquid-fuel booster, leaving the IUS as the only alternative. Since the space shuttle is no longer allowed to launch satellites, the IUS is on inactive status and unlikely to return to America's upper-stage booster inventory.

Figure 2.2. Lift-off of a Trident sea-launched ballistic missile (SLBM) following its initial push from the submarine to the surface by compressed gas. Source: U.S. Navy Archives.

LIQUID FUEL ROCKET ENGINE—PRIMARY COMPONENTS

The typical liquid fuel rocket engine is a complex design product with a dizzying number of variables. Even so, the basic operating principles of the liquid chemical rocket engine can be introduced using three basic components: the combustion chamber, the exhaust nozzle, and the turbopump.

Combustion (Thrust) Chamber

A rocket combustion chamber is designed to accommodate stable, high-temperature, high-pressure combustion, and allow efficient propellant mixing in a violently turbulent environment, while removing excess heat. Cooling of the combustion chamber walls is critical in high-thrust engines because of the extremely high combustion temperatures that can reach 3,870°C (6,000°F). High-thrust combustion chambers must also be designed to allow efficient propellant mixing, yet shaped to suppress acoustic standing waves that contribute to combustion instability. The combustion chamber must also include baffling near the injectors to avoid gas swirling that can also contribute to unstable combustion. As liquid rocket motors become larger, these instabilities become more important since they can lead to degraded thrust performance or to extinguished combustion or even to the damage or destruction of the engine.

The basic combustion chamber design is somewhere between spherical and cylindrical in shape, with an inner wall and an outer wall to allow cooling fluid—one of the propellants, usually the fuel—to circulate between before being forced through the injector system. Heat removed from the engine in the process is added to the propellant before injection, adding to the propellant's heat energy. The process is called regenerative cooling and first appeared on the small Soviet GDL engines in the 1930s, and later on the German V-2 engines. Regenerative cooling is used on medium-to-large liquid-fuel engines and is well suited to the space shuttle main engines (SSMEs) since engine cooling is required, and because liquid hydrogen fuel is the coldest cryogenic propellant available and must be heated before reaching the injectors. Regenerative cooling on the SSME is one of the most extensive circulation cycles of any engine. The cryogenic hydrogen fuel is routed through the SSME nozzle, combustion chamber jacket, and injector head, before entering the injector tubes that are also cooled in the process.

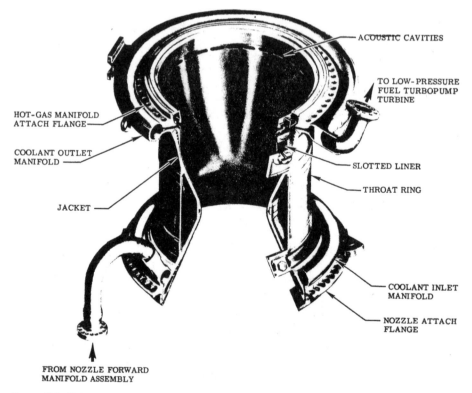

Figure 2.3. Cutaway drawing of the space shuttle main engine combustion chamber and exhaust nozzle throat. Not shown but attached to the hot gas flange at the top is the injector head assembly and the nozzle extension that is attached to the nozzle attach flange at the bottom. Liquid hydrogen is circulated between the chamber wall and the outer jacket and through the coolant loop shown at the upper right. Source: NASA SOFM.

Combustion chamber design is also aimed at confining the hot combustion gas for optimal pressures and temperatures to maximize exhaust velocity. This is accomplished with a constriction at the combustion chamber exit that is also the entrance of the exhaust nozzle, and a bell-shaped nozzle exit fairing to couple the internal pressure to the ambient (outside) pressure with a specific expansion factor called the expansion ratio or area ratio.

Exhaust Nozzle

A rocket exhaust nozzle's primary function is to accelerate the exhaust gas coming from the combustion chamber to a designed maximum velocity. This is accomplished by first constricting the exhaust from the combustion chamber to increase the gas pressure internally, while restricting the internal exhaust flow to subsonic speeds. The exhaust gas is then expanded beyond the nozzle's narrow throat to supersonic speed, pushing the exhaust gas to its maximum velocity. Ideally, the gas exiting the nozzle is at the same pressure as the outside gas pressure, which, for the vacuum of space, is approximated with an extended nozzle. The bell-shaped nozzle extension offers a better compromise than the extended conical nozzle because of its shorter length and its higher expansion performance.

Unusual as it may seem, supersonic gas flow through a divergent cone will increase the flow speed, while subsonic flow will decrease in speed. This is due to the expansion waves created in supersonic flow. A sketch of the components will help to identify the flow associated with the two basic nozzle conical shapes. In subsonic gas flow (less than the speed of sound), the gas will accelerate if constricted in a converging, open-ended cone as sketched in figure 2.4a. Conversely, subsonic gas flow through a divergent cone will decrease gas flow speed. For supersonic flow, the opposite relationship applies. Supersonic gas flowing through a convergent nozzle will decrease in speed, but will increase when flowing through a divergent cone, as shown in figures 2.4b and 2.4d.

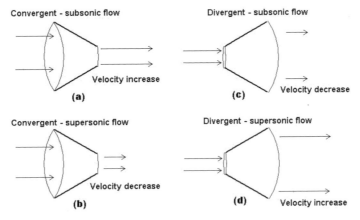

Figure 2.4. Sketch of the relative change in gas velocity flowing through divergent and convergent open-ended conical geometries for both subsonic and supersonic flow.

For a simple nozzle, two sections from those shown above would be arranged to do two things:

1. Generate the highest exhaust velocity exiting the nozzle
2. Match the pressure flow from the combustion chamber to the ambient atmosphere, or the vacuum of space

The first criterion is accomplished by forcing subsonic flow from the combustion chamber to the nozzle throat to increase the exhaust gas pressure, then transition to supersonic exhaust flow at the wide nozzle exit in the divergent section. Not surprisingly, this arrangement is shaped similarly to any conventional rocket nozzle, from model rockets to the space shuttle main engines.

Two important nozzle design variables dominate the performance of the rocket motor:

1. Nozzle throat size: The nozzle throat diameter is chosen to restrict (choke) the flow and set the mass flow rate of the motor with a targeted speed of sound (Mach 1). The product of mass flow rate and exit velocity determines engine thrust.
2. Area ratio (nozzle exit area divided by the nozzle throat area): Maximum exhaust exit velocity is determined by the gas expansion, which is determined by the area ratio. The expansion (area) ratio also determines the gas exit temperature and the resulting exit gas pressure. Exit gas pressure should match ambient pressure, so other design variables come into play. As mentioned above, the product of the maximum exit velocity and mass flow rate determines engine thrust, which makes these two variables primary in determining engine thrust.

A simple conical shape for the divergent nozzle is not the ideal shape for maximizing exhaust velocity and optimizing exhaust exit pressure. Moreover, matching a conventional bell shape of the exit of the nozzle in order to produce an exhaust pressure equal to atmospheric pressure is important, but only applicable for a specific altitude or atmospheric pressure. Nozzle design for a vacuum or for sea level operation is a relatively simple process. Designing the nozzle for a rocket traveling from low altitude (higher ambient pressure) to high altitude (lower ambient pressure) becomes a compromise in the nozzle shape and the expansion area ratio, which introduces more variables.

Other important nozzle design considerations are the active cooling of the nozzle and the flange extension. Even modest-sized liquid engines have far more combustion heat than can be removed from the nozzle by radiation cooling. Hence, an active cooling system is used to circulate one of the liquid propellants around the nozzle for cooling. Like the combustion chamber, nozzle regenerative cooling removes the heat from the inner nozzle surface by conduction. This makes high thermal conductivity important in the selection of materials for the nozzle, as it does for the combustion chamber.

The earliest rocket engine with active cooling was also the first liquid-fuel rocket launched—Goddard's simple LOX and gasoline engine. His active cooling design used coolant tubes wound around the exterior of the nozzle, which also strengthened the nozzle. This technique was duplicated in the German V-2 engine, but using alcohol

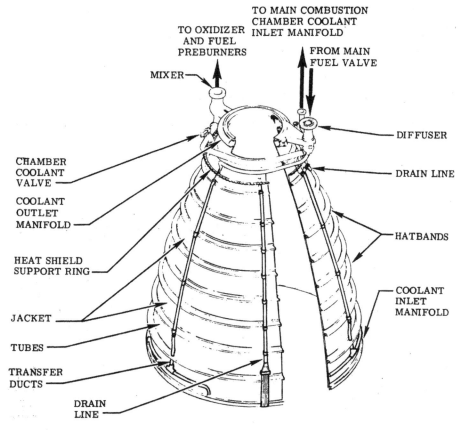

Figure 2.5. Sketch of the space shuttle main engine nozzle extension, including cooling components and support structures. Source: NASA SOFM.

fuel and a combustion chamber cooling jacket. A significant improvement in cooling efficiency was later made by brazing the cooling tubes to the inner nozzle wall parallel to the exhaust gas flow. Reaction Motors Inc. (RMI) is credited with this technique first used in the late 1940s. The success of this innovation is demonstrated by the fact the nearly every production liquid-fuel rocket engine since has employed some variation of RMI's "spaghetti construction" coolant flow method, including the space shuttle main engine design. To overcome the weakness in the parallel channels against the high-pressure stress from combustion, the nozzle is reinforced with external circular bands or hoops (see figure 2.5 above). Interior surfaces of these nozzles are often high-temperature, high-conductivity copper or nickel alloys.

Turbopumps

Propellant flowing from the storage tanks through the injectors into the combustion chamber can be driven either by pressurizing the tanks or by forcing the propellants through the injectors with high-pressure pumps. In general, smaller engines benefit from pressure-fed propellant flow because of the lower weight and the simplified design. High-thrust, high-efficiency liquid engines require high-pressure turbopumps to feed the liquid propellants at the proper pressure, and in the proper ratio.

The flow rate for liquid propellants is representative of the exhaust gas mass flow rate, which, along with exhaust velocity, determines thrust. Propellant flow rate can be expressed by the same equation for specific impulse.

Propellant weight flow rate = Thrust/I_{sp}

For a 100,000 N (245,000 lb$_f$) thrust engine with an I_{sp} of 400 s, 250 kg/s (550 lb/s) of propellant would have to be fed into the combustion chamber through the injector system. Calculation of an approximate exhaust velocity for this engine would be:

$V_{exhaust}$ = g I_{sp}, where g is Earth's surface gravity constant of 9.81 m/s^2. Hence
$V_{exhaust}$ = 9.8 m/s^2 × 400 s = 3,920 m/s.

Propellant turbopump systems use several methods to drive the turbines, including the two most common—gas generators and staged combustion systems. Gas generators create expanding hot gas from the combustion of monopropellants such as hydrazine or hydrogen peroxide to push the turbine blades. Staged combustion systems burn a small amount of engine propellants with the exhaust directed past the pump turbine. Modern engine designs route the spent exhaust from the turbine into the main engine injector system, making it a staged, or two-stage, combustion cycle.

The three most common types of rocket engine turbopump systems are:

- Gas generator—separate propellant driving turbopump (Goddard's P-series rockets, V-2, Redstone are examples)
- Expander cycle—fuel used to cool the combustion chamber and nozzle is heated in the process and then passed through the turbine to drive the turbopump
- Two-stage (staged) combustion—combustion of a small amount of engine propellants is used to drive turbopumps (the space shuttle main engine is one example)

Early Rocket Development: From World War II to Apollo

Germany's defeat in World War I severely limited the scale and number of its military weapons until the Nazis began preparations for the domination of Europe, and ultimately World War II. Since rockets were not a threat in World War I, the Versailles Treaty that prohibited Germany from rebuilding its armies did not specifically ban rockets. As the Nazis began a weapons buildup ahead of World War II, Hitler's interest in long-range rockets as weapons led to the formation of a missile group directed by Walter Dornberger and including a number of scientists from the German VfR rocket club. Development of the new rocket technology began in earnest in 1932 with plans for a much heavier warhead and a much greater reach than any artillery piece.

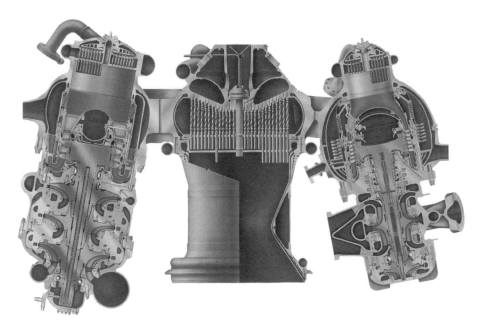

Figure 2.6. Cutaway detail of the space shuttle main engine high-pressure turbopumps, along with the injector head and injector pins, ignitors, and the main combustion chamber. Both the oxidizer high-pressure turbopump on the right and the high-pressure hydrogen turbopump on the left are fed with low-pressure pumps (not shown) and driven by their own smaller gas generator combustion chambers located on the top of each assembly. The multistage turbine drives for both are located near the center, with the impeller located at the bottom of the turbopumps used to drive the propellants into the injector head assembly at top center. Injector pins are shown at the top (head) of the combustion chamber, as well as in the small preburner combustion chambers at the top of each turbopump. Exhaust gas from the preburners is routed through the injector manifold that physically connects the top of both high-pressure turbopumps to the engine injector head assembly. Each of the three combustion chambers includes a spark ignitor system at the top center. Source: NASA SOFM.

By 1933, a liquid-fuel engine had been designed to power the first of the Nazi A-series ballistic missiles, the A-1. Parallel development of the rocket, the guidance and control systems, and the high-explosive warheads took nearly a decade to complete. Preliminary test flights were made with the long-range A-4 missile in 1942, followed by the V-2 production model launched against nearby Allied cities. A decade of delays in the development of a hurried but well-funded weapons project attests to the myriad problems that had to be solved. One of the most difficult of these problems was controlled supersonic flight, which had never before been attempted. Another was the nozzle design for exhaust gas velocities that could push the V-2 to Mach 4.

Even though Germany's long-range missile project went unnoticed during its development, Allied intelligence quickly pinpointed the German missile test launches from Peenemunde, located on the North Sea coast of northern Germany. Soon after, Allied bombers attacked the A-4/V-2 center, which forced the Nazi's V-2 program underground in the nearby Harz Mountains. To avoid destruction of the V-2 production line, individual missiles and launchers were transported by rail from the Mittelwerk facility beneath the Harz Mountains to the launch sites around western Germany.

V-2

Germany's A-4 design challenges included an inertial guidance system, patented by Robert Goddard, that was only crudely effective and required augmentation with ground-based radio signals to improve accuracy. The project was costly and too late in the war to make a significant difference in the outcome. But from the V-2 project came a research tool that both the United States and the Soviet Union exploited in their Cold War civil and military rocket programs.

As Germany collapsed at the end of the war in 1945, both the United States and the USSR dispatched military units to quickly collect as many of the V-2 weapons, missile and production materials, documents, and personnel as possible. The Americans got the lion's share of the V-2 equipment and personnel, although the Soviet Union captured a number of V-2 engineering staff and some of the missile equipment and documentation. Britain was also involved with the V-2 weapons collection, but on a smaller scale. Britain launched several of the V-2 missiles after war's end, and even used the basic V-2 design for their Blue Streak and Black Knight missile projects. The French reportedly also used captured V-2 technology for their first long-range missile programs.

As director of the Nazi V-2 missile development program, Wernher von Braun also led various teams in the planning stages of the more powerful, intercontinental A-9, A-10, and A-12 heavy-lift, winged weapons. The much larger A-10 and A-12 boosters also had space exploration applications, which were far more interesting to von Braun. Rumors of his interest in space travel reached the Gestapo, who questioned von Braun about his attitude toward the Nazi war effort. Von Braun was also perceived as a flight risk because he was an active pilot, which led the Gestapo to jail von Braun for several weeks until his commander intervened directly through Hitler and had him released.

Germany's inevitable defeat in 1945 drove von Braun to lead a team of his closest missile designers and scientists on a roundabout route to surrender to the Americans in Bavaria. His alternatives were to surrender to the Soviets and their brutal dictator, Joseph Stalin, or be recaptured by the SS and likely face a death squad. Von Braun and his team were successful in their surrender and even more successful in their work for the U.S. missile programs afterward.

Tremendous advances were made in rocket technology with Germany's V-2 missile during World War II, while America's efforts were limited to small missiles built for the Army by the Jet Propulsion Lab. Almost overnight, the captured V-2s offered a reliable platform that could reach altitudes never explored before and that far surpassed both the U.S. and Russian rocket programs. A few of the more important technology advances of the V-2 are listed below.

- The V-2 engine had a dual-wall steel alloy chamber regeneratively cooled with the alcohol fuel before being injected into the combustion chamber. The dual-wall design proved successful enough to be adopted in many subsequent engine designs, including the space shuttle main engine.

Figure 2.7. Allies view captured V-2 missiles at the Mittlelwerk underground facility in the Harz mountains. Source: U.S. Army archives.

- Broad "shower head" injectors were used with burner cups to improve dispersion and cooling of the combustion chamber walls, while improving fuel and oxidizer mixing.
- The fuel and oxidizer turbopump was driven by a hydrogen peroxide gas generator that supplied 600 hp to the propellant turbine. The V-2 turbopump machinery was constructed mostly of light-weight aluminum alloys for the housing, impellers, and turbine.
- The aerodynamic shape of the missile and the nozzle design pioneered the first supersonic and hypersonic flight vehicles.

Guidance during the V-2's ascent phase was accomplished by placing graphite vanes in the exhaust flow, a concept first designed and patented by Goddard. V-2 guidance was augmented with rudders on the outboard corner of the four guidance fins. Two independent gyroscopes drove vane and rudder control surfaces to command the trajectory program that could also counter in-flight oscillations and prevent adverse yaw rotation.

As innovative as the V-2 was, its design was not without its flaws. One of the V-2's greatest limitations was a minimum ballast mass requirement for stable flight. This meant that the payload was only part of the non-propulsion mass, which reduced both its range and altitude. The V-2's turbopump was also a problem because of the weak turbine blades, which compromised the missile's reliability. Yet the importance of the V-2 was not in its original use as a long-range weapon but in its later application as a research platform.

- 5,000 built
- 3,200 launched
- 7,200 Allied fatalities
- 20,000 prisoners died in the production program

Table 2.6. V-2 Specs

Fuel	75% ethyl alcohol, 25% water
Oxidizer	Liquid oxygen
Thrust	24,958 N (55,000 lb)
Turbopump	Gas generator driven Tanks pressurized by nitrogen gas
Propellant	Hydrogen peroxide (80 %) and a mixture of 66% sodium permanganate with 33% water
I_{sp}	239 s
Empty weight	4,539 kg (10,000 lbs)
Launch weight	12,700 kg (28,000 lbs)
Maximum velocity	4,400 ft/sec (1,341 m/sec)
Maximum altitude	83 to 93 km (52 to 60 mi)
Maximum range	321 to 362 km (200 to 225 mi)

V-2 as a Research Tool

After Germany's defeat in World War II, its captured V-2 missiles and program engineers were transported to the United States and the USSR for their respective missile and space research projects. Improvements began almost immediately with extensive instrumentation and inertial navigation system enhancements. The original V-2 rockets launched from White Sands could reach an altitude of 97 km (60 mi) on an optimized suborbital flight with its single-stage. To increase its altitude, a second stage WAC Corporal liquid rocket built by the Jet Propulsion Lab was attached to the V-2. The combined WAC/V-2 booster, called Bumper, was first launched in February 1948, reaching an altitude record of 402 km (250 mi) and a speed of 8,050 kph (5,000 mph). But with the increased range of the new Bumper, the White Sands missile range was stretched to its limits of safety. In May 1947, a wayward V-2 overflew El Paso, Texas, and landed near Juarez, Mexico, creating an international incident. A Joint Chiefs of Staff study led to the decision to create a long-range missile range on federal property at Cape Canaveral, Florida, with the Atlantic Ocean serving as a buffer between the launch site and the nearest inhabited land. Approval of the site in 1949 was followed a year later by launch of the first rocket from the range—the Bumper 8—on July 24, 1950.

Before the end of World War II, German missile designers began experimenting with a winged upper stage on the V-2 that could extend the range of a warhead to 500 km (310 mi) or more. The program progressed through the development stage and the launch of two A-9 vehicles. Although both flights ultimately failed, the second winged two-stage V-2 did succeed in reaching Mach 4 before the second-

Figure 2.8. Launch of the Bumper research rocket from the White Sands missile range in New Mexico. Source: U.S. Army archives.

stage breakup. While the A-9 project did not involve von Braun directly, he did outline plans for an intercontinental rocket, incorporating a much larger booster with a winged capsule capable of carrying a pilot and warhead from Germany to the United States with the intention of bombing New York. That multistage winged rocket later surfaced in a series of related, but far less diabolical, concepts of the first manned rockets to the Moon and Mars that appeared in *Collier's* magazine between 1952 and 1954.

ARMY HERMES MISSILE

Project Hermes was a joint venture of the Army and the General Electric Company that also encompassed two V-2 flight development contracts, as well as several V-2 Bumper flights. Hermes was not just a research program but a collection of suborbital rockets that resembled German missiles designed by Wernher von Braun and Walter Dornberger before and during World War II.

Hermes A-1 was the first rocket in the series and was small in size and based on the German Wasserfall surface-to-air missile converted into a test vehicle for flights at

Figure 2.9.
Launch of a sin-
gle-stage Hermes
A-1 test vehicle
at White Sands.
Source: U.S. Army
archives.

the White Sands missile range. Fuel and oxidizer for the smaller engine were liquid oxygen and alcohol, but the altitude and range were far less than the V-2. Follow-up designs included the Hermes A-2 that would have employed a solid-fuel engine, but was never flown, and the Hermes 3A and 3B which had a range of approximately 241 km (150 mi). Although only a handful of these vehicles were ever built and tested at White Sands, the Hermes 3A and 3B furnished a useful guidance platform and radio guidance system that were adopted for longer-range missiles that included the Army's Redstone and Sergeant (Bullard 1965).

Three other Hermes prototype models were launched at White Sands including possibly the most unusual V-2 derivative, the Hermes II. The 2-stage research plat-form employed a V-2 as the first stage to power a second-stage booster that included a ramjet engine. None of the four Hermes II test launches succeeded in proving the ramjet technology, however.

Hermes C was a large three-stage rocket that consisted of a six-engine first stage, and a third-stage glider similar to the A-9 design. The boost-glide system had a planned range of 3,200 km (1,988 mi), but was never built. Hermes C1 was a more advanced version of the single-stage V-2 missile, but with a planned range of 805 km (500 mi). Arguably the most important of the Hermes series, the C1 was used as the prototype for the Redstone and the Sergeant missiles. The Army's transition to the Redstone and other missile programs led to the cancellation of the Hermes program in December of 1952 (Bullard 1965).

REDSTONE MISSILE

The U.S. Army's tactical Redstone missile that originated as the Hermes project in 1949 was one of America's most productive boosters in the early space race with the Soviet Union. Redstone was developed as an intermediate-range nuclear weapons delivery system within the Army Ordinance Guided Missile Center. An important transition took place in 1956 with the conversion of the Army Ordinance Guided Missile Center into the Army Ballistic Missile Agency (ABMA). Located in Huntsville, Alabama, and headed by Major General John Medaris, the ABMA was centered on a research and development team led by Werner von Braun, a fact that accounted for many of the similarities in the V-2 and the Redstone designs. Although the Redstone was a product

Figure 2.10. Launch preparation of the Hermes II with the second-stage winged ramjet atop the V-2 first stage. Source: U.S. Army archives.

of the ABMA, the engine contract was awarded to the North American rocket group (later becoming Rocketdyne) that chose to use the same propellants and many of the features of the V-2 engine. V-2 technology was also used as a template for the Redstone's guidance system, using exhaust vanes and aerodynamic fins, but with a number of improvements. Liquid oxygen and alcohol propellants were also used for the Redstone engine, as was the gas generator turbine for the turbopump. There were major performance differences in the Redstone and V-2 missiles, including a 40% thrust improvement in the Redstone engine that was due, in part, to a new injector mechanism that improved combustion stability.

One of the Redstone's inherent weaknesses that had little impact on its later mission as a space launcher was the use of cryogenic liquid oxygen. As with other cryogenic-fueled ballistic missiles, like the Atlas and the Soviet R-7, the cost of loading, unloading, and refurbishing the booster for the next propellant load limited the missile's use to a fraction of its launch preparation time. Even so, the Redstone intermediate-range ballistic missile (IRBM) was enough of a success to be deployed as a tactical nuclear weapon throughout the United States and Western Europe until 1964.

Initial development of the Redstone by the Army Ordinance Guided Missile group was turned over to the Chrysler Corporation for production as a tactical IRBM in 1952. First launch of the Redstone in August 1953 was followed by the first American deployment of the nuclear IRBM in 1954. Two years later came one of the most important launches of the Redstone in a three-stage booster called Jupiter C (C for composite reentry capsule) that carried a reentry vehicle. The September 20, 1956, launch of the three-stage Redstone/Jupiter C payload reached an altitude of 1,095 km (680 mi) and a downrange distance of 5,310 km (3,300 mi).

Like his Soviet counterpart Sergei Korolev, Wernher von Braun was far more interested in space exploration than missile-making, and like his Soviet counterpart, he was in the fortunate position of being able to do both, and with more accessible military funding. The Redstone project was already underway when the United States announced its intention to launch a research satellite in support of the International Geophysical Year in 1957. Von Braun convinced his superiors to develop a multistage Redstone to launch the first satellite, and, within a year, spurred his team to prepare and launch a three-stage Redstone called Jupiter-C to an altitude of more than 1,000 km. Everyone in the project knew that this launch achieved the near-orbital velocity of Mach 18, and that a satellite could have been orbited simply by adding a single rocket as a final fourth stage. But since the ABMA was directed not to launch a satellite before the Naval Research Lab's Vanguard satellite, the first successful Redstone satellite launch was delayed by nearly eighteen months. When the directive came to launch Explorer following the Vanguard failure, the ABMA Jupiter C/Juno 1, with a fourth stage and a satellite payload added, was prepared and launched within a month. Had the Redstone-Explorer been allowed to launch at its first opportunity, there is little doubt that the United States could have put a satellite in orbit more than a year before the Russians.

Table 2.7. Redstone Missile Specs (Initial)

Length	21.12 m (69 ft 4 in)
Width	1.78 m (70 in)
Weight, empty	7,489 kg (16,510 lb)
Weight, loaded	27,826 kg (61,345 lb)
Payload	2,858 kg (6,300 lb)
Range	92.5 to 323 km (57.5 to 201 mi)
Altitude	63–106 km (34–57 mi) (single stage)
Flight time	288–375 sec
Engine	Rocketdyne NAA 75-110 (S-3) models A-1 to A-7
Fuel and oxidizer	Ethyl alcohol and liquid oxygen
Thrust	35,380–37,650 kg_f (78,000-83,000 lb_f)

Mercury-Redstone

Following the emergence of the National Aeronautics and Space Administration (NASA) on October 1, 1958, its agency managers took over the tasks and responsibilities of its predecessor, the National Advisory Committee for Aeronautics (NACA). The NASA leadership directed its research center personnel to begin organizing and executing America's first space exploration initiatives. One of the most important of those was the manned Mercury project that was already in the planning stages. Since the preferred Atlas booster was not yet tested or man-rated, the Redstone was adopted as the launcher for the initial Mercury suborbital manned space flights scheduled for 1961. However, conversion of the Redstone missile to the manned Mercury-Redstone booster needed over 600 component and system additions or improvements for its man-rating. A partial list of the major changes included (NASA 1964):

- Booster stage
  - Lengthened 6 ft to provide 20 s additional thrust from the larger tanks.
  - Total weight increased to 66,000 lb.
- Engine
  - Thrust was increased to 78,000 lb.
  - Improvements were made in the hydrogen peroxide turbopump.
  - Anti-fire squelch components were added.
  - Improvements were made in engine reliability and operational stability.
- Instrumentation
  - Control sensing unit was added to signal errors and malfunctions.
  - Telemetry was added to provide readings on attitude, vibration, acceleration, temperatures, pressures, thrust level, etc.
- Flight control
  - A simpler, more reliable flight control unit was incorporated to increase stability and reduce vehicle drift.
- Abort
  - Abort management instrumentation and control components were added in order to identify problems in thrust levels, engine vibration, electrical failure, etc.
  - Control functions were expanded for engine shutdown, booster separation, and escape rocket activation, either from an automated sequencer or from manual commands from the pilot-astronaut in the Mercury capsule.

Redstone proved its worth as an important tool in America's early space exploration efforts, from the first satellite and interplanetary launchers, to the first manned boosters. Redstone's legacy extended well beyond its many space program successes and its launch capability. One of its greatest contributions to rocket technology was the Redstone engine, which was redesigned by Rocketdyne and used on the Thor, Atlas, Viking, and Saturn I/IB boosters. Less spectacular but still important were the Redstone propellant tanks, which were lengthened and adapted for use in the Apollo Saturn I and Saturn IB launchers.

Figure 2.11. Photograph of the Jupiter-C assembly plant at the Army Redstone facility showing the Juno-I launcher being assembled in the center, with the narrow Explorer-1 satellite and its fourth-stage Sargent solid rocket motor mounted at the top (lower right). Explorer-1's aerodynamic shroud is located just below and to the left of the Explorer-1 spacecraft and third-stage booster ring. Source: U.S. Army archives.

NAVAHO CRUISE MISSILE

Prior to World War II, America's rocket programs consisted of small missiles and scientific sounding rockets sponsored by the military. An early and successful contributor to these early projects was the Jet Propulsion Lab that began as a U.S. Army–funded propulsion research laboratory. Another successful American rocket builder was the aircraft designer North American Aviation (NAA), which later became North American Rockwell and would eventually win the contract to build NASA's space shuttle orbiters. North American Aviation also included propulsion expertise and was awarded its first military and civil rocket projects in the late 1940s. In what would later become Rocketdyne, the NAA propulsion development group expanded quickly because of the escalating Cold War with the Soviet Union, and the need for more advanced missiles.

North American Aviation's decision to gather the best rocket propulsion designers and managers to expand into missile and space launcher projects helped them

Figure 2.12. Photo of the V-2 56,000 lb thrust engine on the left and North American Navajo 75,000 lb thrust engine on the right. Note the smaller, cylindrically shaped combustion chamber and simpler injector platform at the top of the Navajo engine. Source: NASA SP-4221.

win the contract to develop high-thrust engines for the Navaho MX-770 cruise missile for the Air Force using an advanced variation of the V-2 engine design. The Navaho booster engines were uprated from 25,400 kg$_f$ to 34,020 kg$_f$ (56,000 lb$_f$ to 75,000 lb$_f$) thrust to bring the vehicle up to operating speed for the cruise missile's ram-jet engine.

North American Aviation's Navaho engine caught the interest of Wernher von Braun while he was working on the Army Redstone booster project. The Army had been restricted to short- and medium-range missiles since separating from the Air Force, limiting the booster range to 322 km (200 mi). Within that constraint, the Army missile group began working on the first tactical intermediate-range nuclear-tipped Redstone missile, named after its home at the Army Redstone Arsenal in Alabama. Substantial improvements in the performance of the Navaho 75 K engine over the V-2 soon led to its adoption as the Redstone power plant. While the Navaho vehicle made contributions to many other rocket and missile system designs, none was greater than the North American Aviation's Navaho engine.

NAA's propulsion group established itself as a leader in rocket engine design and production almost as soon as the design improvements on the V-2 engine were adopted for the Navaho and the Redstone. NAA's Redstone 75 K engine, which was also its thrust rating, was used to place the first U.S. satellite in orbit and to place the first American astronauts in space after extensive man-rating of the engine and booster. Improvements in the engine and vehicle came quickly. Second- and third-generation Navaho cruise missiles called for an increase in the engine performance from 37,648 kg$_f$ (83,000 lb$_f$) to 54,430 kg$_f$ (120,000 lb$_f$) for each of the booster's two engines. This

Figure 2.13. Navajo G-26 cruise missile mounted on the lower booster unit in preparation for test launch. Source: U.S. Air Force archives.

dramatic increase in thrust required significant design improvements in the engine, including, for the first time, the "spaghetti" cooling design mentioned earlier. Increased strength, more efficient heat transfer, and a lighter-weight engine allowed a higher concentration of alcohol fuel and a higher combustion pressure to reach the desired thrust of 54,430 kg_f (120,000 lb_f). NAA's engine thrust was later increased to 61,235 kg_f (135,000 lb_f) for the later Navaho missiles (Gibson 1996).

The U.S. Air Force was assigned responsibility for long-range ballistic missiles soon after its separation from the Army in 1947, an assignment that would lead to America's first intercontinental ballistic missile, the Atlas ICBM. Atlas had little in common with the Army Redstone and smaller ballistic missiles, except for the engines. North American Aviation was also contracted to develop and produce the Atlas main engines for the USAF project. Originally designated the MX-774, the Atlas project languished because of underfunding until the Soviets detonated their first hydrogen bomb in 1954. Almost overnight, the Atlas was given top priority with precedence over other intermediate-range ballistic and cruise missiles, including the Navaho. Nevertheless, components and systems from some of those projects were integrated into the Atlas by the prime contractor, Convair, then a division of General Dynamics. NAA's rocket propulsion group, which was spun off to become Rocketdyne in 1955, furnished the prototype third-generation Navaho engines for the early Atlas design. In the process, Rocketdyne's 61,235 kg_f (135,000 lb_f) thrust Navaho engines were converted from alcohol to RP-1, then modified for an increased thrust rating of 68,040 kg_f (150,000 lb_f).

Several far-reaching recommendations made in 1955 led to proposals for a number of satellite launchers, intermediate-range missiles, and several smaller military launchers that would employ existing Atlas and Redstone technology. From these recommendations came two intermediate-range ballistic missile projects that were started almost immediately. The first was the Thor missile—a derivative launcher that used a single Atlas engine. Douglas Aircraft was awarded the booster contract, while the engine contract went to the NAA/Rocketdyne. Thor had immediate applications as a Cold War ballistic missile, and, soon after, as a reconnaissance satellite launcher. Thor's single S-3D, 68,040 kg_f (150,000 lb_f) thrust engine was augmented with the same duo of small LR101-NA vernier guidance engines used on the Atlas. A two-stage Thor satellite launcher soon emerged with an upper stage Agena developed for DoD reconnaissance payloads. A variety of other second- and third-stage boosters would also be adapted later for use on the Thor.

ARMY-NAVY-AIR FORCE JUPITER MISSILE

Interagency cooperation in the early missile and launcher programs was uncommon until formal and informal agreements for large manned projects were ushered in by NASA and its space programs. Although rare, there were examples of collaborative projects that preceded NASA, one of which was the Jupiter missile. By combining the Army's tactical missile needs with the Navy's need for a submarine-launched intermediate-range missile, the shortened and widened Redstone design, called Jupiter, was intended to reduce costs and facilitate missile deployment to counter the spread of Soviet ballistic missiles. Yet the shorter, larger-diameter missiles were incompatible with the Navy's submarines and were never used, leaving the Army with a short, fat booster it had little use for. Instead, solid-fuel Polaris missiles were developed for integration into the Navy submarine fleet. Polaris and similar solid-fuel missiles have been used for this purpose since.

Soon after the development of the Jupiter, the Army was assigned its short-range missile charge, which led to the deployment of the Jupiter IRBM for a brief period in Europe in the early 1960s. The Jupiter missile also inherited the 150,000 lb_f thrust Navaho engine, which gave it an impressive payload capacity. And because of that capability, a new agreement between the Army and NASA led to the adoption of the Jupiter as a booster for NASA's interplanetary missions, but with the new name Juno II. The Redstone-derived Jupiter-C, which was unrelated to the Jupiter missile, was the first launcher called Juno, or Juno I.

Jupiter left behind a record of modest utility as an IRBM and an interplanetary rocket, but it did furnish a number of innovations that included gimbaled engines for active guidance during launch. Previous thrust-vector designs used less efficient exhaust vanes invented by Robert Goddard, and later employed on the V-2 and Redstone.

Figure 2.14. Launch of a prototype Jupiter IRBM missile from the Redstone missile range on September 16, 1959. Source: U.S. Army archives.

USAF THOR IRBM

Rival programs in all three of the Armed Services during the late 1950s introduced a number of ballistic missiles with nuclear capability. The Army's Redstone missile success was soon followed by the combined Army-Navy Jupiter IRBM missile project. The Air Force was in the development phase of their Atlas long-range ICBM missile program when the decision came to expand the number of intermediate-range missiles that could also launch reconnaissance satellites into orbit. These new intermediate-range missile efforts employed technology similar to the Atlas and Jupiter, although simpler. Because the Thor was thinner than Jupiter, it was also capable of much wider deployment as an IRBM, since it could be transported in the fleet of American C-124 Globemaster cargo aircraft.

Thor's first stage consisted of a single Rocketdyne LR-79 (S-3D) engine, developed for the Atlas, that could maintain 667,000 N (150,000 lb$_f$) thrust. Two smaller vernier engines with 1,000 lb$_f$ thrust rating were used for ascent guidance. All three engines used the same LOX/RP-1 propellants, making it difficult to retain as a strategic missile because of the problems supporting cryogenic LOX launch systems.

Thor's payload capacity was later augmented with upper-stage boosters known as the Thor-Able and Thor-Agena. With a propulsion system originally developed to power weapons on the B-58 Hustler, the Agena second stage was adopted in a wide variety of applications including the Naval Research Lab's Vanguard upper stage, the Gemini rendezvous vehicle, the Corona/Discovery spy satellites, and an upper stage for Atlas and Titan boosters. Additional thrust was added to the Thor first stage by attaching solid rocket motors externally to increase payload capacity. These simple yet powerful solid rocket motors also proved useful on a wide variety of launchers, including the Delta, Titan, space shuttle, and, more recently, the Atlas V.

Thor was only a modest success as a strategic missile, even though it was the first non-Soviet IRBM deployed in Europe. Its utility in the European theater came from its range, which could reach Moscow with a nuclear warhead from the UK. Thor also proved useful as a launcher for both military reconnaissance satellites and space exploration flights, beginning with the early Pioneer missions to the Moon. Even earlier flights of the Thor-Able included suborbital reentry tests and biological specimen research. Thor, with its upper-stage booster Able, increased the single-stage IRBM's 2,000 km range to over 9,000 km and allowed it to reach altitudes of 1,600 km. But Thor was never seriously considered an ICBM candidate because of the much higher payload range and capacity of the Atlas.

Thor's legacy was considerable, and even flies today under a different name. The Delta launcher used in NASA's early interplanetary missions was originally a Thor

Figure 2.15. Thor IRBM in preparation for a launch at the Cape Canavera Range. Source: U.S. Air Force archives.

four-stage launcher called Thor-Delta, a name that was later contracted to Delta. The Thor legacy also extended into NASA's Saturn booster project. Rocketdyne's S-3D engine, which was developed for the Thor missile, was upgraded to the H-1 engine, which was used in both the Saturn I and the Saturn IB first stage.

Thor-Agena

Thor-Agena launchers, identified as A, B, and D models, were also capable of carrying a fourth-stage solid booster. An Agena booster was used on the Thor first stage for military payloads between 1959 and 1972. The Thor-Agena low-Earth-orbit (LEO) payload of 1,200 kg was nearly ten times the Thor-Able's LEO payload mass.

The upper-stage Agena was not only a booster, but a multipurpose spacecraft used for Earth-orbit military and civil satellites and for interplanetary exploration payloads. Agena D was perhaps the most versatile of all upper-stage boosters, serving as payload and booster for the Thor, Atlas, and Titan missiles. The last of the series was the Agena D, used as the rendezvous vehicle for the Gemini program. The Agena proved to be one of the most successful booster spacecraft, with a total of 269 successful launches.

Figure 2.16. Orbiting Agena-D rendezvous vehicle used for NASA's Gemini program that prepared the Apollo astronauts for the difficult rendezvous and docking maneuvers on their lunar missions. Source: NASA GRIN.

Table 2.8. Thor IRBM Specs

Airframe manufacturer	Douglas Aircraft Company
Liftoff thrust (sea level)	667 kN (150,000 lb$_f$)
I$_{sp}$ (vac)	282 s
Diameter	2.44 m (8.0 ft)
Total Length	19.82 m (65 ft)
Weight (loaded)	49,800 kg (110,000 lb)
Empty weight	3,125 kg
Maximum range	2,400 km (1,500 miles)
Ceiling	480 km (300 mi)
Main (first stage) engine	Rocketdyne LR79-NA-9 (Model S-3D)
Turbopump	Gas generator
Propellants	Lox/RP-1
Vernier engines	Dual Rocketdyne LR101-NA engines; 4.5 kN (1,000 lb$_f$) each

Thor-Agena was also used to launch the first U.S. Corona spy satellites in 1959 under the Discovery program name.

AIR FORCE ATLAS ICBM

Distrust between the USSR and the United States after World War II—the Cold War—accelerated advanced weapons development of long-range bombers and long-range missiles to unprecedented levels. Experiments with German V-2 missile technology led the development of long-range ballistic weapons for both superpowers, but neither country could reach the other with those missiles—until the launch of the first intercontinental ballistic missile by the Soviet Union in August of 1957. The successful Soviet ICBM launch was well ahead of the efforts by the U.S. military branches in developing long-range missiles of their own. The Army was developing the Redstone IRBM under the guidance of Wernher von Braun, while the Air Force was assigned the first American ICBM called Atlas.

The Atlas project began as a long-range guided missile in 1946, but sputtered in the following years, changing designation and capabilities, with the important structural components established by the mid-1950s. Spurred by the USSR's hydrogen bomb detonation in 1953, the program became an urgent priority with healthy funding within a year.

A number of important innovations in the design of the Atlas made it ideal as an accurate, long-range nuclear delivery system. Its light-weight construction was accomplished by utilizing thin sheets of stainless steel for the missile frame. Welded together and sealed, the stainless steel sheets formed both the primary structure and the tanks in a simple design that replaced traditional cylindrical frame sections made of ribs and stringers. The fuel and oxidizer tanks, called balloon tanks, comprised the rocket structure that was made rigid simply by pressurizing the tanks. Without internal pressurization, the structure could have easily buckled, since the thin stainless shell contained no supporting members except at the top and bottom attachment rings. The Atlas booster was not just unique; it remains the rocket with the lowest empty-weight-to-loaded-weight ratio on record.

The Atlas was also the first booster to have multiple-gimbaled engines. Rocketdyne's three first-stage engines adopted from the Navaho program had a dual configuration unlike any previous launch vehicle design. Both outboard booster engines had a separate fairing that released with each outboard engine after fuel exhaustion. The slightly smaller central "sustainer" engine supplied thrust until preparation for payload separation—originally a warhead. This configuration was termed "stage-and-a-half" since the booster engines and sustainer engines were started at the same time. Initial prototypes were first launched as the Atlas A model in 1957 but included only the two outboard boosters and a dummy warhead. The first successful launch of the complete Atlas, built by Convair (earlier called Consolidated Vultee Aircraft), was the B model in November 1958, with a downrange flight of 9,660 km (6,000 miles) (Walker 2005).

Following its development and testing, the Atlas ICBM was deployed as a nuclear delivery system, first in above-ground hardened buildings, then in below-ground silos. But even with the advantage of Atlas's 10,000 km range, the ICBM was hampered by its liquid oxygen propellant. Like other cryogenic-fueled ICBMs, the Atlas would eventually be replaced by solid-fuel ICBMs like the Minuteman and Peacekeeper. Regardless, the Atlas introduced an impressive number of booster innovations and became a versatile space launcher that still flies today.

Atlas was the first American launcher with a completely autonomous guidance system, although the first models were guided by ground-based radio commands. Preprogrammed targets and accurate guidance gave the Atlas the capability to destroy many of the early below-ground silos that became the mainstay of Soviet and U.S. missile deployment. Atlas was also the first to incorporate onboard digital computers for its guidance, control, command, and systems operations. Atlas E models were the first to use an accurate, inertial guidance system, which made it immune from the earlier models' susceptibility to radio-frequency (RF) jamming.

NASA had initially planned using the Atlas rockets for their manned Mercury program, but adopted a two-phase approach when the Atlas launch tests were delayed beyond the first flights scheduled for Mercury. To keep pace with the Soviets, the Redstone was adopted for the initial Mercury suborbital launches, with the Atlas integrated into the project for later manned orbital missions. By 1961 and the official announcement of the Apollo program, it was clear to NASA managers that something larger than Atlas would have to be used for the Gemini flights that would transition from Mercury to Apollo.

Atlas's contribution to NASA's manned program included the Mercury orbital missions and the Gemini Agena launches, though its use in unmanned exploration missions was far more extensive. Atlas and Atlas-Agena were used for NASA's first successful interplanetary missions, named Mariner. Atlas II, III, and V launchers con-

Figure 2.17. Mercury-Atlas 9 (MA-9) shown in preparation for launch of astronaut Gordon Cooper for the last Mercury mission on May 15, 1963. Source: NASA GRIN.

tinued in their role, including the New Horizons mission to Pluto and the Kuiper Belt, launched January 19, 2006, using an Atlas V. The heavy-lift Atlas V has an entirely new look since it is no longer a stage-and-a-half booster, and no longer uses the balloon tanks. Atlas V is powered with a single Russian RD-180 engine using LOX and RP-1 propellants, with as many as five optional solid rocket strap-on boosters, in addition to a three-engine Centaur upper stage.

Atlas technology included:

- Light-weight structure that employed a thin-wall stainless steel monocoque tank and body structure that was kept rigid by the internal tank pressure (introduced on the Viking rocket)
- Gimbaled rocket engines for effective and efficient ascent guidance (originally patented by Robert Goddard)
- Detachable payload/warhead section
- Stage-and-a-half method for jettisoning the booster engines during the ascent
  - Both booster engines and center/sustainer engine ignited at liftoff
  - Boosters jettisoned at cutoff
- Onboard digital computer for advanced functional controls

Table 2.9. Atlas D Specs

Diameter	3.05 m (10 ft (16 ft at base))
Length	23.11 m (75.8 ft (85.5 ft in ICBM configuration))
Weight	117,900 kg (260,000 lb) maximum at launch
Engines	• 2 Rocketdyne LR105-NA strap-on boosters 154,000 lb$_f$ thrust • 1 Rocketdyne LR89-NA-3 sustainer 57,000 lb$_f$ thrust • 2 small vernier rockets for attitude correction 1,000 lb$_f$ thrust
Engine thrust at launch	163,290 kg$_f$ (360,000 lb$_f$)
Propellants	• Fuel: RP-1 • Oxidizer: LOX • Consumption: ~680 kg/s (1,500 lb/s)

U.S. NAVY VIKING AND VANGUARD

Weapons and weapons projects flourished in the early Cold War on both sides, although the Soviets pursued missile technology far more aggressively than the United States and entered the space race in 1957 with mature, heavy-lift launchers and advanced spacecraft. Fortunately, the two engineers responsible for designing the Cold War missiles for each side were far more interested in building space exploration hardware than weapons. While the Soviet programs were well funded and under the command of a single design bureau, the U.S. effort was splintered by the competitive efforts of the three armed services.

Nonetheless, the divided missile efforts were reconciled for a brief period when the Eisenhower administration made the decision to launch America's first satellite using the Naval Research Lab's Viking missile. Viking began as an intermediate-range missile developed for improving V-2 technology. Smaller than the Redstone, but with a number of significant improvements, Viking's first stage was powered by an engine similar to the V-2 and using the same propellants. The Viking engine, built by the Reaction Motors Company, was the first missile to use a gimbaled engine, and the first to use turbopump exhaust to steer the rocket instead of exhaust vanes. This was an important innovation, since it allowed active steering after engine cutoff, which provided greater accuracy to its target. Viking also was designed with the first reaction control system that is now common on launchers, boosters, and many spacecraft. Although not as successful as the Rocketdyne/North American Aviation propulsion group, Reaction Motors later merged with Morton Thiokol and successfully developed a number of the NACA/NASA X-series spacecraft engines, including the XLR-11 for the Bell X-1, and the XLR-99 for the North American X-15.

Viking's structure was unlike the V-2 in several important ways. First, the Viking missile was the first to integrate the tanks as part of the vehicle's structure, which, along with the use of aluminum alloys, reduced vehicle weight considerably. The integrated tank and vehicle structure predated the Atlas ICBM that is often listed as the first to use the monocoque tank-structure design. Viking's thin, cylindrical shape and triangular fins were also unique for the rockets and missiles of the era. Even though the Viking missile played an important role as a starting point for other projects, it was replaced by lighter, cheaper sounding rockets that included the Aerobee and Nike.

America's commitment to launch its first satellite in support of the 1957–1958 International Geophysical Year came with the stipulation that the program not interfere with the IRBM and ICBM development efforts being carried out by the Army and the Air Force. Yet all three armed services submitted proposals for the launcher and the scientific satellite because of its importance. Naval Research Lab's (NRL) Viking proposal consisted of a Viking first stage and a liquid-fuel Aerobee-Hi second stage—a combination later called the Vanguard. Vanguard was not a true civil booster, but was a more attractive project to the Eisenhower administration in light of the Cold War. Even though the Vanguard was untested and had a limited payload capacity, it was selected to launch America's first satellite for a variety of reasons that are still debated today.

Vanguard

The Navy's Project Vanguard was a two-stage booster comprised of a Viking first stage and an Aerobee-Hi sounding rocket for the upper stage. Vanguard's inaugural launch used a small, solid-fuel third stage that was responsible for the final boost of the 22.7 kg (50 lb) payload into orbit. The Martin Company that was contracted to build the Viking missile was also contracted to build the Vanguard and integrate the upper stages with the satellite.

Figure 2.18. Launch of a Viking test vehicle at White Sands. Source: U.S. Navy archives.

Table 2.10. Viking Specs (RTV-N-12A)

Inaugural launch (Viking-1)	May 3, 1949
Length (final version)	13.7 m (45 ft)
Weight	6,700 kg (14,800 lb)
Payload	450 kg (1,000 lb)
Engine	Single XLR10-RM-2 Reaction Motors liquid fuel engine
Thrust	91,200 N (20,500 lb$_t$)
Propellants	Alcohol and liquid oxygen
Turbopump	Hydrogen peroxide gas generator
Maximum altitude	254 km (158 mi)
Guidance	Active—gimbaled engine Passive—4 tail fins
Manufacturer	Glen L. Martin Co. (airframe) Reaction Motors, Inc. (engine)

Vanguard's first satellite payload was a three-pound, grapefruit-sized spherical payload also called Vanguard. The satellite was designed with a battery-operated internal transmitter used solely for tracking because of the satellite's small size and the difficulty of tracking the small satellite by radar or optically beyond several hundred kilometers. The first launch attempt to orbit the satellite was made on December 6, 1957, with the Vanguard TV3 (Test Vehicle 3) launcher that resulted in a spectacular failure on the launch pad. America's response to the Soviet's impressive Sputnik 1 and 2 satellite launches left the NRL's Vanguard program with a tarnished history that is often portrayed as an embarrassing disaster. While it's hard to argue Vanguard's disappointing early record, Vanguard did launch a 1.47 kg (3.25 lb) satellite into a 3,966 km (2,465 miles) × 653 km (406 mi) orbit successfully on March 17, 1958, following a second failure. That satellite, designated Vanguard-1, is the oldest artificial satellite in Earth orbit. Six more launch failures and two successful satellite launches closed the Vanguard Program in September 1959.

Beyond its televised and spectacular first launch failure, the Vanguard did leave behind a worthy legacy in the Naval Research Lab's engineers and scientists who worked on the project. Soon after NASA's creation in 1958, the space flight office for the Mercury program was organized at the new Goddard Space Flight Center in Maryland. It was the NRL staff that were chosen as the core personnel for the first manned space flight program activities.

USAF TITAN ICBM

Like the Atlas, Thor, Viking, and Saturn boosters, the Titan ICBM began with engine technology derived from the Navaho program. Titan was originally created as a

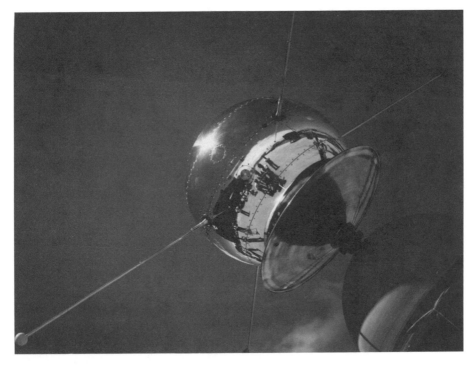

Figure 2.19. Photograph of the Vanguard satellite being prepared for mounting on the upper stage of the Vanguard booster circa 1958. Source: NASA GRIN.

backup ICBM for the Atlas, while later models involved significant improvements over the Atlas that included more secure silo-based storage, storable propellants to avoid fifteen-to-thirty-minute delays in cryogenic fueling, and a larger payload that could carry a four-megaton warhead over 12,000 km. Titan's larger engine thrust and more advanced guidance systems were adapted for space launchers in the Titan II, followed by the Titan III and the Titan IV models.

Titan I

Titan I's first launch in 1959 was followed by a number of development tests until its acceptance for duty as a USAF ICBM in 1962. Although the Titan I was not a space launcher, it was the booster designated for the Air Force's Dyna Soar/X-20 project. Propellants for the Titan I were RP-1, a highly refined kerosene, and liquid oxygen which made it difficult to maintain for ICBM-ready operations.

Titan II

Titan II was the USAF's workhorse ICBM and space launcher, with over 100 launches made between its first flight in 1962 and its last in 2003. Propellants for the Titan II were the storable (non-cryogenic) but highly toxic nitrogen tetroxide (NTO) and un-symmetrical dimethyl hydrazine (UDMH). Both of these propellants were stable over a useful range of storage and operating temperatures, which was a great advantage over the cryogenic liquid oxygen used in earlier ICBMs. Several federal agencies contracted with Lockheed to provide Titan II launchers for a variety of missions, including meteorological satellites, defense reconnaissance satellites, and NASA's Gemini program.

Titan III

Titan III, which was designed as a stretched version of the Titan II, was the first U.S. heavy-lift military booster. Later versions of the Titan III were augmented with optional solid rocket boosters for added payload capacity and range. Titan IIIA was first launched in 1964, followed by the Titan IIIB, which carried an Agena third stage, giving it even greater payload and range capability. Titan IIIC was the largest of the Titan III series, with two strap-on solid-fuel boosters. The solid and liquid engines on the Titan IIIC gave it more than four times the thrust of the Titan I. Titan IIIs could be outfitted with a variety of upper-stage boosters, including the Centaur, which was used to launch the Voyager and Viking deep-space probes.

Titan IV

Operated in the era between the Apollo Saturn V and the new Ares V launchers, the Titan IV was America's heaviest-payload expendable launch vehicle, even rivaling the space shuttle in lift capacity. The comparable payloads of Titan IV and space shuttle were not coincidental, but by design, since the Titan IV was developed to replace the space shuttle as the Air Force's heavy satellite launcher. Payload limitations of the space shuttle, as well as its high costs and restrictions after the *Challenger* accident, ended the Air Force's participation in NASA's Space Transportation System program in an expensive twist of irony. It was the Air Force that saddled the space shuttle design with a huge cargo capacity, a large crew capability, and wings for gliding to a landing after reentry.

Table 2.11. Titan IVB Specs

Manufacturer	Lockheed Martin
Diameter	3.05 m (10 ft) core
Length	44 m (144 ft)
Weight	943,050 kg (2,079,060 lb)
Engines	*1st stage*: 2 Hercules USRM 15 solid boosters, 12 MN (3,400,000 lb$_f$) thrust, I_{sp} 286 sec *2nd stage* (core): 2 LR-87 NTO + UDMH engines, 2,440 kN (548,000 lb$_f$) thrust, I_{sp} 302 sec *3rd stage*:1 LR-91 NTO + UDMH engine, 467 kN (105,000 lb$_f$) thrust, I_{sp} 316 sec
Payload	21,680 kg (47,790 lb) to LEO
Stages	3–5

Improvements in the Titan IV involved larger-thrust solid boosters and a larger Centaur third stage. A fourth stage was also available as the Boeing Inertial Upper Stage (IUS). However, the introduction of the Atlas V rocket and the Delta IV heavy rocket boosters made the Titan obsolete—a legacy that ended with its retirement flight in April 2005 from Cape Canaveral.

Sounding Rockets

Exploration of the atmosphere began with the launch of the first simple instruments on small, solid rockets in the early 1900s, followed by science experiments first flown by Robert Goddard on his P-series liquid-fuel rockets. His first instrument launch in 1931 carried an aneroid barometer, a thermometer, and a camera focused on the two instrument readouts (SP-4401). These early rocket experiments were small in scope but were important for two reasons. This was the beginning of scientific research

Figure 2.20. Launch of the Cassini spacecraft to Saturn on a Titan IVB from Cape Canaveral, October 15, 1997. Source: NASA GRIN.

centered on reaching space, and it also represented the broad interests of the three pioneers of rocketry from Russia, Germany, and the United States. Still, progress in these first rocket experiments was slow because funding was difficult to find in the era of the 1930s Great Depression.

On the other side of the world, and at about the same time that Goddard was flying his early rocket instruments, the GDL Russian rocket club, headed by Fredrik Tsander, began launching instrumented sounding rockets in 1933, with the first successful flight reaching 11 km several years later. Tsander's liquid-fuel rockets were being built by the first generation of rocket scientists that would later lead Russia in the space race against the United States.

The first American high-altitude sounding rockets that were designed specifically to carry instruments for upper-atmosphere research were scaled-down Army Corporal missiles called Wac Corporal that were built by the Douglas Aircraft Company. The Army Corporal was designed by the Jet Propulsion Lab, known at the time as the Guggenheim Aeronautical Laboratory at Cal Tech (GALCIT). Founders of GALCIT, JPL, and the Aerojet Corporation, which built the Wac Corporal propulsion system, included a number of prominent scientists, including Theodore von Karman and Tsien Hsue-shen. Tsien Hsue-shen was not only one of America's leading rocket scientists, he co-founded JPL and advised the Pentagon on missile technology. The importance of his position allowed him a temporary rank of colonel in the Air Force to evaluate German V-2 rocket scientists at the end of World War II. Not long after, Tsien was deported to China in the hysterical Joseph McCarthy anticommunist purges, and within only a few years, he became the director of China's space program. Almost single-handedly, Tsien took the Chinese space program from a collection of primitive Soviet launchers to an aggressive exploration program that now includes plans for a space station and a lunar outpost within the next decade.

Also in the Army's inventory at the time were the Private, Corporal, and Sergeant surface missiles. While wholly unoriginal in name, these missiles were small, efficient, and tailor-made for atmospheric research. The Corporal and Sergeant proved effective in their role as small missiles, but were even more useful as second- and third-stage boosters for atmospheric research rockets. Most notable was the solid-fuel Sergeant rocket that made up the two upper stages of the Jupiter-C, which launched the Explorer-1 satellite into orbit.

WAC CORPORAL

Arguably the most successful of these rocket soldiers was the Wac Corporal, originally designed for the Army Signal Corps as a meteorology probe. Soon after its production began, the Wac Corporal became a staple of upper atmosphere research projects, since it could be manufactured inexpensively, quickly, and reliably. Wac Corporal was later improved to become an even more successful Aerobee high-altitude sounding rocket that is still in use today.

The two-stage version of the Wac Corporal consisted of a solid-fuel 50,000 lb_f thrust first-stage booster named Tiny Tim, and the second-stage Wac Corporal

Table 2.12. Wac Corporal Specs

Length	4.9 m (16 ft 2 in)
Weight	300 kg (700 lb)
Thrust	6,700 N (1,500 lb)
Propellants	Nitric acid (oxidizer)
	Aniline and furfuryl alcohol mixture (fuel)
Manufacturer	Douglas Aircraft Corp. (airframe)
	Aerojet Engineering Corp. (2nd-stage engine)
	California Institute of Technology/JPL (1st-stage booster)

liquid-fuel rocket. As an upper stage for the Bumper, only the upper stage of the Wac Corporal was used, along with its guidance fins, which appear in the figure below.

The first Wac Corporal was launched on September 26, 1945, from White Sands, New Mexico, roughly two years before launch of the first V-2 from that site. Soon after the V-2s arrived at the White Sands range for flight testing, a second stage was added to the converted V-2 missiles to increase the payload altitude. The Wac Corporal was selected for its size and performance, extending the V-2's maximum altitude from approximately 60 mi to 250 mi with a payload mass of 11 kg (25 lb). This was the first American hypersonic vehicle, reaching more than 5,000 mi/hr during atmospheric reentry. Altitude and range of the two-stage Bumper vehicle became a problem

Figure 2.21. Two-stage Wac Corporal shown here on a portable launcher at the White Sands missile range. Source: U.S. Army archives.

Table 2.13. Bumper Specs

Length	17.25 m (56.6 ft)
Diameter	V-2: 1.65 m (5.4 ft); Bumper Wac: 30.5 cm (12 in)
Weight (total)	12,800 kg (28,300 lb)
Speed (reentry)	5,260 km/h (3,270 mph)
Maximum altitude	393 km (244 mi)
Propulsion	1st stage—V-2 Thrust 267 kN (60,000 lb_f) 2nd stage—Bumper Wac Thrust 6.7 kN (1,500 lb_f)

for the limited White Sands missile range, which hastened the move to the new Cape Canaveral missile test range in Florida. Bumper-8 was among the first launches at the Cape Canaveral facility.

AEROBEE

Aerobee, a contraction of Aerojet and the nickname "Bumblebee" given to the Wac Corporal, was a sounding rocket developed for the Naval Research Lab by Aerojet, the same firm responsible for the Wac Corporal propulsion system. Aerojet's Aerobee consisted of a first-stage solid-fuel booster with a second-stage liquid-fuel engine powered by nitric acid and furfuryl alcohol mixed with amine (an ammonia-like compound). Aerobee's second stage was based on the Wac Corporal's design, enlarged, and with numerous improvements. Over a thousand launches of the Aerobee and its variants began with the first flights in 1947 as the Wac Corporal program was ending. Aerobee's first stage developed 8,165 kg_f (18,000 lb_f) thrust, with a burn time of approximately 2.5 seconds. Aerobee's second-stage liquid-fuel engine developed 1,815 kg_f (4,000 lb_f) thrust, boosting small payloads to an altitude of 230 km. A larger Aerobee with larger propellant tanks, the Aerobee-Hi, attained an altitude of 400 km and was last launched in 1985. The Navy's Vanguard Project used the Aerobee upper stage to launch several of the early Vanguard satellites, one of which is still in orbit.

Early Aerobee rocket flights provided some of the first data on solar emissions since their reach was well above the altitude reached by balloon-borne instruments at that time. These early flights carried ultraviolet spectroscopy experiments used for solar research and brought back data on the cosmic ray particles that could not be measured at Earth's surface. Recovery of the suborbital flight instruments was made by parachute. The Aerobee sounding rocket was also used to launch the first nonclassified radar built by the Jet Propulsion Lab to an altitude of 162 km (NASA 1971).

Figure 2.22. Aerobee (also known as RTV-A-1 and the X-8) rocket mounted on the transport cradle at the White Sands missile range. Source: U.S. Air Force archives.

Sounding rockets were the first commonly available tool for space exploration because of their simple technology and because they preceded the much larger IRBMs and ICBMs. Their success continues today because of their simplicity, low cost, and high operational altitudes for their size. A list of U.S. and foreign sounding rockets still in use is given below.

Sounding rockets currently in use (NSROC n.d.):

- Aerobee 100
- Aerobee 150/150A
- Aerobee 170
- Aerobee 200
- Aerobee 300
- Aerobee 350
- Arcas
- Arcon
- Aries
- Argo D-4 Javelin
- Argo D-8 Journeyman
- Argo E-5
- Astrobee D
- Astrobee F
- Astrobee 1500

- Black Brant IIIB (U.K. production version of the Aerobee)
- Black Brant IV
- Black Brant IX
- Black Brant X
- Black Brant XI
- Black Brant XII
- Black Brant VC
- Bullpup-Cajun
- Hawk (Orion)
- Iris
- Nike-Apache
- Nike-Asp
- Nike-Black Brant V

- Nike-Cajun
- Nike-Hawk (Orion)
- Nike-Javelin
- Nike-Malemute
- Nike-Tomahawk
- Skylark
- Terrier-Lynx
- Terrier-Malemute
- Taurus-Orion
- Taurus-Nike-Tomahawk
- Terrier-Orion
- Taurus-Tomahawk
- Viper-Dart

ELECTRIC PROPULSION

Electric rocket propulsion comes in a wide variety of engines, and an even greater variety of propellants. What is common among the electric propulsion (EP) systems is the electric power used to generate the thrust. Simplest of these devices is the electrically heated gas thruster that increases the thrust and efficiency of a compressed (cold) gas thruster simply by heating the gas. Heating improves both performance and efficiency of compressed gas by giving it a greater exhaust velocity and a higher mass flow. The resulting thrust efficiency multiplication can be considerable since I_{sp} for neutral gases is roughly proportional to the gas temperature. In addition to electric power, gas heating can be generated by direct solar heating, and by microwave (RF) heating, similar to a microwave oven. The highest-temperature sources for heated-gas thrusters are radioactive reactor cores that do not require electric power.

Electric Propulsion Categories

The wide variety of electric thrusters can be categorized by the method used to accelerate the propellants, from simple heating systems to complex induced field accelerators. Five basic electric propulsion categories are outlined below.

1. Electrothermal
    - Resistojet thruster
    - Arcjet thruster
    - Solar thermal thruster
    - RF-heated thruster
2. Electrostatic
    - Ion thruster
3. Electrodynamic
    - Magnetoplasmadynamic thruster (MPD)
    - Hall-effect thruster
    - Pulsed-plasma thruster
    - Variable specific impulse magnetoplasma thruster
4. Photon
    - Solar sail
    - Laser momentum delivery
5. Nuclear
    - Nuclear propulsion—thermal
    - Nuclear propulsion— thermal ion

   ***1. EP—Electrothermal* Resistojet thruster**. Cold gas thrusters powered by compressed gas are simple, inexpensive, low-thrust propulsion systems that also have notoriously low I_{sp} ratings. Yet these simple systems can be improved dramatically by heating the gas directly or indirectly. The simplest type of electrically heated gas propulsion entails heating the gas with electrical current flowing through a resistive heating element, analogous to electric range coils or an electric room heater. Expansion of the heated gas is in rough proportion to the temperature increase that is responsible for increasing the exhaust gas velocity and thrust efficiency. The corresponding I_{sp} increase may be a factor of 100 or more with this or other types of gas heating. The same concept is employed in the nuclear thermal engine, but with the use of a much higher temperature reactor core instead of an electric-resistive heater.

   Resistojet thrusters can be used with nonreactive gases such as hydrogen or nitrogen or employ reactive gases such as hydrazine for even greater expansion and propulsion efficiency with I_{sp} values that range as high as 300 seconds or more. Resistojet thrusters were planned for use on Space Station Freedom to reboost the station using methane gas generated in the carbon dioxide reduction system. That propulsion technique was scrapped, along with a number of other systems and components that did not reach the development phase for use on the International Space Station.

Arcjet thruster. A more efficient method of heating and expanding gas that can improve both thrust and efficiency uses a high-current arc in a conical chamber, similar to the expansion nozzle of a rocket motor. The arcjet thruster produces higher chamber temperatures than the resistojet, but also requires significantly more electrical power. Improvements in efficiency for the arcjet can reach as high as 600 s for hydrazine propellant, and 2,000 s for hydrogen. Arcjet thrusters require power in the 1–100 kW range, which makes them suitable for moderate-sized propulsion units on interplanetary missions.

Solar thermal thruster. Solar thermal propulsion is similar to other heated gas thruster concepts, but uses concentrated solar energy focused on an expansion chamber to heat the propellant as it exits through an expansion nozzle. The increase in I_{sp} over a compressed (cold) gas system is a function of the internal chamber temperature, a value determined by the solar concentrator efficiency and the collection area. Although not flown in space, the solar thermal engine would likely yield an I_{sp} of 500–800 s at modest power and thrust levels.

Recent experiments on solar thermal thrusters by the Air Force Research Laboratory have reached over 2,204°C (4,000°F) within the absorber/heater unit. A photo of the collector/absorber is shown in figure 2.24.

RF-heated thruster. Microwave energy can be directed into a heating chamber similar to the solar thermal engine, producing the same gas expansion effect. And, like similar heated gas thrusters, the RF thruster could use either inert propellants (H_2, N_2), or reactive propellants such as hydrazine and ammonia. Propellant heating from RF/microwave excitation is in proportion to the microwave energy available in the thruster, and the efficiency of energy absorption by the specific propellant.

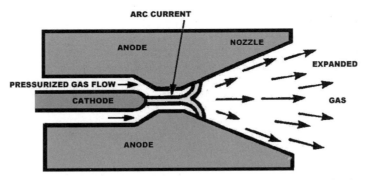

Figure 2.23. Simplified arc jet thruster showing the off-axis propellant gas input to circulate the gas and the ionized plasma through the electrical arc for high-temperature expansion. The arc current represents the current flow between the cathode (-) and the anode (+). Source: NASA TM-4322A.

Figure 2.24. Solar collector and absorber elements of a prototype solar thermal thruster shown during development and testing. Source: U.S. Air Force archives.

2. EP—Electrostatic **Ion Thruster**. Unlike cold gas and heated gas thrusters, the ion (plasma) engine does not use gas expansion for thrust. Instead, ionized gas is accelerated out of the ionizing chamber using electrostatic and/or magnetic fields. A negatively charged grid at the exhaust end of the thruster pulls high-mass positive ions toward the mesh. Positively charged ions pass through the mesh as an ionized gas exhaust at very high velocity. But because the ion/plasma density is extremely low, the thrust available from the ion/plasma engine is also low. These two basic characteristics distinguish the ion/plasma engines from the other propulsion units—very high efficiency, and very low thrust.

High-mass ions used for ion/plasma engine fuel are created in several ways, one being the bombardment of neutral gas atoms by high-energy electrons. A good example of plasma fuel is xenon, a noble gas, which is first ionized, then accelerated through a negatively charged grid. Exiting positive ions are then neutralized with a beam of electrons to keep them from returning to the negatively charged engine and its screen since opposite charges attract. Without neutralization, the charge separation of the positive gas and the negative vehicle would simply build a charge-difference condition that would be self-attractive and supply no net force—like putting a fan on a sailboat. A second low-voltage grid can also be placed outside of the high-voltage screen to repel electrons that may stray from the neutralizing electron beam.

Calculation of the velocity or the energy of the ions is straightforward for a simple system. The exhaust velocity is equal to the net electrostatic field (E_{net}) times the charge (q) of the accelerated ion divided by its mass. Energy would be the square of the exhaust velocity times the exhaust mass.

An approximation of the exhaust velocity can be made using:

$$V_{exit} = 2qE_{net}/m$$

where V_{exit} is the ion exit velocity, E_{net} is the net electrostatic field, q is the charge in electron/proton units, and m is the mass of the ion.

This electrostatic ion thruster has extremely high efficiency but suffers from very low thrust. And because of the low thrust, the applications are limited to long-duration, low-thrust operations such as geostationary satellite station-keeping, or small satellite orbital boost from LEO to GEO, or extended-duration interplanetary flights. The very high efficiency of the ion engine makes it ideal for deep-space missions because of the low propellant mass required compared to chemical thrusters. Typical exit velocities for the ions are 30,000 m/s (67,000 mph), which is some 50–100 times that of conventional chemical rocket exhaust gas. Since the exhaust velocity can be approximated by the specific impulse times g (9.8 m/s²), the efficiency is also 50–100 times greater. More important, the fuel required for the same ΔV boost is 50–100 times less than a typical chemical rocket, although efficiencies for both types of thrusters vary widely with system design and propellants used.

3. EP—Electrodynamic Electrodynamic engines use both electromagnetic fields and electrical currents to accelerate ions in similar fashion to the electrostatic ion engine, but with different techniques and propellants. Magnetoplasmadynamic engines, for example, generate thrust by the interaction of ionized propellant gas acting as a current that interacts with its own induced magnetic field; a process that increases acceleration.

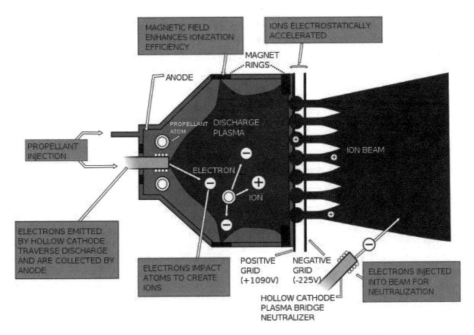

Figure 2.25. Functional diagram of an ion engine showing propellant injection and electron ionization on the left of the confinement chamber. Positive ions are accelerated to the right by a high-voltage grid (inner) and a lower-voltage screen to the right for repelling neutralization electrons. Acceleration of the ions is augmented with magnetic fields in this idealized sketch. Source: NASA DS-1.

Magnetoplasmadynamic (MPD) thruster. By adding an external magnetic field to the basic ion/plasma engine, the small magnetic field induced by ion flow through the exhaust (an ion current) will be enhanced, further accelerating the plasma particles. The net effect is an increase in the exhaust velocity and the I_{sp}, as well as an increase in thrust, since there is greater plasma mass flow. These thrusters require high currents but also generate the highest thrust of the EP engines, making them ideal for larger interplanetary spacecraft.

The MHD engine accelerates ionized gas as it passes through concentric electrodes at high voltage, which creates a plasma current. That current self-induces a radial magnetic field, which creates Lorentz force ($j \times B$, or current perpendicular to the magnetic field) acceleration of the plasma. In this geometry, both the electric field and magnetic field accelerate the plasma current because it is perpendicular to both. An external magnetic field can also be superimposed to increase the acceleration even more. The relatively high currents produce sufficient thrust for use on larger spacecraft, while the high I_{sp} values offer significant weight savings over chemical thrusters.

Hall-effect thruster. The earliest ion/plasma engine used in space was the Hall-effect thruster (HET) used on Russian satellites in the 1960s. The Hall-effect thruster employs both electric and magnetic fields to accelerate ions to high velocities. And although the thrust efficiency of the HET is generally less than most other ion engine types, its thrust range is greater. This makes the HET more useful for near-Earth applications where station-keeping and orbit operations require greater thrust.

Pulsed-plasma thruster. Pulsed-plasma thrusters generate high-specific-impulse propulsion with relatively low power consumption, although thrust levels are also low. This makes the pulsed-plasma thruster best suited for applications on small spacecraft that include attitude control, station-keeping, and low-thrust maneuvers. Solid propellants can also be used with the advantage of greater system simplicity and lower cost, while retaining a relatively high specific impulse.

Variable specific impulse magnetoplasma thruster. One of the newest and most efficient electric propulsion thrusters is the variable specific impulse magnetoplasma rocket (VASMIR). The VASMIR has no conventional electrodes to generate thrust but instead employs RF/microwave energy to ionize the propellant and magnetic fields to accelerate the plasma. The device, sometimes called the electro-thermal plasma thruster, can generate constant thrust propulsion at variable efficiencies, hence its name. Because VASMIR has relatively high thrust levels and very high efficiency, and because it can also use hydrogen rather than the typically expensive propellants used on other EP engines, it is being researched by a number of labs and agencies worldwide. The combination of relatively high thrust (compared to other electric propulsion engines) and high efficiency makes the VASMIR engine suitable for most, if not all, space propulsion applications.

4. EP—Photon **Solar sail**. Even though photons have no mass, they do have momentum and can exert a very slight pressure on any surface they impact. For spacecraft outside of the atmosphere and in zero-gravity, photons can exert a small force capable of propelling a small payload, provided there is a large surface area to reflect or absorb the photons. Since particle or photon elastic recoil is more efficient in transferring momentum than inelastic absorption, the optimal solar sail

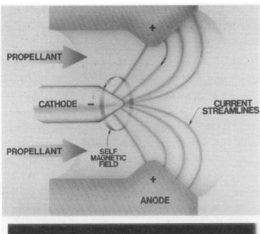

Figure 2.26. (Top) Sketch of the basic operations of magnetoplasmadynamic thruster showing the acceleration of the plasma gas perpendicular to both the electric and induced magnetic fields. (Bottom) A composite image of a thruster and a cutaway of the engine design. Source: NASA NIX.

is reflective. The limit of solar sail propulsion is roughly the orbit of Mars because the sunlight intensity decreases at $1/r^2$ (r is distance/separation). While prototypes of the solar sail have been launched, none have been successful. However, photon pressure has been used for attitude control and orbit changes for spacecraft including the Mariner 10 mission to Mercury. Photon pressure can also produce a slight but measurable force that requires correction in near-solar missions, or perturbations that may need cancellation for spacecraft that require high accuracy in orbit position or trajectory.

Laser momentum delivery. Intense laser light has been proposed for pushing a spacecraft away from Earth without the need for a large sail structure, although there are a number of limitations to such a process. First, laser light is collimated, which is a distinct advantage since it does not diverge as $1/r^2$, but intense laser light has propagation anomalies in the atmosphere that can disperse the collimated beam. Second, the thrust is limited to the momentum of the laser light, which is small. Third, while long exposure time offers the same advantage as it does for solar sailing, the Earth's rotation is a significant complication to continual thrust and precise trajectory control. Laser momentum propulsion experiments have so far been limited to small-scale terrestrial concept demonstrators.

5. EP—Nuclear Nuclear energy is released by either the fusion or the fission of atomic nuclei in a process that provides millions or even billions of times more energy released than in chemical reactions. Nuclear fusion has been regarded as the ultimate source of energy generation on Earth, but has a long list of technological hurdles to overcome before it becomes available commercially. Nuclear fission reactors are far easier to build, but have their own list of problems related to the extremely hazardous radiation by-products that make them difficult to manufacture, handle, and isolate from the environment. Nuclear reactors also have applications in space, primarily as electrical power sources and as propulsion systems, but have additional risks from launch, reentry, and Earth orbit collisional accidents. A brief summary of the two types of nuclear propulsion are outlined below.

Nuclear propulsion—thermal. Nuclear rocket propulsion employs a nuclear (fission) reactor to heat pressurized gas to very high temperatures and pressures, which can generate some of the highest exhaust velocities of any rocket engine. Radioactive materials used for the nuclear heating must be of a specific type that releases neutrons as part of the fission of the nuclei. Compared to the typical radioisotope that decays at a specific rate—like those used in radioisotope thermoelectric generators—the nuclear reactor fissile material releases neutrons that stimulate the fissioning of nearby nuclei. These cascading fission reactions can be moderated by introducing absorbing material throughout the reactor core, which moderates the reaction rate, or can be enriched and slammed together to create a runaway chain reaction—the atomic bomb. Absorbing materials are typically used in power reactors to regulate heat generated, while power levels in the propulsion reactors are regulated by controlling the outflow of neutrons released in the radioactive decay process. Spacecraft reactors commonly use reflectors on the exterior of the core instead of absorbers in the interior. Extending or retracting a reflective shield surrounding the reactor increases or decreases the neutron flux in the core, respectively. Beryllium, which has one of the highest neutron reflection coefficients (cross-sections), is used in alloy form for these reflectors. The advantage of using the more-hazardous fissile materials for nuclear propulsion comes from the extraordinarily high energy content compared to nuclear thermal RTG sources. Energy released from fissile reactor materials can be as much as 500,000 times greater than from the same mass of simple radioisotopes that decay at a fixed rate.

In the era of heightened Cold War weapons buildup of the 1960s, nuclear reactors were used to power prototype propulsion systems in both the United States and the Soviet Union. The U.S. NERVA nuclear rocket engines were developed and tested in this period, although none were launched. Likewise, the Soviet Union's RD-0410 nuclear rocket engine prototypes were tested for several decades, but never flown. However, nuclear reactors were launched into orbit for supplying electrical power to military spacecraft that continue to present a significant and continuing radioactive debris hazard, mostly from their liquid-metal coolant.

Optimum propellants (not the reactor radioisotope) for this system are the lowest atomic mass material available, which is hydrogen. Storage of hydrogen for these rockets must be as a cryogenic liquid to reduce volume to a minimum, since its density is already low.

Nuclear propulsion—thermal ion. As nuclear reactor propellant temperature reaches beyond the ionization temperature of the gas, the ions can be further accelerated in the same process as ion engines—electrostatic and/or magnetic fields. The combination of the nuclear thermal engine and ion acceleration generates even higher specific impulse and higher thrust levels. While these engines could be used for upper-stage launchers, their radioactive hazards make them unsuitable for anything but interplanetary boosters.

Nuclear Propulsion Programs

Between 1955 and 1972, the Department of Defense, the Air Force, the Department of Energy, and, later, NASA sponsored research into nuclear reactor propulsion for moderate-to-high-thrust engines. These designs were to be used for a variety of proposed applications that covered defense, space exploration, upper-stage boosters, and even surface launchers. Reactor fuel for the propulsion units was concentrated ^{235}U, with liquid hydrogen as the propellant. Reactor cooling was accomplished with internal liquid loops, while power levels were regulated by neutron reflection/absorption devices. Thrust levels were 11,340 to 113,400 kg_f (25,000 to 250,000 lb_f), with the thermal power generated in the largest reactors reaching 4,000 MW.

As with any advanced propulsion system, problems surfaced in the reactor engine's technology. For the reactor thrusters, there were major complications with corrosion and with hydrogen fuel brittlizing the metal components it contacted. Similar problems with later turbopump designs exposed to liquid or gaseous hydrogen were solved by using suitable alloys. In fact, later reactor designs that included the NERVA series required high-pressure, high-speed turbopumps to force the hydrogen propellant through the engine. Even though the propellants were not the same, these early efforts led to the solution to the problems that faced designers of the huge F-1 engine turbopumps used on the Apollo Saturn V. This was not a coincidence, but an advantage that resulted from Rocketdyne's contract to develop both the NERVA and the F-1 engine turbopumps (Gunn 1989).

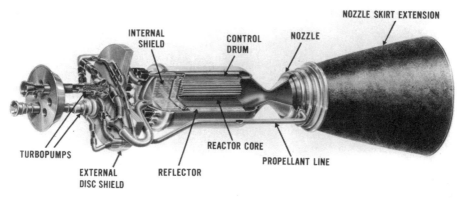

Figure 2.27. Cutaway diagram showing the components of a nuclear reactor engine. Source: NASA-GRIN.

Pulsed Thermonuclear

In the period between 1946 and 1957, the Air Force, the Department of Defense's Advanced Research Projects Agency (ARPA), and NASA funded studies on a thermonuclear propulsion system called Project Orion. The basic concept was to use small atomic bombs to propel the leading vehicle through space by the impulse from the blast waves from the nuclear explosions. Project Orion was conceived not only for space exploration, but manned missions to the Moon, Mars, and beyond. Tests of the shock impulse propulsion took place with chemical explosives, but little hardware was actually developed for the diabolical project. Although it was reported to have the support of Wernher von Braun, the future of the project was doomed by a wide variety of forces that included the nuclear test ban treaty, NASA's reluctance to pursue such a radical and risky program, the expensive and burgeoning Apollo program, and the power of the rational mind.

Project Prometheus

NASA managers have long wanted to improve propulsion technologies for interplanetary missions because of the enormous energy needed to reach the outer planets and because of the inefficiency of chemical rockets supplying that energy. The efforts of earlier nuclear propulsion systems were dusted off when the second Bush administration began reconsidering nuclear propulsion as a way to further advance space exploration. Project Prometheus, a vehicle designed with dual reactors for power and propulsion, was promoted quietly with modest congressional backing because of its ambitious and attractive exploration mission to study the outer solar system moons, and because of its advanced technology that could also be used for Mars cargo missions. Project Prometheus, initially called the Nuclear Systems Initiative and later known as the Jupiter Icy Moons Orbiter (JIMO), soon encountered stiff headwinds, as congressional budgets were slashed before substantial resources were committed to the program. The project's high cost and its unpopular space nuclear reactor technology led to the project cancellation in 2006. Prometheus's final funding was allocated to terminate uncompleted contracts.

Project Pluto

An air-breathing version of a nuclear propulsion system was under development in the late 1950s and early 1960s with the project name Pluto. A ramjet-fueled reactor would power the cruise missile to altitudes of 10.5 km (35,000 ft) at Mach 4 or greater, although propulsion efficiency and reactor cooling suggested a lower-altitude operation while in cruise. The concept, born in the era of the Cold War nuclear weapons expansion, was fraught with problems that included the extreme heat generated in the reactor, radiation products dispersed by the reactor and exhaust along the cruise missile's path, the hazards from the release of extremely radioactive reactor materials as the cruise missile impacted its target, destructive sonic booms along its path, and a host of other serious issues. Together with the development of the long-range, hypersonic ICBM, the Pluto project (also known as SLAM) was terminated before its full-scale prototypes were built. A scaled nu-

Figure 2.28. Artist's sketch of the nuclear-powered Project Prometheus vehicle that was proposed for exploring Jupiter's moons. Source: NASA JPL.

clear reactor power plant was built and tested, but the missile was never completed before its cancellation in 1964.

Mission Propulsion Requirements

Space mission planning is guided by a multitude of objectives, most of which match the details of the spacecraft and its propulsion systems with the mission's purpose and timeline, often orchestrated by the position or changing features of the exploration target. An overwhelming influence on the planning process for space missions is cost, simply because of the typically enormous costs for space exploration programs. These costs can range from tens of millions of dollars for small projects to more than $100 billion for ambitious manned exploration programs. The inherent inhospitaly of space environment also contributes to the need for reliable, durable, and, hence, expensive spacecraft. In a balancing act between cost, performance, and versatility, mission planners and spacecraft designers try to reduce the mass of systems, components, and the overall spacecraft to an absolute minimum. One of the larger mass variables on a spacecraft is its onboard propulsion system. Assuming other variables fixed, a spacecraft's propellant mass is inversely proportional to the propulsion system's specific impulse. As suggested earlier in the chapter, the ideal specific impulse for a spacecraft propulsion system would be as high as possible using reachable technology. However, the high specific impulse of an ion/plasma engine also requires a large electrical power supply that may have a corresponding high mass. A compromise for electric propulsion system design is found in two dimensions of mass. A higher propulsion efficiency reduces propellant mass, but the high

Figure 2.29. Photo of the nuclear reactor-powered ramjet engine planned as the power plant for the Pluto/SLAM cruise missile project. Source: USAF archives.

I_{sp} ion/plasma engine requires greater electrical power mass. Figure 2.30 shows the power mass requirement for a given I_{sp} plotted against the propellant mass needed for the mission (also based on I_{sp}), which crosses at some point. Assuming a fixed thrust output for the booster during the mission, that intersection is the optimal compromise of the payload mass and the I_{sp}.

Another important consideration in the choice of propulsion systems for deep space missions is the balance between propulsion technology, cost, and mission flight time. Chemical rocket propulsion with its higher thrust offers shorter transfer times, but with a lower specific impulse than electric propulsion systems. The result is a greater propellant mass for chemical rockets. However, electric propulsion systems require nuclear power beyond Mars, which means that the mass of the power supply does not increase with I_{sp} as it does for photovoltaic power systems. Even so, more distant exploration targets result in a decrease in payload mass available for both electric and

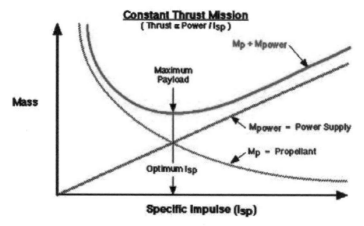

Figure 2.30. Plot of the power supply mass and fuel mass versus Isp. Note the declining propellant mass and increasing power supply mass with increasing Isp. Optimal payload mass and Isp is the intersection of the two curves. Source: NASA Glenn Research Center.

chemical propulsion boosters. But an important distinction arises at large distances. The higher specific impulse of electric propulsion improves the available payload mass significantly above chemical rocket boosters, and that difference increases with distance. Severe limits are placed on the mass of payloads to the outer planets with chemical rocket propulsion, ending at roughly the distance of Jupiter without gravity assist. In contrast, current technology would allow small spacecraft to reach interstellar distances with electric propulsion even without gravity assist maneuvers.

It is possible to augment interplanetary propulsion with planetary flyby gravity-assist maneuvers that can pull a spacecraft in the direction of the planet, either adding or subtracting velocity depending on the relative vector of the spacecraft and the planet during the flyby. This "free" propulsion can take payloads to the outer planets, or even out of the solar system. Using multiple encounters with Earth and Venus, the same technique can be used for an even more demanding voyage to the Sun. Jupiter offers the greatest possible ΔV gravity assist of any planet, although it is not the only planet or moon that can be used for gravity assist. Any payload that can reach Jupiter can be accelerated to any of the distant planets or even toward the nearest stars. Spacecraft reaching Jupiter or Saturn can also use the gravitational tugs from their largest moons to steer the spacecraft trajectory through the family of moons for close flyby exploration that would otherwise require propulsion systems far larger than any available today, or in the past, or expected in the future. Gravity-assist maneuvers that began with the Mariner 10 mission to Mercury have become an accepted practice in multiplying the available energy from conventional launchers and boosters. A graphic that compares the payload limitations for chemical and electric propulsion systems is shown below, plotted as payload mass ratio versus energy. These values do not contain any additional propulsion from gravity-assist maneuvers.

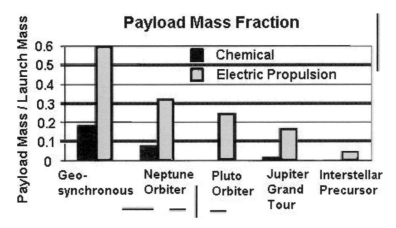

Figure 2.31. A plot of payload mass ratio (payload mass over launch mass) for destinations that range from geosynchronous orbit (36,000 km) to distant interstellar missions. The distinct advantage of electric propulsion system over chemical boosters is lessened by the use of gravity assist maneuvers, but not for orbital missions to the distant planets where a retro booster is needed to slow the spacecraft to orbit-entry speeds. Source: NASA Glenn Research Center.

References

Bullard, John. 1965. *History of the Redstone Missile System*, Washington, DC: U.S. Army Missile Command.

Gibson, James N. 1996. *The Navaho Missile Project: The Story of the "Know-How" Missile of American Rocketry*. Atglen, PA: Schiffer Military/Aviation History.

Gunn, Stanley V. 1989. "Development of Nuclear Rocket Engine Technology." AIAA-1989-2386. The American Institute of Aeronautics and Astronautics.

NASA. 1964. *The Mercury-Redstone Project*. TMX 53107. Huntsville, AL: NASA Saturn/Apollo Systems Office.

NASA. 1971. *Sounding Rockets, 1958–1968: A Historical Summary*. NASA SP-4401. Washington, DC: NASA Scientific and Technical Information Office.

NASA. 1981. *Shuttle Flight Operations Manual*. Volume 8A, *Main Propulsion System*. Washington, DC: NASA.

NASA. N.d. *Deep Space 1 Technologies*. DS-1 NASA-JPL. Retrieved from: http://nmp.jpl.nasa.gov/ds1/

NASA. N.d. *NASA Reliability Preferred Practices for Design and Test*. TM 4322A. Retrieved from: http://www.nasa.gov/offices/oce/llis/0736.html

NASA JPL. 2004. "NASA Selects Contractor for First Prometheus Mission to Jupiter." News release, September 20. Retrieved from: http://www.jpl.nasa.gov/news/news.cfm?release=2004-232

NSROC (NASA Sounding Rocket Operations Contract). N.d. Retrieved from: http://www.nsroc.com/front/html/mmframe.html

Sutton, George P. and Biblarz, Oscar. 2001. *Rocket Propulsion Elements*. 7th ed. New York: Wiley.

Walker, Chuck. 2005. *Atlas: The Ultimate Weapon*, Burlington, ON: Collector's Guide Publishing.

NASA GRIN (Great Images in NASA). Retrieved from: http://grin.hq.nasa.gov/

U.S. Air Force image archives. Retrieved from: http://www.af.mil/photos/

U.S. Navy image archives. Retrieved 2/30/2009 from: http://www.navy.mil/view_galleries.asp

U.S. Army image archives. Retrieved from: http://www.redstone.army.mil/history/photo/welcome.html

NASA NIX (NASA Image Exchange). Retrieved from: http://nix.nasa.gov/

NASA Glenn Research Center image archives. Retrieved from: http://grcimagenet.grc.nasa.gov/home/scr_main.cfm

CHAPTER 3

Planets

Planets have an invariable bond with stars since both form in the same process, and from the same materials. Yet the two share few characteristics, since their definitions are separated by size and mass. However, the range of planet and star masses does converge where the smallest stars meet the largest planets, but without significant overlap. Based on this division, a practical definition—and surprisingly close to the actual definition of a star—is that stars are large enough to generate energy by hydrogen fusion, while planets are not.

Formation of a star begins with the gravitational collapse of a gas cloud, often fragmenting into multiple stars, but favoring mass accumulation in the star instead of the planets. Stars condensing out of a gas cloud invariably leave remnants in the turbulent collapse process. Gravitational dominance of the central star induces fairly simple infall of the gas cloud toward the center, while planet formation involves fragmentation, condensation, and accretion of the gases that have a small ice and dust component. The size of these remnants quickly grows from particles and clumps to mountains and moons, in a collisional maelstrom that ends only when the largest survivors sweep their orbit paths clear of debris and gas. The largest of these survivors are the planets, and the largest of the planets are the gas giants, since more than 90% of a gas cloud collapsing to form the star is composed of hydrogen and helium.

Smaller in size and generally lower in density than the planets are the planetary moons that are formed by the same collisional accretion process. These satellites range in size from small fragments only a few kilometers in diameter, to moons that can be larger than small planets.

Smaller yet are the asteroids and comets orbiting the Sun, called planetesimals, that have recently been designated small solar system bodies (SSSB). The planetesimals that remain, so named because they represent the original building blocks of the planets, have survived gravitational scouring by the dominant planets and their moons for more than 4 billion years. Even smaller fragments locked in meteorites that survive atmospheric entry provide a wealth of information on the solar system's origin and formation process.

Smallest in the hierarchy of the solar system's members is the gas and dust that has been almost completely swept out by the Sun, planets, and their moons. What particles remain are diffuse but still detectable, and occasionally visible as meteors.

Planet Names

Names of the planets, the brightest stars, and the star constellations come primarily from the Greek, Roman, and Arabic cultures, although different names for the same objects can be found in the historical records from the Indian, the Chinese, and the North and South American native cultures. A common feature threading through these names are the mythological gods, characters, and creatures of the respective cultures. For the Western world, these originated in Ancient Egypt during more than five thousand years of cultural continuity. While many of the deities changed in name and meaning, the final identities of their gods were absorbed, along with the Egyptian sovereignty, by the Greeks and Romans. Prominent gods of Egyptian mythology became the elder gods of Greece that were coupled to the prominent celestial bodies. Egypt's supreme gods were replaced by the Greek and Roman gods and titans, with Zeus (Jupiter) reigning as king. Although these were not the only gods in Greek mythology and its heritage, the highest-order deities were assigned to the known planets. Lesser gods and characters that were a part of the rich mythological jewelry that crowds the sky as stars and the constellations we still use today.

Table 3.1. Mythological Origin of the Planet Names

Planet	Greek Name	Roman Name
Mercury	*Hermes*: The messenger god	*Mercury*: The messenger god
Venus	*Aphrodite*: Goddess of Love	*Venus*: Goddess of love
Earth	*Gaea* or *Terra*: Represented Earth. Created the Universe and gave birth to the first race of gods (the titans), and to the first humans	*Terra*: Goddess of Earth
Mars	*Aires*: God of War	*Mars*: God of war
Jupiter	*Zeus*: King of the gods and son of *Gaea*	*Jupiter*: King of the gods
Saturn	*Cronus*: God of agriculture	*Saturn*: God of agriculture
Uranus	*Uranus*: God of the sky	*Uranus*: God of the sky
Neptune	*Poseidon*: God of the sea; brother of *Zeus*	*Neptune*: God of the sea
Pluto	*Hades*: God of the underworld	*Pluto*: God of the underworld

Note: Uranus, Neptune, and Pluto were not known to the ancients, but their names were taken from the Greek and Roman mythology after their discovery.

NAMES OF THE SMALLER BODIES

The introduction of the telescope in the sixteenth century furnished astronomers with their most powerful tool and quickly led to the discovery of more planets, moons, asteroids, comets, and stars. As technology advanced, so did the number of cataloged celestial objects. In order to maintain some continuity in the naming convention, major mythological gods were assigned newly discovered planets, with moons and asteroids taking on the names of other gods and characters from the mythological past. But with the construction of more powerful telescopes came the discovery of many more celestial bodies, which placed the naming process on a more prosaic path. Classical literature names were initially used for small bodies, as were popular names, though those were soon exhausted, and a numbered listing was born out of necessity. With hundreds of thousands of listed and partially listed asteroids and comets, the procedure has diverged somewhat between the comets, which are assigned the name of the discoverer(s) and the year of discovery, and the asteroids, which are named with an assigned number and/or year. These numbers are often appended to a name that is later established by a formal committee of the International Astronomical Union (IAU).

Moons

Names of the natural satellites of the planets were originally derived from Greek mythology, but with the Romanized name. One exception to this rule is the moons of Uranus, named after Shakespearean literary figures (e.g., Cordelia, Ophelia, and Bianca). Hundreds of moons and moonlets discovered by robotic spacecraft in the past decades have led to the adoption of a numbering system similar to that used for asteroids and comets.

Asteroids

Names of the asteroids (also called minor planets or planetoids) were originally derived from Greek mythology.

- Trojan asteroids—found at Jupiter's Lagrange L4 and L5 stable points
  - Greek camp—L4
  - Trojan camp—L5
- Near-Earth asteroids
  - Apollo—Earth-crossing asteroids with a semimajor axis greater than one Astronomical Unit and perihelia less than Earth's aphelion. The Astronomical Unit (AU) is the mean distance between the Sun and Earth.
  - Atens—Earth-crossing asteroids with a semimajor axis less than one AU and its largest reach from the Sun (aphelion) greater than Earth's closest distance from the Sun (perihelion)
  - Amors—Non-Earth-crossing asteroids with a semimajor axis between the orbits of Earth and Mars and perihelia slightly outside Earth's orbit (1.017–1.3 AU). These asteroids can cross the orbit of Mars, but not Earth.
- Centurion—distant asteroids located between Uranus and Neptune

Comets

Comets are named for the discoverer (Halley, Shoemaker-Levy, etc.), with the year and sequence in the year of discovery appended as a suffix. A recent discovery of a comet with the unexciting name of 2006 SQ372 was the first detected from the inner Oort cloud, with an orbit period estimated at over 22,000 years. Since it was identified from sky survey images, it does not carry the name of a discoverer.

Planet Classification

Planets have traditionally been defined by their size and their physical makeup. But as exploration technology improves, those definitions are expanding to include the properties of planets outside of our solar system, called extrasolar planets, or exoplanets. However, the discovery of small planets in our own solar system introduced inconsistencies in the conventional definition of a planet. To avoid creating new and conflicting categories, efforts are underway to categorize planets using definitive features as well as orbital characteristics. Pluto is an example of a planet that has gone from the status of a simple planet to a dwarf ice planet, yet is still recognized as having planetary features. As more planets are discovered, the classification categories could undergo redefinition.

Planet types are still based on the size, orbit, and composition of the planets, but now include the requirement that the surrounding material be cleared from the planet's orbital path. Unfortunately, the inconsistencies in the new definitions established by IAU members in 2006 have left open the question of actual membership, as well as the classification of the more distant objects in the solar system. These unresolved issues for smaller objects have prompted the suggestion of new categories that better describe the dim, distant celestial objects in our solar system that could also be found orbiting other stars. More quantitative methods for classifying planetary bodies are available, but agreement among the astronomical community on the exact classification is arduous. For now, the eight planets recognized by the IAU are Mercury, Venus, Earth, Mars, Jupiter, Uranus, and Neptune. Each of the ordained planets satisfies the three following criteria:

1. Orbits the Sun
2. Has sufficient mass to overcome material forces, which results in a spherical shape (hydrostatic equilibrium)
3. Has cleared its orbital neighborhood of other objects

The third criterion is by far the weakest of the newly established standards, since Jupiter, the Earth, and Neptune have material in or near their orbital paths. Jupiter's questionable qualification as a planet under this category is complicated by the innumerable asteroids and comets in or near its orbital path. Both Neptune and Jupiter are instrumental in clearing out and collecting small bodies in the solar system, roughly between two and sixty Astronomical Units (AU). In fact, small bodies have been, and continue to be, pulled toward and pushed away from Jupiter and Neptune through the natural Lagrange L1 and L2 gravitational gateways, making rigorous classification of new types of solar system bodies more difficult.

Pluto, which does orbit the Sun and is large enough to be spherical in shape, was not considered to have cleared its neighborhood of other objects, since it lies in a resonance region of Neptune with the same exact semimajor axis as the other "Plutinos." For this reason, it was changed to dwarf planet status. Although objects larger than Pluto have been discovered recently, their classification was not established under the IAU accord. Ironically, those new objects were responsible for precipitating the debate and redefinition of planets but were not given a meaningful classification. It was these icy dwarf planets that increased the planet rolls by dozens and placed Pluto in a new category of planet.

Another method of planetary classification employs the physical characteristics using two fundamental properties. The first is the three basic composition differences that exist between the planets: (1) terrestrial (rock and metal), (2) gas, and (3) ice. The second basic difference is the much more subjective range in three sizes for the planet population: (1) dwarf, (2) intermediate, and (3) giant. This classification serves to delineate the planetary bodies by their physical character and their size, both of which are closely related to their origin, as discussed later.

GIANT GAS PLANET OR FAILED STAR?

Exploration of other star systems has uncovered large planets ten times the mass of Jupiter or more, but smaller than the smallest stars. Largest of the neither-planet-nor-star objects found so far are the "brown dwarfs" that have the same approximate composition as the giant gas planets. Although too small to be a star, brown dwarfs have sufficient mass to create temperatures and pressures in their central core that can fuse

Table 3.2. Planetary Types within our Solar System Listed by Approximate Size and Composition

Terrestrial (rock, metal)		
Dwarf	*Intermediate*	*Giant*
Ceres, Vesta (asteroids)	Mercury, Venus, Earth, Mars	None
Gas (H, He)		
Dwarf	*Intermediate*	*Giant*
None	None	Jupiter, Saturn
Ice		
Dwarf	*Intermediate*	*Giant*
Pluto, Quaoar, Sedna, Eris, etc.	None	Uranus, Neptune

the small amount of deuterium in their cores into helium 3, and, if slightly heavier, fuse the tiny fraction of lithium into helium 4. Smaller than the brown dwarfs are the largest giant gas planets, with an internal energy source powered only by the contraction of the interior mass. Aside from the residual heat created in the planets' formation, internal contraction and its heat release can take place by condensation. One of the contraction mechanisms is thought to be the precipitation of liquid helium, which then sinks and converts gravitational potential energy into kinetic energy, called Kelvin–Helmholtz contraction. While brown dwarfs could also generate contraction energy internally, the giant gas planets are not massive enough to kindle deuterium fusion. If positioned at the right viewing angle in its orbit around the parent star, the more massive brown dwarf has a more pronounced perturbation on the position of the central star. Brown dwarfs, often called failed stars, are warmer than the Jupiter-sized planets and therefore easier to detect with infrared-sensitive telescopes.

Observation techniques that measure oscillations in the position of a parent star, or the dimmed light of the star as a large planet passes in front, have led to the discovery of hundreds of extrasolar giant gas planets and brown dwarfs. Statistics coming from precise measurements of extrasolar planet positions and spectra reveal several fundamental differences in the two categories of these gigantic planets. First, a mass roughly fifteen times the mass of Jupiter (M_J) permits the fusion of deuterium (^{2}H) in the planetary core, which can be detected with infrared instruments as a warm glow at the outer shell. Thus, 15Mj approximates the maximum mass of the giant gas planets and the minimum mass for the brown dwarfs. Both theory and spectral observations of brown dwarfs show that lithium fusion begins at a mass of approximately 65Mj, although lithium is also detected in very young stars and cannot be used to identify brown dwarfs without another method to corroborate the data.

Surprisingly, many of the giant planets are found closer to their host stars than Mercury is to the Sun. Survival of a giant gas planet in such a high-temperature environment is still puzzling, but one observational consequence is well established. Planets closer to their stars are hotter and more easily detected than the same planet at greater distance. There is a selection effect that biases the detection of extrasolar planets in favor of close-in giants. This is accentuated by the stronger gravitational tug of a nearby planet on the parent star than from the same planet in a more distant orbit.

An even more important mass boundary exists in the giant planets between the maximum mass of the brown dwarf and the minimum mass of a star, which is approximately 0.1 solar masses (the Sun is defined as one solar mass, or $1\ M_\odot$). This mass is the minimum needed to generate temperatures and pressures high enough to fuse hydrogen into helium in a star's central core. Objects below this mass can barely kindle hydrogen fusion, leaving them as hot planets without the characteristics of the bright emissions and the stellar winds of a hydrogen-burning star.

TERRESTRIAL EXOPLANETS

Giant gas planets and brown dwarfs comprise nearly all of the exoplanet discoveries because of their large mass and their warm emissions, both of which make them far easier to detect than smaller, cooler planets. A snapshot of our own solar system from a nearby star would show the same results. Jupiter would be observed tugging on the Sun, and almost solely responsible for the slight movement in the Sun's position. Terrestrial planets are far smaller than the giant gas planets, even the giant ice planets Uranus and Neptune. However, overcoming the detection bias is difficult with current instruments and methods for the same reason that distant comets are more difficult to detect than nearby comets. In spite of the difficulty, finding terrestrial planets outside our solar system is an ongoing effort because of the greater likelihood of finding life on solid rock planets than on ice or gas planets. A terrestrial planet detection satellite called Kepler that was launched on March 6, 2009, uses a 1 m wide-field telescope and photometer to detect planets as they transit their parent stars. The slightly dimmed light for some 100,000 stars in Kepler's field of view is used to identify a wide variety of planets, large and small. Those measurements will then be used to determine the statistics for the terrestrial planet population in a small region of the Cygnus constellation. Another of NASA's future missions called the Terrestrial Planet Finder (TPF) is part of their Planet Quest program that will target exoplanets in the ultimate search for life outside the solar system

Planetary Formation

Formation of the planets was a minor but important sequence in the Sun's creation; minor, because more than 99.9% of the mass in the solar system was locked in the Sun, and important because life as we know it cannot exist on a star. One crucial step in the creation of planets during formation of a star is the rotational momentum of the collapsing gas cloud. Because angular momentum must be conserved in the collapse, the increasing cloud rotation with its shrinking diameter will form a disk aligned perpendicular to the rotation axis, and centered on the newly forming star called a protostar. This pancake-shaped disk gathers the orbiting material into a dense, flattened region, enhancing the collisions among the small solid clumps that begin forming from smaller particles. These collisions set the stage for building larger and larger solid bodies, and eventually the planets.

A rapid collapse of the solar nebula concentrates mass at the very center, quickly leading to a high enough pressure and temperature to initiate the fusion of deuterium, lithium, and later hydrogen, at the core. This early fusion process generates intense heat, resulting in dramatic changes in the inner solar system. Mass infalling toward the center is stalled by the outward radiation pressure from the young protostar, which establishes the approximate mass of the star for much of its life. UV heating from the young protostar also evaporates the ices and breaks down the volatile materials (molecular nitrogen, methane, water, ammonia, etc.) into constituent gases. In our solar system, the Sun's newly created magnetic field is believed to then begin slowing the nearby orbiting gas and dust. This permits a more direct pull inward, which increases the Sun's mass by infall accretion. However, materials that remained inside this hot ionization zone of the inner solar system were only the dense, high-temperature (refractory) materials comprised of rock and metals. This high-wind outflow and heating phase of the early Sun, called the T-Tauri stage, is also seen in other modest-sized stars in their early formation stages.

The young planetary disk that reached to the outer solar system likely remained unchanged as larger and larger ice clumps and comets formed. A similar collisional accretion process took place in the inner solar system, but dominated by rock and metal clumps that formed the asteroid planetesimals. Yet even with the separation of the ices and the rock that formed in the planetary disk, the rocky asteroid and icy comet building blocks are found scattered throughout the disk, presumably by the violent, chaotic collisions. As planetesimals grew into planetary cores, and cores grew into planets, the collisional process that formed the final planets came to an end. Most of the planetesimals were pulled into the planets and their moons during their formation, and most of that in the first billion years. The small amount of material we find today as SSSBs, the comet and asteroid planetesimals, remain scattered throughout the solar system, or stabilized in various orbits.

PLANET FORMATION SEQUENCE

1. Gas Cloud Collapse with Rotation

A protoplanetary disk is created as a large molecular gas cloud collapses, which is characterized by high-speed infall and turbulent flow. A flattened disk forms due to centripetal forces created by increasing angular rotation speed as the cloud, called the solar nebula when referring to our solar system, shrinks toward the stellar core. While disk formation is not universal, it is common for small to moderate mass stars. The disk structure forms with a higher density of material in the central plane of the rotating disk.

2. Condensation and Crystallization

In the disk formation process, dust grains and heavy molecules and ices begin to condense out of the nebula gas and dust. A general migration begins due to cooling and collisions, while electrostatic attraction creates the first small clumps that will create the planetesimals.

3. Pre-Stellar Heating

The central star continues to grow larger by gravitational attraction and infall toward the central mass. High temperatures and pressures at the center create a hot core, which generates intense UV radiation and a magnetic field from ionization and deuterium fusion, although the central core is not yet hot enough for hydrogen fusion. Radiation heating and the magnetic field slow the rotation speed of the inner disk material, which then infalls to the protostar center, clearing out much of the nebula gas and dust in the inner region and heating the ices and dust, while creating crystalline materials. The remaining high-density, high-temperature refractory materials are the rock, metal, silicate, and oxides that later form the asteroids and rocky planets.

4. Planetary Core Formation

Outer regions of the solar system are not heated or stripped of gas like the inner region, allowing the small particles to quickly form larger and larger particles, until kilometer-sized and larger planetesimals begin concentrating with the aid of turbulent flow. The result is a violent, runaway collision and a buildup of larger planetesimals. These rocky and ice cores begin to accumulate gas atmospheres after reaching Earth-mass and larger. The planetary cores continue accreting solid material by gravitational attraction, but more slowly than gas accumulation.

5. Gas Dispersion and a Transparent Disk

Central star heating and gas dispersion of the planetary disk depletes the mass reservoir responsible for building the outer, giant planets after several million years. This ends their rapid growth, although accretion continues at a much slower pace until the surrounding materials are either scattered or pulled into the planets and their moons. The race between the rapidly growing outer giant planets and the diminishing disk material ends with the stabilization of the planet masses, but this does not stabilize their orbits.

6. Planet Stabilization

Smaller terrestrial planets form more slowly because of the lower-density environment in the inner-disk region, and the more rapid removal of the gas reservoir during their formation. The size difference between the terrestrial planets and the Jovian (giant) planets is dramatic, although the cores are probably similar in composition (core composition of the giant planets can only be inferred because of the uncertainty in the density gradient and actual conditions within their cores).

7. Outer Solar System Remnants

The distant, outer solar system retains the original solar nebula makeup, although the composition of the smaller objects consists only of ices and dust grains because their mass is too small to collect and retain gas atmospheres. These objects in the outer reaches of the solar system include the comets and cometary debris observed in the Oort cloud, which is a large spherical region postulated to lie well outside of the planetary disk containing comets and comet/ice fragments. Another region thought to contain large comets and icy planets has been found in the same plane as the planets, but extending beyond Neptune and Pluto. This region, called the Kuiper disk or Kuiper belt, was named after Gerard Kuiper, who was one of several astronomers that proposed an enhanced population of large comets and ice planets beyond Neptune.

Terrestrial Planets

Each of the four rock and metal terrestrial (meaning Earth-like) planets differs widely in its physical features, although there are important similarities in the members. Similarities can also be found between the terrestrial planets and the asteroids, and even between the terrestrial and the Jovian (giant) planet cores. As a group, the terrestrial planets share three fundamental features; a metal core, a rock mantle, and a rigid crust. Like the larger gas and ice planets, the mass and size of the planet is related to its distance from the Sun by several variables. For the rocky planets, the Sun's hot outflow and magnetic field in its early T-Tauri phase depleted gases and volatile materials, leaving behind mostly high-temperature materials within several AU of the Sun. As a result, the material density that should have been greatest closest to the Sun was depleted, with only the densest and highest temperature particles remaining. Instead of high-density gas near the Sun, the most dense materials—metals and rock—persisted closest to the Sun, but in low abundance. Based on this interpretation, a rough model of terrestrial planet formation would result in the lowest-mass, highest-density planets found closest to the Sun, and increasing in size but decreasing in density outward. Measurements of the terrestrial planets show they roughly follow the argument that terrestrial size should increase with increasing distance from the Sun, and density should decrease with increasing distance from the Sun, although there are some glaring differences. As shown in the figure below, the size of the terrestrial planets does increase with distance, until Mars. Rock and metal planetesimals created during the Sun's early heating would be expected to decline at some point, but without a more comprehensive model of planetary formation for the solar system, the decreased mass of Mars is indicative only of the outcome of the formation process. Also, the assumption of declining density with distance is correct, except for the Earth, which has the highest density of any planet. Compression from the Earth's gravitation accounts for Earth's higher density, a phenomenon which is supported by laboratory experiments on the formation of minerals thought to make up the Earth's deep interior at high temperatures and densities.

TERRESTRIAL PLANET DENSITY

Mercury, which should have the highest density according to the simple model of decreasing density with increasing distance from the Sun, has a density of 5.54 g/cm^3 (5.54 times the density of water). Likewise, Venus should have a lower density than Mercury, and does, at 5.24 g/cm^3. Earth's density, which should be even less, is actually the highest observed in the solar system at 5.50 g/cm^3. Mars should have the lowest density, and does at 3.94 g/cm^3. If Earth's density is corrected for the compressibility of the interior magma, it would have an average value of 4.07g/cm^3.

Another important characteristic of the terrestrial planets is the metal core size and mass compared to the planet's overall size and mass. The simple model that posits higher density materials closer to the Sun also suggests a higher core-to-mantle mass ratio and a higher core-to-mantle size-ratio closer to the Sun, which is true for Mercury, Venus, and Mars as shown in figure 3.1. A discrepancy in Earth's higher density and elevated core-to-mantle ratio attributed to gravitational compression of Earth's mantle is also likely influenced by changes in the mineral and metal composition of the interior.

An important comparison in the terrestrial planet interiors is Earth's core size compared to the Moon's core. At roughly 30% of the total planet's mass, Earth's core-to-total mass ratio is about three times that of the Moon. This disparity is important,

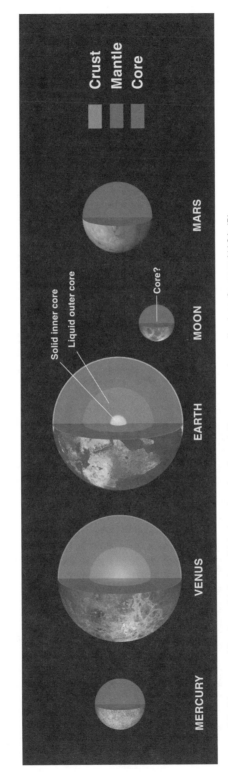

Figure 3.1. Diagram of the terrestrial planet interiors showing the relative core and mantle sizes. Source: NASA JPL.

since Earth and the Moon cannot have formed from the same material (or equivalently, cannot have formed by accretion in the same region). In addition, the average composition of Earth is not the same as the average composition of the Moon, which further limits the possibilities of the formation of the Earth-Moon system.

PLANETARY MASS

Gravitation played the leading role in building planets by mass accretion, and in shaping the terrestrial planets and their moons, as well as forming the asteroids and comets. Internal pressure due to gravity can be used to model the shape and the material composition within a celestial body, and is also one of the criteria for determining the class of a planet. In general, bodies as large as or larger than our moon, or the largest asteroid, Ceres, take on a spherical shape because of the enormous central force of gravity overcoming the material rigidity of the body. Planets or moons with sufficient mass have a spherical shape in which the central force of gravity creates hydrostatic stability. Smaller bodies, like asteroids and small moons, show less of the spherical influence of gravity. Our own Moon has a nearly perfect spherical shape, while the two moons of Mars, Phobos and Deimos, have irregular shapes because of their much smaller mass. Asteroids, which are generally much smaller in mass than prominent moons, are still shaped by gravity, but are not massive enough to conform to a spherical figure. Ceres,

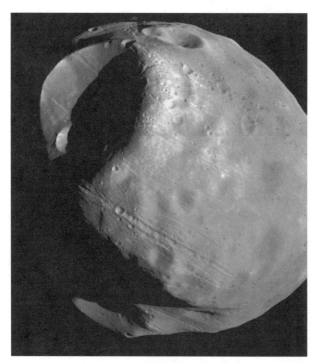

Figure 3.2. Image of the Mars moon Phobos taken with the Viking orbiter. The irregularly shaped Phobos has a mass less than one-millionth that of the Moon ($1.46 \times 10^{-7} M_{moon}$). Source: NASA JPL.

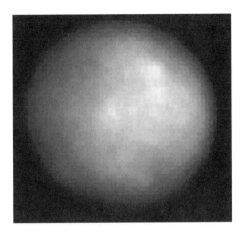

Figure 3.3. Hubble Space Telescope image of the main-belt asteroid 1 Ceres that is now classified as a dwarf planet. Ceres has a mass that is 0.13% the Moon's mass, but is still spherical in shape ($1.23 \times 10^{-2} M_{moon}$). Source: NASA STScI.

which is the largest asteroid in the main asteroid belt, is the one exception, and large enough to be classified as a dwarf planet.

Mass is important not only in determining the shape of a solid planet, but in determining the long-term geological activity of a terrestrial planet, as well as differentiation of the high- and low-density materials. A good example is the Moon, which is too small to be a planet, but is nevertheless terrestrial in composition. It has sufficient mass to have gone through partial differentiation of the interior during its early formation stages, since heat from collisional accretion was enough to melt much of the mantle during formation. This allowed the partial separation of the densest materials, which sank toward the core, from the least dense materials, which floated toward the surface. While the Moon was massive enough to have generated a weak magnetic field, it was much smaller than the terrestrial planets, which forced the smaller Moon to cool quickly. The upper mantle solidified, which froze the incompletely separated materials in place and stopped geological activity roughly 3 billion years ago. In contrast, Earth's mass, which is eighty-one times greater than the Moon's, has undergone complete differentiation of the rock and metals. Lighter rock floated to the surface to form the crust, overlaying a dense rock mantle layer. At the Earth's core are the heaviest metals, dominated by iron.

Heat retained from the early formation process and radioactive isotopes contained in the rock and metal have kept the core and mantle molten in the two largest terrestrial planets. The liquid interior permitted extensive geological activity on the three largest terrestrial planets, although rotation also induced activity in the planetary surface and crust. Venus with a mass 81% of the mass of Earth has a molten interior, but rotates very slowly. As a result, Venus has an entirely different record of geological activity than that of Earth. Without significant rotation and the induced circulation of the liquid mantle, Venus's upper mantle undergoes mostly convective motion that is not capable of moving the relatively thin crust of Venus in the same way that the Earth's crustal plates shift. Because Venus's stationary crust evolved differently from Earth's dynamic crust, their planetary features bear little resemblance to one another.

TERRESTRIAL PLANET ATMOSPHERES

Atmospheres of the terrestrial planets are entirely different from the structure and composition of the Jovian planet atmospheres, and even show characteristics that vary widely among the terrestrial group. There is consistency, however, in the original primordial atmospheres of these inner planets. Evidence locked in the earliest surface rocks and upper mantles shows an early buildup of nebular gases, predominately hydrogen and helium. These primary atmospheres were soon lost from the intense heat of formation and replaced with a secondary atmosphere of gases released from the hot, active surface and crustal rock, and from the impacting comets, whose ices vaporized on impact. Carbon dioxide remained common to all three of the largest terrestrial planets, since its high molecular mass was easily retained by the planets' gravity. These secondary atmospheres also evolved with time and settled into the quasi-stable terrestrial planet atmospheres we see today.

Because the thermal (kinetic) velocity of a gas is inversely related to its mass, the higher-mass gases were easier to retain because higher-mass gas molecules have lower kinetic velocity compared to lighter gases at the same temperature. Lighter gases can easily escape a planet's gravity if their thermal velocity is greater than escape velocity in the upper atmosphere. The actual details of which gases are lost or retained in an atmosphere are far more complex than suggested by this simple velocity comparison, but the concept is useful as a starting point.

The escape velocity for gases near a planet's surfaces is not an accurate representation of the escape condition at the top of the atmospheres where the gas actually escapes, but in some cases can be reasonably close. Also, the temperature at the planet's surface is not a realistic value for upper-atmosphere temperatures on planets with large atmospheres, but it is roughly so for objects with thin or no atmospheres. Using surface temperatures to represent gas temperatures is inherently inaccurate, but can be instructive. With those caveats in mind, a plot of the planet masses, temperatures, and retained gases is shown in figure 3.4, along with three moons.

The original atmospheres of the terrestrial planets were composed of the gases contained in the original nebula, but were stripped away by the high temperatures during the early formation stages of the planets. Secondary atmospheres rich in nitrogen, carbon dioxide, and water, with some methane, were then generated by gases escaping from the rock and collected from comet impacts. While Venus, Earth, and Mars began with roughly the same primary and secondary atmospheric compositions, much of the water contained in the accreted rock, asteroids, and comets was lost from both Mars and Venus, but not from Earth. Part of the reason for the different atmospheres is seen in figure 3.4. The heaviest gases, including CO_2, can be easily retained in the atmospheres of all three planets. However, atmospheric water and Earth's active, thin crust soon led to a significant divergence in the three largest terrestrial planets' atmospheres.

The geological record of the terrestrial planets bears the evidence of volcanic activity, at least in their early history. This is important, since geological activity has one of the most dramatic effects on the chemistry of the terrestrial planets' atmospheres. Earth's geological activity is unique in several ways, as is its atmosphere. First, Earth's volcanism has been almost continuous since its formation 4.5 billion years ago. Second, the Earth's rapid rotation allows active crust motion in the form of plate tecton-

Atmospheric Gas Escape/Retention

$$V_{gas} = \sqrt{3kT \Big/ m_{gas}}$$

Let's begin with a simple expression for the gas velocity relation to the molecular/atomic mass as shown in the Boltzmann equation of gas velocity. Here, V_{gas} = gas velocity, k = Boltzmann constant, T = characteristic temperature, and m_{gas} = molecular mass of the gas.

For a planet to retain a gas, its atmospheric temperature must be sufficiently low so that the gas velocity is less than escape velocity. But because the actual gas velocity is a Boltzmann distribution with two tails, one at decreasingly higher velocity (or, equivalently, temperature), and one at decreasingly lower velocity (temperature), the retained gas condition expressed as $V_{gas} < V_{escape}$ is more accurately expressed as $V_{gas} < 1/6\ V_{esc}$. This is because atmospheric gas losses, gains, or changes have occurred over roughly 4 billion years, and the small tail of the Boltzmann distribution of gas velocities contributes significantly to the losses over that period of time.

$$V_{esc} = \sqrt{2GM \Big/ r}$$

A second important expression for atmospheric gas retention is gas escape velocity from a planet, represented here as the radial velocity required to overcome the gravitational pull at distance r from the surface of a planet. In this equation, G is the gravitational constant, and M is the mass of the planet.

For a rough calculation of the gases that could be retained in a specific planetary atmosphere, first combine the two equations and solve for molecular mass.

$$m_{gas} \approx \sqrt{\frac{54\,kTR}{GM_{planet}}} = C\sqrt{\frac{TR_{planet}}{M_{planet}}}$$

This expression shows that the molecular mass of the retained gas is affected by the atmospheric temperature T, by the planet radius R (assuming a thin atmosphere), and by the planet mass M. With an increased atmospheric temperature, only higher-molecular-mass gases would be retained. For a planet with the same mass but a greater radius, lower-molecular-mass gases would be lost. As the planet mass increases, lower-molecular-mass gases can be retained. In a nutshell, heavier planets retain lighter gases and have larger atmospheres. Smaller planets such as Mercury and the planetary moons have thin atmospheres, or only trace gases, with Saturn's moon Titan a notable exception.

ics. Together, the plate motion and active volcanism release significant gases, even today. Third, water covers much of the Earth's surface. The combination of water, crustal motion, and volcanism actively transforms the surface rocks and atmospheric gases, unlike on Venus and Mars.

Release of gases from the outflow of molten rock has enhanced atmospheric CO_2 and H_2O over the Earth's history. Reactions between the surface rock, atmosphere, and early oceans have produced a variety of compounds, including carbonates. Today, as in the past, the oceans' shell life also produces abundant calcium carbonate, which

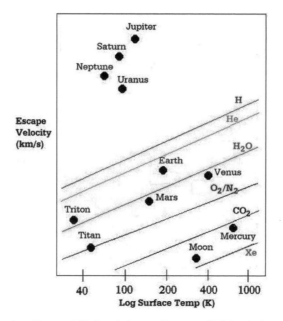

Figure 3.4. Plot of a planet's capability to retain specific gases. Retained atmospheric gases are plotted on diagonal lines, along with the representative temperatures and escape velocities of the eight planets and several moons. The gases that are retained by a planet or satellite have diagonal lines below the planet or moon. Both Venus and Mars, for example, can retain CO_2, O_2, and N_2 that lie below their position, but not H_2O. The Earth can retain all gases shown, except for hydrogen and helium. All Jovian planets can retain all gases because of their large masses. Titan can retain O_2 and N_2, although only marginally. Titan's atmosphere is dominated by methane (NH_3) which is just below the O_2/N_2 line, but methane can easily escape. Because methane is observed in Titan's atmosphere, it must have a mechanism to replenish the gas from within, and does. It is important to note that neither Mars nor Venus can retain the water molecule. Note also that neither Mercury nor the Moon can retain gases lighter than xenon (Xe).

sinks to the ocean floors in sedimentary layers. Some of the layers composed of carbonates and silica compounds are also forced below the crust by sinking (subducting) continental plates. These rock layers can return much later in the form of volcanic magma and gases, but with radically altered composition and chemistry. For the Earth, these processes are important in the absorption and removal of CO_2 from the atmosphere. For both Venus and Mars, there are no similar CO_2 removal mechanisms, which results in an almost pure CO_2 atmosphere on both planets.

In addition to the removal of molecular CO_2, the Earth's geological and biological activity evolved nitrogen and nitrogen compounds, which today make up 78% of the atmosphere. Although oxygen was not originally a significant gas in the secondary atmosphere, it is now 21% of the total by mass, with contributions coming from both early and later biological life. Evolution of the Earth's atmosphere has been far more dramatic than the atmosphere of either Venus or Mars, due to the influence of surface water, and the major physical differences in the planets—primarily planetary mass and rotation. Active plate tectonics and volcanism are both responsible for the makeup of

Earth's unique atmospheric composition, along with the contributions from water and biological life. This raises the question, would there be life on Earth today if it rotated as slowly as Venus? The answer is probably not.

Greenhouse Gases

Terrestrial planet atmospheres contain various amounts of the gases that convert the solar UV, visible, and near IR energy into lower frequency (far) infrared radiation that heats the lower atmosphere. Greenhouse gases that consist of CO_2, H_2O, O_3 (ozone), N_2O (nitrogen dioxide), CH_4 (methane), and NH_3 (ammonia), contribute to planetary atmosphere heating in different amounts because of the different atmospheric compositions, and because of the varied distances of the planets from the Sun. For the Earth, nearly all of the greenhouse gas heating comes from CO_2, H_2O, O_3, and CH_4 in descending order. Total heating from these gases in their low abundance (H_2O ~ 1%, CO_2 ~ 0.04%) is approximately 35°F, which keeps the Earth from a perpetual ice age. In sharp contrast, Venus's atmosphere, consisting almost entirely of CO_2 and N_2, has a greenhouse heating contribution of 500°F from these gases, giving it the hottest average surface temperature in the solar system. Mars, even with its 96% CO_2 composition, only has a 5°F heating from greenhouse gases. This is thought to originate from Mars's early oceans, which likely removed most of the original CO_2 but left the planet frozen thereafter.

MAGNETIC FIELDS

Mass plays another important role in the physical characteristics of planets and large moons as a determinant in the strength of their internal magnetic field. Larger planets

Table 3.3. Composition of the Terrestrial Planet Atmospheres

Gas	Mercury	Venus	Earth	Mars
CO_2	Trace	96.5%	0.038%	95.7%
N_2	Trace	3.5%	78.1%	2.7%
O_2	42%		21.0%	0.2%
Ar	Trace	0.007%	0.9%	1.6%
H_2O	Trace	0.002%	1%	0.03%
CH_4				Trace
H_2	22%			
He	6%	Trace		
Other	29% Na, trace Xe, Kr, Ne	0.0017% CO, trace Ne		0.07% CO, trace Ne, Xe, Kr

have large, high-temperature metal cores under high pressures that promote the circulation of the highly conductive metal cores. Any weak external magnetic field (the Sun's, for example) can induce electrical currents in the conductive metal surface of the liquid metal core. The electric currents can, in turn, generate stronger magnetic fields. Since the magnetic field strength is roughly determined by the core conductivity (related to the core composition, pressure, and temperature), and by the planet's rotation rate, faster rotation induces higher Coriolis circulation forces and potentially higher electric currents. This model of terrestrial planet magnetic fields, called the geodynamo model, is a reasonable approximation in predicting terrestrial magnetic field strength, but is too simple a model for the massive Jovian planets and their magnetic fields, at least beyond hinting that planets with the largest mass and fastest rotation have the strongest magnetic fields.

According to the geodynamo model of planetary magnetic fields, Mercury and Venus should have little or no magnetic field. Mercury is too small and also rotates too slowly, while Venus is massive enough, but rotates only once every eight months. The Earth's magnetic field should be relatively strong, since it has a large terrestrial mass and rapid rotation, and it is. Mars's magnetic field should be very small or negligible because of its small mass, even though it rotates nearly as fast as Earth. The exception to this simple theory is Mercury, which has a very small mass (6% of the Earth's mass) and a slow rotation rate of fifty-nine days. Yet, it has a small, residual magnetic field, most likely induced by the intense magnetic field of the nearby Sun, with a core kept liquid by the Sun's tidal flexing. The Mercury Surface, Space Environment, Geochemistry and Ranging (MESSENGER) spacecraft, which will settle into orbit around Mercury in 2011, should resolve much of the uncertainty of the internal and surface character of the planet.

MERCURY

Mercury is the closest planet to the Sun, and also the smallest, with only 6% of the mass of Earth and one-third the diameter. Mercury is one of the solar system's two inferior planets (meaning inside the Earth's orbit) and is only seen in the morning or evening twilight, since its orbit is so close to the Sun. The maximum separation angle (elongation) of Mercury from the Sun as viewed from Earth is 28°.

Mercury's more notable features are similar to the Moon's: its surface is almost completely cratered; its surface has a very low albedo (percentage of reflected light), which is 10% vs. the Moon's 12%; it has newer and relatively smooth lava-filled basins; and it has only a trace-gas atmosphere. One substantial difference between the Moon and Mercury is the magnetic field. Mercury is almost five times more massive than the Moon, which has no residual dipolar magnetic field, but it rotates twice as slowly as the Moon. The dynamo theory suggests that Mercury should have little or no magnetic field, yet measurements taken during MESSENGER spacecraft flybys show a much larger-than-expected field strength, which is roughly 1% of Earth's.

Mercury's geology shows younger lava-filled basins created by the ancient impacts during its formation, much like the Moon's mare. The largest of these craters are also the oldest, and reach a width of 1,600 km. Mercury's surface also has numerous narrow ridges and scarps that were likely formed by the cooling, shrinking interior fracturing the thinner, outer solid crust. Some of the most unusual terrain on the planet is seen at the antipode (opposing position) of the enormous Caloris crater that stretches 1,300 km in diameter. The fractured landscape suggests that crushing seismic waves from the impact at Caloris traveled around the planet's crust and converged at the antipode. Diversity in the surface geology of Mercury is much greater than for the Moon or Mars, but, at the same time, features from both Mars and the Moon are found on Mercury. These include plateaus that may be volcanic in origin, and folds crisscrossing the planet from earlier stages of cooling and shrinking.

Mercury's density of 5.43 g/cm³ is the highest in the solar system, with the exception of the Earth. However, if the material densities were corrected for gravitational compression, Mercury's mean density would be greater than Earth's. Mercury also contains a greater percentage of iron than any of the other terrestrial planets. Based primarily on density and measurements of the flybys of Mariner 10 and MESSENGER, the metal core is approximately 1,800 km thick, while its mantle and crust are 600 km and 100–200 km respectively. The core is thought to be still molten in spite of its small mass because of the Sun's continual tidal flexing, a force more than ten times that of the Sun-Earth tides.

Temperature extremes on Mercury's surface are greater than on any other planet or moon in the solar system, with a sunward maximum of approximately 700 K (+800°F) and a night-side minimum of approximately 90 K (-298°F). In spite of the extreme temperatures, Mercury is believed to have ice in its polar craters due primarily to its 0.01° axial tilt. This obliquity (tilt) angle is by far the smallest of any planet

Figure 3.5. Image of the north pole region of Mercury taken by the MESSENGER spacecraft in January 2006. In addition to Mercury's heavily cratered surface are a wide variety of geological features. This is the first image taken of this quadrant of Mercury's north pole. Source: NASA/Johns Hopkins University APL.

in the solar system. As a consequence, basins within the deeper craters near the poles could be shielded from sunlight for billions of years. Data from Earth-based radar signals reflected off Mercury show high reflectivity at the poles typical of water or ice. Other mechanisms could be responsible for the high reflectivity on Mercury's surface, however, including sodium ions condensing near the cooler polar regions.

Mercury's orbital eccentricity is the highest of any planet in the solar system at 0.2056. Enormous tidal forces from the Sun contribute to both the noncircular orbit and its tidally locked spin-orbit resonance of 3:2. Mercury has a rotation period of 57 days (sidereal), and an orbit period of 88 days. This means that the nearby Sun would be immense in the daytime sky, with an apparent retrograde motion for a portion of the 88-day orbit. The planet's slow rotation, which is close to its orbit period, stretches the solar day to 176 Earth days. This is precisely two Mercury years (orbits) per solar day.

Mariner 10 was the only spacecraft to visit Mercury until the flyby encounters of the MESSENGER spacecraft twice in 2008 and once in 2009. The next encounter will be orbit insertion of the planet, scheduled for 2011. MESSENGER's twelve-hour orbit will provide high-resolution imaging of Mercury's entire surface to provide composition and topographic maps. Data from the mission is also expected to provide an accurate profile of volatile compounds and elements at or near the surface, especially in the permanently shadowed polar craters. Accurate measurements are also planned to build a detailed three-dimensional model of Mercury's surprisingly large magnetic field.

Mariner 10's original mission plan was for a single flyby encounter of the planet, which was expanded to three flyby encounters by using its solar panel extensions to change the spacecraft's trajectory around the Sun. For the first time ever, solar radiation pressure was used to change the orbital flight of a spacecraft. These multiple flyby opportunities required Mariner 10's solar orbit to be twice Mercury's orbit period to provide sunlit encounters. But with a two-year orbit, only one side of the planet could be imaged. As a consequence, only 45% of the planet was mapped by Mariner 10.

Measurements of the atmospheric gases show small amounts of potassium, oxygen, sodium, argon, and helium released from radioactive decay in the surface rock. Other gases measured by MESSENGER in trace amounts include calcium, magnesium, silicon, and water, along with water ions including the hydroxyl radical (OH^-), and ionized water (H_2O^+). Magnetic field measurements made by both Mariner 10 and MESSENGER indicate a stable field strong enough to deflect some of the Sun's intense solar wind.

VENUS

Venus, named for the Roman goddess of love (Aphrodite in Greek mythology), is often called the Earth's sister planet because it is 95% of Earth's radius and 82% of Earth's mass. Like Mercury, Venus's orbit inside Earth's makes it an inferior planet and visible only near sunset and sunrise. Maximum elongation for Venus is 47°, which permits viewing the planet in the early morning and early evening. Because of its size and nearby orbit, Venus is bright enough to be seen in daylight in the right Sun-Earth-Venus alignment.

Table 3.4. Mercury Statistics

Mass	$3.302 \times 1{,}023$ kg (0.0553 M$_{Earth}$)
Radius	2,439 km
Mean density	5.43 g/cm^3
Escape velocity	3.70 m/s^2
Orbital eccentricity	0.2056
Orbit inclination	7.0°
Semimajor axis	0.387 AU (Astronomical Unit) (5.791×10^7 km)
Orbit period	87.969 days
Rotation period	58.646 days (2:3 orbit-to-rotation resonance period)
Rotational axis tilt (obliquity)	0.01°
Magnetic field	0.0033 Gauss (1% of Earth's)
Albedo (% reflected visible light)	10.6% (Earth = 37%, Moon = 12%)
Atmosphere	Trace (approximately 1,000 kg total) (includes K, Na, Ar, O, O$_2$, He)

One of the many dissimilarities between Venus and Earth is the difference in the planets' rotation rate. Venus's 243-day rotation period is close to its 225-day orbit period, but retrograde—opposite in direction to all other planets except for Uranus. The slow rotation is thought to be responsible for the lack of a significant magnetic field within Venus. The planet's slow rotation also does not allow significant lateral circulation in the upper mantle needed to move the overlying thin crust. Unlike the stationary crust on Venus, Earth's crust is pushed from underneath by the mantle, which is responsible for the plate tectonics and the dynamic geological activity.

Features similar to Earth's include Venus's interior, a molten or partially molten metal core with a radius of roughly 3,000 km. Overlying the metal core is a rock mantle and crust that together make up the planet's overall density of 5.24 g/cm^3.

Venus' surface geology is an unusual aggregation of large volcanic plateaus, mountainous regions, large and small pancake-shaped domes, and long channels, but with surprisingly limited cratering. Unlike Earth and portions of Mars, water played no role in sculpting Venus's surface features, although it could have in its infancy. Large cuts or rills that are seen in the broad plateaus have been attributed to lateral shifts in the crust instead of being cut by erosional forces. One of the most unusual characteristics of the planet's surface is the random distribution of craters, which are smaller than the largest found on the Moon and Mercury, but larger than roughly 35 km. Small craters missing from the surface are believed to be the result of Venus's dense atmosphere breaking up or burning up smaller meteoroids before they could impact the surface. The largest craters found on Venus are significantly smaller than the largest craters on Mercury and the Moon, which limits their age to approximately 300–500 Myr (million years) (Schaber et al.).

Two highland areas on Venus dominate its surface features; the Ishtar Terra highlands (Ishtar is the name of the Babylonian goddess of love) in the northern hemisphere, which contains the largest mountain on the planet, named Maxwell Montes, and the Aphrodite (Greek goddess of love) Terra in the southern hemisphere. In keeping with Venus being the only planet named after a female goddess, almost all of Venus's features are named for female characters and mythological goddesses.

Even more interesting than the unusual surface features on Venus is its geological history. Estimates of the age of the surface of Venus that are based on the distribution of crater sizes are inferred to be the same age as the largest and oldest craters—roughly 300–500 million years. One proposed mechanism for modifying the surface of Venus is the remelting of the lower crust, allowing outflow of the upper mantle through fissures and craters in a thinning crust. Subduction of much of the crust in a 500-million-year remelting cycle could also be responsible for erasing a large portion of Venus's surface features. A lack of crustal motion to dissipate the mantle's heat, like the Earth's continental plate shift, is thought to be responsible for the heat buildup in Venus's upper mantle and lower crust.

Thick cloud layers of hydrogen sulfide and hydrosulfuric acid droplets in Venus's dense carbon dioxide atmosphere make it impossible to view the surface in visible light.

Figure 3.6. Three-dimensional image of the prominent Maat Mons (Maat mountain, named after the Egyptian Goddess of truth and justice). Data from Magellan's synthetic aperture (imaging) radar and its radar altimeter were used to project surface features with a vertical scaling magnified by a factor of 10 to emphasize surface details near the now-extinct volcano. Light-colored regions are the lava outflow onto the surrounding plains. Source: NASA JPL.

The thick atmosphere also creates a pressure ninety-two times the Earth's atmosphere at the surface, where temperatures reach 740 K (865°F). Even though Venus is 28% closer to the Sun than Earth, its extreme temperatures are relatively even at all latitudes and longitudes because of the high-density atmosphere and the efficient absorption of solar heat in the upper atmosphere. In spite of the extreme surface temperatures, the wind speeds at the surface are only a few meters per second because of the very high atmospheric density. Global winds at high altitudes reach nearly 90 m/s, accompanied by lightning. For this reason, and because of the sulfur compounds in the atmosphere, Venus is thought to have active volcanoes even though none have been observed directly.

Exploration

Exploration of Venus began with the launch of the Mariner 2 on August 27, 1962—the first successful interplanetary explorer. The Mariner 2 flyby mission discovered low temperatures in the upper atmosphere, but a surface temperature of about 450°C (723 K, 842°F), and no detectable magnetic field or radiation belt. Russia's Venera exploration spacecraft series began with the launch of the first successful Venera 2 mission on November 12, 1965. The first successful atmospheric entry probe was the Venera 4 probe launched June 12, 1967, although it did not reach the surface as planned. Onboard instruments measured atmospheric CO_2 levels at 90–95% of the total atmosphere. The U.S. Mariner 5, originally scheduled as a complement spacecraft to Mariner 4 to Mars, was launched in June 1967 on a Venus flyby mission. Mariner 5 had improved instrumentation that returned significant data on the Venusian atmosphere and environment, including radio occultation measurements of the atmospheric density, and UV emissions.

Russia's dual-mission Venera 5 and 6, which were orbiter and atmospheric probe pairs, were launched just five days apart in January 1969. Both probes were separated from their orbiters after their arrival four months later to enter Venus's atmosphere just one day apart. Data from both probes was relayed through their respective orbiters before being crushed by higher-than-anticipated atmospheric pressure. It was not until more accurate estimates of Venus's atmospheric pressure and density were made and the Venera lander strengthened sufficiently that the first landers could succeed in reaching the surface (Mitchell 2003). Venera 7 launched August 17, 1970, was the first successful Venus lander, and the first spacecraft to land on another planet, touching down on December 15, 1970. Venera 8, an advanced orbiter and lander of similar design to Venera 7, included sophisticated instrumentation and a camera to measure lighting conditions during its descent through the atmosphere in preparation for the next lander mission, which was programmed to photograph the surface. The low light levels measured on Venera 8 led Soviet scientists to add halogen lamps to Venera 9 and 10 for their photographic mission.

First Images of Venus's Surface

The Soviet's Venera 10 and 11 spacecraft launched in June 1972 were also orbiter and lander pairs. Data from the Venera 9 and 10 landers confirmed the 480–500°C

(896–932°F) surface temperatures and 92 bar surface pressure, but included much more information on the atmospheric composition and cloud layer particles, as well as surface soil/rock composition, and, for the first time, close-up images of the surface of Venus. Protective optical window covers failed to separate on one of each of the landers' dual cameras, but image details from the working cameras showed a relatively smooth solid rock surface strewn with sharp, broken rocks that indicated little surface erosion. Spectroscopic data confirmed the surface rock was basaltic lava. To shorten the descent through the dense lower atmosphere and save battery life, the landers separated from their parachutes 50 km from the surface, then slowed the vehicles to touchdown using a built-in aerodynamic speedbrake flange.

Venera 11 and 12, launched in December 1978, were almost identical to the Venera 9 and 10 spacecraft and with similar mission objectives, but they failed to return image data because of optical cover separation failures on both spacecraft. Venera 13 and 14 missions were launched five days apart in the 1981 Venus launch opportunity. The successful arrival of the dual lander-orbiter spacecraft was followed by the successful landing of both probes, which sent back the first color images of Venus's surface, in addition to more detailed data on the atmospheric and surface rock composition. Both spacecraft used the same structures as the Venera 9–12, but with advanced instrumentation.

The last of Russia's Venera series were the dual-spacecraft Venera 15 and 16 missions launched one day apart in June 1983. The spacecraft bus was similar to the Venera 9–14 vehicles, but had a completely different payload that consisted primarily of synthetic aperture radar instrumentation. The first comprehensive views of the Venus surface topology were generated from the data sent back by the duplicate spacecraft on their one-year mission.

The most detailed images of Venus's surface were returned by NASA's Magellan spacecraft, deployed from space shuttle *Atlantis* on its STS 30 mission launched May 4, 1989. Magellan's synthetic aperture radar imaging system provided surface resolution down to 150 m. Magellan's one-Venus-year (eight Earth months, or 243 days)

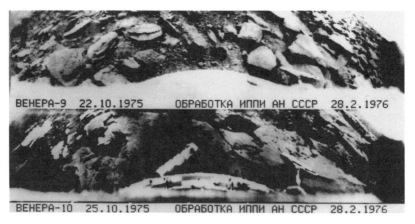

Figure 3.7. Panoramic images from the first successful Venus landers, Venera 9 (top) and Venera 10 (bottom) showing the small, fragmented rocks strewn across the flat, solid rock landscape. While not identical, the images show a surprising similarity in the fragmented rock on a surface taken by landers separated by roughly 2,000 km. Source: NSSDC.

Table 3.5. Venus Statistics

Mass	4.869×10^{24} kg (81% M_{Earth})
Radius	6,052 km (95% of Earth's)
Mean density	5.24 g/cm³
Escape velocity	10.4 m/s²
Orbital eccentricity	0.0067
Orbit inclination	3.39°
Semimajor axis	0.723 AU
Orbit period	224.7 days
Rotation period	243.0 days (retrograde)
Rotational axis tilt	177.4° (2.64° from vertical)
Magnetic field	$<10^{-5}$ Earth's
Albedo	75% (Earth = 37%)
Atmosphere	92 bar (92 Earth atmospheres)
Composition	96.5% CO_2, 3.5% N_2

mission was extended to three cycles because of the vehicle's operational success and the extraordinary quality of the data. In addition to the SAR surface images, the Magellan returned accurate radar altimeter data for the surface observations that allowed three-dimensional imaging of 84% of Venus's surface. After completing its imaging mission, Magellan was placed into low orbit for accurate measurements of Venus's gravity field using precise Doppler tracking of its communications signal. On its final directive, Magellan de-orbited into Venus's atmosphere to measure the density of the upper regions, and to remove the spacecraft from orbit around Venus before the attitude control and orbit propellant were exhausted.

EARTH

Largest of the four terrestrial planets is the Earth, with a mass sufficient to create its unique internal dynamics and surface geology, and with an atmosphere unlike any other in the solar system. Moreover, the Earth's rapid rotation differentiates it from the other terrestrial planets by powering a dramatic interaction between the upper mantle, the crust, and the atmosphere. A third important difference between the Earth and the other terrestrials is the Earth's oceans, which are lacking on Mercury, Venus, and Mars. Compared to the static atmospheres of Mars and Venus, the Earth's atmosphere is continually undergoing changes from the chemical and dynamical interaction of the thin, moving crust with the oceans and the atmosphere. Like both Venus and Mars, the Earth's atmosphere would have been dominated by carbon dioxide except for its oceans. The Earth was also created in a zone between the extreme heat near the Sun and the frozen world beyond Mars. By virtue of our planet's mass, its rapid rotation, its place in the solar system, and the watery surface overlying its thin, active crust, the Earth inherited the ideal conditions for spawning and maintaining life.

Formation

Because of the forces of erosion, the Earth's formation record is difficult to track using its crustal rocks, but it can be reconstructed from asteroid samples and lunar materials brought back during the Apollo program. The reason for this is the Earth's dynamic surface activity, which transforms surface rock almost continually. Radioisotope dating of surface samples shows that nearly 80% of Earth's surface rock is less than 200 Myr old, while more than 70% of lunar surface is older than 4 billion years (By). Nevertheless, difficulties in correlating the formation events in the terrestrials can be overcome by careful analysis of the composition and the element ratios found in rock samples. A simple assumption can also be made that the Earth's formation began no later than the oldest rock samples from the Moon, since both were formed at the same approximate time, and no earlier than the oldest meteorite samples, since those are the building blocks that formed the terrestrial planets. These bracket the Earth's age at 4.54 By.

Similarities in rocks found on the Earth's crust and in lunar and asteroid samples are seen in some of the most abundant metals, called siderophiles (iron-loving materials which include nickel, cobalt, platinum, tin, and tantalum) and in several specific oxygen isotopes. Similarity in the $^{18}O/^{16}O$ ratio compared to the $^{17}O/^{16}O$ ratio indicates the same origin for both the Earth's and the Moon's crusts, since the isotope compositions seem to vary with positions in the solar system. Further, these ratios strongly suggest that the impactor that recreated the Earth and produced the Moon had oxygen isotopic ratios that were consistent with the merged values found in Earth's crust. In contrast, meteorites identified as originating from Mars do not show the same ratio variations. This alone makes a sister of Mars an unlikely candidate as the impactor that blasted much of the material from the Earth to form the Moon (Jones and Palme 2000).

Along with the important similarities in the terrestrial planets are composition differences between the Earth and the other terrestrials. One important distinction is the iron content of the crust of the Earth and Mars. Even though Mars has a higher iron content in the surface material samples than the Earth, it has a smaller iron core. Computer models confirm that additional heating from accretion bombardment of the Earth, resulting in its larger mass, also raised the surface temperature of the upper mantle in its early formation process. This allowed metallic iron to separate and sink to the core, while Mars was too small to reach temperatures where heavier iron oxides would easily sink through the semiliquid mantle.

Simulations of terrestrial planet formation suggest that the accretion of numerous large cores the size of the Moon, or even Mars, are responsible for creating the planets, rather than the formation of a few planets starting with their cores building from many smaller planetesimals. The cataclysmic collisions responsible for building the many protoplanetary cores that shaped the final four terrestrial planets in these simulations take place in a relatively brief episode of violent activity spanning roughly 10^5 yr (Lin 2008). Planetary formation resulting from violent collisions rather than gradual accretive formation is also attractive as a descriptive cause-and-effect relationship because of the unusual features of the terrestrial and even Jovian planets. Venus's slow, retrograde rotation and lack of a moon, as well as Uranus's 98° rotational axis tilt are commonly attributed to violent collisional events during the planetary formation sequence. The Earth-Moon formation has also been shown to be most likely the result of a violent Mars-sized object blasting a portion of the Earth's crust and mantle, which created an Earth-orbiting cloud that formed our Moon.

Structure

The Earth is a well-differentiated planet composed primarily of rock and metal, meaning that its mass is segregated into layers of increasing density toward the core. In fact, all of the terrestrial planets have sufficient mass for the collisional heat energy absorbed in their violent formation to allow separation of the high-density core materials from the medium-density rock mantle and the low-density crust. How complete the differentiation of the various materials is in rough proportion to the object's mass. For example, Venus and Earth are completely differentiated, while Mars and the Moon are only partly so. Small and modest-sized asteroids are thought to be undifferentiated. Material found in the terrestrial region inside the orbit of Jupiter, a zone that excluded gases and low-temperature volatiles early on, consists of high-temperature silicates, metals, and metal oxides that are dominated by iron. This is reflected in the terrestrial planet model consisting of a metal core in the center of the silicate rock outer mantle. The metal core, which makes up roughly 50% of the terrestrial planet's radius, is thought to be a significant part of Jupiter's moon Io, although much smaller in proportion, like the Earth's Moon.

Segregation of the various rock and metal materials in the Earth resulted in distinct layers that vary in density and chemical composition. The primary layers consist of the thin, rigid crust composed of lighter silicate rocks, some 10–40 km in depth. The thin fragile crust is actually comprised of several rigid plates that float on the upper mantle. Plate tectonics, the motion of those continent-sized plates, determines much of the shape and activity of the Earth's surface, including mountain building and the shrinking and expanding ocean floors and deep trenches.

Below the thin crust are several layers of mantle, or molten rock, that are higher in density than the crust layer, but lower in density than the central metal core. Included in the uppermost semirigid mantle is the lithosphere—a combined layer of the thin crust and the uppermost mantle—reaching approximately 60–70 km in depth. Below the lithosphere is a more dense layer called the asthenosphere which is less rigid than the lithosphere, but less fluid than the upper mantle that it floats on. The semirigid asthenosphere that extends in depth from about 100–200 km can and does tear, as it is pulled by the circulating, dense mantle below while encountering the friction of the rigid lithosphere above. Structure and activity of the Earth's crust—the lithosphere—and the asthenosphere are measured by the seismic wave propagation generated by Earth's internal activity.

Earth's central core is a dual-phase metal sphere composed primarily of iron, along with nickel and minor elements. The outer liquid-metal core surrounds the solid

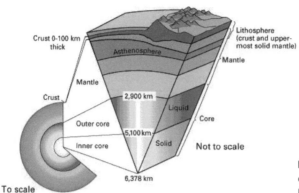

Figure 3.8. Structural layers of the Earth's interior. Source: USGS.

central metal core, which together make up approximately 35% of the Earth's mass. Circulation of the highly conductive outer liquid core is believed responsible for the Earth's magnetic field in the geodynamo mechanism. More complex theories of the core are needed to model reversals in the Earth's magnetic field polar orientation found in the geological record with a period of less than 1 million years. The last reversal of the Earth's magnetic field occurred some 770,000 years ago, while the next reversal cycle is currently underway.

Atmosphere

Geological evolution has had a dramatic impact on the Earth's early atmosphere because of the heating and differentiation of the materials in the upper mantle and crust. At the same time, the Earth's atmosphere also affected the planet's early surface activity. Had there been little or no atmosphere in the early stages of the Earth's formation, cooling of the partially molten surface would have been rapid, inhibiting the smooth differentiation of the interior materials. Also moderating the heat loss in the Earth's early formation stages were the crust's insulating layer and, based on computer simulations and analysis of zircon (zirconium silicate) grain deposits in the oldest rock, a significant ocean. These zircon granules dated at 4.4 billion years require a significant water reservoir for formation, which is corroborated by oxygen isotopic ratios that confirm the Earth's early oceans (Wilde et al. 2001).

Unlike the smaller Mars and the hotter Venus, liquid water on Earth combined with several types of rock and sediment, which locked up some of the carbon dioxide in the young atmosphere as carbonate rock. Life, which began as simple bacteria some 3.3 billion years ago, began converting carbon dioxide into oxygen for the first time. Later and more abundant plant life employed photosynthesis to convert the CO_2 into oxygen, changing forever the Earth's atmosphere and removing any similarities to the atmospheres of Venus and Mars. Adding to the oxygen content of Earth's early atmosphere was the photodissociation of water in the upper atmosphere.

Earth's atmosphere, like its interior, is layered in higher-to-lower density shells that also contain different gas-abundance ratios. The lowest of the layers, and the densest, is the troposphere, containing nearly all of the water in the atmosphere. The troposphere, which ranges in height from 17 km at the equator to 6 km at the poles, also contains roughly 80% of the mass of the entire atmosphere.

The next layer in ascending order is the stratosphere, which overlays the troposphere—the two separated by a boundary layer called the tropopause. The altitude range of the stratosphere reaches from the top of the troposphere to approximately 50 km, with wide variation in season, in latitude, and in solar activity. Unlike the troposphere, temperatures are inverted in the stratosphere, meaning they increase with altitude. The inverted temperature lapse is due to the Sun's UV radiation heating the ozone molecules concentrated in the troposphere. Above the stratosphere is the mesosphere, which ranges from the stratosphere near 50 km, to a height of approximately 80–85 km. The mesosphere has only minor molecular heating from sunlight, hence a normal temperature lapse that decreases with altitude. The mesosphere has two transition boundaries near the lower (stratosphere) and upper (thermosphere) layers, called the stratopause and mesopause, respectively. Above this is the thermosphere, which is very low in density but encounters intense solar heating of the diffuse oxygen and nitrogen molecules. This layer, which ranges from approximately 80 km to 650 km, also exhibits an inverted temperature lapse that increases with altitude. Atop the thermosphere is the exosphere, which is the top layer of Earth's atmosphere, and the most ionized and least dense gas, consisting mostly of hydrogen and helium. As well as receiving direct heating from the Sun, the exosphere interacts with the Earth's magnetosphere, and the solar wind, since it can extend as far as 10,000 km during the Sun's active phase.

Table 3.6. Earth Statistics

Mass	5.974×10^{24} kg ($1/333{,}000$ M$_{Sun}$)
Radius	6,378 km (equatorial)
Mean density	5.52 g/cm^3
Escape velocity	11.20 km/s^2
Orbital eccentricity	0.017
Orbit inclination	0.0° by definition (solar system reference orbit plane called the plane of the ecliptic)
Semimajor axis	1.00 AU
Orbit period	365.24 days (1.00 yr)
Rotation period	23 hr 56 min (sidereal)
Rotational axis tilt	23.5°
Magnetic field	0.308 Gauss
Albedo	37%
Atmosphere	1 bar (1 Earth atmosphere)
Composition	78.1% N$_2$, 20.9% N$_2$, 0.9% Ar

Exploration

Space exploration began with the launch of the first artificial satellites designed to explore the space environment surrounding Earth. By reaching above the atmosphere, the first satellites could observe the Sun's extreme radiation, the Earth's magnetic field, and the dust and debris orbiting the planet. Sputnik and Explorer were the first in the satellite series that began space exploration by orbiting Earth.

NASA's Explorer series, which began before NASA was created with the Explorer 1 launch, includes more than 75 missions that bridged nearly a half century and extend as far as Lagrange-point observations of the solar and interplanetary influences on Earth. Explorer 7, better known as ACE (Advanced Composition Explorers), is one of the newer Explorer spacecraft, launched in 1997. ACE orbits the Sun near the Earth at the Lagrange L1 point of equal gravity between the Earth and the Sun, some 1.5 million miles from Earth. The position of ACE on the Sun-Earth line allows the continual measurement of the solar wind composition and particle energies and the high-energy cosmic rays coming from supernovas and the galactic center, as well as continual communications with Earth-based ground stations.

A new class of Explorer spacecraft sponsored by NASA is made up of four mission classes, all of which are aimed at the exploration of our space environment. Those are (NASA Goddard Space Flight Center n.d.):

- Small Explorers (SMEX)—cost projections < $120 million
- Medium-class Explorers (MIDEX)—cost projections < $180 million
- University-class Explorers (UNEX) / STEDI—cost projections < $15 M
- Missions of Opportunity (MO)—cost projections < $35 M

THE MOON

Our Moon is closely linked to the Earth and its surrounding space environment for several reasons. First, the strong gravitational attraction between Earth and the Moon has locked the Earth-Moon system into the highest angular momentum coupling in the solar system. Gravitational effects of this bond include strong tides, coupled rotation, and the Moon's stable rotational axis. Another close relationship is the common origin of the Earth and the Moon thought to be the result of a cataclysmic impact that blasted a quarter of the Earth's crust and mantle into an orbiting cloud of debris that formed the Moon. In addition to the close physical and dynamical relationship of the Earth and the Moon, our understanding of the solar system and its formation has come primarily from exploration of the Moon, the closest celestial object, which retains one of the oldest surfaces in the solar system.

Lunar Structure

The Moon's 1,734 km radius includes a 300–425 km iron-rich core, a 1,000 km mantle, and a crust that ranges from tens of km to 100 km in thickness, with an average depth of 45 km. Interior estimates of these major layers are based on seismic data collected during and after the Apollo missions and from orbital satellite data. Seismic wave propagation and refraction measurements used to estimate the depth of the Moon's layers were also used to constrain density and pressure values of the interior, as well as the likely chemistry located at or near the discontinuities between the layers.

The Moon's density of 3.34 gm/cm³ is nearly 40% less than Earth's, which is the largest in the solar system. And while the composition is similar to the Earth's, the abundances of most materials are distinctly different. The Moon has a mass large enough to allow some differentiation of the rock and metals. Like the terrestrial planets, the Moon's core is composed of the highest density materials, although the iron and nickel are not believed to be condensed into a pure metal core. For comparison, small and large asteroids exhibit only partial to negligible differentiation, which is dictated by mass. For the Moon, the upper mantle experienced incomplete segregation during its early molten phase. Irregular-shaped blocks of lighter material rising and heavier material sinking were frozen in place, which created regions of varying density within the upper mantle. These irregularities, called masscons, a contraction for mass concentrations, were discovered early in the lunar exploration program by Soviet and American orbiting spacecraft.

Lunar Interior

Similar structural boundaries exist in the Moon as they do in the Earth. The crust, mantle, and core regions have different densities and compositions, which are measured by the refraction of seismic waves, which is straightforward, although not simple. Seismic and gravity studies collected during the Apollo program found that the core region was displaced toward the Earth. The Moon's synchronous rotation during its semimolten state was responsible for the drift of the core toward Earth. The process also carved a thinner crust on the side facing Earth, and a thicker crust on the far side. The lunar core also provides an important distinction between the Earth and its moon. The mass of the lunar core, which is roughly 8% of the total mass, is far less than the Earth's core, which makes up about 30% of its total mass. This mass ratio difference is essential in the formulation of a theory for the origin of the Moon.

Residual heat from the Moon's interior suggests a higher level of radioactivity than estimated for the Earth, although the measurements were taken at the Apollo landing sites, which were found to be in regions of elevated radioactive emissions. Temperature estimates of 830°C (1,530°F) at the Moon's center to 170°C (340°F) near the surface are likely skewed by the location of the Apollo thermal probes. Spectral imaging from *Clementine*, the Lunar Prospector, and the Lunar Reconnaissance Orbiter also shows that thorium and other isotope levels are significantly higher in the equatorial regions visited by the Apollo astronauts.

Models of the formation and evolution of the lunar surface and interior indicate a complex layering or stratification of materials in the crust and mantle, but probably not in the core. Theories of a mantle and crust composed of molten magma are supported by the discovery of KREEP (an acronym for potassium [K], rare-earth elements [REE], and phosphorous [P]) rock generated by the slow, uneven separation of higher- and lower-density materials as the cooling crust suffered heavy impact fracturing and cratering in its youth. Shifting of the underlying magma, due to the pressures of the crust and the infusion into the fractured, cratered solid crust, followed by impact mixing of the crust and mantle and the changing heat flow from impacts and radioactive decay are generally accepted models of the Moon's early interior, but the accuracy of the available data is limited.

The structural model of the lunar interior was pieced together from seismic data generated by quakes from within the Moon, and from Apollo Lunar Module ascent stage and S-IVB booster impacts, as well as from Apollo seismic explosive experiments. Seismic data from inside the Moon comes from moonquakes originating at two different depths—the first roughly 50 km at the base of the crust, and the second roughly 500 km deep. The deeper quakes occur with the same period as the lunar orbit, making them tidal in origin. These make up most of the recorded quakes, which occur roughly 500 times per year, followed in number by the shallow quakes triggered by the slow shrinking of the Moon as it cools (Eckart 1999).

Rock samples and magnetometer measurements from orbital satellites show weak, localized magnetic fields on the Moon today. Presuming a stronger dipolar field existed in the past based on measurements of weak, irregular magnetic remnants within ancient surface rocks, the chronological record and geophysical features should corroborate the model of a primordial molten metal core. However, a molten metal core would retain more heat than is seen flowing from the interior. A metal core would also require a greater overall density than is observed. Another essential condition for an early geomagnetic core is sufficient heat to generate convective heat transfer upward from the Moon's hot core. When compared to the Earth's crust, samples from the Moon show a definitive lack of volatiles with the exception of the KREEP rocks frozen in the highlands as the crust and upper mantle solidified. The Moon's incomplete differentiation and depleted volatiles in surface samples would preclude convective heat and material transport and a molten metal core, and thus, a global magnetic field. Nonetheless, slight magnetization appears in some of the boundaries between the maria and the highlands, likely due to impact shock. Lunar samples show small but measurable magnetic fields that existed at various times during the crustal rock formation. As on Earth, the magnetized lunar rock helps establish the time of crustal formation and the chronology of crystallization of the surface rock.

Upper Mantle and Crust

Lunar Highlands Conditions on the surface of the newly formed moon are often referred to as the "magmatic ocean" because of the molten condition of the upper mantle that cooled to form the crust. Lighter rock that cooled first contained abundant plagioclase, a silicate rock rich in aluminum, calcium, silicon, and oxygen. As the lighter plagioclase crystallized some 4.5 billion years ago, it rose to the surface in irregular shapes, creating a bedrock frozen in place permanently. However, the condition of the crust would change rapidly from impacts from the same planetesimals that created the Earth. Other light-colored rock consisting mostly of feldspar plagioclase also shaped the lunar crust, which dominates much of the original lunar surface as anorthocite rock. The original plagioclase crust, commonly called highlands because of its generally higher elevation, experienced an intrusion of slightly heavier magnesium-rich magma containing less plagioclase and more olivene and pyroxene, which are both silicate rocks rich in magnesium and iron. The Moon's first phase of crust formation ended as the lunar surface cooled approximately 4.1 billion years ago, a solidification that reached into the upper mantle. These later intrusion formations constitute about one-quarter of the highland regions, yet provide important insight into the formation and early evolutionary processes of the Moon.

Lunar Maria Heavy bombardment of the lunar surface continued until approximately 3.9 billion years after its formation when most of the planetesimal material had been swept out by the growing planets. Asteroid and comet impacts not only cratered the lunar surface permanently, but also created a layer of rubble known as regolith that reached deeper and became more fragmented with time. As planetesimal impacts began to wane, the thermal insulation of the crust and continuing radioactivity in the Moon's mantle began heating the interior, raising the upper-mantle/lower-crust temperatures to the molten state. At this point, the molten magma began to flow upward through the broken, cratered crust, which would change the lunar surface for the last time.

Remelting of the Moon's lower crust initiated a profound change in the lunar surface that we can see today with the naked eye. While the solid crust was thinning from within, the fractured, cratered crust still put enormous pressure on the underlying liquid mantle. The solid crust's tremendous pressure forced the molten magma to flow upward

Figure 3.9. Photograph of the lunar far side taken by the Apollo 16 astronauts showing fewer and smaller maria than on the Moon's near side that faces Earth. Source: NASA NIX.

through the fissures into the lower basins of the largest and deepest craters. Darker, denser magma colored by olivene and pyroxene quickly cooled after filling the lowest basins and fractures in the crust, outlined by the largest and deepest craters. These vast areas, called maria, or seas, have distinctly different geological appearance and composition from the original highlands. These darker basins are now only approximations of circular craters because of billions of years of bombardment by comets and asteroids. From a distant vantage, the maria appear smooth compared to the rough valleys, mountains, and craters that dominate the highlands. In addition, crater features on both highlands and maria establish that larger and older impacts are found on the highlands, a fact more firmly established by the lunar rock samples returned by the Apollo astronauts.

Lunar Composition

Estimates of the Moon's composition and the relative abundance of elements are based on a variety of sources, including the mean density, the surface geology, spectral surveys from orbiting satellites, comparative analysis between lunar samples, asteroid samples, and Earth rocks, as well as Apollo rock samples. Even though the Apollo rock and soil samples are the most revealing, they represent only a tiny fraction of the surface located near the lower latitudes on the Earth-facing side. Uncertainty in the actual lunar composition is greater for the interior because of the lack of bedrock samples, and because it is no longer active. Nevertheless, the basic physics and chemistry associated with the formation of the minerals and rock recovered from the lunar surface during Apollo offer many clues to the lunar makeup. While the Apollo samples are invaluable in establishing the chronology of the formation and evolution of the Moon, significant improvements in lunar models will require more in-place (in situ) measurements and samples from the Moon's surface, especially from the many regions not visited on the Apollo missions.

Minerals found in the two primary rock types that consist of basaltic lavas and anorthocites represent the two different densities of rock from the crust and the upper mantle. The oldest, lightest material is the anorthocite, predominantly aluminum calcium silicates ($CaAl_2Si_2O_8$). Both calcium and aluminum are more abundant in the lunar crust than on Earth, and are potentially useful for construction of structures or facilities on the Moon. The same is true for other materials that could be extracted from anorthocite, including silica glass (silicon oxides), calcium oxide (lime), and alumina (aluminum oxide).

Basalt, in contrast, is composed of a broad combination of silicates and oxides rich in magnesium, iron, and titanium. These minerals are commonly metal oxides (MgO, TiO, FeO, and Fe_3O_4) combined with silica. One of the more common metal oxide silica minerals in the lunar basalts is olivene, a combination of magnesium oxide (MgO) and silica (SiO_2) that forms Mg_2SiO_4. A very important mineral found in the lunar basalts is ilmenite, $FeTiO_3$, important because of its oxygen content, which could be used for a propellant or for producing water or for breathing. Ilmenite's titanium content could be used for high-temperature and lightweight structural metal.

Many other minerals and potential resources are available from the lunar rock and soil, but extraction of the elements and/or minerals requires extensive processing.

Table 3.7. Surface Rock Composition Based on Apollo Samples (Permanent n.d.)

Component	Apollo-11	Apollo-12	Apollo-14	Apollo-15	Apollo-16	Apollo-17
SiO_2	42.47%	46.17%	48.08%	46.20%	45.09%	39.87%
Al_2O_3	13.78%	13.71%	17.41%	10.32%	27.18%	10.97%
TiO_2	7.67%	3.07%	1.70%	2.16%	0.56%	9.42%
Cr_2O_3	0.30%	0.35%	0.22%	0.53%	0.11%	0.46%
FeO	15.76%	15.41%	10.36%	19.75%	5.18%	17.53%
MnO	0.21%	0.22%	0.14%	0.25%	0.07%	0.24%
MgO	8.17%	9.91%	9.47%	11.29%	5.84%	9.62%
CaO	12.12%	10.55%	10.79%	9.74%	15.79%	10.62%
Na_2O	0.44%	0.48%	0.70%	0.31%	0.47%	0.35%
K_2O	0.15%	0.27%	0.58%	0.10%	0.11%	0.08%
P_2O_5	0.12%	0.10%	0.09%	0.06%	0.06%	0.13%
S	0.12%	0.10%	0.09%	0.06%	0.06%	0.13%
H	51.0ppm	45.0ppm	79.6ppm	63.6ppm	56.0ppm	59.6ppm
He	60ppm	10ppm	8ppm	8ppm	6ppm	36ppm
C	135ppm	104ppm	130ppm	95ppm	106.5ppm	82ppm
N	119ppm	84ppm	92ppm	80ppm	89ppm	60ppm

**Table 3.8. Abundance Estimates of Elements
by Weight for the Moon, Asteroids, and the Earth**

Element	Lunar surface	Asteroids (average)	Earth surface
Oxygen	42%	36%	47%
Silicon	21%	18%	27%
Aluminum	7%	<2%	8%
Iron	13%	26%	5%
Calcium	8%	<2%	4%
Sodium	<1%	<1%	3%
Magnesium	6%	14%	2%

Thus, the attractiveness of lunar resources is dependent on the cost of processing at the Moon. Table 3.7 lists the various lunar sample compositions, averaged at sampled sites for each of the Apollo missions. All but Apollo 16 were in mare regions. Samples from the Apollo 16 site near the Apennine highlands ridge reflect the anorthocite character with higher Al and Ca content. Basaltic samples from all but Apollo 16 show a higher iron, titanium, and magnesium content. Volatile gases and carbon measured in parts per million are embedded in the surface rock and soil grains over time by solar wind and cosmic rays. Vertical core samples of the lunar surface made by the Apollo astronauts show progressive churning of the regolith by the continual impact of micrometeoroids. Larger impacts have churned deeper layers of the pulverized regolith, while smaller, more recent micrometeoroids have mixed shallower layers. Thus, the depth and composition of the layers provide a chronology of the rock fragmentation, as well as a record of the Sun's solar wind and the cosmic rays embedded in the rock and soil.

WATER ON THE MOON?

The hopes of finding water ice on the Moon from eons of comet impacts was dashed by the analysis of rock and soil samples brought back in the Apollo program. No evidence of water in any of the lunar samples suggested that the extreme conditions on the Moon, including temperature extremes and the vacuum of space, would have sublimated any residual ice over time. But a preliminary finding that water may be present on the Moon came from the *Clementine* spacecraft that orbited the Moon for several months in 1994. Radar data returned from the spacecraft indicated volatile ices may be present, including water. Four years later, NASA's Lunar Prospector returned neutron absorption data that inferred abundant hydrogen near the polar regions of the Moon, strengthening the argument for water ice on the lunar surface, especially near the shadowed polar regions. Lunar Reconnaissance Orbiter (LRO), launched in June 2009 with its companion spacecraft LCROSS, discovered emissions from water and the related hydroxyl radical (OH^+). While spectral emissions were detected at higher levels near the polar regions, the same emissions came from the low and intermediate latitudes. The impact of LCROSS in October 2009 in the Cabeus crater confirmed the presence of the water and hydroxyl molecules, although not necessarily in the form of pure water ice. NASA as well as international agencies are planning lunar lander projects to determine the presence of water on the Moon which include International Lunar Network landers proposed for launch in 2016 and 2017.

Lunar Formation

The Moon is an ideal laboratory for studying the solar system origins since it is our closest celestial neighbor, and it is large enough to attract and survive large planetesimal impacts, maintaining a nearly complete record of bombardment. Moreover, the Moon is sufficiently small to retain its early surface history because the internal heat was not great enough to either remelt some or all of the surface (as with Venus), or to sustain a molten interior that creates a thin crust that can easily shift, like the Earth. It is also too small and too close to the Sun to amass a significant atmosphere that could erode the surface record over time.

Valid theories of lunar formation and evolution must take into account the physical properties of the Moon, and the characteristics of the Earth-Moon system, including angular momentum and orbital features. Whether detailed surface composition data or bulk lunar characteristics, a robust theory of the Moon's formation must match observational data. Observations and measurements used to constrain the model of lunar formation include but are certainly not limited to the following abbreviated list:

Lunar Formation Constraints

- Ratios of oxygen isotopes ($O^{16}/O^{17}/O^{18}$) in the Earth and in the Moon are the same.
- The Moon and the Earth have differences in various other isotopic ratios.
- The Earth's density is 5.5 g/cm^3, while the Moon's is 3.3 g/cm^3.
- The Moon's crust contains roughly 12% iron while the Earth's crust contains 4%.
- The mantles of the Earth and the Moon have distinctly different iron/nickel/cobalt metal (siderophile) makeup.
- Refractory (high-temperature) element concentrations are higher in the Moon than in the Earth; however, their ratios are the same.
- Angular momentum of the Earth-Moon system is higher than any other planet-satellite pair in the solar system.

Four main theories of lunar formation have survived through the years, although more have been proposed. To be considered sound, a lunar formation theory would require validation by measured physical properties, including dynamical measurements, and by numerical simulation. The formation theory considered the most probable, and best supported by physical observations (see constraints listed above) and numerical modeling, is the collision/impact theory. The four main theories are briefly discussed below for comparison.

1. Lunar Capture Assuming both the Earth and the Moon are in orbit around the Sun in close proximity, a gravitational capture of the Moon by the larger-mass Earth is possible, provided some process can either remove some of the Moon's orbital energy or pull the

Moon through the Earth's L1/L2 regions. Even though portal capture may be more likely than outright orbital capture, removing energy in order to bind the Earth and its moon requires a passing planet or moon just as the Moon passes the Earth. Such an encounter is highly improbable, but not the greatest weakness in the theory. Any capture scenario would result in a very small binding energy and a correspondingly small orbital angular momentum of the Earth-Moon system that is actually the largest in the solar system.

- Attractive because of its simplicity
- Easily dismissed because of the different composition abundance of Earth and Moon (especially iron)
- Greatest weakness is the actual binding energy being much greater than possible in a capture scenario
- Also improbable because a large number of large planetesimals required to have one passing by a close Earth-Moon encounter at the exact time of capture

2. Coaccretion During the Earth's formation by collisional accretion, a second, smaller body could be formed close to the Earth from the same material, and with little binding energy. Neither of these characteristics is supported by observational data.

- Attractive because of simplicity and the lack of the need for a catastrophic event
- Highly improbable because the actual composition differences between Earth and the Moon are not accounted for
- Also improbable because the orbital angular momentum is much too small compared to the actual Earth-Moon angular momentum
- Also improbable since a merger is more likely than a separate, lightly-bound Earth-Moon pair

3. Fission of a Rapidly Spinning Earth George Darwin, son of the noted Charles Darwin, published a number of papers on the theory of the Moon's origin based on a rapidly rotating molten Earth spinning off a portion of its lighter outer layer to form a satellite. Assuming that the Earth were spinning fast enough, and that resonant modes in the molten Earth would be sufficient to help heave a significant mass outward, a molten satellite could break away from the Earth and form the Moon. Lighter materials would be concentrated in the upper mantle of the Earth that are reflected in the actual composition of the Moon. However, the extreme angular momentum of the spinning Earth required for the fissioning process is far too great when compared to the rotating Earth and Earth-Moon orbital angular momentum. Other weaknesses in the theory include a high refractory material abundance on the Moon that is not supported by a rapidly spinning Earth. The high inclination of the lunar orbit to the Earth's equatorial plane is also unsupported by the fission model.

- Attractive because it can explain composition differences in the Earth and the Moon
- Validation of a fissioning planet is difficult to model numerically without an external impactor
- Highly improbable because the angular momentum required to spin off the Moon is not observed today

4. Collision of a Small Planet with the Earth The theory of a colliding body forming the Moon first proposed by William Hartmann and Donald Davis begins with a Mars-sized body impacting the Earth off-center. The violent, oblique impact would produce a ring of debris that would coalesce and accrete to form the Moon out of what remains of the orbiting debris. Numerical simulations indicate a mass ratio of roughly 10:1 (Earth-to-impactor), and a majority of the material forming the Moon coming from the impactor.

- Attractive because it can account for composition differences between the Earth and Moon
  - Resulting Moon would be less dense than Earth overall since the lower-density mantle of the Earth would make up a large part of the ejected material
  - Measurements of the Moon show a reduced iron/metal core (7% vs. 30% for Earth)
- The edge-on collision produces an uneven ring of debris that rapidly forms a diffuse, then dense central core orbiting the Earth
  - The newly-formed moon, along with the Earth, sweeps out the rest of the material
- Details of the theory have been supported by numerical simulation, although several questions remain
  - Circular, inclined orbit requires specific impact parameters
- This is the most probable of formation theories, considering:
  - Moon's higher iron composition surface
  - Low lunar core mass and density compared to Earth
  - Oxygen radioisotope similarities between the Moon and the Earth
  - High orbital angular momentum of the Earth-Moon system

Lunar Topology

Major lunar surface features were formed in three stages, beginning with the rocky crust solidifying during a period of intense bombardment by asteroids and comets. The Moon's battered crust is preserved in the highlands regions with a topography shaped by cratering. What appear to be mountains are actually formed as crater rims connected randomly and sometimes dramatically into a surface relief that is twice Earth's Mount Everest. Impact basins on the Moon are the largest in the solar system, with the Aitken Basin that spans the lunar southern pole extending roughly 2,500 km in diameter. The South Pole-Aitken crater has the lowest elevation (-6 km) and some of the highest rim elevations (+8 km) on the Moon.

Erosional forces on the Moon's surface are entirely from small and large impacts that have rounded the sharp features and layered the surface with the rubble and dust of the impact fragments. This regolith has a variable depth from tens of meters to kilometers if fractured rock is included, with a layer of fine dust several centimeters in depth coating the surface. The electrostatically charged dust contains fine particles of lunar and impactor rocks, as well as fine micrometeoroid dust.

Other lunar topological features include ridges and valleys, called rills, that form with the uneven shrinkage of the Moon. More spectacular rills were formed from lava flows in the early lunar remelting epoch when lava tubes collapsed to create features resembling river beds. The intriguing possibility of an ancient river of lava led NASA

Figure 3.10. Photograph of a large region surrounding the Apollo 15 landing site that contains complex geological features including a large, flat lava-filled mare surrounding older mountainous highlands. This mountainous terrain is actually ancient crater rim remnants called the Apennine mountains which envelop the thin, winding valley named Hadley Rille. Source: NASA NIX.

to assign the Apollo 15 mission exploration site to the Hadley Rille region near the boundary of the Mare Ibrium (Sea of Rains) and the Apennine mountain range. Hadley Rille, which runs along the Apennine boundary, reaches some 4,600 m in height (15,000 ft), with averages of 1.5 km in width, and 400 m (1,300 ft) in depth.

A number of volcanic lava domes are found on the lunar surface that resemble the shield volcanoes on Earth and the pancake-shaped lava outflows on Venus. A few small cinder cone–like volcanic vents are also found on the Moon, but the features are much smaller than those on Earth, which are powered by entirely different mechanisms. Volcanic glasses from impact-melted silicas and from volcanic vents are also found on the lunar surface. Excitement over the discovery of orange soil by Harrison Schmitt on Apollo 17, the only scientist to visit the Moon, confirmed the expectation that glasses could be found near ancient volcanic vents. By pure coincidence, an impact that created the young crater named Shorty lifted the orange-colored silica glass from deep within the crust to the surface.

MARS

The basic character of the surface of Mars has a number of important parallels to Earth's geology, making it one of the most attractive planets in the search for extraterrestrial life, meaning life beyond Earth. But the extreme conditions on Mars make finding life unlikely because of the cold temperatures, the high radiation levels, and the very low pressure atmosphere composed of almost pure carbon dioxide. Even with these extremes, the possibility of life on Mars cannot be ruled out since biological life is found in conditions on Earth that are just as extreme. Biological life has been found in the extremely cold, barren Antarctic regions, in the high-temperature environments

Table 3.9. Lunar Statistics

Mass	7.349×10^{22} kg (1/81 M_{Earth})
Radius	1,738 km (equatorial) (0.27 R_{Earth})
Mean density	3.35 g/cm³
Escape velocity	59.5 km/s²
Orbital eccentricity	0.055
Orbit inclination	5.14° (from ecliptic)
Semimajor axis	384,400 km
Orbit period	29.5 days solar (27.3 days sidereal)
Rotation period	29.5 days
Rotational axis tilt	1.54° (to ecliptic)
Magnetic field	< 0.0001 Gauss
Albedo	0.12
Atmosphere	Trace amounts of helium, argon, sodium, and potassium

near underwater volcanic vents, even in highly acidic hot springs rich in heavy metals. Moreover, the discovery of past water deposits and water flows on Mars make the possibility of life in subterranean layers in its warmer past even greater.

Mars has many of the same geological features found on Earth, which include seasonal variations in the northern and southern hemispheres, polar ice caps, deep valleys cut by erosional forces, high mountains, broad plateaus, and large volcanoes. One of the major differences in Martian and Earth surface features is Mars's heavily cratered surface, although less extensive than cratering on Mercury and the Moon. But with all of its diversity, Mars's geological features show little evidence of recent geological activity on a large scale. Data from recent rover and orbiter missions to Mars, however, do show recent water flows, landslides, and subsurface water ice deposits.

Arguably, Mars's most unique geographic feature is the enormous valley cut through a high, broad plateau, called Valles Marineris, extending 20% of the length of the entire surface. Even more intriguing are the two small moons Phobos and Demos, which are both too small to have a spherical shape. Studies of the surface features and spectra indicate the moons bear closer resemblance to asteroids than their parent planet; not surprising, since Mars lies on the edge of the asteroid belt that contains as many as 1.5 million members bunched between the orbits of Jupiter and Mars.

Structure

Like the other terrestrial planets, Mars's primary structural features include a rigid crust, an interior rock mantle, and a central metal core. And like the other terrestrial planets, the composition, mass ratios, and geological activity of Mars are different from any other planet. Mars, which has an average radius of 3,386 km—just half the size of Earth—includes a metal core with a radius of 1,480 km, a 1,550-km rock mantle, and a crust 50 km thick. Mars's core consists of primarily iron, iron compounds, and nickel that is at least partially molten. Simulations of the Martian core using iron and siderophile materials at the core temperatures and pressures in the planet's center indicate a lower density than Earth's core, and approximately 14% sulfur by weight (Jacqué 2003). The crust of Mars, which is one-third greater in thickness than Earth's (roughly three times Earth's crust-to-diameter ratio), also varies with longitude. The depth of the southern hemisphere of Mars, as measured by the Mars Global Surveyor and other orbital spacecraft, ranges over 100 km in depth, while the northern hemisphere is only half that. The crust also exhibits features that indicate an active period of plate tectonics that ended some 4 billion years ago, in part explaining the buildup of enormous stationary volcanic plateaus and the highest mountain features in the solar system. While internal activity on Mars is possible, NASA's Viking landers failed to register any seismic activity.

A dipole magnetic field is not found associated with the core of Mars, even though significant paleomagnetic fields have been measured in the crust and surface rocks. Measurements of the Martian surface show periods of alternating reversals in polarity that are frozen into the rocks, similar to the record written into the surface rocks found on Earth. By comparison, the crust and mantle of Mars contain 18% iron compared to 8% on Earth, an indication of the lower temperatures in the early formation period of Mars. Its smaller size did not allow the complete differentiation of the higher and lower density materials. As a result, less iron sank to the core and more remained in the mantle and crust. This excess iron on the surface of Mars is not only found in the spectral data and surface samples, but can be seen as a reddish hue from the rich iron-oxide soil.

Surface

The surface of Mars is mostly volcanic basalts, with some silicate rocks and rock formations. Smoother volcanic flow has covered much of the northern hemisphere, although craters still litter its surface. The southern hemisphere is dominated by a highlands region with more volcanic features, mountains, and ancient craters than found in the northern hemisphere. The enormous Olympus Mons shield volcano found in the southern hemisphere rises 35 km from the surface, nearly three times the height of Mount Everest. Valles Marineris, the largest valley in the solar system, extends 4,000 km and cuts 7 km deep into the surface, more than three times as deep as the Grand Canyon.

High-resolution images from the recent Mars Global Surveyor and the Mars Reconnaissance Orbiter show evidence of unusual surface erosion of the surface of Mars. Erosion channels and gullies appear on a scale of several kilometers, but small features like tributaries feeding larger channels are missing. Long-term erosion from wind and small impacts, which are both likely responsible for erasing the smaller features, date most of these features to a much earlier period. Since the only possible agent that could make these flow features at the atmospheric temperature and pressure on Mars is water, the lack of significant surface ice suggests large-scale subterranean ice deposits. Large flows of liquid water were likely produced in the past by episodic heating of stored ice from below. Melted ice flows are also seen in images of craters

Figure 3.11. Mars Global Surveyor Image of the Martian southern hemisphere showing the dominant Valles Marenaris (Mariner Valley) stretching more than 4,000 km across the surface. Source: NASA JPL.

and canyon walls on Mars, and recently beneath shallow soil covering by the Phoenix lander. While Mars's gravity is too weak to retain water in its atmosphere, ice can be stored in the cold polar regions and in subsurface deposits for long periods.

Atmosphere

The atmosphere of Mars is much smaller than the Earth's because of the planet's smaller mass. Like Venus, the atmosphere of Mars is composed of 96% carbon dioxide. The loss of Mars's magnetic field some 4 billion years ago allowed the solar wind to ionize and strip off the upper layers, further decreasing the atmosphere which ultimately resulted in a pressure less than 1% of the Earth's sea-level pressure of 1013.3 mb (101.3 kPa). Without oceans, a magnetic field, and active crust tectonics, Mars's atmosphere, like Venus's, will remain under the rule of carbon dioxide.

Martian Moons

Both of the Martian moons, Phobos and Deimos (named after the Greek god of fear and the Greek mythological character representing panic, respectively), have spectral signatures of carbonaceous asteroids rather than the rocky-metal composition of Mars. For this and other reasons, they are more likely captured asteroids than accreted moons. The probability that both moons are captured asteroids is enhanced by Jupiter's influence over many asteroids in Mars-crossing orbits, which introduces a chaotic nature that promotes travel through the Mars L1 and L2 passageways. Phobos is currently orbiting inside the synchronous orbit radius of Mars, a condition that removes orbital energy from the moon and decreases the satellite's orbital radius gradually until it impacts Mars in an estimated 30–90 million years. The dense field of nearby asteroids also has populated the Mars orbit with at least one Martian Trojan asteroid.

Exploration

Exploration of Mars by robotic spacecraft had a spotty record until the last two decades when more experience, more advanced technology, and broader international participation improved the overall mission success rate. That is not to say that recent efforts have not had problems, such as America's Mars Observer that was launched in 1992 and ended communications before entering Mars orbit, and the British Mars lander named Beagle 2 that ended communications before atmospheric entry in 2003.

The first successful Mars robotic explorer was Mariner 4, launched on a flyby mission in 1964. Mariner 4 showed for the first time that Mars was a barren wasteland covered with craters, and not the canals, deserts, and forests suggested by Percival Lowell in several publications from the early 1900s. The Soviets were no less determined to reach Mars, and between 1960 and 1996, launched nineteen spacecraft to Mars on flyby, orbiter, and lander missions. Of the nineteen Soviet missions launched, three orbiters succeeded in their orbital exploration missions, along with one lander that returned data for twenty seconds.

Considered the most successful of the early Mars missions were the dual U.S. Viking spacecraft launched in 1977 as orbiter-lander pairs. Following orbit insertion and a period of evaluation for prospective landing sites using images from the Viking orbiters, the two Viking landers were deorbited and soft-landed on opposite sides of Mars. The dual landers were the first spacecraft to return comprehensive measurements of Mars's atmospheric temperature and density, wind speed, diurnal and seasonal variability, as well as atmospheric composition, and dust storm phenomena. Additional instruments onboard the landers measured the chemical character of the surface soil in three separate tests for possible life, but without confirmation.

Data from the Viking orbiters and landers provided the most complete understanding of the Martian interior, surface, atmosphere, and environment until the arrival of the Mars Global Surveyor (MGS) in 1996, and the Mars Exploration Rovers Spirit and Opportunity (MER-A and MER-B) in 2003. Since then, our understanding of the composition, character, and evolution of Mars has expanded dramatically. Proof of the existence of periodic flooding of the craters and valleys by subsurface water is visible in the MGS images, as well as in newer images from ESA's Mars Express Orbiter, launched in 2003, and the U.S. Mars Reconnaissance Orbiter, launched in 2005. Evidence has been recently made available on lava pooling, similar to features on the Moon, and the identification of hematite minerals indicative of iron compounds processed in stagnant water. Recent data from the host of Mars explorers show evidence of water-rich mineral formation coming to an end less than a billion years after its formation, along with the end of major geological changes.

Future missions are planned for Mars exploration by the United States and several international partners, as well as a number of projects under development by other coun-

Table 3.10. Mars Statistics

Mass	6.421×10^{23} kg (0.11 M$_{Earth}$)
Radius	3,397 km (equatorial) (0.53 R$_{Earth}$)
Mean density	3.93 g/cm^3
Escape velocity	5.02 km/s^2
Orbital eccentricity	0.093
Orbit inclination	1.85°
Semimajor axis	1.5237 AU (227,939,100 km)
Orbit period	686.971 day (1.8808 yr)
Rotation period	24.623 hr (sidereal)
Rotational axis tilt	25.2°
Magnetic field	No central dipolar field, but areas of weak, banded, localized magnetization of the iron-rich crust
Albedo	0.15
Atmosphere	0.006 bar 95.2% CO_2, 2.7% N_2, 1.6% Ar, 0.2% O_2

tries. These missions are scheduled for launch opportunities that correspond to the period of repeated relative position between the two planets, known as the synodic period. For Earth and Mars, the synodic period is 780 days, or 2.14 years. Recent and future launch opportunities to Mars include 2007 (NASA Phoenix), 2009 (NASA Mars Science Laboratory), and 2012 (NASA Mars Scout mission is planned for launch in 2013).

Jovian Planets

The largest planets in our solar system, called giant or Jovian planets (named for Jupiter, the Roman king of the gods), are all located beyond the orbit of Mars and the terrestrial asteroid belt. The composition of the giants also reflects their greater distance from the Sun than the rocky planets because of their gas and ice content, which resembles the original solar nebula rather than the rock and metals that dominate the inner solar system. Jupiter, the largest planet and the closest giant planet to the Sun, is also most like the Sun in its makeup of 95% hydrogen and helium. Beyond Jupiter, the H and He abundance decreases, while the ice content increases.

Jovian planets follow a more accurate classification in pairs according to their dominant composition. The two giant gas planets, Jupiter and Saturn, are found inside the orbits of the two giant ice planets, Uranus and Neptune. Both Jupiter and Saturn contain more gas than ice and rock, while Uranus and Neptune contain more ice than gas and rock, although the term "ice" is not an appropriate description for the materials within the giant planets. Beyond Neptune lies a frozen world of small and large comets composed primarily of ices. Some of these are large enough to be classified as dwarf planets, including Pluto.

Major differences in the four giant planets are as numerous as the major features that are common to the giants, including a rock and metal core and an outer envelope of hydrogen and helium gas. Both of these properties suggest not only a similar formation process, but a similar mixing process that brought some of the rocky materials from the terrestrial zone to the outer planetary disk. A similar scattering process was also responsible for carrying cometary ices from the outer regions to the inner solar system even though the bulk of the rock and ice planetesimals remain distributed closer to their origin. This imbalance in asteroids and comets in the inner and outer solar system can be modeled, at least in general terms, as the consequence of solar wind pressure and heating during the Sun's T-Tauri phase, which would create more asteroids and fewer comets in the inner solar system and just the opposite toward the giants.

FORMATION

Mathematical models of particle and gas dynamics are the basis for studying planetary formation, whether terrestrial or Jovian, in addition to numerical simulations of the physical process of cohesion, collisions, and turbulence that create the first particles and clumps. Corroborating these modeling tools are physical samples that contain primordial, planetesimal, and planetary materials that can be precisely measured for abundances, composition changes, and age. Yet even with these tools available to construct robust theories of planetary formation, questions remain concerning a number of important details on particle formation, planetesimal formation, and planet formation. Details of the formation processes that distinguish between building small planets, large planets, and their moons are also challenging. Questions related to building giant planets far outnumber the questions of forming smaller terrestrial planets because of the complexity and diverse composition of the giants. Problems are also encountered in simulating the rapid accretion of material needed to build giant planets, and in shutting off the mechanisms that feed material into the growing planets.

PLANET COMPOSITION

A quick glance at the planets in our solar system suggests the greatest dissimilarities are between the terrestrial and Jovian planets. Less obvious but more dramatic are the physical differences found among the four giant planets due mostly to their enormous masses and their gas and ice makeup. Unlike the terrestrial region that was swept clear of the gas and ices during the Sun's early T-Tauri phase, the outer solar system contained the complete solar nebula mix that was enhanced near Jupiter by the gases and vaporized ices from the Sun's radiation outflow. A secondary effect came from impact and slingshot scattering of silicate rock outward from the terrestrial region, with a counterflow of icy planetesimals scattered inward from the outer solar system. The scattering process, which continues today but on a much smaller scale, provided

Table 3.11. Jovian Planet Characteristics

	Jupiter	Saturn	Uranus	Neptune
Mass	1.8986×10^{27} kg ($317.8 \times M_{Earth}$)	5.6846×10^{26} kg ($95.15 \times M_{Earth}$)	8.6832×10^{25} kg ($14.54 \times M_{Earth}$)	10.243×10^{25} kg ($17.147 \times M_{Earth}$)
Density	1.326 g/cm^3	0.687 g/cm^3	1.318 g/cm^3	1.638 g/cm^3
Radius (equatorial)	$71,492$ km ($11.2 \times R_{Earth}$)	$60,268$ km ($9.45 \times R_{Earth}$)	$25,559$ km ($4.01 \times R_{Earth}$)	$24,764$ km ($3.88 \times R_{Earth}$)
Rotation period	9.925 hr	10.783 hr	17.240 hr	16.110 hr
Orbit period (sidereal)	11.86 yr	29.46 yr	84.07 yr	164.88 yr
Semi-major axis	5.203 AU	9.537 AU	19.191 AU	30.069 AU
Eccentricity	0.04839	0.05415	0.04717	0.00859
Core temp (estimate)	$36,000$ K	$12,000$ K	$7,000$ K	$7,000$ K
Obliquity (tilt) angle	$3.13°$	$26.73°$	$97.77°$	$28.32°$
Notable features	Four bright Galilean moons, Metallic hydrogen interior	Complex ring system, Metallic hydrogen interior	High tilt angle produces retrograde rotation	Near-supersonic wind speeds in upper atmosphere reaching 2,000 km/hr

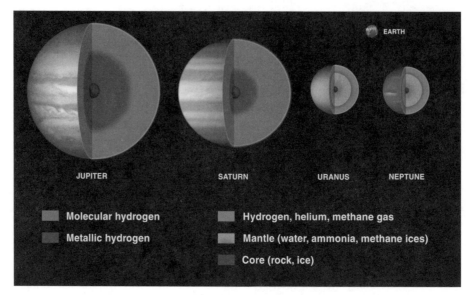

Figure 3.12. Graphic comparison of the Jovian planets' interior and composition illustrating the differences in the gas giants (Jupiter, Saturn) and ice giants (Uranus and Neptune). Earth is shown to scale on the upper right. Source: NASA JPL.

rocky cores for the gas and ice giants and their moons, and a shower of gas and ices that helped build the atmospheres of the newly formed terrestrials. In consequence, the general composition of the Jovian planets includes a large gas envelope with a liquefied ice mantle surrounding a rocky core many times the mass of the Earth. This created two classes in the Jovian family; the giant gas planets closest to the Sun and the giant ice planets beyond Saturn. Also created in the planetary disk but scattered even wider were the dwarf ice planets and the comets.

The atmospheres of the Jovian planets are dominated by hydrogen and helium, which were also the most abundant gases in the original solar nebula. But the striking colors seen circulating through the Jovian atmospheres in varied patterns are not from the hydrogen and helium, which are colorless. The spectacular spectrum of colors of Jupiter and the other giants come from the trace gases, which include methane, ammonia, acetylene, hydrocarbons, and possibly phosphorous and sulfur compounds. More muted colors and less turbulent circulation patterns are seen in the upper atmosphere of Saturn. Uranus has little variation in its methane-colored green-blue haze, while Neptune, which is also colored by trace methane gas that is responsible for absorbing more of the red light in the visible spectrum, is a deeper blue. Lower internal temperatures in Uranus and Neptune also reduce the convective upwelling that is the source of the spectacular circulation patterns seen in the atmospheres of Jupiter and Saturn.

Significant differences in the Jovian family are also found in the liquid mantles. In both Uranus and Neptune, a mantle of liquid ice compounds surrounds a dense rock core, while Jupiter and Saturn contain a dense liquid mantle in layers of liquid hydrogen, liquid helium, and metallic hydrogen. Temperatures and pressures beneath the massive liquid layers in the two largest planets force the electrons in the hydrogen into degeneracy, creating superconductive metallic hydrogen. As pressure increases deeper in the planet's mantle, a point is reached at which higher pressure does not increase compression, but produces a higher energy state in the material—the degenerate condition. For Jupiter and Saturn, metallic hydrogen within the inner cores represents unique conditions, including a superconductive shell of metallic hydrogen thought to be responsible for the large magnetic fields located in the mantles instead of the cores in both planets.

JUPITER

Jupiter, named for the mythological king of the Gods, is the largest planet in our solar system, and also one of the brightest objects in the night sky because of its enormous size and light-colored upper cloud layers. The first recorded use of the telescope to observe Jupiter was made by Galileo, who made detailed observations of the four inner moons of Jupiter called the Galilean moons, and the Great Red Spot that is still seen today. More detailed observations of the planet continued using larger and larger telescopes, but all were surpassed by the first spacecraft to visit the planet, Pioneer 10 and 11 launched in 1972, which returned spectacular details of Jupiter, its moons, and its magnetic field. Dual Voyager spacecraft launched in 1977 improved the quality and detail of exploration data even more during their flyby missions that extended to Saturn, Uranus, and Neptune. A dedicated orbiter launched in 1989, named Galileo, spent eight years exploring the plant and its surroundings, which furnished the most detailed information on Jupiter and its moons to date.

Jupiter's strong magnetic field and intense radiation belts have been measured over the years in the decametric band (tens of meters in wavelength), which led to the discovery of lightning in Jupiter's upper atmosphere. These low-frequency emissions were also used to study early details of the Jovian magnetic field and led to the discovery of a subtle mismatch between the rotation rate of Jupiter's core and its rotating surface features.

Magnetic Field

Jupiter's mass is more than twice the rest of the planets combined and has the fastest rotation rate of any planet at 9.925 hr. Structural effects from the rapid rotation are Jupiter's oblate shape which has an equatorial diameter more than 9,200 km greater than the polar diameter. Dynamical effects of the massive planet's rapid rotation include an enormous dipolar magnetic field that harbors the highest radiation levels of any of the planets. Jupiter's magnetic field is also the largest structure in the solar system, excluding the Sun's magnetosphere. Jupiter's toroidal (donut–shaped) magnetic field interacts strongly with the upper atmosphere, and with the inner moon, Io. Recent findings from Cassini show similar ion and magnetic field interactions with the next moon out, Europa. Even farther out, the magnetic field is reshaped by the solar wind particles that reduce the sunward boundary, and extend the lee-side magnetosphere to reach as far as the orbit of Saturn, some 5 AU away.

Figure 3.13. Jupiter and its four Galilean moons are shown to scale in this mosaic of images from the Galileo spacecraft. From top to bottom in the same order of distance from Jupiter are Io, Europa, Ganymede, and Callisto. Source: NASA JPL.

Atmosphere

One of the unique characteristics of Jupiter is its composition, which is a close representation of the primordial solar system. Jupiter's huge gravitational mass allowed it to sweep up material from a wide path around its orbit in a region enriched by the Sun's outward push of the gases in the inner solar system. Similar to mass differentiation in terrestrial planets, Jupiter's upper atmosphere contains more of the lighter hydrogen gas than found in lower layers. The denser interior is still dominated by hydrogen and helium, but in a different ratio, and with an increased proportion of heavier gases. Measurements of Jupiter's upper atmosphere show a ratio of 75% hydrogen and 24% helium by weight, with a slight but important deficit in the noble gases compared to the Sun's makeup due to precipitation.

The most prominent feature in Jupiter's atmosphere is the Great Red Spot first noted by Galileo in 1610, and later recorded by Cassini and Hooke. Coriolis rotation in the southern hemisphere directs the circulation of the Great Red Spot, which is a high-pressure (anti-cyclonic) storm that extracts its energy from Jupiter's internal heat. This huge feature, three times the diameter of the Earth, is in some ways similar to the tropical cyclones on Earth. Jupiter's internal heat exceeds heat absorbed from the Sun, creating a net heat flow outward and a source of energy for the anti-cyclonic storms on the planet. Similar internal heating is also found on Saturn and Neptune, and thought to be generated by the same condensation mechanism involving the noble gases helium and neon. As these stable gases condense under extreme pressure, their fall through the upper and middle mantle creates a slow, slight contraction of the planet's mass. In response, heat is released as the potential energy of the infalling mass is converted into kinetic energy.

Interior

Jupiter's structure and dynamics are based on models that use a number of assumptions about the composition and mass of the material making up the planet. Jupiter's interior is modeled, in part, by the character of its magnetic field, with a density profile provided by Doppler measurements from the Pioneer, Voyager, and Galileo spacecraft. Estimates based on these models indicate Jupiter is comprised of three separate regions that begin with a dense rock and metal core fifteen to forty-five times the Earth's mass (M_{Earth}). Central core temperature and pressure for Jupiter are estimated at approximately 25,000 K and 10^6 bar, respectively. Jupiter's largest structure is the metallic hydrogen mantle that extends approximately 75% of the Jovian radius, although there is considerable uncertainty because of the variables introduced by the compounds formed in a range of uncertain temperatures and pressures. A lighter, liquid hydrogen mantle overlays the metallic hydrogen interior, as shown in figure 3.12. Overlaying that is the light hydrogen-helium gas atmosphere that would be transparent, as would be the liquid hydrogen and liquid helium layers beneath it, except for the colorful trace gases and the cloud droplets that make all but the top of the upper atmosphere opaque.

Rings

Ring features around Jupiter were first discovered in the Voyager 1 image data, and studied more completely by instruments onboard the Voyager 2 and Galileo spacecraft. Jupiter's faint ring system consists of four components. Unlike Saturn, Jupiter's rings are made up primarily of dust instead of ices, and thought to originate from the small moons orbiting in or near the rings. The closest ring structure is a faint halo that extends from Jupiter's cloud tops, halfway to the main ring. The densest ring is roughly 7,000 km wide. Outside the main ring is a pair of very faint "gossamer" rings bounded by two small moons, Amalthea and Thebe (Throop et al. 2004).

Galileo Spacecraft

The first spacecraft dedicated to exploring Jupiter and its environment was Galileo, launched October 18, 1989 on space shuttle *Atlantis*. A three-year launch delay due to the space shuttle *Challenger* accident was compounded by a switch to a lower thrust booster, since the Centaur upper stage booster could no longer be carried on the

Table 3.12. Jupiter Statistics

Mass	1.8986×10^{27} kg (317.8 M$_{Earth}$)
Radius	71,492 km (equatorial)
Mean density	1.326 g/cm^3
Escape velocity	59.5 km/s^2
Orbital eccentricity	0.048775
Orbit inclination	1.305°
Semimajor axis	5.204267 AU (778,547,200 km)
Orbit period	11.85920 yr
Rotation period	9.925 hr (sidereal)
Rotational axis tilt	3.13°
Magnetic field	10–14 Gauss (poles)
Albedo	0.52 (geometric)
Atmosphere	89.8% H$_2$, 10.2% He, 0.3% methane, 0.026% ammonia, 0.003% deuterium, 0.0006% ethane, 0.0004% water

space shuttle. Instead, Galileo used a multiple Earth-Venus flyby gravity assist called VEEGA (Venus-Earth-Earth Gravity Assist) to compensate for its smaller Inertial Upper Stage solid booster. The delay may have also contributed to its high-gain antenna failure, which reduced the total number of images that were returned from Jupiter, although a clever data compression algorithm compensated for much of the reduced transfer rates from the smaller antenna that was substituted. In spite of Galileo's operational wrinkles, which included problems with the mass memory tape drive, the dual-craft mission was an overwhelming success.

Galileo's atmospheric probe was released five months before the main spacecraft arrived at Jupiter in December 1995. The atmospheric probe relayed data through the Galileo orbiter about the conditions in Jupiter's upper atmosphere before being crushed by the high pressure encountered in its descent. Following the probe's record 230 g, 48.2 km/s (108,000 mi/h) entry, data on Jupiter's atmospheric composition, lightning, temperature, density, and wind speed showed a turbulent but dry atmospheric layer. Water vapor measured initially at only 1–2% relative humidity was later found to be more than 20% by the Galileo orbiter. It was coincidence that the probe descended through a dry spot in the atmosphere. Combined data from the probe and the orbiter identified trace amounts of methane, water vapor, ammonia, carbon, ethane, hydrogen sulfide, silicon-based compounds, neon, oxygen, phosphine, and sulfur in the upper atmosphere of Jupiter. Although the colorful features on Jupiter remain puzzling because of the temperature in the uppermost atmosphere, their likely origin is from hydrocarbons such as acetylene, as well as phosphorous and sulfur rising from a warmer, lower layer of the atmosphere.

Galileo's many discoveries include huge thunderstorms observed in Jupiter's upper atmosphere, although the frequency was less than expected. Jet streams reaching 725 km/h (450 mi/h) were found at lower levels, confirming interior heat-driven circulation. In addition, auroras were seen at Jupiter's poles, along with large ion currents generated in Jupiter's magnetosphere by its interior moons, especially Io. The Galileo orbiter also found weak magnetic fields on Ganymede, Callisto, and Europa, probably originating in conductive oceans beneath the solid ice crusts interacting with Jupiter's extended magnetic field. Galileo's many successes led to the extension of its prime mission three times. Even though more could have been completed, the spacecraft was sent on its final assignment to burn up in Jupiter's atmosphere before it depleted all of its thruster propellant, and to avoid contamination of Jupiter's moons. Galileo's end came on September 21, 2003, when its signals stopped at the Goldstone Deep Space Network tracking station.

JUPITER'S MOONS

Io

Jupiter's most prominent satellites are the four Galilean moons named Io, Europa, Ganymede, and Callisto, in order of distance from the planet. Io, being the closest, also exhibits the greatest tidal influence from Jupiter's enormous mass. Since Io follows a slightly eccentric orbit, its varying distance from Jupiter results in repeated tidal stretching, much like the lunar tides on Earth, but much stronger. This tidal flexing generates sufficient heat within Io to melt at least part of its interior, resulting in continuous volcanic activity on the surface. Io's volcanoes were first discovered in the Voyager 1 flyby images and found to be composed of sulfur compounds and silicon. Continual volcanism on Io, which makes it the most geologically active object in the solar system, not only resurfaces the moon on a regular basis, but also spews sulfur, sodium, and other ions into Jupiter's magnetic field. Those ions furnish a conductive path for currents to flow between Io and Jupiter along the magnetic field lines, while generating radio frequency emissions. Io is in a resonant orbit with its two neighbors Europa and Ganymede (1:2:4) around Jupiter, which is the driving force behind Io's eccentric orbit, which would otherwise circularize over time. Io's semimajor axis is 422,000 km, about 10% greater than our Moon's semimajor axis around Earth, but it orbits in only 1.7 days compared to our Moon's 28-day orbit. This is a clear indication of how much greater Jupiter's mass is than Earth's.

Io's surface is devoid of craters because of its prolific volcanism, yet, over time, the volcanic activity has produced mountains higher than Mount Everest, as well as deep pits and depressions, and extensive lava flows. Io's interior appears to resemble the terrestrial planets, with significant iron at the core and a silicate rock mantle. About 5% larger than our Moon, Io has the highest density of any of the major moons in the solar system, at least to date. Like the rest of the Galilean moons, Io was discovered by Galileo Galilei in 1610, when he first used the telescope for astronomical observations, although under ideal conditions the satellites can be seen with the naked eye.

Figure 3.14. False-color image of Io taken by the Galileo orbiter showing the many volcanic flows and features on the solar system's most active body. Source: NASA JPL.

Europa

Europa, which is the smallest Galilean moon, lies beyond Io in a 3.55-day, 671,000 km orbit. Like Io, the resonance orbits of nearby moons force Europa into a slightly eccentric orbit. Still within the strong gravitational pull of Jupiter, the eccentric orbit generates tidal flexing that heats the interior, which is likely responsible for the liquid upper mantle. Europa's rotation period is just slightly faster than its orbit, making it nonsynchronous by definition, but only by a hundredth of a degree per year, based on Voyager and Galileo images. The slightly faster crust rotation offers supporting evidence of a liquid water ocean beneath the surface crust because of the uncoupling of the rotation of the solid rock and metal interior from the solid ice crust. Atmospheric gas was detected on the surface of Europa, consisting primarily of molecular oxygen.

Europa's structural features include a metal core, a silicate-rock mantle overlaid with ice, and a liquid water layer beneath a solid ice crust. Like Ganymede and Callisto, Europa contains a weak magnetic field induced by its conductive liquid ocean passing through Jupiter's enormous magnetic field. Galileo's images of Europa show a smooth but cracked surface that is likely an ice crust fractured on a regular basis by the tidal tugs from Jupiter. This allows the interior liquid water to flow upward to the surface because of the pressure from the overlying ice crust, which shapes a smooth surface as it freezes. The possibility that the subsurface water on Europa could harbor life has created a steady stream of exploration program proposals, although as of 2010, none have been approved for funding.

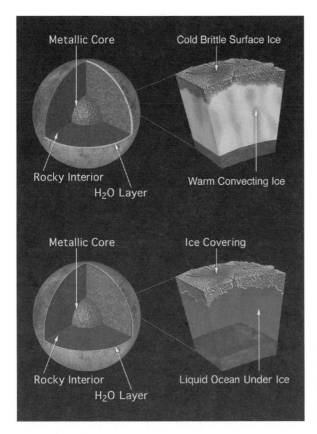

Figure 3.15. Artist's sketch of two possible structures making up the interior of Europa and the liquid/solid water ice ocean beneath the moon's icy crust. Source: NASA JPL.

Ganymede

Beyond Europa is the Galilean moon Ganymede with a 7.15-day, 1.07 million km orbit of Jupiter. Ganymede's orbit resonance features are many, and include Io (4:1) and Europa (2:1), as well as synchronous orbit-rotation with Jupiter. Ganymede is the largest moon in the solar system and larger than the planet Mercury, although half its mass. Like Europa and Callisto, Ganymede's surface is composed of ices, with a silicate rock mantle and a metal core. Surface features include dark-colored, ancient, cratered regions with unusually flat rims, in addition to newer, though still very old, lighter colored areas with extensive grooves and ridges.

Spectral data from Galileo's instruments and the Hubble Space Telescope (HST) were used to identify a thin atmosphere on Ganymede composed primarily of water vapor sublimated from the surface. Photodissociation of the water generates a thin oxygen layer near the surface, with lighter hydrogen rising to higher altitudes before escaping the weak gravity of Ganymede.

Galileo also detected an internal magnetic field on Ganymede, which is unique in all of the moons in the solar system. The weak magnetic field is thought to have its origin in a liquid metal core dynamo like Earth's. Data from Galileo's sensitive fluxgate magnetometer also showed an induced magnetic field component in Ganymede that orbits within Jupiter's strong magnetic field. This suggests a conductive layer of water under the icy crust, similar to Europa.

Callisto

Callisto is the outermost Galilean satellite locked in synchronous rotation with Jupiter in its 16.69-day, 1.88 million km orbit. Callisto is the third-largest moon in the solar system and nearly the same size as Mercury, but far different in composition. Models based on Galileo's flyby encounters indicate that the composition of Callisto is dominated by roughly 40% ice and a 60% rock and metal mix, but not differentiated like the other Galilean moons. Callisto's ice and rock surface has a liquid layer beneath that contributes to its average density of 1.83 g/cm³—the lowest of the four Galilean moons. Although there is no internal magnetic field, a small field induced by motion through Jupiter's strong magnetic field was found, indicating a conductive liquid ocean beneath the icy crust like Europa and Ganymede. This conductive layer is most likely water kept in the liquid state by internal heat and the pressure of the overlying crust.

Callisto's surface features are dominated by heavy cratering to the point of saturation. This would imply an ancient surface with little of the crustal activity that resur-

Figure 3.16. Cutaway drawing of Jupiter's four Galilean moons Io, Europa, Ganymede, and Callisto (left to right) from top right to bottom left. The metal cores in the three inner moons do not appear in the Callisto sketch because of the more uniform mix of the moon's interior which had been modeled with data from the Galileo flyby encounters. Source: NASA JPL.

faces the other Galilean moons. Callisto's surface composition has been estimated as a 2:3 mixture of rock and ice based on Galileo's image data. Surface features on the moon are dark, with an albedo of only 20% reflectivity. In addition to the pervasive cratering, Callisto's surface contains large, concentric ring patterns surrounding impact basins extending as much as 3,000 km in diameter.

Smaller Satellites

Pioneer flyby missions augmented with Earth-based telescopes were used to identify thirteen Jovian satellites by the mid-1970s. These included four smaller moons that reside inside the orbit of Io: Metis, Adrastea, Amalthea, and Thebes. Metis, the closest satellite to Jupiter, has a semimajor axis of 128,000 km and an orbital period of only 7 hr. 4.5 min. With the arrival of the Voyager spacecraft at Jupiter in 1979 and Galileo in 1995, the total satellite count reached sixty-three. The most distant of Jupiter's satellites, called S/2003 J23, is one of the smallest identified so far. The tiny moon is in a retrograde orbit that has a semimajor axis of more than 30 million km and an orbit period of nearly three years. Jupiter's remaining moons reside outside the orbit of Callisto.

SATURN

Saturn, the father of Jupiter in Roman mythology, is the second-largest planet in the solar system, located twice the distance from the Sun as Jupiter. Typical of gas giants, Saturn's composition is dominated by hydrogen and helium, although less than Jupiter's. Like Jupiter, Saturn's structure model includes a rock and metal core, and a superconducting metallic hydrogen mantle overlaid with a liquid hydrogen layer. Saturn's atmosphere, which is composed primarily of hydrogen and helium, also contains trace amounts of methane, water, ammonia, and hydrogen sulfide. Comparison with Jupiter, however, fails to explain important details of Saturn's interior, since the period of rotation of the magnetic field and the associated radio emissions are not the same as the rotation period of its metallic hydrogen mantle. An assumption that the metallic hydrogen mantle is responsible for the magnetic field is unsupported, since the rotation periods are not locked together. Saturn's rapid rotation rate of 10.5 hr creates an oblate shape like Jupiter, with the equatorial radius more than 5,900 km wider than the polar radius.

On the largest scale, Saturn's most unique feature is its low density which is the lowest of any planet or moon in the solar system. Saturn's density of 0.69 g/cm³ is lower than the specific density of water at 1.00 g/cm³. The extended atmosphere of Saturn, which contributes to its low density, is cool enough for ices to form and circulate in its high-velocity winds. Even with its low density, Saturn's huge mass creates conditions extreme enough to form metallic hydrogen in its mantle and generate a core temperature that reaches an estimated 12,000 K (21,140°F).

Saturn's atmosphere lacks the color depth in its features because of its deeper atmospheric layers, atmospheric haze, and storm depths, all of which wash out the distinctive patterns and colors seen on Jupiter. Even so, Saturn's circulation bands are similar to

Jupiter's, as are the complex eddies and storms concentrated along the boundaries of the countercirculating bands. These are not unlike the low-pressure and high-pressure systems on Earth. Some of the most distinctive of Saturn's atmospheric features are the unusual polar circulation patterns. One of the most puzzling of these features is the set of hexagonal rings and the strong vortex that circulate at the north pole. Adding to their peculiar behavior is the fact that they do not have a differential rotation period like the circulation bands. Both unusual features have the same period as the magnetic field and the radio emissions of the planet. Saturn's north and south polar regions also have elevated temperatures, but the south pole circulation differs from the north pole in many ways, including more moderate circular winds. At the center of Saturn's south pole is a unique circulation feature found only on Earth—an eye wall. Infrared imaging from the Cassini spacecraft shows deep atmospheric storms located far from the poles that produce impressive radio bursts characteristic of lightning.

Images of Saturn's outer atmosphere sent back from Voyager and Cassini show wind speeds near the equator approaching 1,700 km/h—much higher than any atmospheric winds found on Jupiter. Spectroscopic measurements of the upper atmosphere identified something just as unusual. Data from Cassini show a distinct helium deficiency (7%) with an excess of hydrogen (93%) compared to Jupiter and the solar nebula model. This deficiency has been attributed to the same mechanism that generates heating in Jupiter's interior, which is the condensation of helium in the cooler, upper atmosphere. Heat from helium condensation and the shift in mass toward the center of the planet, known as Kelvin–Helmholtz contraction, are thought to be the source of Saturn's internal heat, which is 2.5 times the heat received from the Sun.

Saturn's most prominent feature is its large and complex ring system, which can be easily spotted with a small telescope or binoculars. Under good conditions and with the correct ring projection angle, Saturn's rings can be seen with the naked eye. Even though the scale of Saturn's ring system is enormous, the broad, thin ring structure is made up of fine ice particles, with some ice clumps ranging in size from sand to gravel. Particle sizes for these and other ring systems were established by radio-signal occultation experiments using the Voyager and Cassini spacecraft transmitters.

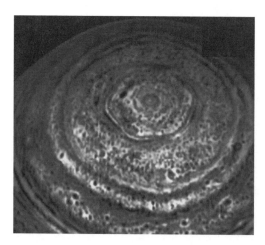

Figure 3.17. Saturn's unusual hexagon-shaped circulation pattern surrounding its north pole shown in this image taken with the Cassini infrared mapping spectrometer. Source: NASA JPL.

Saturn's complex ring system is actually composed of hundreds of individual bands that are sculpted by the orbital dynamics of the material orbiting the massive planet, along with divisions created by several small satellites interspersed within the rings that scatter particles near their orbit with their slight gravitational pull. The ring system begins about 10% beyond Saturn's atmosphere at 67,000 km, identified as the D ring, and extends to roughly 480,000 km at the outer edge of the E ring. The major ring regions in order of increasing distance from the planet are identified as the D, C, B, A, F, G, and E rings.

The small ice particles that make up Saturn's rings show evidence of strong electrostatic charge, possibly from inter-particle collisions. This charge buildup allows the orientation and position of the particles to change slightly when changes in the Saturnian magnetic field pass through. Dark outlines that look vaguely like rotating spokes within the rings were spotted in the Voyager flyby images, and can be seen more clearly in the Cassini data. Visible, infrared, and ultraviolet observations returned from Cassini instruments also show a variety of large- and small-scale structures within the rings. Unexpectedly, oxygen was found elevated from the particle disk plane, likely from photodissociation of the water ice particles that releases both hydrogen and oxygen. The oxygen ions could be repelled from the overall charge of the nearby ring material, which would result in the slight levitation of the ionized oxygen (Waite, et al. 2005).

Saturn's magnetic field is weaker than Jupiter's, and about one-fifth its diameter at the magnetopause—the radial distance of the outer reaches of the magnetic field. Saturn also has a much lower charged particle density within its radiation belt. In contrast to Jupiter's energetic, intense particle belt that is fed by ions from Io's volcanic spray, Saturn's only interactive moon orbiting within its magnetosphere is Titan. The magnetic field–charged particle interaction does exist, but on a much smaller scale than in the Jovian system. Saturn's radiation belt, which extends from the inner A ring to Titan, is roughly the same particle density as Earth's van Allen belt. What is perhaps most unusual about Saturn's magnetic field is its orientation, which is almost perfectly oriented with its rotational axis, and its axisymmetric shape. Saturn's magnetic field features, which are more like Earth's than Jupiter's, include bright auroras that appear in visible and UV images of the polar regions (see figure 3.20).

Figure 3.18. UV image of Saturn's main ring system outlining the position of the A, B, and C ring bands. Source: NASA JPL.

Figure 3.19. Negative mosaic from Cassini's visible and infrared spectral imaging cameras of the southern hemisphere of Saturn shows the deeper layers of the atmosphere outlined by the internal heat in the upper hemisphere, with a brighter northern hemisphere below illuminated by the Sun. Source: NASA JPL.

SATURN'S MOONS

Saturn's moons range in size from moonlets that are only meters in diameter, to its largest moon, Titan, which is larger than Mercury. The moons extend from inside the inner rings, to as far as 377,000 km, nearly ten times the Earth-Moon distance. As of 2009, the total count of Saturn's moons with confirmed orbits is 61, and roughly 200 unconfirmed. Seven of these moons are considered the major moons of Saturn and can be viewed on Earth with a modest telescope. The innermost of these is Mimas, a 400 km satellite orbiting 185,000 km from Saturn's center. As a small satellite, Mimas contains an impressive crater with the largest crater-to-moon diameter ratio in the solar system. Inside Mimas's orbit are seven smaller satellites, with the 30 km diameter Pan the closest at 133,600 km from Saturn's center in a 13.8 hr orbit. Outside of Mimas are the other major moons, Enceladus (504 km dia, 237,950 km orbit), Tethys (1,066 km dia, 294,610 km orbit), Dione (1,123 km dia, 377,396 km orbit), Rhea (1,529 km dia, 527,108 km orbit), Titan (5,151 km dia, 1,221,930 km orbit), and Iapetus (1,472 km dia, 3,560,820 km orbit). Mixed with the major moons are seven smaller moons, which include four Trojan members belonging to Saturn's L4 and L5

Figure 3.20. Hubble image of the aurora phenomena seen on both of Saturn's north (top) and south poles. Source: NASA STScI.

Lagrange stable regions. Beyond the orbit of Iapetus lie 38 smaller moons, some just a few km in diameter, and most discovered in the Cassini image data.

Titan

Titan is Saturn's largest satellite, and the solar system's second-largest moon, with a diameter greater than Pluto and Mercury. Titan is also the only moon in the solar system with a significant atmosphere. Its unusually dense, opaque atmosphere has

Table 3.13. Saturn Statistics

Mass	5.6846×10^{26} kg (95.15 M_{Earth})
Radius	60,268 km (equatorial)
Mean density	0.687 g/cm^3
Escape velocity	35.5 km/s^2
Orbital eccentricity	0.055723
Orbit inclination	2.485°
Semimajor axis	9.58201720 AU (1,433,449,370 km)
Orbit period	29.657296 yr
Rotation period	10 hr 32–47 min (sidereal)
Rotational axis tilt	26.73°
Magnetic field	0.2 Gauss (measured at equator)
Albedo	0.47
Atmosphere	96% H_2, 3% He, 0.4% methane, 0.01% ammonia, 0.01% deuterium, 0.0007% ethane

puzzled astronomers since the first spectra were taken with Earth-based telescopes. Details of the moon were even more puzzling when spectral and image data came back from the Pioneer and Voyager spacecraft that showed Titan with a thick, opaque, yellow-brown atmosphere consistent with a complex chemical process not seen on any other moon or planet.

The first opportunity to examine Titan closely was on the dedicated Saturn exploration spacecraft Cassini, named for the seventeenth-century astronomer who was one of the first to study Saturn and Jupiter in detail. Titan's atmospheric probe was designed to be transported by the Cassini orbiter for delivery to Saturn's moon. On entry, the probe Huygens, named after the contemporary of Cassini who discovered Titan, Christiaan Huygens, would measure the atmospheric profile and sample the gases as it descended, and relay images through the Cassini orbiter as it slowed to a soft landing.

The Huygens lander was built for the joint exploration missions by the European Space Agency to return images of the cloud layers during its descent to its landing site, as well as to relay data on Titan's atmospheric makeup and winds. Data returned during the successful Huygens descent show that the unusually thick atmosphere is composed mostly of nitrogen (87–97%), methane (1–6%), and argon (1%). What gives Titan its opaque, colorful character is the roughly 10% hydrocarbons and organic gas content consisting mostly of hydrogen (2,000 ppm); ethane (20 ppm); acetylene (4 ppm); and ethylene, propane, and hydrogen cyanide, each at approximately 1 ppm (ESA Science 2005).

Clouds that dominate Titan's visible atmospheric features are composed of methane and ethane droplets, which precipitate as rain. Convective storms in Titan's atmosphere were observed by the Cassini instruments, as well as seasonal weather patterns. Titan's dense atmosphere, which has a surface pressure 1.6 times Earth's, is composed of mostly molecular nitrogen. This makes Titan an attractive model for studying our early atmosphere, although there are significant differences. Since Titan cannot retain nitrogen in its atmosphere, nitrogen is likely resupplied from an interior reservoir of methane through volcanic vents.

Temperatures on Titan's surface range near -180°C (-292°F), which is close to the triple-point for both ethane and methane, meaning that it is possible to have both gas and liquid phases of ethane and methane in Titan's atmosphere and on its surface. Clouds over the north pole indicate a seasonal variation on Titan that allows liquid and gas methane, therefore methane rain. Clouds resupplied by the evaporation of methane and possibly ethane lakes are not just possible, but recently observed at polar latitudes. These alcohol lakes appear to shift to the south polar region in that hemisphere's winter. Few craters have been observed with Cassini's synthetic aperture radar instrument, which is an indication that cryovolcanism is still active on Titan.

Surface features on Titan are rich in geological diversity and more closely resemble Earth than any other planet or moon. Although Huygens's descent images were limited in coverage, subsequent synthetic aperture radar images from Cassini's close flybys of Titan show large, smooth regions that appear to be ocean beds or lake beds. Hills at higher elevations on Titan are carved with river channels that lead to basins—strong evidence of periodic methane/ethane rain. Dune features on Titan are indicative of significant surface winds as well as dry climate regions.

Based on recent Cassini and Huygens data, Titan's interior appears to be similar to several of Jupiter's moons, with a rocky core overlaid with a mantle that

Figure 3.21. Synthetic aperture radar image of the Titan surface taken by the Cassini space-craft that clearly outlines the hilly regions cut with drainage channels and the dark, flat basin that is more reflective and hence returns less signal to the radar sensor since the flat areas are off-per-pendicular to the surface. An image taken directly above the surface would show a circle of bright re-flection directly below the instrument. The dark basin is not filled with liquid methane/ethane, but has been in the past. Similar images from Titan's polar regions do indicate liq-uid-filled lakes or oceans. Source: NASA JPL.

is composed of water and ammonia ice layers. Roughly two thirds of the moon is rock/silicate/metal that remains warm, or perhaps hot, from tidal flexing by Saturn, which would likely create conditions for a liquid water layer in the upper ice mantle. Surface rocks formed of water ice and volcanic methane plumes indicate a liquid reservoir of methane and water.

Enceladus

Enceladus was discovered to be one of the more interesting of Saturn's moons follow-ing the analysis of Voyager 2's flyby images. Enceladus is located in the dense region of Saturn's diffuse outer E-ring. Speculation about the ring's source of fine-particle ice quickly focused on Enceladus because of its light-colored, icy, highly reflective surface. Enceladus also shows two different surface characteristics between the more north-erly latitudes and the southerly latitudes. Voyager 2 images show a sparsely cratered northern region compared to the young, uncratered southerly zone. The difference is dramatic, and indicative of an active moon. Images from more recent Cassini flybys show even more intriguing features and an active surface that is likely due to tidal stretching by Saturn's strong gravity. Even though the majority of the moons and the rings of Saturn are in circular orbits, Enceladus has a relatively high eccentricity of

Table 3.14. Titan Statistics

Radius	2575.5 km
Density	1.88 g/cm³
Orbit period	15.945 days
Rotation period	15.945 days (synchronous)
Albedo	0.21
Surface pressure	147 kPa
Surface temp	94 K (-290°F)

0.0047, which is driven by its 2:1 resonance with the orbit of Dione. The tidal heating of the interior generates liquid cryovolcanoes near the south polar region that spew frozen particles into the E-ring. Meteoroid impacts are also likely contributors to the E-ring particles, as with other moons in the same region, including Mimas, Thehys, and Dione.

URANUS

Uranus, named for the father of Saturn (Greek god Cronos), is not visible to the naked eye and was therefore unknown to the ancient astronomers. Since it is the closest of the giant ice planets, it was discovered before the larger, but more-distant Neptune. William Herschel's discovery of Uranus in 1781 made it the first planet to be discovered with a telescope, although other astronomers noted it earlier without identifying it as a planet. Uranus's most unusual feature is its rotation axis that is tilted more than 90° from its orbit plane. The only other planet with a retrograde rotation, Venus, has its rotation axis nearly 180° from the rotation axis alignment of the rest of the planets. Although not conclusive, supporting evidence of a violent collision turning Uranus on

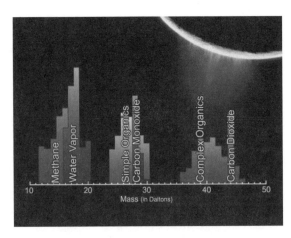

Figure 3.22. Spectral histogram plot of the materials detected by Cassini's instruments as the spacecraft passed through the ice plumes in Enceladus's southern region in a July 2005 flyby. Daltons are mass units commonly used for molecules, expressed in unified atomic mass units. Source: NASA JPL.

its side can be found in the chaotic features of its moon Miranda. A 90° reorientation of the planet would make orbits previously aligned with the planet's equatorial plane unstable. Along with the inevitable shower of debris from an impact with Uranus, any satellites formed before the impact would likely not have survived intact, or in Miranda's case, may have been broken apart to later reform as a jumbled mass.

Because of Uranus's 98° obliquity (rotation axis tilt), seasonal variations would be the same as if the Sun appeared to rise and set to the north and south instead of the east and west once a day. At polar latitudes, the Sun would not set for half of the Uranian year, and would not appear for the other half of the year. Near the equatorial zone, the Sun would rise and set each day, but would not disappear in the summer or winter seasons. The Sun would only reach down to the east and west horizon. Intermediate latitudes would experience intermediate daily and seasonal variations between the polar and equatorial extremes. Uranus's 19.2 AU orbit with a corresponding 84.07-year orbit period would translate into a 42-year day and a 42-year night at the poles.

Models of Uranus's interior and atmosphere are based almost entirely on the Voyager 2 flyby data returned during the latter part of 1985 and early 1986, and from later observations with the Hubble Space Telescope. Sequential images of Uranus's upper atmosphere show indistinct cloud features within the upper atmosphere mixed with haze layers circulating in three bands in each hemisphere. Polar and equatorial band circulations in the southern hemisphere during its summer captured in the Voyager 2 images were divided by a narrow circulation band described as a collar. Nothing similar has been found in the northern hemisphere by the Hubble Space Telescope during that hemisphere's winter season, although the expectation is that the collar will reform in the opposite hemisphere as the seasons reverse.

Uranus is the lightest of the giant planets with a mass roughly 14.5 times that of the Earth. Its diameter is slightly more than four times Earth's, which gives it a mean density of only 1.27 g/cm³. This makes Uranus the second-least-dense planet in the solar system. Unlike Jupiter's and Saturn's dedicated exploration mission spacecraft, the single flyby visit of Voyager 2 provided only approximate limits on the distribution and makeup of the Uranian interior.

The structural model of Uranus consists of three primary layers centered on a rocky core approximately ½ M_{Earth}, which is thought to make up 20% of the radius of Uranus. Above the core is the lower-density liquid mantle composed of ice materials under high pressure and temperature. This massive liquid layer would not be separated into layers of traditional ice compounds, but mixtures of radically different compounds originally consisting of water, carbon dioxide, methane, and ammonia altered by the extreme pressures and temperatures. This layer makes up most of the planet's mass, estimated at 10 to 14 M_{Earth}. Since the mass of the planet is dominated by the liquid mantle that itself consists of heated ices under pressure, both Uranus and Neptune are classified as giant ice planets.

Atop the liquid mantle is the atmosphere, composed primarily of hydrogen and helium, estimated at 1 M_{Earth}. Voyager 2 measurements of the upper atmosphere composition were 83% H, 15% He, 2% methane, 0.01% ammonia, with smaller amounts of ethane, acetylene, carbon monoxide, and hydrogen sulfide. Because there is almost no internal heating like that found in Jupiter and Saturn, there is little vertical mixing

Figure 3.23. Voyager 2 image of Uranus's south polar region showing the planet's indistinct atmospheric features. Source: NASA JPL.

in the upper atmosphere. The lack of convective mixing gives Uranus its characteristic greenish-blue hue, a condition promoted by the absorption of red light from the methane in the upper atmosphere.

Characteristics of the Uranian magnetic field are unusual in two ways. First, its dipolar axis orientation is aligned 59° from the rotation axis. Second, the dipolar magnetic field axis is displaced along its rotational axis by about 30% of its full radius. Because Neptune also has a magnetic field aligned off-axis and displaced while Jupiter and Saturn do not, these peculiarities may be common to giant ice planets. A rough "texture" in Uranus's magnetic field detected by the Voyager 2 magnetometer during its flyby indicates that the magnetic field is generated closer to the surface, compared to the giant gas planets, or to the Earth. This corroborates the model of a conductive water-methane ocean in the upper part of the Uranian liquid mantle being responsible for the magnetic field. Even though the magnetic field is comparable in strength to Earth's but with a much stronger dipole moment, the mechanism that generates Uranus's magnetic field is not understood at this time. Particles trapped in Uranus's magnetic field are almost exclusively protons and electrons whose origin is Uranus's ionosphere. Helium and other heavy nuclei that could be created by orbiting moons and possible ring interactions are not present as they are on Jupiter and Saturn. Dark surface features on several of Uranus's moons and in the ring system confirm the proton-electron composition of the radiation belt.

The first reliable observations of the ring system surrounding Uranus were made in 1977 during occultation observations on NASA's Kuiper Airborne Observatory—KAO. (KAO has since been replaced by the B-747 Stratospheric Observatory for Infrared Astronomy—SOFIA.) The small ring system was found to be in the same approximate plane as Uranus's equator and moons. Later images from the Voyager 2 flyby in 1986 were used to identify several more rings, which brought the total to 11. More recently, HST observations located two additional small rings twice the distance from the planet than the others.

Table 3.15. Uranus Statistics

Mass	8.681×10^{25} kg (14.54 M$_{Earth}$)
Radius	25,559 km (equatorial)
Mean density	1.27 g/cm^3
Escape velocity	21.3 km/s^2
Orbital eccentricity	0.0044405
Orbit inclination	0.773°
Semimajor axis	19.2294 AU (2,876,679,080 km)
Orbit period	84.3233 yr
Rotation period	17.2900 hr (sidereal)
Rotational axis tilt	97.77°
Magnetic field	0.1–1.1 Gauss
Albedo	0.51 (geometric)
Atmosphere	83% H$_2$, 15% He, 2.3% methane, 0.007% deuterium, 0.007% ethane, 0.0004% water

Discovery of the moons Umbriel and Titania in orbit around Uranus were first made by William Herschel, the same astronomer that discovered Uranus. These are the largest of the five major moons of Uranus, with diameters only 10% smaller than the Earth's moon. In addition to the five major satellites, there are twenty-two smaller moons located mostly inside the orbit of Miranda, the innermost of the major satellites, with a 470 km diameter, and a semimajor axis of 129,000 km, which corresponds to an orbital period of 1.4 days.

Miranda's surface is best described as chaotic. Its surface features are dominated by an unusual patchwork of broken terrain and enormous canyons as deep as 20 km—more than ten times as deep as the Grand Canyon. Large racetrack-shaped ridges suggest that icy (cryo) volcanoes from liquid layers beneath the icy crust are responsible for filling lower elevations on the broken surface. The few craters that appear on its surface indicate that the violent activity that shaped Miranda came well after its formation.

Beyond Miranda are the four other major satellites of Uranus: Ariel (1,160 km dia, 191,000 km orbit), Umbriel (1,170 km dia, 266,000 km orbit), Titania (1,580 km dia, 436,000 km orbit), and Oberon (1,520 km dia, 584,000 km orbit). Features on these are generally older than Miranda and without the broken terrain. Common to the major moons of Uranus is a low albedo, or dark surface, the darkest of which belongs to Umbriel, which has an albedo of 0.16. Radiation trapped in Uranus's magnetic field is thought to be responsible for darkening the methane on the moons' surfaces.

The majority of the Uranian moons lie inside the orbit of Miranda, with Cordelia the closest, orbiting 49,770 km from the planet with a period of 8.03 hours. Beyond Oberon are nine small satellites, the most distant of which is Ferdinand at 20,901,000 km.

Figure 3.24. Voyager 2 image of Miranda showing the unusual and irregular surface features that include prominent racetrack-shaped ridges called coronae. Source: NASA JPL.

NEPTUNE

The most distant of the giant planets is Neptune, and, because of that distinction, it was the last of the giants to be discovered. Neptune's discovery as a planet was the first that involved the calculation of orbit perturbations to extract the position of an unknown object. The gravitational tug that could be measured on Uranus's orbit, and to a lesser extent on Jupiter's and Saturn's, led astronomers to the unknown planet's position.

Neptune's distant orbit also influenced its makeup, as with Uranus. Its structure is dominated by an ice-compound mantle, with a rock and metal core, and a large atmosphere that is much smaller in proportion to its mass than with Jupiter and Saturn. And like Uranus, Neptune has no metallic hydrogen within its mantle because of its modest mass compared to the giant gas planets. Although Neptune is larger in mass than Uranus (17 times M_{Earth}), it is 3% smaller due to its slightly higher density at 1.64 g/cm^3.

Neptune's atmosphere has a similar composition and temperature profile to Uranus, but varies in abundance. Measurements by the Voyager 2 instruments and ground-based observations show Neptune's atmosphere dominated by hydrogen (80%) and helium (19%), with methane (1.5%), and trace amounts of deuterium (2H—0.02%), and ethane (0.00015%). A variety of hydrocarbon and organic and inorganic compounds are observed in the atmospheric ices and droplets, ranging from acetylene and hydrogen sulfide to carbon monoxide and water (Lunine). The bluish color of the atmosphere is due primarily to the red and IR absorption by methane, while the whitish clouds are likely droplets of ammonia, ammonium sulfide, and hydrogen sulfide.

Figure 3.25. Full-disk view of Neptune taken during the Voyager 2 flyby in 1989 shows the major atmospheric features that include the Great Dark Spot (center left), the high-altitude, high-speed white clouds, a variety of circulation bands, and several atmospheric storms. Source: NASA JPL.

Winds in Neptune's upper atmosphere are the highest in the solar system, with cloud features tracked in Voyager 2 images at 2,100 km/h (1,300 mi/h). The driving forces for the high winds on Neptune are thought to be internal heating and the very cold upper atmosphere temperatures, which are comparable to Uranus's minimum atmospheric temperature of 50 K roughly at the same pressure level. Circulation bands in the upper atmosphere of Neptune are far more dramatic than on Uranus. A dark blue spot called the Great Dark Spot, seen in detail from the Voyager 2 flyby images taken in 1989, shows a strong circulation feature similar to Jupiter's Great Red Spot, and to the anticyclonic patterns seen on Saturn. The Great Dark Spot was not long-lived, however, and may have been a region of depleted gas rather than a heat-induced storm. Hubble Space Telescope images fifteen years after the Voyager 2 flyby no longer show the large dark spot on Neptune.

Neptune's magnetic field is offset and tilted similar to the magnetic field of Uranus. The magnetic field axis is some 47° from the rotational axis, and displaced more than halfway to its surface—roughly 14,000 km off-center. This has led to the speculation that Neptune's magnetic field, like Uranus's, is generated in a conductive layer, or layers, in the liquid mantle rather than in the core. Neptune's magnetic field strength, which varies widely because of its alignment and offset, is comparable in strength with the Earth's and Uranus's at the surface.

A faint ring system significantly smaller than Uranus's rings was discovered around Neptune's equator in the Voyager 2 images. The outer ring is composed of three arcs rather than a complete ring, which cannot be duplicated with current simulation techniques. Just as intriguing is the short lifetime of Neptune's rings, discovered in recent observations with the Keck Observatory, which shows a weakening in the ring structure compared to the Voyager 2 images.

Table 3.16. Neptune Statistics

Mass	1.0243×10^{26} kg (17.147 M_{Earth})
Radius	24,764 km (equatorial)
Mean density	1.638 g/cm³
Escape velocity	23.5 km/s²
Orbital eccentricity	0.011214
Orbit inclination	1.305°
Semimajor axis	30.10366 AU (4,503,443,660 km)
Orbit period	164.79 yr
Rotation period	16.110 hr (sidereal)
Rotational axis tilt	28.32°
Magnetic field	0.14 Gauss (equator)
Albedo	0.41 (geometric)
Atmosphere	80% H_2, 19% He, 1.5% methane, 0.019% deuterium, 0.00015% ethane

Of Neptune's thirteen known moons, the largest and arguably the most unusual is Triton. The moon's rarely seen retrograde orbit indicates that it is a captured object rather than shaped from the same accretion and collision events that formed Neptune. Triton's retrograde orbit is even more unusual, since the moon is locked in synchronous rotation in its orbit around Neptune. Physical characteristics of Triton also strongly suggest the moon originated beyond Neptune, since its surface composition, density, and even size are similar to Pluto and many of the other trans-Neptunian objects. Density and mass measurements constructed from Voyager 2's flyby data point to a core of rocky material making up roughly two-thirds of the moon's mass, with the other third composed primarily of water ice.

Triton has a diameter 78% of the Earth's moon, but only 28% of its mass, which is reflected in its relatively low density of 2.05 g/cm³. This low density is typical of many of the Jovian moons, and of Pluto and its relatives in the outer solar system. The surface of Triton shows complex features, with few craters and a very young surface. Spectral imaging data show Triton's surface covered with nitrogen ice, water ice, carbon dioxide ice, methane ice, and ammonia ice. Active geysers/cryovolcanoes of nitrogen and water also indicate a geologically active past and present, which corroborates the young age of the surface, estimated at 50 million years for the oldest features. Some of the terrain on Triton has the appearance of a cantaloupe skin, complete with flat outlines surrounding shallow depressions. Triton also retains a cold, thin atmosphere composed mainly of nitrogen, with some ammonia, similar to Pluto. While extremely thin, Triton's atmosphere has winds strong enough to carry small particles of frozen nitrogen ejected from cryovolcanoes far enough to coat many areas of Triton with light-colored ices.

Figure 3.26. Voyager 2 flyby image of Neptune's largest moon Triton showing the widely varied surface features. Source: NASA-JPL.

Neptune's twelve other known moons are all smaller than Triton, with four inside Triton's orbit and eight outside. Eight of these moons were discovered in the Voyager 2 image data.

Trans-Neptunian Objects

For nearly a century after the discovery of Neptune, astronomers searched for an enigmatic planet called Planet X that appeared to have a slight influence on Uranus's orbit. Even though the search was successful in finding the outermost planet, Pluto, the calculations used for the orbit perturbations were in error. Pluto's discovery in 1930 by Clyde Tombaugh opened the door to the icy world orchestrated by the gravitational influence of Neptune. Pluto's discovery also closed the door on the search for Planet X, which was postulated to explain the incorrectly calculated motion of Uranus and Neptune. Pluto's discovery further discredited the Titus-Bode Law that predicted Pluto's position at 77 AU (Pluto's semimajor axis is 39.5 AU), but could not account for Neptune.

Recent discoveries by the powerful Hubble Space Telescope and several new ground-based adaptive-optics telescopes reveal hundreds of Pluto-like bodies roughly in the same plane as the major planets. This region that extends beyond Neptune, called the Kuiper Belt after astronomer Gerard Kuiper, is populated with moon-size comets that resemble Pluto but are generally much smaller. The disk population is confined to a semimajor axis of about 55 AU, and clustered at specific positions with respect to Neptune's orbit period (see figure 3.27). Not only does Neptune's gravitation direct the motion of these outer bodies, Neptune has stabilized most of these icy bodies inside or outside the orbital resonance positions. Because of Neptune's fundamental role in shepherding these icy satellites, including Pluto, they are classified as Trans-Neptunian Objects (TNOs).

Data from the HST and Earth-based telescopes like the Keck Observatory have uncovered more than a thousand of the TNOs, which show a distribution pattern remarkably similar to the asteroids between Mars and Jupiter. An intriguing example is the handful of TNOs at the same exact semimajor axis as Pluto, called Plutinos, that are in the 2:3 resonance point with Neptune's orbit period. This means that for every three orbits of Neptune, these objects orbit the Sun two times.

Resonance orbits are important in the interactions of the giant planets and their smaller neighbors, as are two of the five Lagrange stable points—L1 and L2. Integer orbital resonances such as 1:2, 2:3, 3:2, 3:5 are orbit period ratios that result in either stabilization or destabilization by the periodic gravitational tugs from larger bodies, or from companions with similar mass. A graphic example of a destabilizing orbit resonance is the Cassini Division in the Saturn ring system created by the 1:2 orbital resonance of Mimas's orbit around Saturn. In contrast, Pluto and the Plutinos are examples of the 3:2 stabilizing resonance with respect to Neptune's orbit period.

Neptune's interactions with innumerable cometlike objects in the outer solar system over a period of billions of years have put a unique fingerprint on the distribution of trans-Neptunian objects. Countless interactions with these small bodies have also increased Neptune's semimajor axis orbit by some 5 AU and shaped the Kuiper Belt object distribution in the process. Numerical simulations have been used to show Neptune's motion related to the migration of the TNOs and Kuiper Belt objects, and the scattering and capture of large and small comets in the process (Levison and

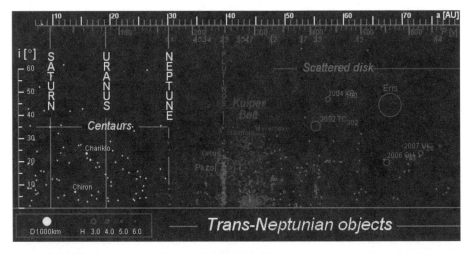

Figure 3.27. Distribution plot of the small objects in the outer solar system including the resonance-distributed Trans-Neptunian Objects that lie beyond Neptune's orbit. Kuiper Belt objects extend from Neptune to about 50 AU, beyond which are the Scattered Disk Objects (SDOs). Centaur objects are comets and asteroids located between Jupiter and Neptune. Source: Eurocommuter.

Morbidelli). Extending beyond the Kuiper Belt are bodies even more widely dispersed, known as scattered disk objects, which include the largest dwarf planet known, named Eris. These scattered objects show a declining population beyond the Kuiper Belt, which reflects, at least in part, the decreasing density of material with increasing distance from the Sun. This effect is also accentuated by the increasing difficulty in detecting small, cold bodies at these distances.

Pluto

The discovery of several Pluto-like planets in the outer solar system opened a contentious debate on the character of planets, followed by several internationally recognized panel discussions intended to define more accurately what qualified as a planet. The outcome was a redefinition of planets in 2006 by a consensus of the International Astronomical Union that was anything but unanimous. Pluto's redefinition was not a demotion in its status, as many believed, but the result of the need to expand the categories of large and small objects inside and outside the solar system. The debate has not ended, however. With the discovery of a variety of other solar and extrasolar planets, the classification of planets is converging on three sizes and three compositions. As such, Pluto and its moon, Charon, are now members of the dwarf ice planet category.

Pluto and Charon are unique in several ways. Because Charon's mass is 12% of Pluto's, the pair are better described as companions rather than a planet-moon pair. The pair also satisfies a more quantitative method of defining companions since their barycenter (center of mass) is outside of Pluto's radius. Both Pluto and Charon are thought to have a rocky core with an ice mantle roughly as large as the core. Since no spacecraft has visited the planet-pair as yet, internal structure and density gradient estimates are mostly educated guesses. Most literature suggests a differentiated silicate-ice core and a water and nitrogen liquid-ice mantle.

The atmosphere of Pluto, which has been examined with occultations of several background stars, shows a very small atmosphere with a pressure of 0.15 Pa (1.5×10^{-6} bar), slightly more than 1 millionth of Earth's pressure. Sublimation of the surface ices and the resulting mass of Pluto's atmosphere varies with the season, which is determined by the orbital distance from the Sun and with location on the surface. Because sublimation increases cooling, a sunward location on Pluto would experience slightly higher pressure, but a slightly lower temperature.

Both Pluto and Charon are locked into tidal synchronous rotation, with the face of one always aligned with the face of the other. As viewed from any position on Pluto, Charon never moves in the sky; only the stars and other planets move in the background. The same applies to any position on Charon, where Pluto is always fixed in the same relative location.

In 2005, two smaller moons, named Nix and Hydra, were discovered orbiting Pluto using images from the Hubble Space Telescope. All three moons orbiting Pluto are in circular orbits and are close to the same orbital plane. This suggests a similar origin for the four bodies, possibly a collisional event similar to the Earth-Moon system formation.

Pluto travels inside the orbit of Neptune for approximately 20 years of its eccentric 248-year orbit. But since Neptune keeps Pluto locked in a 3:2 resonance orbit (three orbits of Neptune to two Pluto orbits), the relative position of the two never comes within less than 52° of each other. Other more subtle orbit resonance effects also maintain long-term separation of Neptune and Pluto. As a result, a collision between the two planets is not possible.

After decades of frustration with unsuccessful planning and funding efforts, NASA finally succeeded in developing and launching a spacecraft dedicated to the exploration of Pluto and the more distant Kuiper Belt. The small 1,000 kg spacecraft called New Horizons was launched on a large Atlas V booster January 19, 2006, propelling it to the fastest Earth-departure speed ever reached. The same three-day cruise to the Moon needed for the Apollo missions was covered in just nine hours by New Horizons. The spacecraft needed the high-velocity boost to reach Pluto before its atmosphere collapsed into a surface ice layer as its eccentric orbit took it beyond 30 AU. New Horizons and its eight major instruments are scheduled to arrive at Pluto in 2015, after just eleven years in cruise and one gravity assist from Jupiter. The normal Hohmann transfer period to reach Pluto from Earth without a gravity assist is forty-five years, although there is no booster available that could place even a small 1,000 kg spacecraft that far from the Sun without a gravity assist.

Table 3.17. Pluto Statistics

Mass	1.305×10^{22} kg (0.0021 M_{Earth})
Radius	1,195 km
Mean density	2.03 g/cm^3
Escape velocity	1.2 km/s^2
Orbital eccentricity	0.24881
Orbit inclination	17.142°
Semimajor axis	39.48168677 AU (5,906,376,272 km)
Orbit period	248.09 yr
Rotation period	6 d 9 h 17 m 36 s (sidereal)
Rotational axis tilt	119.59° (to orbit)
Magnetic field	Unknown
Albedo	0.49–0.66
Atmosphere	nitrogen, methane, and carbon monoxide (surface ice sublimation)

Dwarf Planets

Newly discovered Kuiper Belt objects and trans-Neptunian objects are interesting in many ways. In general, the orbits of these objects have a wide variation in eccentricity and inclination, though all are more eccentric and inclined than the major planets. A wide range of albedos and colors is also seen in these distant icy bodies, which reflects variations in formation, in surface evolution, and even in dynamics. In addition, at least one of these newly discovered dwarf ice planets is larger than Pluto. Eris, the largest of the dwarf planets found so far, lies beyond the Kuiper Belt in a 557-year, 67.7 AU orbit. First known as SDO 2003 UB313, then Xena, distant Eris has a moon named Dysnomia that was used to establish an accurate mass for Eris of 1.67×10^{22} kg, some 27% greater than Pluto. Because of its size and its spherical shape, and because it travels in a unique orbit around the Sun, Eris is classified as a dwarf ice planet. Eris's highly inclined (44.18°), highly eccentric (0.44177) orbit is typical of the SDOs seen outside the Kuiper Belt.

So far, the most distant dwarf ice planet found is Sedna, with a semimajor axis of 525.9 AU (7.867×10^{10} km) that corresponds to an orbital period of approximately 12,000 years. Sedna is typical of SDOs, with a high orbital inclination (11.93°) and high eccentricity (0.855) orbit. Even though Sedna is classified as a trans-Neptunian object, its orbit is too distant to fit comfortably into the category of SDOs, which may prompt a new classification scheme for the more distant cometlike dwarf planets. In fact, Sedna's discovery in 2003 was partly responsible for the debate over redefining the planets, and eventually the compromise classification of dwarf planets. Sedna has a reddish hue like Mars, but it is created by entirely different surface materials and conditions.

Other TNOs recently discovered by the Hubble Space Telescope and powerful Earth-based telescopes include Quaoar, which was originally thought to be the next-largest object beyond Neptune, second only to Pluto. Recent measurements show Quaoar, with its approximately 1,260 km diameter and 43.6 AU orbit, to be half the size of Eris (~2,600 km, 67.7 AU) and smaller than Sedna (~1,800 km, 526 AU), 2003 EL61 (~1,500 km, 43.3 AU), and 2005 FY9 (~1,500 km, 45.8 AU) (Grundy et al. 2005). Future higher-resolution images of these distant dwarfs could rearrange the order of their size, and help identify any orbiting moons belonging to these dwarf ice planets.

Small Solar System Bodies

A similar crisis in defining what constitutes a planet occurred two centuries ago, and for much the same reason and with a similar outcome. Discovery of the largest asteroid, Ceres, in 1801 soon led to its adoption as a planet. One prevailing argument for the decision was that Ceres was large enough to validate the Titus-Bode Law that predicted a planet between Mars and Jupiter. Discovery of several large asteroids in the vicinity of Ceres, which is within the main asteroid belt, expanded the number of planets in the solar system until it was clear that hundreds of the newly discovered small bodies should not have the same classification as planets. This led to the inevitable debate on what qualified as a planet and introduced the terms *asteroid* and *minor planet*, both of which are in common use today. Two centuries later, the 2005–2006 debate and the redefinition of most objects in the solar system simplified the classification of comets, asteroids, minor planets, and planetesimals into a single category called Small Solar System Bodies, while expanding the general classification of planets into the categories of dwarf, giant, and terrestrial planets.

In order to separate planets from other objects in heliocentric orbits, the IAU committees established measurable criteria in 2006 that defined planets as satisfying three basic criteria. Accordingly, a planet must:

1. Orbit the Sun;
2. Sweep out interplanetary materials surrounding its orbital path; and
3. Be large enough to have a spherical or near-spherical shape (a condition of hydrostatic equilibrium).

Dwarf planets such as Pluto and Eris do not satisfy the second and most subjective criteria, yet are still formed from the same planetesimal building blocks that formed the largest planets. These building blocks, previously known as asteroids, comets, and minor planets, but excluding dwarf planets, are now called Small Solar System Bodies. SSSBs contain the earliest solid materials that formed in the solar system, which makes them much more than a scientific curiosity. Broken bits and pieces of these that enter the Earth's atmosphere and survive as meteorites can be analyzed for composition, origin, age, even parent bodies in some cases. Insight into planetary formation, as well as the process of condensation in the solar nebula, can also be garnered from these asteroid/SSSB samples. In addition, timelines are available from these samples that show when and how the Sun and planets cleared material from the planetary disk. That same dispersion process is seen in nearby early stars that are just forming planets in a swirling, turbulent disk of gas and dust surrounding the young star. The various stages of planet creation, breakup, and equilibrium seen in the extrasolar worlds are contained in the record of our own solar system, which is locked in fragments varying in size from dust to failed planets.

What appear to be random collisions of asteroids and comets with the planets and moons may not be random at all, but rather events that are modulated by the tidal influence of our solar system traveling through the disk of the Milky Way galaxy. Our solar system orbits the galaxy center in a period of approximately 225 million years, but it also oscillates vertically through the plane of the galactic disk in a period of about 30–35 million years. That oscillation period may have an important connection to the impact events on Earth in the past several hundred million years, which average roughly 30 million years between major impact events. Notable examples are the impacts in the Chesapeake Bay on the East Coast of the United States, and at Popagai in Siberia, both occurring 35 million years ago. More familiar, perhaps, is the Chicxulub impact crater on the Yucatán Peninsula 65 million years ago. Although correlation cannot be construed as cause, the case can be made that the solar system passage through the central portion of the galactic plane could disrupt the cometary debris in the surrounding Oort Cloud and create the conditions for a significant increase in comet-planet collisions. Nevertheless, this may be little more than speculation, since the Chicxulub-Yucatan impact was from an asteroid and not a comet. Evidence of the type of impactor is found in the high concentrations of iridium deposited around the world at the same time, characteristic of asteroids and not comets. It could also be argued that the perturbations from passage through the galactic plane could perturb the sensitively balanced chaotic orbits of many of the Jupiter-dominated or even Neptune-dominated asteroids, which would correlate with the same 30-million-year period of catastrophic cometary collisions.

Asteroids

William Herschel coined the term *asteroid*, meaning "starlike" after the discovery of the first and largest asteroid in 1801, named Ceres. Typical asteroids range in size from fractions of a kilometer to hundreds of kilometers in diameter, and are composed of the same silicate rocks and metals that make up the terrestrial planets. The largest collection of asteroids is located in a band between Mars and Jupiter. However, such generalizations offer little insight into the origin and early evolution of the rock and metal asteroids and meteorite samples of these asteroids that make it to Earth's surface. Meteorites, as opposed to meteors, which are the events we see in the night sky as "shooting stars," and meteoroids, which are debris in space, have a broad range of composition. Because of the diversity in their makeup, and because each fragment has at some time originated as part of a larger asteroid, they are one of the most powerful tools available for studying the solar system's content and formation.

The oldest and most abundant of the asteroids and meteorites are the chondrites, which make up more than 85% of the meteorite samples, and about 75% of all known asteroids. Chondrites are generally very small and show no evidence of being heated by compression or of being changed by differentiation, which occurs in large asteroids and planets. Chondrites generally represent the earliest solid asteroid materials, although the meteorite samples are altered by the heat of atmospheric entry. Rich in ferrous metals (Fe, Ni) and metal oxides (Ca, Al), chondrites also contain presolar grains that condensed before planetesimals began forming in the solar system.

CLASSIFICATION

Classification of asteroids runs somewhat parallel with meteorite classification, which roughly entails composition, spectra, and albedo. A simple grouping of four asteroid categories is given below, along with a brief discussion of their characteristics. More detailed asteroid classification schemes have been created that include multispectral characteristics, composition differences, albedo features, and membership in families and groups with other asteroids. These more comprehensive classification schemes offer greater insight into the origin and evolution of the thousands of observed asteroids, but extend well beyond this brief discussion.

C-Type (Carbonaceous)

Carbonaceous chondrites are similar in makeup to the chondrites but are formed far-ther out in the main asteroid belt. The cooler environment beyond 3 AU is thought to be responsible for the enriched organic (carbon) content and the water compounds found in these asteroids and meteorites, as well as amino acids. The least-changed of these meteorite samples contain as much as 20% water, but represent only a few per-cent of the chondrite samples. These are some of the darkest asteroids, with an albedo range of 0.03 to 0.10.

S-Type (Silicaceous)

Silicon-rich asteroids are the next most common type, making up nearly 17% of the observed asteroids. The silica-rich S-type asteroids are concentrated near the central portion of the main asteroid belt. Their composition is dominated by metal silicates, including iron and magnesium, which correlates well with a relatively high albedo range of 0.10–0.20.

M-Type (Metallic)

Metal-rich asteroids, which have similar spectral features to iron meteorites, often contain a mixture of rocklike silicates. Yet, along with a wide range of spectral features, some of the M-type asteroids appear to contain little or no metals. Like the S-type, M-type asteroids have a moderate albedo range of 0.10–0.20.

D-Type (Dark)

Darkest of the asteroid types are the D-type, which typically reflect a reddish color. Because of their dark spectra, and their location in the distant region of the main belt of asteroids, the D-type are thought to contain organic silicates, with some being anhydrous (dry). However, there is also evidence that some may contain water ice in their interior.

GROUPS AND FAMILIES

Ceres is the largest of the main belt asteroids, and was considered a small planet when it was discovered in 1801. After nearly a century of measurements, Ceres has been found to be a protoplanet, containing more than 30% of the mass of all other asteroids between Mars and Jupiter. The perturbations from Ceres measured on nearby asteroids indicate Ceres is composed of a rock and metal core surrounded by a large ice mantle. The ice mantle, which is predominately water ice, is thought to contain a liquid ocean beneath the icy crust. A closer look at Ceres and the other asteroids between Mars and Jupiter shows a complex clustering due to Jupiter's enormous mass and nearby orbit. Peaks and gaps that are clearly outlined in figure 3.28 show evidence of the resonant features in

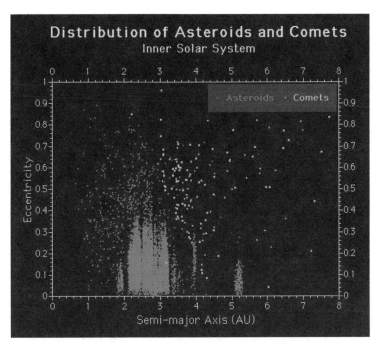

Figure 3.28. Plot of the distribution of asteroids and comets in the asteroid belt that shows clear influence of Jupiter's orbit and gravity. The asteroid population shows more circular (smaller eccentricity) orbits than for the comets, yet does not have an even distribution. Instead, these asteroids show gaps that correspond to Jupiter's orbit period resonance integers. One of the most distinctive gaps is located near 2 AU which corresponds to Jupiter's 4:1 orbital resonance. Another large gap can be seen near 3.3 AU which is at the Jupiter 2:1 resonance. The isolated group located just beyond 5 AU is in the same orbit as Jupiter (5.2 AU) but located at the L4 and L5 Lagrange stable points. These are the Trojan asteroids that follow Jupiter in a region 60° ahead of and 60° behind Jupiter. Mars, which is located at 1.5 AU, has only a minor influence on the inner asteroids which does not appear on this plot. Source: NASA JPL.

Jupiter's orbit. The gaps, commonly called Kirkwood gaps, define the distinct distributions within the overall collection of asteroids in the main belt. The inner band, called the Inner Main Belt, is located inside the prominent gap found at 2 AU, which is the 3:1 Jupiter resonance position. This includes the second-largest asteroid discovered so far—Vesta. The Middle Main Belt lies between the 3:1 gap at 2 AU and the 5:2 resonance gap located at 2.8 AU, which contains Ceres. The Outer Main Belt, which lies between Jupiter's 5:2 and 2:1 orbital resonances, extends from 2.8–3.3 AU.

These divisions are significant, not just because of their dynamic origin, but because the asteroids show different characteristics from band-to-band due to the variation in solar heating at increasing distances. As an example, Vesta is smaller than Ceres, but it formed in a warmer environment closer to the Sun and evolved as a dry, differentiated, large asteroid. Vesta's violent impact history produced a number of recovered meteorite samples that are categorized as part of the Vesta family. In contrast, Ceres formed in a cooler region, beyond the "frost line," which allowed water ice to be retained in its upper layers, creating a light-colored, frost-coated, dwarf planet. NASA's Dawn asteroid exploration mission, launched in 2007, is programmed to visit both Vesta in 2011 and Ceres in 2015 to help unlock some of the mysteries of planet formation.

Two intriguing asteroid groups bound by Jupiter's gravity are the Trojan asteroids located at the L4 and L5 Lagrange positions. For Jupiter, these are located at 5.2 AU. Even more intriguing is the Haida group, which is scattered at mild inclinations but bound to Jupiter's 3:2 resonance point near 3.9 AU. The fascinating Haida group follows complex orbits that extend both inside and outside of Jupiter's orbit of the Sun, with the Haida members transiting between the interior and exterior orbit regions through Jupiter's L1 and L2 passageways.

Two asteroid groups that are of interest to our survival are the near-Earth crossing asteroids called the Aten asteroids, which have a semimajor axis less than 1 AU but an aphelion greater than 1 AU, and the Apollo asteroids, having a semimajor axis greater than 1 AU but perihelia less than 1 AU. No Trojan asteroids that would normally be located at the Earth-Sun L4 and L5 stability zones have been observed, although several have been found in the Mars L4 and L5 regions. The two moons of Mars, Phobos and Deimos, have spectral and composition characteristics that are consistent with captured asteroids rather than moons that could have formed from the same or similar Mars materials, like the Earth-Moon system. The likelihood of Mars capturing asteroids is far greater than for Earth or Venus, since its L2 and L1 Lagrange points, which serve as entry portals to Mars, are close to the rich field of asteroids inside Jupiter's orbit.

Asteroid families refer to the parents and offspring in a grouping of asteroids, which is based on spectral features and orbital elements. The two or three dozen recognized families of asteroids in the main belt are related through the breakup of a parent by collision(s), or by cratering of the parent that results in small-fragment offspring. Some families exhibit diverse composition and/or spectral features that imply that the parent was differentiated before it was broken up.

One of the lesser known asteroid groups is the Centaur asteroids, which reside between Jupiter and Neptune. These appear to be in a current or previous phase of migration between the two Giant planets.

METEORITES

Even though robotic spacecraft have yet to bring back samples of asteroid materials, there are samples of the close relatives of small asteroid fragments called meteorites. These fragments, which also include material from Mars and the Moon, are composed of the same materials that make up the asteroids, but they can be altered by the extreme heat of atmospheric entry. Meteorite classification is generally simpler than for asteroids and often based on composition. Beyond classification, however, spectral analysis plays an important role in correlating the meteorite samples with the spectral features of the unsampled asteroids.

About 94% of meteorite samples are stony meteorites, a name suggesting a composition dominated by silicate rock. Actually, the stony meteorites are subdivided into the chondrites and achondrites, chondrites being the most primitive products formed in the early solar system. Chondrite meteorites consist of small spherical inclusions called chondrules, hence the name. Chondrites are also subdivided into five subclasses according to their makeup, one of these being carbonaceous chondrites. The unusual carbonaceous chondrites contain organics and water, and some also may contain amino acids, but these represent a small percentage of the chondrites. Achondrite meteorites are more terrestrial-like and contain silicate rock that is likely fragmented crust from large, differentiated asteroids. Examples of achondrite meteorites can be found in samples originating from the Moon and Mars.

Lower in abundance are the iron meteorites that make up roughly 5% of the meteorite samples. Iron meteorites, which have their own subdivision based on composition and structure, are composed almost entirely of iron and nickel, although some include rocklike silicates. Meteorites that are a combination of metal and rock, called stony irons, have two subcategories and make up only 1–2% of the meteorites found on Earth.

Average Composition

Although meteorites vary widely in their composition, an average of the elemental abundance in a broad selection of meteorites provides insight into the composition of the meteoroid and asteroid population. This average should also be representative of the terrestrial planets, although the meteorite samples generally exclude the volatile materials burned off by the meteorite's atmospheric entry. An average must be used with caution, since chemical and physical changes are introduced by ballistic heating and surface impacts. With these caveats in mind, the average composition of meteorite samples has found to be as shown in table 3.18 (Davis 2005).

ASTEROID/METEORITE IMPACTS

The crater of a comet or meteorite/asteroid impact and the energy released is roughly proportional to the object's mass, since a large meteorite's impact velocity is closer to the average hyperbolic velocities of 20–40 km/s than the much smaller terminal velocity of small meteorites. But the Earth's crater record is sparse, since surface erosion effectively erases small and large craters over time. Impact craters that do exist show the

Table 3.18. Average Composition of Meteorite Samples

Oxygen	36.3%
Iron	25.6%
Silicon	18.0%
Magnesium	14.2%
Aluminum	1.3%
Nickel	1.4%
Calcium	1.3%
Sodium	0.6%

tremendous destructive energy of impact, just as they do on the Moon and other planets. These impact craters also contain the original impactor material, with a few notable exceptions. One important trace element unique to asteroids is iridium, which can be used to distinguish between asteroid impacts and comet impacts because of the dispersal of the dust in Earth's atmosphere. A thin layer of iridium-rich material deposited around the globe is a useful fingerprint for identifying a large asteroid impact. While the lack of iridium is less definitive, its absence can corroborate comet impacts such as the Tunguska event in 1908 that left a wide area of damage in the Siberian forest. The Tunguska blast created a light display that could be seen in Europe for weeks from the forest fires, but it left no crater, nor did it deposit impact material. A comet could have been responsible for the cataclysmic air burst several kilometers above the surface that left no significant material and left no crater, but the same evidence would support the disintegration of a structurally weak meteorite such as a carbonaceous chondrite.

Impact Examples

- Tunguska Siberia (1908)
  - Comet/carbonaceous asteroid estimated at 50 m diameter
  - No crater created, meaning it had to be a weak (carbonaceous) meteorite or a comet disintegrating in an air burst
  - 15 megaton of TNT energy equivalent
  - 1,000 sq km surface devastation
- Arizona crater (50,000 years ago)
  - 1.3 km crater
  - 50 m diameter iron/nickel meteorite
  - 200 megaton energy equivalent
- Yucatan impact (65 M years ago)
  - 70% species exterminated on Earth
  - 13 km diameter asteroid
  - 130 km crater diameter
  - 100,000,000 megaton energy equivalent

An estimate of the impact energy and damage using hyperbolic orbit speed, a high-incidence angle, and typical asteroid density as shown in table 3.19 provides a useful comparison of the potential damage of various-sized impactors.

Table 3.19. Impact Diameter, Energy, and Cratering Potential

Impactor diameter	Energy of impact (Megatons of TNT)	Crater diameter (km)
20 m	5	0.2
100 m	100	1.0
1 km	10,000	10
10 km	10,000,000	100

Comets

Comets are often described as dirty snowballs because of their composition consisting of ices and dust. Ices that make up the bulk of the comets are primarily carbon dioxide, ammonia, methane, and water ices. Spectral signatures of other compounds include ethane, hydrogen cyanide, formaldehyde, long-chain hydrocarbons, and amino acids. Dust-grain spectra vary widely from silicates to organics to metals. Even though comets are primarily ice bodies, they are also some of the darkest objects in the solar system, with albedos measured as low as 0.03 for comet Borelli. The dark surface is due to the progressive concentration of the dust and organic molecules as the outer ice layers evaporate from solar heating and sublimation. This leaves behind the dust grains and the complex and organic molecules in increasingly thicker layers.

The simplest comet classification is divided into two simple categories: short-period and long-period orbits. Short-period comets have an orbit period of less than 200 years, while long-period comets have an orbit period greater than 200 years. Comets measured with hyperbolic orbits suggest their origin is in the distant Oort cloud comet reservoir, well beyond the planetary disk. Closely related to their orbit semimajor axis and perihelion are the expected lifetimes and evolution of comets. Comets reaching within the orbit of Jupiter are thought to undergo significant scattering in hundreds or certainly thousands of orbits. It is also likely that comets reaching Jupiter's orbit, and especially as close as Earth's orbit, would lose a significant amount of surface and interior ice over time. It is likely that multiple encounters within several AU of the Sun would transform a comet into an asteroid-like body with a carbonaceous signature containing an ice mantle instead of rock.

COMET TAILS

Comets consist of an undifferentiated solid-ice core called a nucleus. If close enough to the Sun for significant heating, a gas envelope called a coma builds around the nucleus, pushed away from the Sun by radiation and solar wind to form an elongated tail. The comet tail does not point opposite to the direction of flight like a contrail, but opposite to the relative position of the Sun—at times toward the direction of flight. Solar heating creates the comet's coma, which in turn reflects the Sun's visible, IR, and UV light, making the comet much easier to spot inside the Earth's orbit. A comet's partially ionized coma and tail sometimes forms a second tail created by the Sun's magnetic field pointing the ions in a different direction from the solar photon pressure.

COMET DISTRIBUTION

As you would expect, comets are rarely found near the Sun because of their volatile nature. Less intuitive is the comet distribution peak just beyond the orbit of Neptune in the Kuiper Belt. As part of the broader Trans-Neptunian Object collection, the dense comet field in the Kuiper Belt exhibits a strong interaction with Neptune, and likely

Figure 3.29. Plot of the outer solar system objects including the planets, comets (plotted as small darts—note Hale-Bopp and Halley), and asteroids shown as smaller dots). Source: NASA JPL, Chodas.

is responsible for Neptune's migration away from the Sun after its formation (Desch 2007). Beyond this is an even broader region of scattered-disk objects, also dominated by Neptune (see figure 3.29). Well beyond these distant regions, hundreds and even thousands of AU from the Sun, lies a hypothetical sphere of comets and ice fragments called the Oort cloud. Because ices easily stick together, or accrete, and because the solar nebula collapse was radially inward, ice clumps that were not pulled into the Sun or later into the planets followed nearly straight-line (highly eccentric) orbits from all directions, producing random orbit inclinations. The Oort cloud, hypothesized by Ernst Opik, and later by Jan Oort, was intended to answer the puzzling appearance of comets from all directions at a higher rate than found orbiting in the visible solar system. This distant, diffuse reservoir of comets that can be perturbed by the motion of Jupiter and Neptune may be responsible for the periodic bombardment of Earth by comets.

Recent data from NASA's Deep Impact mission shows a strong presence of both carbon dioxide and water ice, as well as hydrocarbons and organic molecules, which corroborates earlier spectral data from ground-based instrument and spacecraft observations. The initial spectral data from the Tempel 1 comet impact, displayed in figure 3.30, shows the dominant constituents of that comet, including water ice (left), hydrocarbons (center), and carbon dioxide ice (right).

Dust and Dust Grains

Minuscule in size but large in number are the tiny dust grains that made up the clumps that formed the planetesimals and planets in our solar system. These pre-

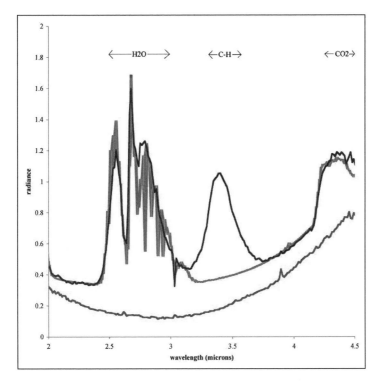

Figure 3.30. Plot of the spectral emission from material released by the Deep Impact probe's collision with comet Tempel 1 on July 4, 2005. The three major spectral peaks clearly show the signature of water (as ice, left), hydrocarbons (center), and carbon dioxide ice (right). Source: NASA JPL.

cursors to the larger debris swept up into planets, asteroids, and comets were first created in supernova explosions and red giant outflows of nearby dying stars. Dust particles in our solar system consisting of high-temperature silicates, metals, and metal oxides, are spread as interstellar dust and mixed with the surrounding hydrogen and helium gas clouds. Gas clouds orbiting the galaxy center that undergo compression from a nearby supernova are seeded with the dust from the blast before collapsing to form a protostar. Long-term mixing and cooling of the cloud of gas and dust introduces ices and other volatiles, which in turn create grains of increasing size and complexity during the cloud's collapse into a star. Changes in these dust grains from compressional heating altered the volatile ices at the center of the collapsing gas cloud, changing also the chemistry of the higher-temperature grains that remained.

Models of the formation and accretion of dust in the solar nebula are based in part on the measurements of chondrite meteorites found on Earth, and recently from the comet debris samples collected by NASA's Stardust spacecraft. Chondrites were the earliest rock materials to form in the solar system and contain some of the original presolar grains embedded within larger silicate and metal particle chondrules. Discovered in Stardust's comet samples were particles that were heated as the Sun passed through its early fusion stage, then later embedded in cometary ices. These samples offer the most comprehensive view of the early stages of solar system formation available. From seemingly insignificant dust grains comes important evidence that comets contain presolar grains scattered throughout the early solar system in addition to the processed grains created only in the inner solar system. Since comets and ices close to the Sun were consumed by its intense heat, those processed grains had to be scattered through the outer solar system in its early formation. This means that the development of planetesimals and the early planet cores was a violent, turbulent epoch. Meteorite samples confirm the picture of a solar system with a violent beginning, since several meteorite families contain interior as well as exterior fragments of large, differentiated asteroids that were blasted apart in violent collisions.

Scattered dust still orbits the Sun in regions that are diffuse but dense enough to be detected by their faint reflected sunlight, and by the emission of a faint infrared glow. Cosmic dust, known as the interplanetary dust cloud, permeates the inner solar system with a diffuse band concentrated on the orbital plane of the planets, which is coincident with the plane of the ecliptic—the Earth's orbit plane. Sunlight scattered by this dust can be seen near sunrise or sunset on clear nights, provided there is little or no background light or moonlight. Scattered light from this band of dust is called zodiacal light because the plane of the ecliptic runs through the zodiacal constellations. However, the scattered light is brightest opposite the Sun, which produces a faint glow slightly brighter than the zodiacal light at the antisolar point, called gegenschein (German for "counter shine").

Particles and dust are also concentrated around the planets and their moons in their respective gravity wells, and in the L4 and L5 Lagrange stability points of many planets. Excluded are the planets nearest the Sun and the Earth-Moon system because of the Sun's strong gravity gradient. Since Mars is more than 50% farther from the Sun than the Earth, Mars's L4 and L5 positions undergo a weaker destabilizing influence from the Sun. Proof of this exists in the few asteroids and dust that have been observed in both of the Mars-Sun L4 and L5 regions. By convention, Mars's captured L4/L5 objects are also called Trojan asteroids. The largest collection of Trojan asteroid dust is found in the Jupiter-Sun system, due to the weak gravity gradient of the Sun and Jupiter's huge mass. Nevertheless, L4 and L5 regions are only "quiet zones" that offer a reduced perturbation environment and a weak Coriolis effect, which retain objects by inducing rotation about the approximate center. The Coriolis effect circulates the objects rather than stabilizes them, leaving the Trojan objects susceptible to perturbations from nearby planets and any noncircular orbital motion of the planet.

DEBRIS

Earth's gravity well shapes a much richer debris field than found in the general interplanetary regions, with the possible exceptions of Jovian planet rings and comas that surround comets. Earth's space environment contains small meteoroids and micrometeoroids, called natural debris, as well as man-made debris from spacecraft boosters, failed and fragmenting spacecraft, pyrotechnic and mechanical spacecraft deployment devices, and fragments from collisions in antisatellite tests. Although functioning spacecraft are not considered space junk, they pose serious hazards to

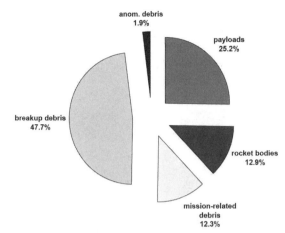

Figure 3.31. Orbital debris break-down of objects tracked by source as of August 2007. Source: NASA Orbital Debris Quarterly.

other spacecraft and are tracked carefully to alert launch and flight operators to the position and potential hazard of objects traveling at relative speeds of 5–20 km/s (11,250–45,000 mi/hr).

Tracking space debris was initiated by the Department of Defense after the launch of the first Sputnik satellite in 1957 to monitor potential threats from Soviet ICBMs. Today, the U.S. Space Surveillance Network that is part of the U.S. Strategic Command monitors over 10,000 cataloged orbiting objects that are over roughly 10 cm in diameter. Those data are passed along and distributed through NASA for planning and operations, since most spacecraft have the capability to change orbits. Objects smaller than 5 cm are difficult to impossible to track with optical and radar

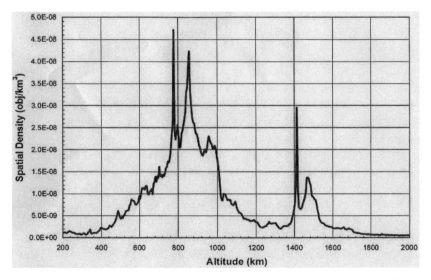

Figure 3.32. Plot of the density distribution of tracked objects in low- and medium-Earth orbit. Source: NASA Orbital Debris Quarterly.

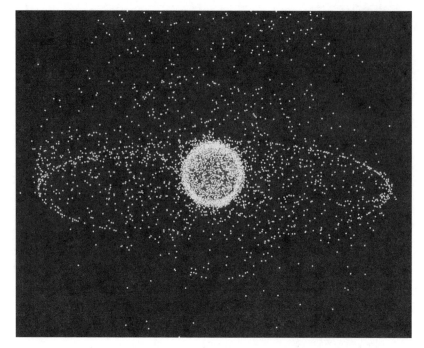

Figure 3.33. Three-dimensional projection of tracked spacecraft and detectable debris showing the large concentration in low- and medium-Earth orbit, as well as the geosynchronous ring. High inclination and high-eccentricity debris/spacecraft appear far from the geosynchronous plane. Source: NASA JSC.

equipment, but still pose a collision risk to spacecraft. The greatest concern over debris collisions is the low-Earth-orbit region, since most spacecraft operate in this region, and even more important, manned space vehicles orbit at these altitudes. Fortunately, low-altitude debris is relatively short lived because of atmospheric drag, depending on mass and size, and orbital altitude.

Atmospheric drag from the Earth's upper atmosphere selectively pulls material out of orbit with a rate dependent on altitude—equivalent to atmospheric density—and the shape of the debris and its density. A peak in the number of trackable debris pieces in Earth orbit appears near 800 km, well above the orbits of the manned space vehicles. While smaller-particle debris is removed from orbit even faster than larger objects, it is also more abundant in number, and more common in low Earth orbit than centimeter-sized or larger material. This sand-grain-sized micrometeoroid material is abundant enough to pit and erode spacecraft that are in orbit for as little as several days. For space stations in orbit for years, the pitting and erosion become a concern because of the potential damage to the vehicle's optical, thermal, and sensor surfaces. NASA's space shuttle orbiters, for example, return with micrometeoroid-pitted windows on each mission without exception. Several of the outer panes from the ten main window sets of the space shuttle orbiters have to be replaced after each flight.

Extrasolar Planets

Recent discoveries of planets outside of our solar system, called extrasolar planets or exoplanets, were first made by careful measurement of the slight periodic motion of the central star pushed and pulled by the gravitational tug from a large orbiting planet. The period of the Doppler shift in the stars' spectra determines the period of the planet's orbit, while the mass of the star is calculated from its spectral features. Both provide an approximate mass for the planet and its semimajor axis, with the greatest uncertainty in the orbit inclination. Similar techniques using accurate measurements of the star's position can be used to detect the planet's gravitational influence on the star when viewed from a polar perspective that has no Doppler signature. Planetary eclipse of the star is also possible provided the star-planet orientation is edge-on, and the planet is large enough to block enough light to be detected as it passes in front of the star. Orbiting high-resolution and high-sensitivity photometer and interferometer telescopes are being prepared for launch that enable the direct detection of terrestrial-sized planets and larger. The NASA Kepler Mission, launched in 2009, and NASA's Space Interferometry Mission, scheduled for launch in 2015, are just two of the exoplanet exploration programs for studying planets in nearby star systems.

Preliminary statistics from exoplanet research are intriguing. As of 2009, nearly 300 extrasolar planets have been cataloged, with most being the size of Jupiter, and orbiting at an average near 1 AU, although many are closer to their parent stars than Mercury is to the Sun, with some orbits as short as one day (Lawson and Traub 2006). Most of the stars chosen for these studies are similar to the Sun, since the environment associated with such stars is more likely to promote planetary formation, including terrestrials. Surveys that include giant and dwarf stars corroborate this bias, since the population of detected planets drops off at the high-mass and low-mass extremes.

The most common planet type found is the low-density Jupiter-like gas giants dominated by hydrogen and helium gas, orbiting much closer to the parent star than Jupiter. The characteristic features of these prevalent exoplanets have led to their informal classification as "hot jupiters." Lower-mass, higher-density planets that range from 5 to 30 Earth masses have been put in a similar category called "hot neptunes," containing more ices and possibly more rocky materials than the hydrogen- and helium-rich Jupiters. At least one Earth-like exoplanet has been discovered—the smallest mass detected so far. Spectra from the planet Gliese 581 c, discovered in 2007, and approximately five Earth-masses, indicates the presence of water at temperatures ranging from 0°to 40° C which suggests a terrestrial structure (Selsis et al. 2007).

RADIOMETRIC DATING

Samples recovered from space, the Moon, and Mars can be dated quite accurately using the same technique for dating Earth rock with a method known as radiometric or radioactive dating. Radioisotopes created within stars and by cosmic ray collisions with surface rock have a fixed rate of decay based on the individual atomic isotope.

This allows measurement of the remaining radioisotope using its activity level. The level of radioactivity is compared to the original amount of radioisotope, measured by the mass of the product(s) that the radioisotope is converted into, called daughter products. The commonly used term radiocarbon dating refers to the use of one specific isotope that has such a short lifetime it can only be used for dating relatively recent fossils. Isotopes with much longer half-lives, or equivalently much slower decay rates, are used for dating meteorites or lunar samples billions of years old. The utility of this process lies in the very stable decay rates for radioisotopes and the wide variety of decay rates available from the radioisotopes in more than a hundred naturally occurring elements.

The type of radioactivity used for radiometric dating is generated by unstable isotopes (isotope meaning a nucleus with greater or fewer neutrons than in the stable atom) breaking apart, or fissioning. The activity level of an isotope is measured by the number of nuclei breaking apart and emitting electromagnetic radiation or particles, or both, in one second. The maximum activity level of an isotope occurs after its creation, with a declining activity count that produces more of the stable by-product. The ratio of these two is an accurate measure of the elapsed time since the fissioning process began. And because the activity level decreases by one-half of its previous level uniformly, the time unit called half-life can be converted into total elapsed time since creation of the radioisotope. The primary limitations to this dating technique are the half-life of the isotope and the level of activity. If more than about ten half-lives have elapsed since formation of the radioisotope, accurate ratio measurements become challenging. Other inaccuracies can be introduced if the sample has gains or losses in either the radioisotope or its daughter product, or if the sample is exposed to elevated temperatures.

The relationship between activity level and elapsed time for a radioisotope contained in a rock or fossil sample is simple, as is the expression relating the two. The decay equation is commonly expressed as:

$N/N_o = \exp[-(A/T_{1/2})t]$ where N/N_o is the ratio of the activity to the original isotope in the same units (normally number of atoms)

N = remaining radioactive isotope atoms
N_o = initial number of radioactive isotope atoms
A = constant = $\ln(2)$ = 0.6931
$T_{1/2}$ = radioactive half-life of the specific isotope

t = time in the same units as $T_{1/2}$
One half-life = time for nuclear activity to reach one-half of its original level, $N/N_o = 1/2$
Two half-lives = one-quarter of activity remains, $N/N_o = 1/4$
Three half-lives = one-eighth of activity remains, $N/N_o = 1/8$
Four half-lives = one-sixteenth of activity remains, $N/N_o = 1/16$

This can also be expressed as $N/N_o = (1/2)^n$ where n = the number of half lives, or $n = (t/T_{1/2})$ (n does not have to be an integer).

A more useful approach is to use the remaining fraction to calculate the age of the sample containing the radioisotope.

One half-life $= 1 \times T_{1/2} = 1/2$ remaining
$2 \times T_{1/2} = (1/2)^2 = 1/4$
$3 \times T_{1/2} = (1/2)^3 = 1/8$
$4 \times T_{1/2} = (1/2)^4 = 1/16$
$5 \times T_{1/2} = (1/2)^5 = 1/32 \ldots$
$10 \times T_{1/2} = (1/2)^{10} = 1/1024$

The maximum practical number of half-lives that can be used for radioactive dating is roughly 10 [note that $(1/2)^{10}$ represents $1/1024$ of the original radioactive level]. Therefore, the maximum range of elapsed time that a radioactive element can provide is roughly ten times its half-life.

Calculation Example 1

What would the remaining activity level be for a thorium 232 (^{232}Th) sample after six half-lives?

$N/No = \exp[-(A/T_{1/2})t]$. This calculation is possible by plugging in the numbers, but it's much easier to use the simple calculation $N/N_o = (1/2)^n$ where n is the number of half-lives. This would be

$N/N_o = (1/2)^6 = 1/64 = 1.65 \times 10^{-2}$, or 1.65%

Calculation Example 2

A rock sample from the surface of Mars is measured to have 0.00103 of the original uranium 236 (^{236}U) remaining. What is the age of the rock?

$N/N_o = \exp[-(A/T_{1/2})t]$ or $\ln(N/N_o) = -(A/T_{1/2})t$ or $t = -T_{1/2}/A*\ln(N/N_o)$ or $t = T_{1/2} \ln(N_o/N)/\ln2$
$t = T_{1/2} \ln(N_o/N)/\ln2 = 2.34 \times 10^7 yr*\ln(1/0.00103)/\ln2 = 2.34 \times 10^7 yr* \times 9.923 = 2.32 \times 10^8 yr$

This is equivalent to 9.923 half-lives.

To date rock samples more than a few million years old, the selection of useful radioisotopes is usually limited to a few thorium or uranium isotopes, or to a number of intermediate daughter-product ratios. Some of the more common isotopes used for radiometric dating are shown in table 3.20.

Table 3.20. Radioisotopes Commonly Used for Sample Dating

Element	Half-life ($T_{1/2}$)	$10T_{1/2}$	Uses
^{14}C	5,730 yr	57,300 yr	Dating prehistoric human remains, mammals
^{236}U	2.34×10^7 yr	2.34×10^8 yr	Dating rock and geological layers on Earth
^{40}P	1.26×10^9 yr	1.26×10^{10} yr	Dating rock and geological layers on Earth
^{232}Th	1.41×10^{10} yr	1.41×10^{11} yr	Dating lunar and the early solar system rocks

References

Davis, A. M., ed. 2005. *Meteorites, Comets, and Planets*. San Diego, CA: Elsevier Science.

Desch, S. J. 2007. "Mass Distribution and Planet Formation in the Solar Nebula." *Astrophysical Journal* 671, 878–93.

Eckart, P. (ed.). 1999. *The Lunar Base Handbook*. New York: McGraw-Hill.

ESA Science & Technology. 2005. "Cassini-Huygens." Retrieved from http://sci.esa.int/science-e/www/object/index.cfm?fobjectid=31187

Eurocommuter. N.d. "GNU Project." Retrieved from http://en.wikipedia.org/wiki/File:TheTransneptunians_73AU.svg

Grundy, W. M., K. S. Noll, and D. C. Stephens. 2005. "Diverse Albedos of Small Trans-Neptunian Objects." *Icarus* 176, July, 184–91.

Jacqué, Dave. 2003. "APS X-rays Reveal Secrets of Mars' Core." Argonne National Laboratory, September 26. Retrieved from http://www.anl.gov/Media_Center/News/2003/030926mars.htm

Jones, John, and Herbert Palme. 2000. "Geochemical Constraints on the Origin of the Moon." In *Origin of the Earth and Moon*, ed. R. M. Canup and K. Righter, ed., pp. 197–216. Tucson: University of Arizona Press.

Lawson, P. R., and W. A. Traub. eds. 2006. *Earth-Like Exoplanets: The Science of NASA's Navigator Program*. JPL Publication 06-5, Rev A. Pasadena, CA: Jet Propulsion Laboratory, NASA.

Levison, Harold F., and Alessandro Morbidelli. 2003. "The Formation of the Kuiper Belt by the Outward Transport of Bodies during Neptune's Migration." *Nature* 426, November 27, 419–21

Lin, Douglas. 2008. "The Genesis of the Planets." *Scientific American*, May.

Lunine, Jonathan I. 1993. "The Atmospheres of Uranus and Neptune." *Annual Review of Astronomy and Astrophysics* 31, 217–63.

Mitchell, Don P. 2003. "Plumbing the Atmosphere of Venus." Retrieved from http://www.mentallandscape.com/V_Lavochkin1.htm

NASA Goddard Space Flight Center. N.d. Explorers Program, code 410. Retrieved from http://fpd.gsfc.nasa.gov/410/missions.html

NASA Orbital Debris Program Office. 2008. *Orbital Debris Quarterly News* 12, no. 2, April.

Permanent. N.d. "Lunar Materials." Retrieved from http://www.permanent.com/intro.htm

Selsis, F., J. F. Kasting, B. Levrard, J. Paillet, I. Ribas, and X. Delfosse. 2007. "Habitable planets around the star Gliese 581?" *Astronomy and Astrophysics* 476, December, 1373–87.

Schaber, G. G., R. G. Strom, H. J. Moore, L. A. Soderblom, R. L. Kirk, D. J. Chadwick, D. D. Dawson, L. R. Gaddis, J. M. Boyce, and J. Russell. 1992. "Geology and Distribution of Impact Craters on Venus: What Are They Telling Us?" *Journal of Geophysical Research* 97(E8), 13,257–301.

Throop, H. B., C. C. Porcoa, R. A. West, J. A. Burns, M. R. Showalter, and P. D. Nicholson. 2004. "The Jovian Rings: New Results Derived from Cassini, Galileo, Voyager, and Earth-Based Observations." *Icarus* 172, 59–77.

Waite, J. H. Jr., T. E. Cravens, W.-H. Ip, W. T. Kasprzak, J. G. Luhmann, R. McNutt, H. B. Niemann, R. V. Yelle, I. Mueller-Wodarg, S. A. Ledvina, and S. Scherer. 2005. "Oxygen Ions Observed near Saturn's A Ring." *Science* 307, no. 5713, February 25.

Wilde, Simon, John Vally, William Peck, and Colin Graham. 2001. *Nature*, January 11, 175–78.

NASA JPL Image Photojournal. Retrieved from: http://photojournal.jpl.nasa.gov (accessed 3/15/09).

NASA STScI (Space Telescope Science Institute). Image gallery. Retrieved from: http://hubblesite .org/gallery/

NASA Johns Hopkins University Applied Physics Lab image gallery. Retrieved from http:// messenger.jhuapl.edu/gallery/sciencePhotos/

NASA NIX (NASA Image Exchange). Retrieved from http://nix.nasa.gov/

CHAPTER 4

Orbits

Celestial orbits describe the motion between two or more bodies interacting through mutual gravitational attraction. Relative motion between the objects plays an important part in the orbit features, as does the relative distance. For two bodies or for thousands, the orbit features of the interacting masses are influenced by the relative motion, the mass, and the separation of each. Nonetheless, the most useful tools for describing the orbital characteristics of multiple bodies are the two types of energy contained in the relative motion of two or more masses and their separation. These two important measures of orbital energy are:

Kinetic energy (E_k) = $\frac{1}{2}mV^2$ where m is the mass of the moving body and V is the relative velocity between the two masses under consideration.

Potential energy (E_p) = -GMm/r where G is the universal gravitational constant, M is the mass of the larger body, and r is the separation between the two bodies.

If one type of energy (these relations apply to each pair of two or more bodies in orbit) is greater than the other, we have entirely different orbit conditions than for the opposite case. Gravitational potential energy is negative by convention, and kinetic energy positive, so adding the two energies to get the total orbital energy $(E_{total} = E_k + E_p)$ will show a positive, negative, or zero net value. Greater binding (gravitational potential) energy than relative velocity (kinetic) energy means that the negative potential energy is greater that the kinetic energy. That total energy is therefore negative and the orbit is bound, since the gravitational pull between the two or more objects exceeds kinetic motion. Faster relative motion, where kinetic energy is greater than potential energy, means that the total energy is now positive, and the orbit is unbound. That is, the two bodies will continue to separate even though they still attract each other. If the energy total is zero, the potential and kinetic are exactly equal, and the two bodies have reached escape velocity.

Most orbits in the solar system can be described reasonably accurately by a two-body orbit definition that excludes any significant external gravitational influence. This is a useful tool for tracing out planetary orbits even though there are a host of other bodies that exert gravitational pull on each planet. The utility of the two-body orbit is its simplicity, but the same virtue of simplicity limits it to an approximation. For any system with more than two bodies, numerical methods must be used to solve the orbital trajectories precisely. In theory, the two-body orbit is simple to define and extremely useful, but in practice, there are no truly isolated systems with two bodies, and accurate orbits almost anywhere in the universe require approximations. In order to simplify the concepts presented, orbits discussed in this chapter will be limited to two-body systems that do not require computed numerical solutions.

The two-body orbit was first defined in the sixteenth century by Johannes Kepler, a student of Tycho Brahe, who was responsible for logging the first precise observations of the planets that were accurate enough to allow Kepler to derive his pioneering orbital equations. For the first time, Kepler's laws of planetary motion defined orbits with elliptical shapes, and established the conservation of orbital angular momentum even though the concept had yet to be introduced by Newton.

Kepler's Three Laws for Two-Body Orbits

1. *The planets orbit the Sun in elliptical orbits, with the Sun at one focus of the ellipse.* An ellipse describes the orbital path of one object around another, without regard to which object is used as a reference. The Earth in orbit around the Sun follows the same ellipse as an orbit projected with the relative motion of the Sun around the Earth, provided the Earth remains at the coordinate center. For planetary orbits, the Sun is an overwhelmingly dominant mass, and each planet follows an independent elliptical orbit with the Sun's position at one of the ellipses' focal points.

Since an ellipse that is not a circle has both a major, or long, axis and a minor, or short, axis, one of the common aspects of an orbit is the ratio of the two axes. Using the convention that the semimajor axis is one-half of the major axis and the semiminor axis is one-half of the minor axis, the orbit geometry is identical to the geometry of an ellipse as shown in figure 4.1.

2. *A line joining a planet and the Sun sweeps out equal areas in equal time intervals.* Planets orbiting the Sun have different velocities, which depend on the planet-Sun separation. For a noncircular solar orbit, the planet has the highest velocity at closest separation, called the perihelion, and travels at the slowest velocity at the most distant point, the aphelion. For circular orbits, the velocity of orbit is constant throughout its path. Because the orbiting body must conserve angular momentum, and because the total energy of an orbit is constant, the continually varying orbital velocity sweeps out equal areas in equal times.

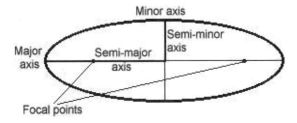

Figure 4.1. Orbit ellipse geometry showing the major axis, minor axis, and focal points.

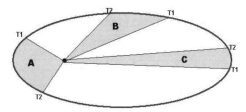

Figure 4.2. Sketch of an elliptical orbit with areas A, B, and C swept out during the same time period between T1 and T2. For a two-body orbit, area A equals area B equals area C.

3. *The period of a planetary orbit squared is equal to its orbital semimajor axis (average orbital separation) cubed, or*

$$p^2 = a^3$$

Kepler's third law defines how the period p of a planet orbiting the Sun increases with increasing distance from the Sun. Since orbit speed and orbit period are inversely related, orbital speed decreases with increasing distance from the Sun. Kepler's third law is simple because it employs convenient units where the period of orbit is in Earth-years and the semimajor axis is in Astronomical Units (AU)—the mean distance between the Sun and the Earth (1.50×10^8km).

The closest planet to the Sun is Mercury, which has the highest orbital velocity and the shortest period of orbit. More distant planets have longer orbit periods and smaller orbit velocities. Pluto, for example, has an orbital period of nearly 248 years, while Mercury has a period of only 88 days. A plot of the orbital velocity versus semimajor axis shows an inverse relation, with the velocity decreasing as $\sqrt{1/r}$. As the separation or semimajor axis increases, the orbit period increases and the rotation velocity decreases. The simple orbit period-separation-velocity relationships are as follows:

$$p = a^{3/2}$$
$$a = p^{2/3}$$
$$V \propto \sqrt{1/r} \text{ where } \propto \text{ is the proportional symbol}$$

To calculate the period of orbit of a comet of known semimajor axis, for example, you would use the equation $p = a^{3/2}$ (p in Earth years and a in AU). For semimajor axis calculations, you would use the equation $a = p^{2/3}$. For these equations to be valid, the objects must be in heliocentric (Sun-centered) orbits.

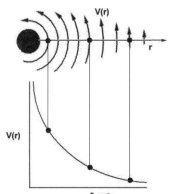

Figure 4.3. Inverse relationship between orbital velocity and separation is shown for Kepler's third law $p^2 = a^3$.

Calculation Example 1

Calculate the period of orbit for an asteroid with a semimajor axis of 3.20 AU.

Kepler's law can be used, since this is a heliocentric orbit. Units for Kepler's period-semimajor axis relationship are Earth years and astronomical units.

$$p = a^{3/2} = 3.20^{3/2} = 5.72 \text{ years}$$

Calculation Example 2

Find the semimajor axis of a comet that has an orbit period of 0.45 years.

$$a = p^{2/3} = 0.45^{2/3} = 0.59 \text{ AU}$$

Orbit calculations for planets, comets, and asteroids are simple because of the dominant mass of the Sun, and because of the convenient units of years and astronomical units contained in Kepler's third law. A more general equation for orbit period and separation was later derived by Isaac Newton and can be used for any body in orbit around another, regardless of mass or mass difference. Newton's generalized expression can be used for satellite orbits around the Earth or the Moon, for star orbits around the galaxy center, or even a galaxy in orbit about the center of mass or a supercluster of galaxies. Newton's general expression of orbit period has two elements in common with Kepler's third law and is written:

$$p^2 = a^3 \frac{4\pi^2}{G(M+m)}$$

where p = period in seconds, a = semimajor axis in meters, G is the gravitational constant = 6.67×10^{-11} Nm^2/kg^2, and M and m are the orbited and orbiting masses in kilograms. Other units can be substituted for meters-kilograms-seconds but must be consistent with the units in the constant G.

Note that the relationship between period and semimajor axis is the same as in Kepler's third law, but the units allow mass and distance values that the Kepler equation does not. In general, Kepler's simple equation can and should be used for anything in orbit around the Sun, including planets, asteroids, and comets, while Newton's equation must be used for any orbit that is not heliocentric. Either equation can be used for heliocentric orbits, although Newton's results in an orbit period of seconds.

Orbit Geometry

The basic two-body orbit geometry is the ellipse and the circle, the circle being a subset of the ellipse. An ellipse includes a major and minor axis, and two focal points.

Calculation Example

Find the period of orbit for a satellite in low Earth orbit with a circular orbit altitude of 250 km.

For this example, the orbit is not heliocentric and Newton's equation must be used. In order to calculate the semimajor axis of the orbit, the Earth's radius of 6,385 km must be added to the orbit altitude. Using 5.97×10^{24} kg for the mass of the Earth, the orbit period of orbits is calculated with:

$$p = \sqrt{\frac{a^3 4\pi^2}{G(M+m)}}$$

where the semimajor axis of the orbit is the sum of the Earth's radius (R_{Earth}) and the satellite altitude (h).

$a = h + R_{Earth} = $ 250 km + 6,385 km = 6,635 km = 7.635×10^6 m (units of meters are used in this example)

$G = 6.67 \times 10^{-11}$ Nm²/kg²

$M_{Earth} = 5.97 \times 10^{24}$ kg

$$p = \sqrt{\frac{a^3 4\pi^2}{G(M+m)}} = \sqrt{\frac{(6.685 \times 10^6 \, m)^3 \, 4\pi^2}{6.67 \times 10^{-11} \, Nm^2/kg^2 \times 5.97 \times 10^{24} \, kg}}$$

(The mass of the satellite is neglected since it is much, much less than the mass of the Earth.)

Thus, p = 5442 sec = 1 hr, 30.7 min.

For the bound orbit, the orbited body is located at one of the two focal points of the ellipse and the other is empty. In the extreme case of the orbited body being much more massive that the orbiting body, the center of mass of the larger orbited body is the focal point. For masses that have smaller differences, say 1,000:1 or less, the focal point is the center of mass of the two-body system. For equal masses, the focal point is located exactly between the two.

Basic Definitions

- Plane of orbit—the plane in which the orbiting object travels.
- Plane of the ecliptic—plane of the Earth's orbit around the Sun.
- Eccentricity (e)—the flatness of the orbit.

Variations on the flatness of the ellipse is defined with the same geometry using the flatness, called eccentricity, that is related to the ratio of the major to minor axis. Orbital eccentricity is expressed as e in the expression:

$$e = \sqrt{1 - (b/a)^2}$$

which can range from zero for a circular orbit, where b = a, to less than one where a >> b. Orbital eccentricity is useful in several applications including the definition of bound and unbound orbits.

- Bound orbits: $0 \leq e < 1$ (e is greater than or equal to zero and less than 1)
- Unbound orbits (hyperbolic): $e > 1$
- Escape (parabolic) orbits: $e = 1$
- Inclination angle—the angle between the orbital plane and a reference plane
  - Earth-orbiting satellites follow elliptical or circular orbits, but require a reference to identify the satellite's position and orbit with respect to a position on the Earth, to other satellites, to the solar system, or to the background stars. The reference is established by defining a fixed plane with a specific orientation. The conventional reference plane is the Earth's equatorial plane (an extended plane passing through the Earth's equator), and the orientation is a point in the sky known as the vernal equinox. This fixed reference plane has several convenient features, the most important being that the stars are fixed in the background sky. In this convention, Earth-orbiting satellites have their orbit plane referenced to the Earth's equatorial plane, and the line of nodes defining the intersection of the two planes.
- Nodes—the nodes represent the two intersecting points between an orbit path and the reference plane.
  - Descending node: point of intersection of the orbit path as the satellite descends through the reference plane.
  - Ascending node: point of intersection of the orbit path as the satellite ascends through the reference plane.
- Line of nodes—the line intersecting the two nodes (also represents the intersection of the orbit plane and the reference plane).
- Periapsis (or periastron)—the closest approach distance within a two-body orbit.
  - Perigee: closest point between the Earth and the object in orbit around the Earth.
  - Perihelion: closest point between the Sun and the object in orbit around the Sun.
- Apoapsis (or apoastron)—the farthest point in orbit between two bodies in a two-body orbit.
  - Apogee: farthest point between the Earth and the object in orbit around the Earth.
  - Aphelion: farthest point between the Sun and the object in orbit around the Sun.

Satellite Orbits

Orbit data for many Earth satellites is available in two general formats that conform to classical orbital elements, called Keplerian coordinates, as well as Cartesian coordinates, both of which can be converted from one system to the other. Each of these orbital data sets has a reference date or epoch that is used to correct for the slowly rotating reference frame of background star positions due to the Earth's rotation axis precessing. These reference epochs are in fifty-year increments that are most accurate

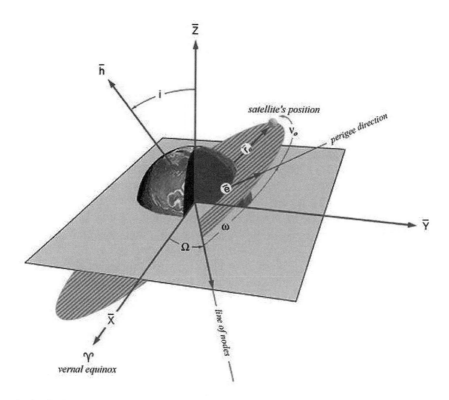

Angles that are defined by the classical orbital elements and shown in the figure above include:

- Ω = Right Ascension (longitude) of the ascending node = angle between X-axis (vernal equinox) and the ascending node of the orbiting body
- ω = Argument (longitude) of the perigee = angle between the ascending node and the perigee
- υ = True anomaly = angle between perigee and position of orbiting body
- T = Time since periastron passage

Figure 4.4. Diagram of the equatorial reference plane of the Earth and a satellite with its orbit plane outlining the conventional (classical) orbital elements. Source: NASA JPL.

for the nearest epoch. For example, positions of the stars and planets corrected to the year 2000 epoch would be more accurate than the 1950 epoch reference, and even more accurate than the 1900 reference.

NASA provides data on many of its satellites in orbit, including the International Space Station and the space shuttle, using the two conventions mentioned above, listed in either the 1950 reference epoch, called the Cartesian Mean Position 1950, or M50, and the Julian Date 2000 precession data, called J2K, meaning Julian date 2000 epoch coordinates. Keplerian coordinates are also available for most satellites. An example of the orbital data for the ISS from 2001 is listed below in English and metric units, with some mixed.

Given:

Satellite: ISS

Position vector time (GMT): 2001/295/09:39:30.000 (295th day of 2001 at 09:39:30 Greenwich Mean Time [GMT])

Coordinates

M50 Cartesian (m)	M50 Keplerian (m)
X = -6577644.96 m	a = 6777727.61 m
Y = 93016.17 m	e = 0.0005995
Z = 1623097.82 m	i = 51.71101 deg
XDOT = 1349.159356 m/s	Wp = 75.92851 deg (argument of the perigee)
YDOT = -4915.420616 m/s	RA = 167.95779 deg (right ascension of ascending node)
ZDOT = 5732.889437 m/s	TA = 301.84176 deg (true anomaly)
	MA = 301.90011 deg (mean semimajor axis)
	Ha = 214.988 nm (altitude of apogee above surface in nautical miles - 1 nm = 1.85 km)
	Hp = 211.192 nm (altitude of perigee above surface)

M50 Cartesian (ft)	J2K Cartesian (m)
X = -21580200.01 ft	X = -6586082.76 m
Y = 305171.15 ft	Y = 19435.18 m
Z = 5325124.08 ft	Z = 1591115.33 m
XDOT = 4426.375841 ft/s	XDOT = 1376.152176 m/s
YDOT = -16126.708058 ft/s	YDOT = -4900.187019 m/s
ZDOT = 18808.692378 ft/s	ZDOT = 5739.510821 m/s

TWO-LINE ELEMENTS

Orbital data for satellites is often available in a modified set of orbital elements called two-line elements. These data contain much of the same information as the traditional orbital elements, but with the addition of orbital decay rates and mean orbital motion in orbits per day. Two-line orbital element sets are readily available for most satellites, but lack accuracy for anything beyond approximate positions with time, and require arcane conversion methods to or from the traditional orbital element formats. The two-line elements have the advantage of providing simple parameter and position information, but are more difficult to use for plotting orbital trajectories. Probably the most useful of the two-line element data are the mean motion (the inverse of the orbital period), and the orbital decay rate.

The following listing is a two-line element set for the International Space Station from the year 2001.

ISS
1 25544U 98067A 01295.46347873 .00057998 00000-0 70152-3 0 9007
2 25544 51.6368 168.5225 0008202 233.9470 126.0925 15.57755304 6968

The legend for this set is as follows:

Table 4.1

Satellite	ISS
Satellite catalog number	25544
International designation	98067A
Epoch time	01295.46347873 = 2001 (yr) 295 (day of year = Oct. 22nd) + 0.46347873 fraction of the day (11:07:25)
Decay rate	5.79980×10^{-4} rev/day2 (1st derivative of mean motion)
Decay acceleration	0.0000×10^{-0} rev/day3 (2nd derivative of mean motion)
Drag term	7.0152×10^{-3} (radiation pressure coefficient)
Element set & checksum	9007
Inclination	51.6368 deg
RA of node	168.5225 deg
Eccentricity	0.0008202
Arg of perigee	233.9470 deg
Mean anomaly	126.0925 deg
Mean motion	15.57755304 rev/day
Period of orbit can be found using 23.93 hr/day/15.57755304 orbits/day = 1.54 hr/orbit	
Revolution at epoch	6968

Orbit Classification

Two-body orbits come in a wide variety of shapes, sizes, and orientations that are typically defined by the three parameters of inclination, eccentricity, and semimajor axis, but may also be classified according to a specific application. The most general orbit categories are just two—bound and unbound. These conditions are determined as described earlier by the sum of the potential and kinetic energies of the orbiting body. A bound orbit has a negative total energy, whereas an unbound orbit has a positive total energy.

Bound (Elliptical) Orbit

- A bound orbit has negative total energy
  - Potential energy is greater than kinetic energy

$$E_p > E_k$$

- Orbital path is described by an ellipse
- Eccentricity is less than one

$$0 \le e < 1$$

Unbound (Hyperbolic) Orbit

- A hyperbolic orbit is unbound with a positive total energy
  - Kinetic energy is greater than potential energy

$$E_k > E_p$$

- Orbital path is described by a hyperbola
- Eccentricity is greater than one

$$e > 1$$

Escape (Parabolic) Orbit

- A parabolic orbit is neither bound nor unbound since the relative kinetic energy is exactly equal to the gravitational potential energy. The parabolic orbit conditions are the same as escape velocity.
  - Kinetic energy is equal to potential energy

$$E_k = E_p$$

- Orbital path is described by a parabola
- Eccentricity is equal to one

$$e = 1$$

Orbit Types

The variations in orbits are generally related to their application, orientation, and eccentricity. A good example of a diverse orbit application is found in the families of communication satellites. Even though circular orbits are employed for most communications satellites, high-eccentricity orbits are found on high-latitude communications satellites placed in highly eccentric, highly inclined orbits called Molniya orbits. Multiple communications satellites in low or medium Earth orbits, such as the Iridium communications satellites, provide broad coverage of the entire Earth, while geostationary communications satellites are more common because of their apparent stationary position on the equatorial plane. More complex orbits can also be found in applications such as remote sensing, a good example being the Landsat satellites. These spacecraft are placed in circular, retrograde orbits to repeat their positions over a given latitude at the same time each day—the sun-synchronous orbit. A list of commonly used orbits is given below.

1. Prograde orbit: A prograde orbit has an inclination angle less than 90°, which means the orbital path is in the same direction as the rotation of the planet or moon.

2. Retrograde orbit: A retrograde orbit has an inclination greater than 90°, which travels in reverse direction to the rotation of the planet or moon.

3. Polar orbit: A polar orbit has an inclination of exactly 90°, which allows global coverage of a planet or moon over a period of hours to days, depending on the orbital altitude. This orbit is commonly used for meteorological and surveillance satellites, for Earth observations, and for planetary exploration and mapping applications.

4. Geosynchronous: A geosynchronous orbit has an orbital period equal to the Earth's rotation period of twenty-four hours (sidereal period must be used which is $23^h56^m4.09^s$). The semimajor axis of the geosynchronous orbit is 42,164 km, with an inclination that can vary from the conventional equatorial orbit of 0°. Orbital tracks from geosynchronous orbits not on the equatorial plane oscillate in latitude over the Earth's surface at a fixed longitude.

5. Geostationary: A geostationary orbit is also geosynchronous, but has an equatorial orbit with an inclination of 0°. The geostationary orbit provides a fixed communications position with respect to the Earth's surface, making it ideal for several applications that include communications satellites, remote sensing satellites, weather satellites, and surveillance satellites. The broad utility of the geostationary orbit also makes it attractive to many companies, agencies, and governments putting the limited number of geostationary orbital slots at a premium.

6. Sun-synchronous: A sun-synchronous satellite orbit maintains a constant orientation between the Sun and Earth, making it useful for several applications including remote sensing applications and astronomical observations. The repeated Sun-Earth orientation creates full back-illumination from the Sun for a consistent illumination angle throughout the year, or complete shadowing from the Sun for astronomical uses during one-half of the satellite orbit. A nearly polar orbit is used for the sun-synchronous orbit to provide a constant Earth-Sun plane. If the orbit were exactly polar, the orientation of the orbit would be fixed with respect to the Sun, but not the Earth. The polar orbit would show a 0.98° per-day change in orientation as the Earth orbits the Sun at a rate of 0.98° per day (360° in 365.24 days). Therefore, the spacecraft needs -0.98° per day of retrograde motion in its orbital plane to counter the Earth's orbital motion around the Sun. To accomplish this retrograde rotation, the Earth's oblate shape is used to apply torque to the spacecraft in its orbit that

will precess the line of nodes (rotate the orbital plane). A range of altitudes and corresponding inclinations are available for this orbit. One example is the Landsat 7 satellite that uses an orbital altitude of 705 km (438 miles—equatorial), giving it an orbit period of 98.9 min. Landsat 7, with an inclination of 98.2°, passes over the equator 10:00 to 10:15 every day on each of its 14 daily orbits. Longitude of passage depends on the plane of the orbit.

7. Molniya orbit: Russia has traditionally used communication satellite orbits with both a high inclination and a high eccentricity for their high-latitude ground stations. These eccentric orbits offer the satellites an extended view of high-latitude stations near their orbit apogee. Molniya's twelve-hour orbit period also places the communications satellite over the North American continent during the second orbit in a twenty-four-hour period, allowing for a secondary objective as a surveillance satellite. Likewise, the United States has used Molniya orbits for its own surveillance programs, primarily for electronic eavesdropping on the Asian continent. Molniya orbit inclination is 63.4° with a semimajor axis of 26,562 km, although varying apogee and perigee values can be used that satisfy the combined period and semimajor axis requirements for a twelve-hour orbit.

8. Tundra orbit: A tundra orbit is an eccentric, high-inclination (63°) orbit similar to the Molniya orbit, but with a period twice as long (one sidereal day). Like the Molniya orbit, this tundra orbit is used primarily for communications at latitudes far from the equator.

9. Parking orbit: A parking orbit is a temporary orbit commonly used for spacecraft checkout operations before departure from Earth.

10. Graveyard orbit: A graveyard orbit is a permanent, higher-than-normal orbit used to remove defective or aging spacecraft from the busy geostationary region (also called supersynchronous orbit).

11. Walking orbit: A walking orbit gets its name from the rotation or precessional motion of the orbit due to the asymmetrical shape of the planet. Torque on an orbiting spacecraft is created by the Earth's oblate shape due to its relatively rapid rotation generating a larger equatorial diameter than polar diameter. The sun-synchronous orbit is an example of a walking orbit.

12. Halo orbit: Halo orbits are not the traditional orbits around a celestial object, but orbits around either the Lagrange L1 or L2 stability regions. These are not true orbits, but circular orbits, or more complex Lissajous orbits, that are permitted in the equipotential regions at the L1 and L2 points. The halo and Lissajous orbits are not stable orbits, nor are L1 and L2 actually stable points, hence spacecraft in these orbits must have thrusters to maintain the spacecraft at these positions.

Transfer Orbits

A transfer orbit is not an orbit classification but an orbit function. Transfer orbits are used to get a spacecraft from lower to higher orbit or vice-versa, or to get from one planet to another, or for an orbit plane change. The transfer process begins typically with a single-thrust departure from an established orbit that provides the correct ΔV to reach the target position or orbit. For a complete transfer from one circular orbit to another such as a flight from Earth to Mars, or a satellite boost from low Earth to high Earth orbit, a transfer sequence is executed in three phases The first phase boost to reach the destination orbit is followed by a coast phase in which the spacecraft follows an elliptical trajectory to the target or desired orbit. A spacecraft with just a single boost would return to the departure orbit if not given a circularizing boost at the target or desired orbit. The third phase is executed as the spacecraft arrives at the target or orbit with a second boost to either circularize the orbit in a higher or lower orbit, or to insert the spacecraft into an orbit around the targeted planet or moon.

A different style of orbit transfer can be made with a continuous-thrust boost from a low-thrust ion or plasma engine. The procedure would not consist of a dual-boost, three-phase transfer, but a continuous thrust flight from departure orbit to arrival, or a significant portion of the departure-to-arrival transfer phase.

Transfer orbits vary not only in function, but also in scale. The smallest orbit transfers are maneuvers used on satellites and spacecraft such as the space shuttle that change orbit character or orbit position using small thruster burns. These minor orbit changes are often employed to complete positioning a spacecraft at a precise rendezvous position with a target using minute changes in ΔV. In contrast, large thruster burns that take interplanetary spacecraft from a parking orbit around the Earth to the outer solar system may last ten minutes or more and add a ΔV to the spacecraft of 18 km/s (40,000 mi/hr). Another related family of transfer orbits is the planetary gravity assists that can propel spacecraft beyond the giant planets. The Voyager 1 and 2 spacecraft are two examples of spacecraft that were given high enough gravity assist ΔV to leave the Sun's gravity and the solar system, and allow them to reach an orbit around the center of our Milky Way galaxy—the ultimate orbit transfer.

Orbit transfers can also be used for changing orbit planes, and for combining semimajor axis changes and orbit plane changes. For these discussions, orbit transfers will be limited to changes in semimajor axis—increasing or decreasing orbital distance—or for gravity-assist maneuvers. Simple orbit transfers that change semimajor axis are often categorized into three types of trajectories that are related to thrust and efficiency. These are:

1. Hohmann Transfer: the most energy efficient coplanar orbit transfer.
2. Spiral (low-thrust) transfer: this is a long-period spiral orbit from departure to arrival that is best suited for high-efficiency, low-thrust ion/electric propulsion systems.
3. Direct (high-thrust) transfer: the direct, short-period transfer requires greater thrust to accelerate the spacecraft quickly to the desired orbit at the expense of greater transfer energy and a larger propulsion system.

Gravity-assist (slingshot) transfer orbits are much more complex and involve a gravitational boost from the close encounter of a spacecraft with a large planet to increase the velocity of the spacecraft with reference to the Sun. This technique can also be used to slow spacecraft to get to a closer orbit to the Sun, or to change the inclination of a spacecraft's interplanetary trajectory.

HOHMANN TRANSFER

The Hohmann transfer requires the least energy of the orbit transfer techniques and is therefore the most efficient transfer method. A two-burn procedure for the Hohmann transfer is started with the first executed 180° from the arrival position. Again, for simplification the assumption is made that the departure and arrival orbits are both circular, and that all orbits are in the same plane. To calculate the semimajor axis of the Hohmann transfer use:

$$a_{transfer} = 1/2(a_A + a_B)$$

where $a_A + a_B$ are the semimajor axis values for circular orbits A and B.

The periapsis of the Hohmann transfer is the circular orbit radius of the inner orbit, the apoapsis is the circular orbit radius of the outer orbit.

Hohmann Interplanetary Transfer Orbits

Hohmann transfer periods for interplanetary missions are usually calculated in Earth years since the transfer phase is within the Sun's gravitational influence. For satellites in other orbits including Earth orbit, the semimajor axis calculations are made with the same technique, but with a period calculated using Newton's period-semimajor axis equation.

To calculate the period of a Hohmann transfer, first determine the semimajor axis of the transfer orbit, then, for interplanetary orbits, use Kepler's law to find the period.

$$P_{transfer} = \frac{1}{2}\, a_{transfer}^{3/2}$$

where the factor of ½ appears because the transfer segment is one-half of the complete orbit.

DIRECT TRANSFER ORBITS

Direct transfer orbits reduce the period of transfer from departure to arrival, but require higher thrust than a Hohmann transfer and are less efficient in thrust and

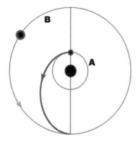

Figure 4.5. Sketch of a Hohmann transfer trajectory departing from circular orbit A and arriving at circular orbit B at an angular position 180° from departure.

Table 4.2. Hohmann Transfer Period from Earth

Planet	Semimajor Axis (AU)	Hohmann Transfer Period from Earth
Mercury	0.39	0.26 yr
Venus	0.72	0.40 yr
Mars	1.52	0.72 yr
Jupiter	5.20	2.73 yr
Saturn	9.56	6.07 yr
Uranus	19.22	16.1 yr
Neptune	30.11	30.7 yr
Pluto	39.55	45.7 yr

energy (ΔV). A direct transfer requires the same two boosts at the departure and arrival positions, but with a larger initial ΔV than for a Hohmann transfer, and a larger projected semimajor axis. The direct transfer segment to the target is only a portion of the transfer ellipse represented by the initial boost. The direct transfer segment that approximates the coast phase of the maneuver traverses less than the 180° Hohmann transfer segment.

Transfer orbits are commonly classified according to the rotation angle between the departure and arrival points. A direct transfer covers less than 180°, while the Hohmann transfer is defined as having a 180° rotation between the first and second burn, at least for coplanar transfers. Orbit transfers that follow a trajectory greater than 180° from the departure point are classified into types that are multiples of 180° (π radians).

Type I (direct transfer)—less than 180°
Type II—greater than 180°, and less than 360°
Type III—greater than 360°, and less than 540°
Type IV—greater than 540°, and less than 720°, and so on.

SPIRAL TRANSFER ORBITS

A spiral transfer implies a Type II or greater rotation using a continuous thrust or multiple thrust burns. The spiral transfer is typically employed for low-thrust, high-efficiency

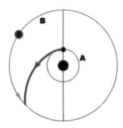

Figure 4.6. Sketch of the direct transfer method also called Type 1.

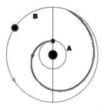

Figure 4.7. Sketch of a simple spiral transfer orbit from orbit A to B.

electric ion engines for small payloads. Travel time for these orbits is typically longer for interplanetary orbits, but may actually be less for distant objectives because of the cumulative ΔV acceleration with continuous thrust propulsion over a long duration.

Mission Planning Using Conic Sections and Patched Conics

Conic sections often used to define the shape of the two-body orbits are so named because the intersection of a cone and a plane at various angles produces the geometrical equivalent of the two-body orbital trajectories. The cone-plane intersection can reproduce circular, elliptical, parabolic, and hyperbolic geometries that can be used to piece together lunar or interplanetary flight trajectories. Two exceptions to the conic shapes are the launch-to-orbit trajectory and the atmospheric entry, which involve continuous force as either thrust or as aerodynamic drag. A collection of these conic sections spliced into a first-order approximation for mission planning is called a patched conic.

The patched conic approach for the preliminary planning of an interplanetary mission is similar to the orbit transfer in many ways, and is also divided into three phases: the departure, the cruise, and the arrival.

DEPARTURE

The departure phase covers operations within Earth's gravity and generally begins with the spacecraft in a circular parking orbit. The spacecraft is boosted into a hyperbolic departure orbit that transitions to a transfer orbit with the Earth at the focus of the departure hyperbola. The gravitational influence of the Sun and other planets are neglected until the spacecraft reaches beyond the planet's gravitational sphere of influence. For a lunar mission, the departure velocity is less than escape speed, hence the departure orbit is a Hohmann transfer ellipse and not a hyperbola.

CRUISE

After departure from the Earth's gravity, roughly 10^6 km from the planet, the spacecraft enters a cruise phase along the transfer ellipse under the influence of the Sun's gravitation. Gravitational influences from the Earth and other planets are neglected. For a lunar mission from Earth to the Moon, the gravity of the Sun and other planets is neglected.

ARRIVAL

The spacecraft arrival phase at the target planet is simply the departure phase in reverse, but with the hyperbolic velocity determined by the mass of the target planet. Without a burn to enter orbit around the planet, the spacecraft would simply fly by the planet and return to the original departure point, although altered by gravity from the flyby. Inside the gravitational sphere of influence of the target planet, the gravitational pull from the Sun and the other planets is neglected.

Example: Flight from Earth to the Moon using patched conics.

Table 4.3

A. Low Earth orbit (LEO) parking orbit	Circular orbit described by the circular conic section
B. Boost and transition from Earth-parking orbit to the lunar transfer (cruise) segment—called the trans-lunar injection (TLI) burn	Hohmann transfer orbit described by elliptical conic section
C. Lunar transfer ellipse (cruise, still in Earth's gravity)	Hohmann transfer orbit described by elliptical conic section
D. Transition to lunar orbit—lunar orbit injection burn	Hyperbolic approximation with respect to the Moon's gravity
E. Lunar orbit	Circular or elliptical, depending on lunar orbit desired

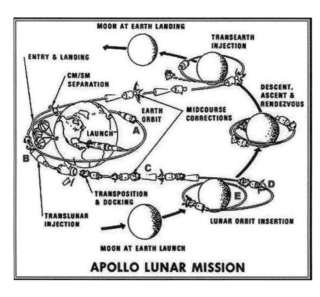

Figure 4.8. Patched conic segments for a lunar mission to a lunar orbit and return. Source: Apollo 11 Press Kit.

Sphere of Influence

Our solar system is dominated by the gravity of the Sun except near the planets or their moons. But the strength of the overall gravitational field in the solar system decreases with increasing distance from the Sun. The same is true for planets, but since planets are much less massive than the Sun, the dominance of a planet's gravity ends at a distance that scales with its mass and its distance from the Sun. A simple measure of the reach of a planet's gravity in the solar system is approximated with the Laplace expression for the sphere of influence of a planet within the Sun's gravity field:

$$r_{sphere} = a \left(\frac{m_{planet}}{M_{Sun}} \right)^{2/5}$$

with r_{sphere} = the planet's gravitational sphere of influence radius, a = the semimajor axis of the planet's orbit around the Sun, m_{planet} = the mass of the planet, and M_{Sun} = the mass of the Sun.

This approximation of a sphere that has equal gravitational pull from the planet and the Sun would extend from the inner L1 Lagrange stable point to the outer L2 Lagrange stable point shown in figure 4.16. Outside the planet's sphere of influence, a spacecraft would be under the influence of the Sun's gravity and in a heliocentric orbit, while inside the sphere of influence a spacecraft would be under the gravitational influence of the planet. From the Earth, this boundary extends approximately 930,000 km in radius. The sphere of influence is also the largest bound orbit radius, and for the Earth, sets a maximum limit to a seven-month orbit period. It is possible for objects that are orbiting the Sun near 1 AU to pass through the Sun-Earth L1 or L2 Lagrange points and transition from a heliocentric orbit to a geocentric orbit, or the reverse, and do so repeatedly. Other approximations exist for calculating the boundaries of equal gravitational force from the Sun and the planets, or between the planets and their moons that include the Hill Sphere, also known as the Roche Sphere.

Planning Interplanetary Missions

Precise orbit and flight calculations for the interplanetary missions are important because of the accurate alignment and orientation necessary for flight to and from the planets, and for gravity assists. A mission to Venus, for example, has a minimum separation distance of 0.28 AU between Venus and Earth, but actually requires a flight distance of 0.83 AU because the flight must take place in an elliptical orbit path around the Sun.

Relative motion and alignment calculations are best done by using a fixed reference frame for the solar system. Although the most common reference for the stars and planets is the Earth's reference, we know that the Earth is both rotating and orbiting the Sun. It is necessary therefore to establish a number of reference frames for interplanetary missions that include the Sun's position and motion, the target planet's position and motion, and the background reference stars' positions.

REFERENCE FRAMES AND REFERENCE TIMES

The most common planetary reference frame used is the Earth-centered coordinate reference called the geocentric coordinate system that has an equatorial plane coincident with the Earth's equator, and poles that coincide with the Earth's north and south geographic (rotational) poles. Stars, planets, the Sun, and the Moon appear to be rotating in the sky around the point in the sky near the star Polaris—the north pole reference point. By using the stars as a fixed or inertial reference in this coordinate system, the geocentric equatorial coordinate system becomes an unambiguous guide to celestial positions in the sky. Originally established as a reference for astronomical observations and for navigation on Earth using stars, the geocentric coordinate system is also used for spacecraft navigation through the solar system and beyond. When carefully corrected for rotation and motion around the Sun, the Earth's motion in the background sky measured in this coordinate system also furnishes an accurate time reference that defines the year. Because the Earth rotates as it orbits the Sun, there are two distinct measures of time that make up one Earth year.

Two methods are available for measuring the orbital and rotational motion of the Earth and the corresponding elapsed time that are based on using either the Sun or the stars as a reference. If we compare the daily motion of the Earth with its annual motion relative to the Sun and stars, an interesting divergence appears. A specific reference point on the Earth with respect to the Sun and a star near the equatorial plane will take slightly longer to return to the same point for the Sun's reference than for the reference star. Using a twenty-four-hour clock, the Sun will repeat its (apparent) position in the sky at the same time on a daily basis, but the same clock will show a slightly slower movement for the evening stars. This is because the Earth is orbiting the Sun once per year in addition to rotating on its axis once per day, which requires one extra day per year to come back to the same position with respect to the Sun (compared to the fixed stars). This time difference of one day per year is the reason we have two basic time references that define the day and the year with respect to the position of the Sun, or solar time, and the day and year with respect to the inertial, or fixed, reference stars, known as sidereal time. Solar time more commonly known as civil time incorporates the Sun as the primary reference, while sidereal time uses the background stars as the primary reference. The background stars are also extremely useful as an inertial reference for precise space navigation and for measuring astronomical positions accurately, since their positions are precisely fixed.

- Solar time—the Sun used as a primary position reference
- Sidereal time—the background stars used as a primary position reference

Solar/Civil Time

Solar time is established by the Earth's periodic motion with respect to the Sun, although it appears that the Sun is also moving with respect to the Earth in a seasonal pattern. Using the passage of the Sun through the southern meridian as noon, we can devise a simple solar time reference. This cycle repeats 365.242 times each year. The Sun's position with respect to the Earth's orbit within the background sky is used as the beginning of the year. For our calendar, that reference is defined as the apparent rise of the Sun through the Earth's orbit plane, the plane of the ecliptic. This position also defines the vernal equinox, also known as the first point of Aries, and the first day of spring.

Sidereal Time

The day on Earth is loosely defined as one rotation of the Earth on its axis and equivalent to an average of 24 hours. But since the Earth also orbits the Sun, there is a difference between a complete rotation with reference to a fixed or inertial background versus the Sun as a background reference. As the Earth rotates and orbits the Sun, it must rotate slightly farther to make up for the 0.99° orbit motion per day. This translates into a difference of 3.94 minutes per day between solar time and sidereal (star) time. Because the Earth's orbit around the Sun also contributes one rotation per orbit, there is a difference of one extra day per year. To calculate the difference per day, use:

$$1/365.242 \text{ days/year} \times 24 \text{ hr/day} \times 60 \text{ min/hr} = 3.943 \text{ min/day}$$

This calculation is also the time difference that represents the apparent motion in the night sky of 3.943 minutes per day, or 24 hours per year. Using 30-day months, the star motion is approximately 2 hours per month (30 days × 4 min/day = 120 min/month), or 1,440 minutes per year = 24 hr. So, sidereal time is shorter than solar time by approximately:

4 min per day
2 hr per month
24 hr per year

Orbit and Propulsion Energy

Calculating the approximate energy required to get from one planet to another is as simple as calculating the gravitational potential energy difference between the departure and arrival planets' orbit positions. To see this in graphic form, a plot of the Sun's gravitational potential energy can be constructed from Newton's expression for gravitational potential energy:

$$E_{potential} = -GMm/r$$

The plot of gravitational potential energy versus distance is clear evidence of the Sun's dominant mass. The spheres of influence of the largest planets are included in the two-dimensional graph shown in figure 4.10, which also provides a key to the booster requirements needed for interplanetary travel. The difference in the Sun's gravitational potential at the departure and arrival planets is the same booster energy required for an orbit transfer between the two planets. Since the energy contained in

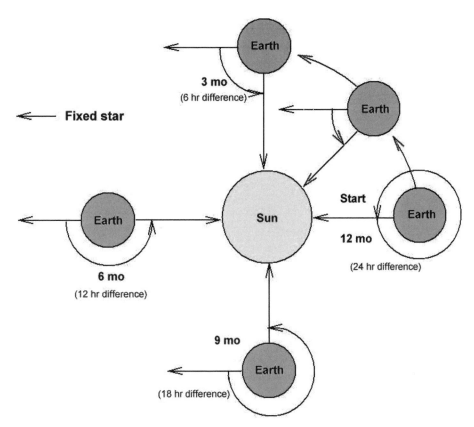

Figure 4.9. The time reference difference between solar and sidereal time is shown here using the Earth's orbital and rotational motion. The two time references begin at the start position on the right with both solar reference and fixed star reference in the same direction. As Earth's orbital position moves from the starting point, the time it takes to complete a rotation and return to the reference point increases for the solar reference (solar time), but remains constant for the fixed reference (sidereal time). The accumulated time difference for one year (one orbit) is one day (one rotation).

a booster is proportional to the booster's propellant mass, the orbit energy difference is also proportional to the size of the booster, which is mostly propellant. To illustrate these points, a plot of the Sun's potential energy is shown in figure 4.10 with the gravity wells of the three inner Jovian planets added.

A similar plot of escape velocity in the solar system can be made by taking the square root of the absolute value of the potential energy, as shown below. Escape velocity which reflects the strength of a planet's gravity decreases with increasing distance.

$$V_{escape} = \sqrt{\frac{2GM}{r}}$$

where M is the mass of the planet, moon, or the Sun, and r is the initial separation distance

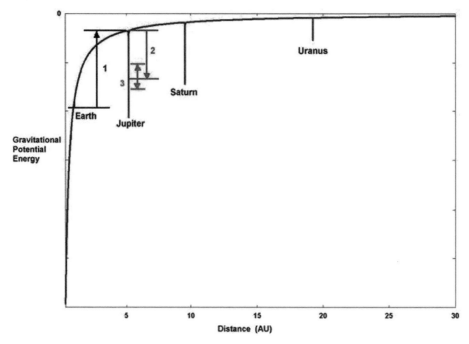

Figure 4.10. Plot of the gravity potential of solar system in arbitrary units on the left and distance in astronomical units on the horizontal scale. Jupiter, the largest planet with the largest gravity well, is followed on the right by Saturn and Uranus. If the negative sign of the potential is removed, the graph would be inverted. Traveling from Earth to Jupiter is a three-step process, although the first is not shown since it represents leaving the Earth's gravity well that is too small to identify on this chart. Second (length 1) is the transfer orbit to Jupiter's sphere of influence in energy units. The third is a retrograde boost to go into orbit around Jupiter. This is a completely arbitrary altitude indicated by the length of arrow number 2. Notice that getting closer to Jupiter requires more energy and a larger retro booster. A compromise would be to reach a lower altitude by making the Jupiter orbit elliptical as shown by arrow 3. The more eccentric the orbit, the closer the spacecraft can get to Jupiter in its periapsis, but the less time it spends in that region.

FLIGHT OPPORTUNITIES

Conceptual planning of an interplanetary mission begins with the choice of the mission's exploration objectives, followed by the selection of orbits needed for the mission's launch, departure, cruise, and arrival phases. The coplanar approximation for the first rough estimate of the spacecraft flight trajectory is not likely to furnish enough information for selecting the launch and booster vehicle, however. The actual three-dimensional orbit geometry of an interplanetary mission from initial departure to final arrival is generally much more complex than can be portrayed in two dimensions. An actual trajectory requires far more ΔV (energy) than a two-dimensional approximation because of the orbit plane changes necessary to transition from one planet's orbit to another. These complications aside, the next step is to calculate the correct alignments of the planets needed for the orbit transfer. Since the arrival planet is not going to be in the correct position for the spacecraft encounter at any arbitrary

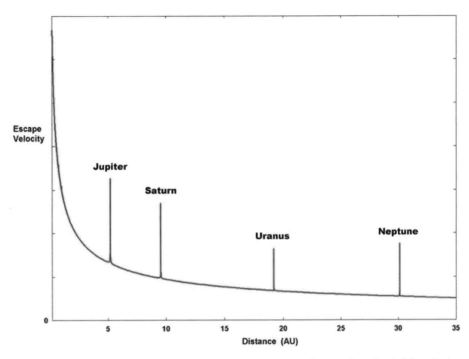

Figure 4.11. Plot of the escape velocity from the Sun and from the gravity wells of all four Jovian planets shown with an arbitrary velocity scale on the left and astronomical units (AU) on the horizontal distance scale. Included are the four Jovian planets in order, with the Earth represented by a very small blip just to the right of the third tic from the bottom. The spikes represent the velocity needed to escape the planets' gravity wells from a specific but arbitrary altitude (higher values represent closer distances to the planets). The baseline curve running from top left to bottom right is the velocity requirement to escape the Sun's gravity.

time, the alignment calculation entails the relative motion and position of the two planets at launch and arrival. For two planets in orbit around the Sun, this repeated planetary alignment is known as the synodic period.

The two basic orbital periods on Earth are the solar year, known formally as the tropical year, which measures the Earth's repeated position with respect to the Sun, and the sidereal year, which is the repeated position of the Earth with respect to the stars. For the Earth, the solar year and the sidereal year differ by one day per year. But if another planet is used for reference, its period of repeating a position in the sky depends on the relative motion of the two planets. Relative motion is slowest for the planets nearest each other since their orbit periods are similar. Therefore, repeated alignment with Earth and Venus, and Earth and Mars, have the longest synodic period, while the shortest synodic periods with respect to Earth are for Neptune and Pluto, although none are shorter than one year.

The calculation of the synodic period is simple, but more than just an exercise in math, since the synodic period is also the period of the launch opportunity from one planet to another. For example, the synodic period between the Earth and Venus is calculated using:

$$\frac{1}{P_{\text{synodic}}} = \left| \frac{1}{P_{\text{Earth}}} - \frac{1}{P_{\text{Venus}}} \right| = \left| \frac{1}{1} - \frac{1}{0.615} \right| = 1.60 \text{ yr} = 584 \text{ days (the negative sign is removed with the absolute value function } |..|)$$

Calculations made for the other planets with respect to the Earth's orbit provide a convenient table of launch opportunities from the Earth, which are the same periods for return flights.

Note that the distant planets approach one year in synodic period since their orbit period is so long—the most distant planets move little in the sky over a period of one year.

Table 4.4. Synodic Period for the Planets with Respect to Earth

Mercury	116 days
Venus	584 days (19.2 mo)
Mars	780 days (25.6 mo)
Jupiter	399 days (13.1 mo)
Saturn	378 days
Uranus	370 days
Neptune	367 days
Pluto	367 days

LAUNCH WINDOWS

Launch opportunities from Earth to the other planets are dictated primarily by the synodic period and the phase position between the two planets (see figure 4.13 below). Since the Earth is in a different orbital plane than all other planets, the alignment of the intersection of the planes at the line of nodes and the phase between the planets becomes an important consideration in launch timing because of added energy needed to transfer to/from non-coplanar orbits. In addition, none of the planets, including Earth, are in circular orbits. This makes the optimum departure and arrival timing a matter of compromise between four primary variables, which are:

1. Orbit phase between launch and arrival planets
2. Line of nodes position
3. Angle between two planetary orbit planes
4. Position in orbit of the two planets

Idealized circular, coplanar orbits have simple solutions for the timing of a minimum energy Hohmann transfer between the planets. But since the planetary orbits are neither circular nor coplanar, finding minimum transfer time or a minimum transfer energy between planets is a compromise. The true planetary orbit variables generate a range of solutions that can be plotted in two dimensions for ease in interpretation.

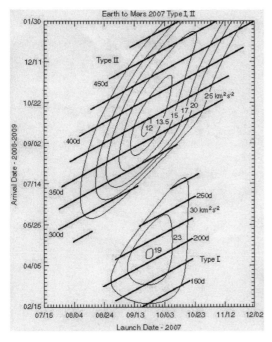

Figure 4.12. Pork chop plot of the launch and arrival solutions of intersecting time of flight (diagonal lines) and energy requirements (closed contours) is shown for the 2007 Earth-Mars opportunity. The energy contours represent total energy in massless units of km^2/s^2 which is ΔV squared. Even though a wide range of departure and arrival times is available, the penalty in delta-V is generally prohibitive beyond the plotted regions on this graph. The minimum energy plotted in delta-V (square root of energy) is found in the center of the closed curves. Minimum flight times appear at the bottom right. Note that the Type II solutions have a lower energy/ solution than Type I. Source: NASA JPL.

The plots, called pork chop plots because of their shape, have two general solutions—a Type I orbit solution and a Type II solution. For missions to Mars, the Type I would be optimal for manned missions because their shortest transfer period would expose the crew to the least radiation and microgravity. The Type II solution would be optimal for cargo or exploration missions since the energy is less than the Type I solutions, but longer in time-of-flight. An example of the two-dimensional solution to planetary transfer in the 2005 flight opportunity from Earth to Mars is shown in figure 4.12.

To understand the consequence of launching in one flight opportunity versus another, the time and energy variables can be better visualized not in a single pork chop plot near the synodic opportunity, called a conjunction, but using a representative plot of the lowest transfer energy values for a series of conjunctions. A histogram plot of the approximate energy minima for future Earth-Mars conjunctions demonstrates this in figure 4.13.

MANNED MISSION TO MARS

Human crews provide mission versatility and problem-solving capabilities that are unrivaled by robotic spacecraft, but at a high cost and with potentially catastrophic hazards. Arguably the greatest problem that human crews face is the physiological and psychological impact of exposure to extended flight in space. Even though the crews would be protected from most radiation by the spacecraft, high-energy radiation and the lack of gravity present a danger to the crews that does not permit human flight beyond the Moon with today's technology and expertise. Accumulated radiation exposure outside of the Earth's magnetic field is a serious problem in the two-to-three-year period

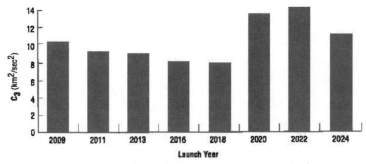

Cargo mission departure energies, 2009–2024

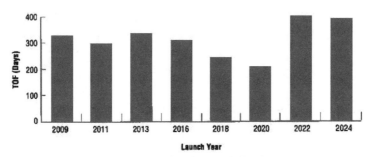

Cargo mission durations, 2009–2024

Figure 4.13. Histogram plot of the minimum energies (top) and corresponding time-of-flight values (bottom) for the Earth-Mars conjunctions listed beneath each plot. Propulsion energy requirements differ because of the changing positions of the two planets in their orbits with respect to their apsides (apoapsis to periapsis orientation) which changes over time. Even though these minima repeat on a 30-year cycle, the repetition is not perfect and long-term effects change the pattern of these values. Source: George and Kos, 1998.

required for a turnaround flight from Earth to Mars and back. Effects of microgravity exposure, which induces progressive bone demineralization and muscle atrophy, has an increasingly adverse impact as the exposure duration increases. And while minimizing flight time is one approach to decreasing the hazards of space flight, those risks are still significant even with an optimistic mission duration as short as one year.

A back-of-the-envelope estimate of the flight time to Mars and back using two-dimensional orbits begins with a simple Hohmann transfer calculation. The transfer orbit semimajor axis is simply the average of the Earth and Mars orbits, or $(1 + 1.52)/2$ AU = 1.26 AU. The corresponding transfer period is $1.26^{3/2}$ years = 1.41 years for the complete orbit. The actual transfer period is one-half of this, or 0.71 years (8.5 months) each way. A Hohmann transfer requires that the target planet be at a position 180° from the point of spacecraft launch from Earth. Thus, the launch can only take place during the correct alignment of Mars and Earth for the spacecraft to reach the target planet Mars, which repeats with the 25.6 month (2.13 yr) synodic period.

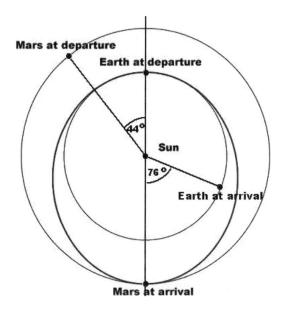

Figure 4.14. Diagram of the relationship between the orbit positions, or phases, of Earth and Mars for an idealized Hohmann transfer of a spacecraft flight from Earth to Mars.

GETTING TO MARS AND BACK

For a rough calculation of the time of flight to Mars and back and the time needed for the two planets to realign correctly for the return trip, begin by finding the mean motion between the two planets in their orbit around the Sun in degrees/day, or similar units. The Earth's mean motion is simply 360°/year × 1 year/365.24days, or 0.986 deg/day. For Mars, this is 360 degrees/1.88 years × 1 year/365.24 days or 0.524 deg/day.

Next, calculate the transfer period flight time to Mars using the circular orbit approximation. This was calculated earlier with a Hohmann transfer period of 0.71 years, or 259.3 days.

The next step is to find how far Mars will travel in its orbit during the spacecraft transfer period from Earth to Mars. This is just Mars's mean motion times the transfer period, or 259.3 days × 0.524 deg/day = 135.87°. Mars must arrive at the 180° position during the transfer orbit; therefore, the spacecraft, which is launched from Earth at the 0° point, would be behind Mars by 180° - 135.87°, or 44.13° (alternatively, Mars would be 44.13° ahead of Earth). This alignment can be approximated from the Earth's and Mars's orbital elements, being careful to not confuse the two-dimensional exercise with the three-dimensional orbital elements.

To calculate the time needed for the phase difference for the two planets to move into a correct alignment for the return trip, use the difference in mean motion between the two planets. This relative motion is just the difference in the mean motion of the two planets, or 0.986 deg/day - 0.524 deg/day = 0.462 deg/day. During the spacecraft transfer that took it to Mars, the Earth has traveled 255.66° in its orbit. The difference in positions would therefore be 255.66° - 180°, or 75.66°, with the Earth ahead of Mars by this value.

For the return trip, Mars must be ahead of the Earth by the same 75.66 degrees (the launch from Mars can be visualized at the 0° position with the Earth-arrival 180° from that launch position, which would begin -255.66° from that point, hence the angular difference between the Earth and Mars would be 255.66° - 180°, or 75.66°). For an estimate of the time required for the Earth to catch up to this position, the angle needed is just 360° minus the current lead of 75.66° minus the lead angle of 75.66°. This is a total of 208.68° (360° - 75.66° - 75.66°). At a closing rate of 0.462 deg/day, this makes the return realignment period = 208.68°/0.462°/day = 451.69 day = 1.24 years.

An important number here is the turnaround or realignment period of 1.24 years. With a transfer period of 0.71 years (times two for return travel time), the total mission time would be 2.7 years for this two-dimensional, circular orbit, Hohmann transfer approximation. Although the transfer orbit period can be decreased by increasing vehicle propulsion thrust, the realignment period is a critical element that can be as long or longer than the travel time to and from Mars. A nearly three-year minimum mission period to Mars and return, of which nearly one and a half years is in zero-g transit, means that the critical human space flight exposure problems must be researched and solved before humans can reach Mars. If the total mission duration were reduced to a minimum with an immediate return—a turnaround time of zero duration—along with high-thrust, direct-transfer orbits, the crews would encounter reduced space exposure but could contribute little, if anything, to a Mars exploration mission.

Gravity Assist

The Sun's enormous gravitational pull makes regions beyond Jupiter inaccessible for everything but the smallest spacecraft, even using the largest boosters available. But that limitation has been overcome with a technique that employs the close flyby of a planet that takes advantage of the gravitational tug of the planet on the spacecraft in the same direction of flight. A gravity assist requires a spacecraft trajectory that passes close to a planet, be it on an outside trajectory with respect to the planet to gain outward velocity or an inside trajectory for a decrease in velocity to reach the inner planets. The first gravitational assist or "slingshot" boost was designed by a JPL graduate student for the Mariner 10 mission to Mercury. The gravity-assist maneuver solved the dilemma of needing a larger booster than was in production to reach Mercury, even though its orbit is as close as 0.7 AU from the Earth.

Both Mariner 10, launched in 1973, and MESSENGER, launched in 2004, employed modest-sized boosters that were augmented by gravity assists from Venus to reach Mercury. Mariner 10 had a single Venus flyby to place it in a heliocentric orbit, while MESSENGER had a total of six gravity assists to reach orbit around the planet. The MESSENGER (MErcury Surface, Space ENvironment, GEochemistry and Ranging) spacecraft executed its first gravity assist with a flyby of Earth in December 2005, followed by two gravity assists from Venus, then three final gravity assists from Mercury. The last swingby of Mercury in September 2009 prepared the spacecraft for orbit insertion maneuver around Mercury in 2011.

As its name implies, a gravity-assist maneuver gains velocity, which is equivalent to propulsion energy, from the gravitational tug of the planet. This makes the mass of the planet one of the primary variables in the magnitude of the gravitational boost. Jupiter

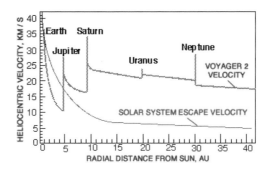

Figure 4.15. Plot of the Voyager 2 velocity changes resulting from the encounters with the Jovian planets. The added velocity from each of the gravity assist flybys except for Neptune contributed to Voyager 2's escape velocity, although the Saturn flyby could have easily generated sufficient velocity for escape as it did with Voyager 1. Figures 4.9 and 4.10 show the possible gravitational influence on a spacecraft in energy and velocity, while this plot shows the resulting velocity change from the encounter. Source: JPL Basics of Space Flight.

is the ideal planet in our solar system for these slingshot maneuvers and has been used more than any other planet for this purpose. Until recently, the smaller planets were not used in gravity-assist propulsion because of their low mass. But with the exception of Neptune, all of the planets have now been used for gravity assist. The Voyager 2 flyby of Neptune was targeted for an encounter with its largest moon, Triton, which carved a trajectory that slightly decreased the velocity of the Voyager spacecraft. Pluto, now classified as a dwarf planet, is scheduled for a flyby of the New Horizons in 2015, although the encounter will have little impact on the spacecraft's velocity.

Augmented propulsion available from a gravity-assist flyby can also be used to change the orbital plane of a spacecraft, as was done with NASA's Ulysses Solar Polar Explorer in 1990. Launched from the space shuttle *Discovery*, Ulysses reached Jupiter in 1992 for its close encounter below Jupiter, which was close enough to gain ΔV to change from a solar equatorial orbit to a solar polar orbit with an inclination of about 80°. The semimajor axis of Ulysses' heliocentric orbit changed little from its Hohmann transfer trajectory, ending with a perihelion of about 1 AU and an aphelion of 5 AU—close to Jupiter's orbital distance from the Sun.

In general, the gravity-assist trajectory is far more complex than the two-dimensional orbit analog, and includes several different orbital planes starting with the departure planet's orbit plane, the orbit of the gravity-assist target, and the orbit plane at the destination.

Because these gravity-assist maneuvers involve many variables that require computed numerical solutions, there is no simple expression to estimate flight time to the various planets as there is with a simple Hohmann transfer. Even so, examples of gravity-assist flight times from past missions are instructive in showing the reduction in flight time to the distant planets as shown in table 4.5.

Several rules-of-thumb are available for planning gravity-assist propulsion maneuvers. The most useful of these are the parameters that generate the greatest ΔV from a gravity assist flyby, which are the following:

Table 4.5. Approximate Flight Times with and without Gravity Assist for Interplanetary Missions from Earth

Mission	Flight Time: Hohmann Transfer	Flight Time: Gravity Assist
Earth–Jupiter Voyager 1 & 2 (no gravity assist)	2.7 yr	1.9 yr
Earth–Saturn Voyager 1 & 2 (Jupiter gravity assist)	6.1 yr	4.0 yr
Earth–Uranus Voyager 2 (Jupiter & Saturn gravity assist)	16.1 yr	8.8 yr
Earth–Neptune Voyager 2 (Jupiter, Saturn & Uranus gravity assist)	30.7 yr	12.0 yr
Earth–Pluto New Horizons (Jupiter gravity assist only)	45.7 yr	9.7 yr

- Larger ΔV available from larger turn angle (approach-to-departure deflection)
- Larger ΔV available from closer approach distance (smaller impact parameter)
- Larger ΔV available from higher planet mass
- Larger ΔV available from higher planet orbital speed
  - The highest gravity-assist ΔV possible is from Mercury, since it has the highest orbital speed, although its mass is small and the flyby periapse must also be small.
  - The highest possible angle of deflection from a gravity assist is from Jupiter because of its enormous mass.

Three-Body Orbits and the Lagrange Stability Points

Two-body orbits are extremely simple but still useful in approximating planetary orbits, satellite and moon orbits, even galactic orbits. One of the greatest limitations of the two-body approximation is that it does not include any influence of other celestial objects. Yet, even with that restriction, the two-body orbit model can be used to build approximation methods for three or more bodies in orbit. By adding a third body, an unusually powerful tool can be fashioned to help model the motion for Earth satellites, and for planets, asteroids, comets, and stars, provided there is a significant mass difference between the first, the second, and the third masses. One of the most important constraints in this three-body approximation technique is that the mass of the first body must be much greater than the second, and the second must be much greater in mass than the third. The second constraint is that the orbits must be approximately circular. For the approximations discussed, the first and second masses are significant, and the third is not.

The third body comes into play, not necessarily as a single mass, but as a probe of the effects of the gravitational field from masses one and two. Between the two dominant masses is a point where gravitational forces balance to zero, hence the relative velocity between objects two and three in this reference frame would also be zero. The same occurs at a point beyond mass two, where the net force is zero and the relative velocity to the second mass is zero. If we look at the velocity values for small masses orbiting mass one throughout this region, three other zero-relative velocity points appear for a total of five. These zero-velocity, or stability, points were researched and published by Joseph Lagrange in the eighteenth century and are therefore given the name Lagrange stability points. The five Lagrange stability points denoted, L1, L2, L3, L4, and L5, are positioned as shown in the figure below. While L4 and L5 are considered true stability points and L1, L2, and L3, are quasi-stable, applications have been found for using each of these stable points for understanding orbits throughout the solar system, and for positioning spacecraft for specific missions.

Orbit Precession

Variations in Earth's orbit around the Sun are both subtle and profound depending on the time frame used to describe the changes. Three important changes in the Earth's orbit have a far-reaching effect on the Earth's climate over hundreds of thousands of years but are not noticeable in a single lifetime. These are the rotation of the orbital apse (orbit major axis), with a period of about 114,000 years; the variation in the rotational axis tilt, called obliquity; and the variation in Earth's orbital eccentricity. All three of these have a direct influence on the seasonal heating from the Sun and can induce extreme climate alterations.

Another important variation in Earth's orbit is the rotation of the spin axis that today points at the North Star. The progressive rotation, called precession, of the equinoxes comes from the Sun's gravitational pull on the Earth's oblate shape, since the Earth is not a perfect sphere but a flattened (oblate) sphere due to its rotation. The pull on Earth's equatorial bulge precesses or rotates the orientation of the Earth's spin axis with a period of approximately 26,000 years. The spin axis, which is tilted at an angle of 23.5° is rotated slowly around the vertical, accompanied by an apparent change in the orientation of the fixed stars with the same period. To accommodate this apparent shift in the background star field, precession corrections are made to the positions of

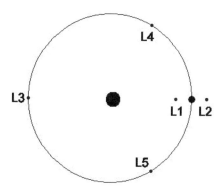

Figure 4.16. Arrangement of the five Lagrange stability points in relation to mass 1 (center) and mass 2 (center right).

stars and planets based on the precessional rate of approximately 0.013°/year. What seems to be a small correction is actually larger than allowable for space flight navigation requirements by a factor of 100 or 1,000, or even more. The correction is also cumulative, in that the precession angle grows with time, and first-order corrections become less accurate with the accumulated error. The correction technique for Earth's precession is best done in small increments and had led to the adoption of the fifty-year epoch for correcting the Earth's axis precession error in order to calculate the true position of stars and planets. The same precession forces act on satellites that are in orbit around the Earth or any other asymmetrical planet or moon, including the Moon's orbit about the Earth. While the Moon's rotation axis does not precess appreciably, its orbital perigee rotates with a period of 8.85 years. A second lunar precession takes place around the ecliptic plane with a period of 18.6 years, and in the opposite direction. This ecliptic period determines the repeated period of lunar and solar eclipses, which repeat twice in this period since the Moon eclipses the Sun at new moon, and is eclipsed by the Earth at full moon.

References

George, L. E., and L. D. Kos. 1998. *Interplanetary Mission Design Handbook: Earth-to-Mars Mission Opportunities and Mars-to-Earth Return Opportunities 2009–2024*. NASA TM-1998-208533. Huntsville, AL: NASA

NASA-JPL. N.d, Basics of Space Flight. Retrieved from: http://www2.jpl.nasa.gov/basics (accessed 4/1/09).

NASA JPL Image Photojournal. Retrieved from: http://photojournal.jpl.nasa.gov

CHAPTER 5

Military Space Operations

First Concerns

Alliances formed to defeat the Nazis in World War II turned from strained to polarized following the detonation of the USSR's first atomic bomb in 1949. Suspicions between the Soviet Union and the United States were aggravated by the defensive posture taken by the Americans because of the attack on Pearl Harbor, and the Soviets because of the Nazi invasion of Russia. As a result, both countries began rearming themselves and their allies with advanced weapons while accelerating spy programs to help determine their adversaries' capabilities. The new world order following World War II was conditional peace, with a deep division between the two new military superpowers. This was the birth of the Cold War.

Secrecy throughout the USSR made America's intelligence gathering difficult, if not impossible. Stalin's brutal purges and cruel dictatorship instilled fear and suspicion in everyone within the USSR, which made personal contacts between the East and the West ineffective in gathering anything more than superficial intelligence. The Iron Curtain, a reference to the Soviet Union's division of Europe after the war, was not a physical barrier, but isolation of the Soviet citizens as well, making conventional spy networks of little use within the USSR. Collecting enough details of the Soviet military programs to make a reasonable evaluation of their military threats was challenging because of Stalin's firm grip on the entire country. And without an effective spy network, the United States would have to rely on new intelligence methods.

Nuclear weapons stockpiles were building rapidly on both sides in the early 1950s, which made military intelligence even more critical. At that time, President Eisenhower had placed emphasis on the Air Force's long-range nuclear bombers for defense and deterrence, and initiated development of nuclear cruise missiles. These programs could not furnish intelligence, while high-altitude aircraft like the U-2, which could evade antiaircraft missiles and interceptors, were only concepts a decade after the end of World War II. One proposed solution to the intelligence-gathering problem looked to the use of entirely new technologies. Captured German V-2 missiles were undergoing modifications for space research and also being improved to help develop long-range ballistic missiles that were attractive for a number of reasons, including the delivery of nuclear warheads over existing defenses. With enough thrust, the missiles could also be used for launching satellites into orbit, more specifically, reconnaissance satellites that could photograph the opposition's defenses and offenses and that would be impossible to intercept.

Although all were under development, President Eisenhower was much more concerned about closing the intelligence gap with a spy satellite in the early Cold War than he was about advancing nuclear delivery missiles or competing with the Soviet Union in launching the first satellite. Eisenhower's secret reconnaissance satellite development program was not revealed until decades later when the collapse of the Soviet Union made continued classification of some of the sensitive documents unnecessary.

The first studies of reconnaissance satellites in the United States were made by the RAND Corporation in 1946. RAND, an acronym for research and development, was an offspring of Douglas Aircraft and continued satellite studies into the early 1950s. These early studies proved timely, since three important new technologies were emerging in the 1950s. First, nuclear weapons designers produced small enough warheads to be carried on long-range missiles. Second, intermediate-range and intercontinental ballistic missiles were progressing toward their initial test flights in 1953. Third, artificial satellites were under development for launch in the 1957–1958 International Geophysical Year. Spurred by the deepening Cold War, President Eisenhower and the Department of Defense began investing resources in all three technologies.

Much of the resources and funding for the Air Force were committed to its strategic air power mandate during the early Cold War, although Air Force planners did begin a slow, deliberate program to develop a nuclear-capable ICBM in 1951. The program, named MX-774 and later renamed Atlas, was reenergized with the detonation of the first Soviet hydrogen bomb in 1953. America's first intercontinental ballistic missile (ICBM) served as a versatile military and civil spacecraft booster and is still in production today as the Atlas family of medium- and heavy-lift launchers.

Soon after the Atlas program was hurried toward completion, a number of other ballistic missile programs were started, including the Titan ICBM, the Jupiter intermediate-range ballistic missile (IRBM), and the Thor IRBM. Other missile projects were also in progress, but, these four were important because of their ability to orbit surveillance satellites simply by adding an upper-stage rocket. Progress on reconnaissance satellite projects trailed work on long-range ballistic missiles in the early 1950s, but these too were also being prepared for the first available launcher.

SPACE MEETS THE MILITARY IN 1955

One of the pivotal years in the development of America's Cold War military strategic policy was 1955. This was the year that both the United States and the Soviet Union had proposed orbiting a scientific satellite in support of the 1957–1958 International Geophysical Year (IGY). President Eisenhower had approved America's first scientific satellite project, but was reluctant to use a military missile for the launch. Even as the launch date approached, he was reluctant to launch a satellite before the Soviets. Both could be interpreted as a provocation toward our Cold War enemy since the laws of spacecraft overflight had yet to be established. Even so, satellites and satellite launchers were attractive to the administration and the Department of Defense, since reconnaissance satellites offered the only meaningful tools of intelligence on the missiles and bombers inside the Soviet Union. U-2 high-altitude reconnaissance aircraft, which first flew inside the Soviet borders in 1956, were successful, but they were limited in coverage; risky; and most of all, a diplomatic nightmare (Day 2007).

The legality of satellite overflights of foreign countries was not a minor point in the Cold War era. In fact, its implications were important enough to establish important U.S. space policy for the following decade. Surveillance of the Soviets in the Cold War years following the division of the East and the West was absolutely necessary, but along with the advantages that came with satellite reconnaissance data were the serious complications directly related to Cold War aggression. A breakthrough came with the release of two reports in 1954 and 1955. The first was the RAND report entitled "Project Feed Back" which supported the concept of satellite reconnaissance. The second was an even more propitious report by the civilian Technological Capabilities Panel appointed by President Eisenhower that recommended development of a scientific satellite in support of IGY. Just as important as these two reports, was the recognition that the problems affiliated with international law and the spy satellite project could be solved by first launching a scientific satellite. A part of that solution was a new policy concept aptly named "freedom of space" that was quickly adopted to counter any future protests from the Soviets concerning satellite launches and overflight. That policy, at least in theory, could tie the first scientific satellite missions to any orbital launches that followed. It was this fundamental connection between scientific orbiting spacecraft and the right of overflight that was essential for planning the first U.S. satellite launch, the first space exploration missions, and the first spy satellites (Day 2007).

RECONNAISSANCE SATELLITES

Basic concepts associated with "freedom of space" and the recommendations of the Technological Capabilities Panel were approved by Eisenhower in 1955. The clever solution of using a preemptive scientific satellite was conceived by Robert Bissell of the CIA and served as the genesis of the military spy satellite project that soon followed. The reconnaissance satellite program, called Corona, and headed by Bissell, began first as the Strategic Satellite System proposal request that was released by the USAF in 1955. Lockheed was the winner of the contract (and most surveillance spacecraft projects since), and in 1956, began work on a preliminary reconnaissance satellite project called Pied Piper (Peebles 1987). While there was interest in the project at the highest levels, progress was slowed by the lack of funding. But the program's fortunes changed abruptly with the launch of the Soviet satellite Sputnik in October, 1957. It is worth noting that a Soviet satellite launch was anticipated in 1957–1958 by the administration and by the intelligence community, as well as by the scientists affiliated with the IGY. What were not expected were the huge ICBM booster and the advanced technology that the Soviet Union displayed on its first and subsequent Sputnik launches. These advances by a nation that was thought to be well behind the Western world in advanced technology were sobering. The response in the United States was covert and swift, and placed a new priority on the reconnaissance satellite projects and launchers. The launch of Sputnik made it obvious that the Soviet nuclear threat was real and that there was little intelligence to construct rational defense strategies with, at least beyond the limited U-2 coverage of the USSR. In a truly interesting twist, these Sputnik launches worked against the perception of President Eisenhower's as a leader concerned about the Soviet space technology advances, yet they freed him to take the necessary steps to launch the spy satellites for assessing Soviet aggression and technology.

Following the launch of the more advanced Sputnik 2 in November, 1957, Eisenhower approved the reorganization of the struggling USAF reconnaissance satellite project identified as WS117-S and the military agencies and committees involved with satellite reconnaissance (Peebles 1987). Sputnik 2 even pushed Eisenhower to designate the Redstone-Explorer vehicle as the backup satellite launcher for the Naval Research Lab's Vanguard, which was chosen as the "civilian" IGY satellite launcher.

To help accelerate the satellite surveillance program, the Air Force's Atlas ICBM, which was still under development, was replaced with an existing Thor booster lower stage topped with an Agena upper stage. Concurrent with the booster preparations was the decision to equip several surveillance satellites with high-resolution film cameras with canister reentry capsules attached, some with electronically transmitted image data. Film canister capsules were constructed for parachute landings to avoid inevitable technology delays in perfecting electronically transmitted images. Electronic cameras had the advantage of offering real-time image data, but the disadvantage of high data rates that required communications links through only a few ground stations (Peebles 1987). The simpler film capsule drop could be made by parachute, or, to prevent difficulty recovery in inaccessible areas, by aircraft capture during the capsule's parachute descent. The canister-parachute capture technique was implemented with several Air Force C-119 "Flying Boxcar" aircraft outfitted with capture devices to snag the capsule and parachute before its touchdown or splashdown.

The Air Force had undertaken its own reconnaissance satellite program, which was more diverse than the film canister deorbit and retrieval technique, but with less success. The USAF Satellite and Missile Observation System (SAMOS) was actually a combination of three film canister reentry systems of varying resolution and design, and an electronic scanning device that generated image data without having to deorbit the capsule and recover the film.

Thor

The early USAF surveillance satellites, code-named Corona, were launched in high-inclination polar orbits to cover all continent and ocean surfaces from pole to pole. This required a new launch facility on the West Coast, since high-inclination orbit launches are not possible from the East Coast. Any launch inclination greater than 57° would not be allowed since the trajectory would pass over heavily populated areas in northeast United States and Canada. The West Coast site chosen was the Camp Cook U.S. Army base in Santa Barbra County, some 150 miles north of Los Angeles, already an active USAF Thor IRBM installation. The new West Coast site, later designated Vandenberg Air Force Base, was constructed specifically for military and high-inclination missions. Vandenberg's first launch after its conversion from a Thor IRBM missile installation was, ironically, a single-stage Thor ballistic missile on December 16, 1958.

The first reconnaissance satellites launched used the Thor for the first stage, since it was already in operation as an IRBM. The second stage was an Agena booster that housed the surveillance camera and film equipment. First launch of the Thor-Agena from Vandenberg on February 28, 1959, was named Discoverer 1 and was part of the early Corona satellite series. This was the beginning of both the Thor's and the Agena's pivotal role in America's first satellite reconnaissance programs and in the early space exploration programs. The first five launches of the Pioneer spacecraft to the Moon were made on the Thor-Able boosters between 1958 and 1960, albeit without success. Agena, which was launched as an upper stage on the Thor, the Atlas, and Titan boosters, was also used as an upper stage and a rendezvous vehicle in NASA's Gemini program. Over 360 Agena vehicles were launched in support of the military reconnaissance projects and the early civil space exploration programs.

Thor's S-3D main engine was adopted from the Jupiter rocket, which gave it the same approximate performance. Thor's structure had a thinner profile than the Jupiter missile, although both were intermediate-range nuclear missiles with origins in Wernher von Braun's Redstone rocket. Thor was also adopted as a tactical nuclear missile for the European theater, since it could reach Moscow from England and could be transported by cargo aircraft instead of by ship (Peebles 1987).

The Thor was later augmented with solid boosters and became an attractive space launcher for NASA's early interplanetary missions because of its low cost, its availability, and its relatively high payload capacity. Thor-Delta, a variation on the Thor-Agena but with four stages, was first ordered as a trial launcher for NASA's early space exploration efforts. Variations in the upper stages, the main engines, and the solid strap-on rockets evolved into the Delta launch family that now includes a much larger Delta IV Heavy launcher.

Thor's S-3D Rocketdyne engine underwent extensive redesign by Rocketdyne and was adopted for use on the first stage of the Saturn 1 and the Saturn IB launchers for the early Apollo program test flights. Even though the H-1 was not used as a prototype for the much-larger Saturn V F-1 engine, its functions and systems proved useful in the F-1's development.

Key Hole

Corona, a satellite family that included a variety of film and electronic surveillance spacecraft, was first launched as the Discoverer series to help obscure its mission. Discoverer's inaugural launch on the Thor-Agena booster January 21, 1959, ended in a failure before launch. A similar rash of failures plagued the Discoverer series in its early hardware and operations test flights, as well as the camera operations. The first booster-satellite success was Discoverer 14, launched August 18, 1960. The satellite was successfully placed into orbit, then deorbited for a successful capture and recovery. Discoverer 14 was also the first to carry a camera and film, and return images of the Soviet Union. These first camera-film devices were later designated Key Hole, which, along with a series number, was contracted to KH, followed by the sequence identification. KH reconnaissance satellites were also affiliated with the name Corona.

Discord among the Department of Defense, the CIA, the USAF, and the White House in 1960 was enough of a problem to assign the high priority reconnaissance satellite program to a new agency called the National Reconnaissance Office (NRO), which still exists today. Concurrent with the creation of NRO was the development of the first electronic imaging satellites under the USAF SAMOS designation. Design of the new electronic imaging device was based on a modified film camera mechanism that generated images from film that were then scanned with an optical device. Scanned data were then transmitted through ground stations as the satellite passed overhead. The SAMOS electronic camera, a derivative of the original Pied Piper system, saw limited success, with only a dozen launches between 1960 and 1972 (Peebles 1987). Nevertheless, its systems worked well on all five of

Figure 5.1. Launch preparation of the Thor-Agena A booster used to launch the Discoverer 6 from the Vandenberg AFB, California, in 1960. The Discoverer satellite was integrated into the Agena upper stage, along with a separate capsule for film recovery. Source: USAF Archive.

NASA's Lunar Orbiter spacecraft flown between 1966 and 1967 that returned high-resolution images of prospective landing sites for the Apollo manned missions.

America's first-generation reconnaissance satellites consisted of the Discoverer-Corona and SAMOS spacecraft outfitted with photographic cameras and electronic scanners. The relatively small Discoverer camera systems were roughly 750 kg in mass and provided the first reliable images of the Soviet Cold War threats. The USAF SAMOS satellites, which began roughly a year after Discoverer, were more than twice the mass of Discoverer and required the larger Atlas-Agena boosters. Both satellites were targeted almost exclusively at Soviet Union missile installations because of the seriousness of the threats. And while the threats were real, Soviet inventories of nuclear-tipped ICBMs were often inflated by American intelligence reports, which served to aggravate the arms race, requiring even greater reconnaissance efforts.

Discoverer flights typically lasted two to four days, which corresponded to the time required to expose the 9,146–10,670 m (30,000–35,000 ft) of film before the canister was ejected for recovery and analysis. But improvements in ground-track coverage and image detail were planned almost from the first flights. Image coverage of

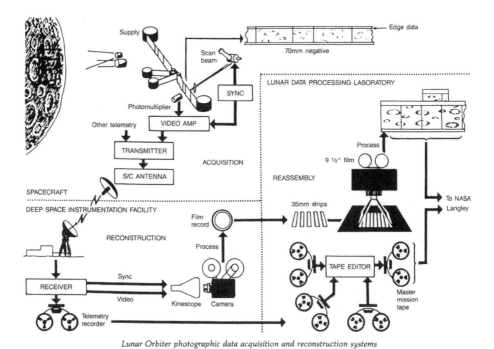

Lunar Orbiter photographic data acquisition and reconstruction systems

Figure 5.2. Functional schematic of the Lunar Orbiter camera and telemetry system that was adapted from the USAF/NRO SAMOS reconnaissance satellite system. Source: NASA SP-480.

the surface area in a given time was a function of launch frequency, which determined program cost. Higher resolution images also required larger-format cameras, since camera resolution is proportional to aperture size—the lens diameter. Aperture size, in turn, dictates the mass of the camera and its supporting film and equipment. Camera mass determines the booster size, which also equates to mission cost, but even more so, since improvements in the needed technologies take time and development funding. Yet, not all resolution improvements were made with larger cameras. Image resolution was improved by taking the satellites to lower altitudes in the eccentric, high-inclination orbits used for surveillance overflights. Improvements also came from advances in optical and electromechanical components.

The growing mass of the American reconnaissance satellites required larger and larger boosters. With the Atlas unavailable until 1962, missile designers turned to upgrading the available launchers. The Thor-Agena pair had a payload of about 1,000 kg (2,205 lb), depending on orbital altitude. By adding strap-on boosters to the first stage, the utility of the Thor-Agena could be improved beyond the increased payload capacity. Since the Thor-Agena combination had already proved itself in its operational success and had low launch costs, the booster-augmented launcher could be, and was, used to expand coverage of the reconnaissance fleet for the growing Cold War threats and the specific, critical events such as the Berlin crisis and the Cuban missile crisis.

Thor was augmented with solid rocket strap-on boosters and renamed the Thrust-Augmented Thor (TAT), which increased its payload capacity for launching larger Discoverer satellites. Solid rocket booster augmentation of the Thor did not end with the

TAT. Thor's strap-on solid boosters proved effective and inexpensive, enough so that even more solid boosters were added, along with larger propellant tanks. This combination is what succeeded in transforming the Thor IRBM into the Delta launcher.

Concurrent with the early Thor-Agena launches were the USAF Atlas trials in 1961 that would prepare the new missile for its multipurpose roles as an ICBM, as an Agena-Discoverer reconnaissance satellite booster, and as a manned launcher for the second phase of the Mercury missions. Test flights of the Atlas-Agena began in March 1962 with the successful launch of the larger 2,000 kg Discoverer from Vandenberg Air Force Base (Peebles 1987).

Discoverer flights ended in April 1964 after seventy-eight missions, but the reconnaissance program continued with changes in satellites and program secrecy. The first Key Hole satellite series that included the KH-1 through KH-4 models was followed by larger, higher-resolution cameras. These reconnaissance satellites incorporated a similar recoverable film capsule, but, because of their greater weight, they were launched on the Atlas and TAT. An even larger camera was developed for launch in 1966 on the largest booster used at the time, the Titan III-B. The Titan-Agena could carry about 3,400 kg (7,500 lb) into orbit with a camera capable of a 0.9 m (2 ft) resolution (Peebles 1987). These third-generation satellites included detachable film reentry canisters, as well as scanned image data transferred electronically through the ground stations. Infrared imaging allowed these satellites to relay detailed nighttime images of China and the Soviet Union from space. An extended version of the Thor-Agena called the Long-Tank Thrust Augmented Thor (LTTAT) booster was also used for these larger satellites.

Evolution of the successful Corona-Discoverer reconnaissance satellite program was a continual process out of necessity because of the increasing numbers and capabilities of Soviet and Chinese ICBMs. Nevertheless, the end of the era of reconnaissance satellite film and retrieval came abruptly with an unsettling compromise in satellite intelligence when Soviet trawlers were discovered under the landing site of one of the Corona film canisters during the recovery process.

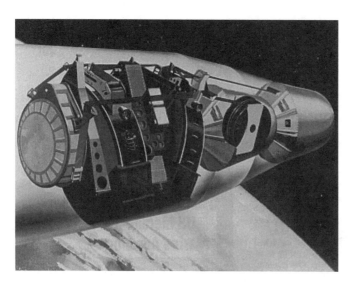

Figure 5.3. Cutaway sketch of a KH-B camera-film reconnaissance satellite showing the multicamera section left of center, with the film storage canister at far left. The exposed-film recovery capsule is shown as the vehicle nose on the right. Source: NRO Archive.

Other models of the Key Hole satellites were launched in the 1960s, 1970s, and 1980s, with various sensor technologies and with varying results. KH-5, code named Argon, consisted of cameras with film-return capsules and electronic scanned film, as well as electronic intelligence missions that were flown between 1961 and 1964. KH-6 (Lanyard) was flown only a few times in 1963, with recoverable film canister payloads. The KH-7 (Gambit) was a slightly more advanced Corona satellite with a camera and film recovery system that was flown between 1963 and 1967. KH-8 was a more prolific reconnaissance satellite series related to the Corona camera and film canister technology, with fifty-four missions launched between 1966 and 1984 on the Titan III booster. KH-8 had the same code name as its predecessor, Gambit (Peebles 1987).

A much larger and longer-lived replacement satellite called KH-9, commonly referred to as Big Bird, was developed to replace the earlier generation spacecraft that required frequent launches. Beginning in the 1960s, the development plan for the new-generation satellite was to incorporate very high-resolution imaging that required only a few flights per year, while using a more versatile real-time electronic data format. Nicknamed for its size and mass, the large KH-9 satellite weighed approximately 13,000 kg (28,660 lb) and carried a 2 m (6.6 ft) diameter mirror, making it comparable in scale to the Hubble Space Telescope. KH-9 "Hexagon" satellites launched from 1971 to 1986 were the first Broad Coverage Photo Reconnaissance satellites that introduced advanced digital imaging technology. Similar designs in later satellites, including the KH-11 (aka, Crystal, Kennan, Big Bird) and KH-12 (unofficial name since NRO removed the conventional KH identification of subsequent reconnaissance satellites for secrecy), were large-optics telescopes with mirror diameters as large as 3.1 m (10.2 ft). The last and largest of the Big Bird series, the KH-12, had an estimated mass of 20,000 kg (44,092 lb), with dimensions of 20 m × 4.5 m (65.6 ft × 14.8 ft), remarkably similar to the space shuttle's payload bay dimensions (FAS 2000).

L-Series

Successors to the massive Big Bird satellites were the advanced technology reconnaissance satellites that no longer carried the Key Hole designation. To avoid compromising mission identity, these and subsequent NRO satellites were simply given a sequential launch number beginning with the NROL prefix (NRO Launch, or L-number). The new satellites are generally smaller than the KH spacecraft and have a longer on-orbit life, but have been plagued with technology and budget problems. The first of these was L-1 (aka Nemesis, USA 179), launched in 2004 as a communications satellite. The spacecraft was placed in a twelve-hour Molinya orbit that lingers over the Asian continent for the longest period of the satellite's orbit. In a twenty-four-hour period, the second twelve-hour orbit has its apogee over the North American continent, conveniently providing that extended period for data download. Perhaps the most notorious of these later satellites was the L-21 (USA 193) advanced reconnaissance satellite launched December 14, 2006. L-21's mission history is a prime example of the complex interplay of advanced technology, politics, and intelligence policy. As a result of the failed launch of L-21, which was part of the NRO 8X program, the satellite wound up in a decaying orbit. Possible compromise of the satellite's technology surviving reentry, if any did, led the defense intelligence agencies and the NRO to plan a missile intercept of the problem satellite before reentry. Destruction of the L-21 satellite by a SM-3 antisatellite missile launched from the U.S. Navy Aegis warship, *Lake Erie*, assured that sensitive pieces of the satellite could not be reassembled or reverse-engineered. A series of press releases claimed the world's population was being protected from 450 kg of toxic propellants surviving reentry, but this was actually a convenient excuse to destroy classified hardware and avoid repeating the calamity of surviving technology, that had occurred with an earlier KH-4A spy satellite reentry mishap in May 1964. After all, similar incidents had taken place hundreds of times earlier with insignificant risk and without the public's knowledge.

America's military reconnaissance satellite naming has gone to a vague designation that is now the USA series, replacing an earlier classification known as OPS. The USA designation continues today and is similar to, and just as ambiguous as, the Soviet/Russian Cosmos spacecraft designation. USA launches began in 1984 with the Navstar-9 GPS satellite, and have totaled 200 military launches as of April 2008.

ELECTRONIC INTELLIGENCE (ELINT) SATELLITES

An important component of defense intelligence is the interception and analysis of electronic communications and related signals using signal intelligence (SIGINT) technologies and techniques. The term signal intelligence involves a broad variety of signal collection and analysis operations that include communications intelligence (COMINT); analysis, measurement, and signature intelligence (MASINT); and several others. Data for these analyses are gathered primarily by low-orbit satellites, although other sources are available from shipborne and airborne platforms.

America's first ELINT satellites were launched for the Office of Naval Intelligence following the shoot-down of the first U-2 spy plane and capture of its pilot Francis Gary Powers by the Soviet Union in 1960. These early electronic-intercept satellite launches employed the GRAB (Galactic Radiation And Background) and GRAB II spacecraft that were announced as astronomical satellite missions. These covert satellites carried additional receivers and data relay equipment to help identify the characteristics of the Soviet radar signals coming from inside Russia. With the transfer of reconnaissance satellite operations to the National Reconnaissance Office in 1962, the early ELINT GRAB and the POPPY satellite were used to relay data from Soviet and Chinese military communications, navigation, and radar signals to secure ground stations. Data were then routed through the NRO to the National Security Agency, then on to the intelligence community.

The National Reconnaissance Office's POPPY satellite system was the successor to the GRAB ELINT satellites that were designed to detect land-based radar installations and to support ocean surveillance. Seven launches of the POPPY on Thor-Agena D boosters took place between 1962 and 1971 from the Vandenberg AFB. While the program was declassified in 2005, its operations and hardware are still classified.

Like the camera/imaging reconnaissance satellites, ELINT and SIGINT satellites have evolved dramatically over the years. Use of these electronic signal reception and relay satellites has blossomed, and they are now used by dozens of nations, both friend and foe. America's ELINT/SIGINT satellites evolved from the GRAB and POPPY spacecraft in the 1960s, to the CANYON, Rhyolite, and Aquacade series, followed by the Vortex, Mag-

Figure 5.4.
Mockup of the GRAB satellite on display at the National Cryptologic Museum adjacent to the National Security Agency at Fort Meade, Maryland. Source: NSA.

num, Orion, and Mentor satellites. These high-orbit and geosynchronous-orbit satellites were succeeded by the JUMPSEAT and TRUMPET satellite series beginning in 1975. TRUMPET and JUMPSEAT were somewhat unique satellites that were placed in Molniya twelve-hour orbits to provide better high-latitude coverage over the Asian continent. The advanced MAGNUM geosynchronous satellites were even more unique in that they were first launched from the space shuttle in 1985 (Richelson 1989).

SOVIET PHOTO RECONNAISSANCE SATELLITES

Extreme Soviet secrecy during the Cold War allowed their spies better access to sensitive information on America's military facilities and operations than the converse. Yet the Soviet leadership pressed for reconnaissance satellites development in parallel with the early civilian orbital satellites. The first of the USSR's reconnaissance satellites was built by Sergei Korolev's design bureau in concert with his manned capsule program because of his command of most space flight projects in the late 1950s and early 1960s. One of his greatest desires was to put cosmonauts in orbit, but clashes with the Soviet military leadership led him to a creative compromise that produced an almost identical design for the Vostok crew capsule and the first reconnaissance satellites called Zenit ("zenith"). The basic Zenit structure consisted of a 2 m diameter spherical capsule with a service module attached for power, thermal control, and communications support while in orbit. In contrast to the Corona canister ejection and recovery technique, Korolev's engineers chose to have the Zenit reentry capsule bring back the film and reusable cameras at the end of its mission. Mission length ranged from four to fifteen days, which was a balance between mission cost and image data return frequency, not the spacecraft's capability.

Combined weight of the spherical reentry capsule and the service module was under 5,000 kg, the approximate lift capacity of the three-stage R-7 ICBM. Five cameras were placed onboard the early Zenits, which included one low-resolution 20 cm (7.9 in) broad-coverage camera and four 100 cm (39.4 in) focal length telescope cameras, as well as ELNIT equipment. To ensure camera orientation during the missions and to reduce blurring, an improved attitude control system was integrated into the spacecraft guidance system (Peebles 1987).

The first Zenit satellites launched were the Zenit-2 series, which totaled eighty-one flights between 1961 and 1970. Later flights switched from Baikonor to the military launch facility at Plesetsk, some 800 km north of Moscow. The classified facility was far enough north to allow high-inclination orbit launches without overflight of major populated areas, although some cities remained at risk. Subsequent Zenit spacecraft were launched on three-stage R-7 derivatives and included the Zenit 2M, 4, 4M, 4 MT, 6U, and 8 models (Peebles 1987). Over 500 of the Zenit reconnaissance satellites were launched between 1981 and 1994, making it the most prolific satellite and satellite launcher in history. Surprisingly, Zenit missions continued for three years after the fall of the Soviet Union in 1991. As with all other military missions and a number of civilian launches, the Zenit series was launched under the obscure Cosmos designation.

Soyuz

Soyuz was designed as a crew vehicle by Korolev and his bureau, first as a manned lunar vehicle, then as space station crew transport. While in its development phase, highly modified versions of the original Soyuz lunar crew vehicle were proposed, then placed under a different branch within Korolev's OKB-1 design bureau. Soyuz P and R versions of the crew vehicle were converted military manned rendezvous vehicles and photo-reconnaissance satellites, respectively. Program changes delayed, and in some cases redirected, these projects until the mid-1970s when a final design of the Soyuz photo-reconnaissance vehicle took shape. In 1975, the first converted Soyuz reconnaissance satellite was prepared for launch with a film return capsule attached, similar in function to the Corona satellites. Like Corona, the multiple film capsule deorbit and capture technique allowed a much longer operational lifetime than the Zenit. This method was attractive to the Soviet planners, since so many of the reconnaissance satellite missions were flown.

Failures plagued the initial Cosmos-Soyuz reconnaissance satellites, although some were successful, such as the Cosmos 905, launched April 26, 1977. This was the first of the Soyuz reconnaissance missions to reach its operational design life of thirty days (Peebles 1987). Mission length increased with newer generations of the Soyuz and other reconnaissance satellites, reaching a maximum with Cosmos 1643 in 1985. This mission lasted over 200 days, comparable to America's Big Bird, while later flights reportedly exceeded one year operational life. Later Soyuz-based military reconnaissance satellites carried the Yantar film and electronic imaging systems that reportedly ended in 2000. The Soyuz-Yantar satellite design was reportedly used for the Resurs DK Earth remote-sensing spacecraft launched in 2006 (Peebles 1987; RussianSpaceWeb n.d.).

SOVIET ELINT SATELLITES

Cold War hostilities spurred the development of new surveillance technologies including electronic listening satellites that could provide a more complete intelligence picture of worldwide military and aggressor activity. For the USSR, only standoff ships and Cuban stations were available to monitor continental U.S. electronic signals. For the U.S., electronic intelligence could be made from nearby friendly nations, but electronic intelligence from the interior of both nations was incomplete. Electronic ears placed on already-orbiting reconnaissance satellites were the earliest and least expensive way to solve this problem.

Listening for specific electronic signals from space is far simpler than acquiring high-resolution images, since the signals are not localized to launch sites, or weapons facilities, or supply caravans. This allows higher orbits to be used for the ELINT satellites than for imaging satellites. The optimal position for electronic surveillance is geosynchronous orbit, although coverage decreases with increasing latitude from the equatorial orbit. Since northern regions of the Soviet Union are not accessible to geostationary surveillance satellites, high inclination orbits are often used for dedicated ELINT spacecraft. Soviet electronic eavesdropping satellites also use the high-inclination orbits, which required a switch from the Baikonor launch site to Plesetsk beginning in 1967. This was the same year that the first of the nuclear-powered Soviet electronic ocean surveillance satellites was launched (Peebles 1987).

New and unique orbit patterns appeared in the Soviet satellite launches in the late 1960s and early 1970s, indicating the ELINT satellites had become operational. Intelligence estimates for these new satellites were three to four launches per year, with an operational life averaging 280 days. Based on U.S. ELINT deployments, it was assumed that the smaller ELINT satellites were primarily used for radar signal reception, while the larger satellites were multireceivers used for other communications and navigation signals. It was also assumed that separate satellites were also developed and launched with infrared imaging systems for early warning of spacecraft and missile launches (Peebles 1987).

SOVIET NAVAL SURVEILLANCE SATELLITES

Ocean surveillance has always been a concern for military tacticians because of the mobility of naval vessels, and because two-thirds of the Earth's surface is covered by water. Ocean surveillance, therefore, requires complete and continual coverage of more than half of the Earth's surface with real-time imaging. Networks of electronic imaging surveillance satellites would seem a reasonable solution except that cloud cover restricts imaging systems to the microwave band. Either side-looking radar or synthetic aperture (imaging) radar can provide cloud penetration, although simpler multiunit radar systems can be used to determine signatures for tracking oceangoing vessels. One of the most serious drawbacks to the radar surveillance satellite system is the high power needed to generate the primary radar signals and relay the high-bandwidth data back to Earth. Another complication with radar data is that the higher orbits needed for greater surface coverage require higher signal power since the signal power required is proportional to the altitude to the fourth power. At low or modest orbital altitudes, tens of kilowatts of electrical power are needed for the signal power, which can be furnished by one of two methods: by large photovoltaic arrays and heavy batteries, or by nuclear reactors. Unfortunately, the Soviets chose nuclear reactors with only marginal safety standards to power their ocean surveillance satellites, some of which are still in orbit today.

Because the USSR is almost completely landlocked, a major effort was made after World War II to expand its presence on the seas with a dramatic increase in naval vessels. Although Premier Khrushchev redirected these efforts toward submarines and smaller vessels after the death of Stalin in 1953, the Soviets wanted to monitor global naval activities to counter-balance its weak navy. The quickest and most direct way to do this was to build an extensive satellite network of reactor-powered radar satellites, beginning with the Topaz reactor prototype first launched in the 1970s.

Topaz was a high-temperature ^{235}U-powered reactor that used thermoionic conversion for generating electrical power. This conversion technique is more efficient than lower-temperature thermoelectric power conversion types, but the high operating temperatures can be difficult to manage. Reactors also employ fissile materials, which generate neutrons in their decay, which accelerates the spontaneous decay rate of the radioisotope. Regulation of the decay rate and the resulting heat produced to generate electricity is accomplished either with absorber material used in terrestrial reactors or with a reflective metal shell typically used in spacecraft reactors. Although the nonfissile radioisotopes have a fixed decay rate and are well suited for small nuclear power sources called radioisotope thermoelectric generators, they have much less energy per unit mass. Fissile materials used in reactors, while extremely hazardous, contain 500,000 times the energy per unit mass than the constant-decay radioisotopes.

Early production Topaz reactors contained approximately 27 kg (59 lb) of 90% ^{235}U as uranium dioxide (UO_2) (Bennett 1989). The uranium fuel was placed in lined tubes within an outer cylinder that allowed ionized cesium gas to flow between the hot interior and the cooler, outer tubes of niobium alloy. Cesium ions would then facilitate the generation of electrical current with their flow between the two conductive tubes. Thermally conductive liquid sodium-potassium circulation loops were used to cool the reactor core and transfer heat to the external copper radiators.

Topaz contained an outer cylinder of beryllium as a reflector, encircled with boron absorbers shaped like rolling pins. These moderator/reflector cylinders were rotated in position outside the reactor core to regulate the flux of neutrons throughout the core, thus regulating its temperature. A critical design flaw in the first Topaz reactors produced an unstable reaction rate that increased with temperature. This made the reactor prone to runaway reaction if temperatures were allowed to increase beyond safe levels. In short, the Soviet Topaz I reactor was susceptible to the proverbial "melt-down." A later Topaz II design was improved in its reliability and stability, and was even considered for integration into the U.S. space power reactor inventory after the fall of the Soviet Union.

Any nuclear reactor in orbit poses a serious atmospheric entry hazard from the release of radioactive debris and gases. Radiation from a deorbiting reactor is more of a concern than the stronger-clad RTG units, since the reactor is not constructed to survive reentry intact, while all but the first RTGs launched are. Reactor breakup during reentry results in at least some of the reactor core materials, as well as the reactor container and secondary radioactive products, surviving reentry. The most infamous case was the Cosmos 954 Topaz reactor reentry accident that spread hazardous radio-active materials across Canada's Northwest Territories in 1978.

To avoid the immediate hazards of reactor reentry, the Topaz reactors that powered the early Soviet ocean surveillance satellites (often referred to as ROSAT for Russian Ocean Reconnaissance Satellites) were designed with three sections that could be separated with pyrotechnics before actual reentry. The first Topaz reactors cores were not only separable, but had their own booster attached to take them to a higher 900 km (559 mi) orbit. At this altitude, the low atmospheric drag would delay reentry for an estimated 600 years, allowing most of the radioisotopes to decay to much safer levels. Even so, the design was not fail-safe, and in December 1977 it became clear that the Cosmos 954 reactor was not going to a higher orbit. Radar tracking indicated that it would probably deorbit sometime in January of 1978 (Peebles 1987). The Nuclear Emergency Search Team (NEST) was alerted for the eventual disaster, then deployed for the January 24, 1978, reentry. Cosmos 954's reactor debris scattered over a region 805 km (500 mi) long in mountainous, snow-covered, inaccessible terrain near the Great Slave Lake in the Northwest Territories. Debris ranging from small particles to large 20 kg fragments required meticulous cleanup, local evacuations, and heavy shielding for transporting the debris to confinement facilities. Cleanup of the nuclear reactor debris was laborious, but mostly complete by April, after which a bill was sent to the Soviet Union for more than $6 million. The Soviets refused to pay for their satellite's cleanup until December 1980, and then only half (Peebles 1987).

Soviet intransigence in its risky nuclear programs continued, and although their nuclear reactor launches were halted briefly, reactor launches were restarted in 1980. Within two years another reactor core separation failure occurred with a reactor reentry incident. The Soviets first denied their Cosmos 1402 was in trouble, then denied that reentry of the core that was separated from the reactor would pose a risk to the general population. In the end, the Soviets accused the West of fabricating the incident to aggravate the already-hostile Cold War. Reentry of the Cosmos 1402 satellite took place on January 23, 1982, landing in the Indian Ocean without further incident, although the NEST team was placed on alert. Because of its lower drag, the reactor core reentry did not occur until February 7, 1982. Fortunately, the impact was in the Atlantic Ocean, which dispersed the radioactive materials over more than 100,000 km² before sinking to the ocean floor (Peebles 1987).

As many as forty of the Soviet reactor-powered ROSAT satellites were launched between the 1970s and the collapse of the Soviet Union in 1991, with the majority of the reactor cores now orbiting near 900 km. These spent 300–600 kg nuclear cores are still highly radioactive and present a potential collision hazard while in space. But these reactors' most problematic feature is the liquid sodium-potassium coolant that leaked out after separation from its host satellite. The large and small droplets of radioactive coolant that are slowly deorbiting present a minor hazard to the public since each droplet will vaporize and disperse on reentry. The greatest risk is that during their lifetime, their decaying orbit and gradual breakup will generate widespread orbital debris. Some of the larger droplets are greater than 5 cm in diameter, and weigh hundreds of grams. As they collide with other debris and themselves, their numbers and their risk to other spacecraft increase exponentially.

The Cosmos 954 reentry catastrophe initiated a number of studies, position papers, and international agreements on space operation safety that centered on nuclear power systems in Earth orbit. Participating nations drafted criteria for space operation safety through the United Nations, and submitted guidelines that addressed agreements made in the United Nations Committee on the Peaceful Uses of Outer Space (COPUOS), and in the Working Group on the Use of Nuclear Power Sources in Outer Space. Safety guidelines for nuclear power sources including both radioisotope and reactor sources have been established by international agreements, with stringent design, operation, and safety criteria. Basic rules for radioisotope sources launched into Earth orbit or into deep space must provide for containment, immobilization, and recovery. Nuclear reactor rules and requirements are far more rigorous since the hazards of reactors pose a far greater risk than radioisotope units (RTGs). Operation and design specifications were established for nuclear power sources to include the following criteria:

- Criticality—power generation in the critical state is allowed only after reaching orbit
- Required boost to high orbit at the end of life, or after an accident/failure to allow radioisotope decay before reentry
- Backup reentry design for dispersing core materials as fine particles that will burn up in the atmosphere
- Contamination levels in the atmosphere and on land below acceptable dose-equivalent limits for the general population of 5 mSv (0.5 rem) per year [the average dose from background radiation is approximately 2.5 milli-Sieverts (mSv)]
- Additional safety evaluation for nuclear-powered spacecraft launches is extensive, and in the United States includes a nuclear safety review process impaneled with representatives from the DoD, the Department of Energy (DOE), and NASA. The safety review also requires a presidential endorsement (Angelo and Buden 1985).

Nuclear powered spacecraft accidents have occurred on several U.S. missions, but generally with only minor environmental impact. Of more than 20 RTG-powered spacecraft launched by the United States, three have reentered Earth's atmosphere. The first incident was a SNAP-9A RTG launched on a Transit 5BN-3 U.S. Navy navigation satellite, the third of three nuclear-powered Transit satellites. A launch abort from Vandenberg on April 21, 1964, placed the vehicle on a suborbital reentry 45–60 mi in altitude. Atmospheric sampling indicated that the RTG unit burned up in the atmosphere as designed. A second RTG accident was on an aborted Nimbus-1 launch from Vandenberg on May 18, 1968. The SNAP-19 RTG survived the abort destruct initiated at 30 km and splashed down in the Santa Barbara Channel. Its recovery five months later showed the RTG to be intact, including the graphite ablator that was used to help survive reentry. No radiation contamination leaked from the RTG unit (Stokely and Stansbery 2008). The same SNAP-9 RTGs were also used on the Pioneer 10 and 11 spacecraft, as well as the Viking I and 2 landers.

The third RTG reentry incident was a SNAP-27 fuel cask (SNAP is the acronym for Space Nuclear Auxiliary Power) that was attached to the Apollo 13 lunar module. The SNAP-27 RTG that was to be inserted into the RTG after lunar landing produced approximately 70 W of electrical power at startup and was used to power experiments placed on the lunar surface by the astronauts. Because the LM was retained on the command and service module as a life boat on Apollo 13 after the oxygen tank explosion, both the service module (SM) and the LM were separated from the command module before reentry. Nearly all of the LM and SM burned up in the atmosphere, but the SNAP-27 RTG survived reentry, as designed. Apollo 13's SNAP-27 RTG now resides at the bottom of the Pacific Ocean south of the Fiji Islands.

One space reactor was launched by the United States on April 3, 1965, and placed into a circular, polar 1,300 km orbit. The SNAP-10A reactor powered the military SNAPSHOT (OPS 4682) reactor test spacecraft with approximately 500 W electrical power for forty-three days. A power failure on the spacecraft initiated a reactor separation and boost to a higher altitude. That orbit will decay slowly, with an expected reentry in approximately 4,000 years. Radar tracking of the SNAPSHOT has identified more than sixty debris pieces from the original spacecraft, substantiating the serious concern that the greatest hazard of reactor-powered satellites is not necessarily the radioactive fallout, but the profusion of spacecraft and reactor debris (Stokely and Stansbery 2008).

DEFENSE EARLY WARNING SATELLITES

While missile-launched weapons are a concern for everyone, peaceful spacecraft launches are generally embraced as worthy scientific endeavors. Nevertheless, identification of launchers and their payloads as friend or foe can approach the impossible, since weapons are launched on the same or similar trajectories as suborbital scientific launches, and may be placed in the same orbit as peaceful satellite missions. The same or similar launcher may also be used for both military payloads and civilian spacecraft, as was common in the Cold War. In addition, the same launch sites, even the same launch pads, have been used for both military and civilian rocket launches. Therefore, trying to identify specific launches as hostile becomes a problem involving a wide variety of variables that include but are not limited to the launch site, the time of launch, launch azimuth or inclination, launch trajectory, and booster signatures. This complex task of analyzing launch events is often accomplished using satellite networks that can furnish three-dimensional imaging for computing missile trajectories, with infrared spectral imaging data to identify rocket exhaust plume features.

The first of the U.S. early warning satellite systems was the MIDAS (Missile Defense Alarm Systems) satellites launched soon after the first reconnaissance satellites as part of the USAF WS-117L project. MIDAS satellites were first flown on the Atlas-Agena boosters in 1960 and became operational in 1963. The active element of the MIDAS was an infrared scanning imaging system placed in the nose of the Agena to detect high-temperature rocket plumes above the limb of the Earth (Peebles 1987). But by 1963, the MIDAS satellites had exposed a number of weaknesses, including the high costs of continual observations of most of the Soviet Union. Some of the most difficult issues were the highly complex systems needed to identify missile plumes, and the inconsistent results analyzing the plume signatures. To help solve these and a host of related problems, improved IR imaging satellites were placed in geosynchronous orbits. By comparison, the early MIDAS operated in an orbital range of 2,000–6,500 km.

New early warning satellites were developed under the name Program 949, which continued to use the Agena booster with a smaller spacecraft and IR imaging sensors cooled with cryogenics for greater sensitivity. An early warning satellite project similar to Program 949 was the Defense Support Program (DSP), also known as Program 647. DSP satellites that were launched on Titan III-C boosters beginning in 1970 used a Schmidt (folded mirror) telescope and a continually rotating spacecraft for improved IR imaging that helped reduce false alarms. Even though the missile gap had closed before the 1970s, the early warning capabilities of the new DPS proved useful for strategic arms limitation compliance, in addition to its primary use as a missile early warning network (Peebles 1987).

Figure 5.5. View from the aft flight deck of Atlantis of the DSP 16 satellite deployment on STS-44. The IR Schmitt telescope shroud is seen at the top of the tubular telescope shroud. Source: NASA GRIN.

DSP satellites, which make up the current Satellite Early Warning System, were also launched aboard the space shuttle using an Inertial Upper Stage (IUS) solid booster, one of which was the DSP 16 aboard STS-44 in 1991 shown in figure 5.5. Geosynchronous orbits for the DSP satellites were somewhat unusual because of their mildly eccentric slightly inclined orbits, which provided increased surface coverage.

The highly successful DSP program has undergone refinements since its first launch more than three decades ago, and it is currently being replaced by a more advanced and more expensive Space-Based Infrared System (SBIRS). The SBIRS system is an integrated collection of IR early warning satellites in geosynchronous orbits, low Earth orbits, and highly elliptical orbits. Some of the DSP technology is being utilized in the new SBIRS spacecraft, along with more advanced systems and sensors that have contributed to the program's cost overruns.

DOD WEATHER SATELLITES

Weather satellites proved to be a powerful tool almost immediately after their introduction in the 1960s. This led to weather data from NASA's spin-stabilized polar-orbiting TIROS weather satellites being shared by civil and Department of Defense agencies—an unusual arrangement that ended in 1965. The split was not due to rival interests but because the broad-area TIROS coverage was useful for synoptic weather forecasting while higher-resolution data were needed for planning military reconnaissance satellite coverage. The two distinctly different program requirements led to the development of the lower-altitude, higher-resolution Defense Meteorological Satellite Program (DMSP) satellites first launched in 1965 by the Air Force. Thor-Altair boosters were used to launch the DMSP satellites into sun-synchronous 483–725 km (300–450 nautical mile) orbits with an inclination of almost 99°. The popular Altair rockets were small solid-fuel upper stages used on the Vanguard, as well as the third stage of the early Delta boosters and the fourth stage of the Scout.

Instrumentation onboard the DMSP included visible and IR image sensors, in addition to gamma-ray detectors and charged particle detectors that were added in support of the nuclear weapons detonation detection program. Space environment measurements of solar activity and radiation belt changes made on the DMSP were also used to evaluate interference in space communications and ballistic missile radar signal propagation. Since the polar-orbiting meteorological satellites operated by the National Oceanic and Atmospheric Administration (NOAA) had similar capability, the separate polar-orbiting satellite operations were merged in 1994. A tri-agency organization was created for the program that included the Department of Commerce (DoC, NOAA's directing agency), the Department of Defense, and NASA. The new program, known as the National Polar-Orbiting Operational Environmental Satellite System (NPOESS), was structured to meet the environmental monitoring needs of all three agencies, while replacing the DMSP program. Operation of the DMSP, previously managed by the USAF, now resides under the Department of Commerce, although program improvements, support funding, and data distribution to the defense agencies remain under the USAF.

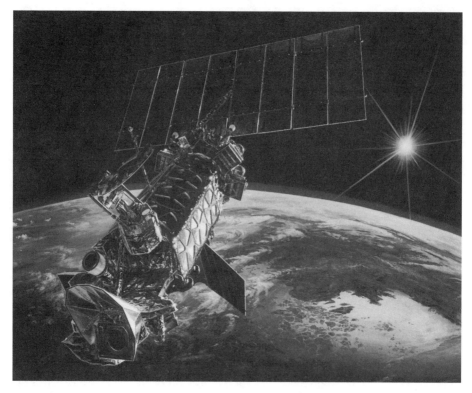

Figure 5.6. Sketch of the DMSP Block 5D-2 meteorological satellite in orbit. Source: DoD.

DEFENSE NAVIGATION SATELLITES

Sputnik 1 had a profound impact on American military planners for several reasons. Probably the most sobering was the realization that the Soviet Union had advanced their missile and weapon systems well beyond any other country, including the United States. Confirmation of the U.S.-USSR missile gap had to wait until surveillance satellites could verify Soviet missile facilities and its missile installations. Along with reconnaissance satellite development projects, the Department of Defense also had to construct its own advanced long-range missile and weapons systems to first match then surpass the Soviets. Supporting facilities for the new American IRBM and ICBM weapons included communications networks and navigation facilities that could furnish tactical data to help counter the Soviet threats.

The first American space navigation system was the Navy TRANSIT system, also known as the Navy Navigation Satellite System (NAVSAT). Triangulation and timing of signals from known satellite positions were the heart of the system—similar to today's military NAVSTAR Global Positioning System, better known as the Global Positioning System, or GPS. The network of TRANSIT satellites supplied worldwide, two-dimensional position coverage using computed ephemeris satellite locations in specially equipped receivers. Submarines that were restricted to slow, shallow-water signal transmissions and reception could use the TRANSIT system for navigation with 200 m accuracy after approximately ten minutes of integration (Banther 2004).

With all its advantages, the space-based TRANSIT system could not provide three-dimensional navigation and was therefore limited to surface operations. That limitation, and its relatively slow operation, spurred research into an advanced space navigation system using precise timing techniques with multiple satellites. Independent of the TRANSIT project, the Naval Research Laboratory and the USAF Space and Missile Systems Organization began development of a more accurate navigation system called TIMATION (TIMe navigATION), using the same timing concept as TRANSIT, but with quartz clocks for more precise signal-timing accuracy (Banther 2004). Still, these early space navigation systems developed in the 1960s could not provide three-dimensional positions.

By 1970, the space-based military navigation problem had become a major concern, and in 1973, the Department of Defense approved the NAVSTAR Global Positioning System. The multiphase GPS program started with proof-of-concept satellite operations using refurbished TIMATION satellites with the first atomic clocks launched in space (Banther 2004). Full-scale development of NAVSTAR/GPS began in 1982, with satellite technology that included:

- Non-military use supplied by an unsecured channel known as Select Availability, or SA
- Stable atomic clocks
- Hardened electronics
- Signal encryption
- Worldwide coverage except for the polar regions (> 80° latitude)

The GPS navigation system's near-global coverage is supplied by twenty-four satellites placed in six orbital planes, each rotated 60° from the adjacent orbit, each at an inclination of 55°, with four satellites carefully spaced in each orbit. Because the four GPS orbital planes have an inclination that is considerably less than 90°, signal coverage for multiple satellites is limited to about 80° in latitude. And because the satellites have an orbital altitude of 20,200 km (12,600 mi), signal reception is limited to orbits below that altitude, since the GPS antennas are directed downward. Backside signals received by deep-space spacecraft would be of little use for navigation because of the small angles separating the satellites at large distances.

The space-based NAVSTAR/GPS navigational system has two types of accuracy: P and C/A. Agreements between the DoD and civilian agencies, including the Federal Aviation Administration, have resulted in the C/A having the same accuracy as the military P-code, although the accuracy of the C/A signal is under the control of the DoD (the C/A signal is the S/A civilian signal that can be turned on and off). Lower accuracy of the C/A signal is created by adding pseudo-random noise to the unencrypted signal known as L1.

Figure 5.7. Artist's sketch of the Block II GPS spacecraft. Source: NASA NIX.

- Military = P (precision) code, which has a rms (root mean square) accuracy of approximately 10 m horizontally.
- Civilian = C/A (coarse/acquisition) code, which has a rms accuracy of approximately 100 m horizontally and 300 m vertically.

Operational frequencies for the military NAVSTAR/GPS navigation signals are in the L-band microwave band, with the primary frequencies identified as L1 and L2.

- L1—1.2276 GHz (P and C/A)
- L2—1.5754 GHz (P only)
- L3—1.38105 GHz (used by the Nuclear Detonation Detection System payload)
- L4—1.379913 GHz (being studied for additional ionospheric correction)
- L5—1.17645 GHz (proposed for use as a civilian safety-of-life signal)

Beyond the GPS's signal coding and the receiver quality, accuracy for GPS receivers is dependent on the relative position of the satellites, on signal propagation through the atmosphere, and on terrestrial noise. Since signals passing through the atmosphere are degraded by a longer path length, signal reliability becomes a problem for a satellite as its apparent position nears the horizon. Because of this, the practical limitation for reliable satellite signals requires a greater than 5° height above the horizon. And, due to triangulation errors, satellites closer to each other furnish less accurate signal data than satellites with large separations. One type of GPS signal accuracy is computed by the receiver for an error estimate of the horizontal and vertical positions. Other types of error and uncertainty are computed by the individual receivers for geometrical configurations, time errors, and receiver and atmospheric errors. Errors due to satellite and receiver positions are often combined into the resulting estimate of uncertainty, called the Geometric Dilution of Precision (GDOP).

Position calculations for GPS and other triangulation navigation systems use signal timing for each of the three dimensions. One time value would describe a sphere representing a solution for the distance from a satellite/transmitter at a specific time. A second timed signal would also provide a spherical solution, which would also supply a circle of intersecting points on the two calculated spheres, meaning that the receiver is somewhere on that circle. A third timed signal from a third navigation satellite/transmitter would again have a spherical solution, and therefore supply an intersection of a sphere with the previous circle—resulting in a solution of exactly two points. Selecting the correct position then becomes a process of computing the correct geometry. For accurate time, velocity, and acceleration information, a fourth GPS is required. Hence, reasonably accurate GPS coordinates require simultaneous signal reception from four GPS satellites. Having more than four reliable satellite signals improves accuracy by reducing the uncertainty in the receiver's position in a process similar to averaging.

GPS information can also be computed and displayed or relayed for velocity and acceleration data. Typical military and spacecraft systems include acceleration data and often have redundant systems for reliability and improved accuracy. Even greater accuracy is available by combining the GPS solutions with data from the Russian GLONASS satellite navigation system. Even higher accuracy three-dimensional navigation, called differential GPS, is available from one or more nearby transmitters that generate accurate timing signals typically used for aircraft precision and nonprecision approaches and landings.

The Russian satellite navigation system known as GLONASS is also in operation, although not fully implemented as of 2008. Like NAVSTAR/GPS, GLONASS is a multiple-orbit plane, multisatellite system that uses timed signal coding, ephemeris satellite data, and atomic clocks for its operation. Hybrid receivers that are capable of receiving both GPS and GLONASS data can obtain higher accuracy position information from signal redundancy.

A European consortium has also begun implementation of a civil space navigation system called Galileo. The three-dimensional satellite navigation system is being designed to (1) have a higher accuracy than either GPS or GLONASS, (2) provide useful signals at high latitudes, and (3) not suffer the risk of military operational control. Galileo is to have a completed constellation of thirty satellites at an orbital altitude of 23,222 km. Galileo's thirty satellites will be placed in three orbital planes with 56° inclination, with ten satellites in each orbit plane (nine operational satellites and one active spare). Galileo is expected to be operational by 2013.

NUCLEAR WEAPONS TEST DETECTION

Widespread proliferation of nuclear weapons after World War II began to slow once the policy of "mutual assured destruction" reached an equilibrium in nuclear deterrence between the United States and the USSR, and later with China. Follow-up agreements introduced limitations on strategic arms known as the Strategic Arms Limitation Treaty, or SALT. These precarious yet important mutual arms limitation agreements and the subsequent Strategic Arms Reduction Treaty (START) would not be practical without reliable confirmation of weapons numbers and weapons testing. Concerns over SALT and START compliance introduced new technologies and techniques to monitor nuclear weapons and detect nuclear weapons tests anywhere—underground or aboveground. Surprisingly, the most difficult to detect were atmospheric tests because they did not generate seismic waves, and particle fallout did not appear in surface monitors for days or weeks.

America's first nuclear space detonation detection program employed both ground-based sensors and satellite sensors, in addition to live atomic bomb detonation experiments in space to determine the effectiveness of the systems. The reality, more akin to lunacy, of detonating nuclear weapons to detect nuclear weapons, accelerated the efforts to limit nuclear weapons tests, although both American and Soviet nuclear testing continued beyond the early test ban treaties.

Detecting nuclear weapons tests in the atmosphere using spacecraft started with Vela (Spanish for "watchman") satellites designed to carry particle, X-ray, and gamma-ray sensors. Vela satellites were first launched in 1963 into the highest orbits of any military spacecraft, roughly 65,000 km (40,370 mi), corresponding to an orbit period of more than 100 hours (Peebles 1987). This was beyond the Earth's radiation belts and almost a quarter of the distance to the Moon. Sensors on Vela at these altitudes were capable of detecting nuclear device detonations in the atmosphere as well as in deep space. Vela, with its X-ray sensors, could also measure solar eruptions and even high-energy bursts from stars and distant galaxies. Advanced Vela satellites developed and launched in 1967 also included flash analyzers to help determine the yield of nuclear tests.

Vela's detector instruments offered much more than tools for verifying nuclear weapons compliance during its program life. Its contribution to the scientific community was impressive, and included monitoring solar flares for the Apollo and Skylab programs, observation of the first suspected black hole in Cygnus X-1, and the first detection of the enigmatic gamma-ray bursters that are found throughout the universe and were initially thought to be clandestine weapons tests. Several events characterized by the Vela instruments as nuclear detonations have been ascribed to weapons tests in Israel and South Africa, but without corroborating evidence such as radioactive fallout or multiple-sensor confirmation. The versatile and long-lived Vela satellites have also been used to study the character of lightning bolts (Peebles 1987).

Vela programs have been replaced with military satellite systems on unrelated missions carrying compact electromagnetic and particle instruments developed for detecting nuclear tests. The nuclear detonation detection program originally carried on Vela was transferred to the Defense Support Program (DSP) satellites, then in the late 1980s to the Navstar/GPS satellites. The program is now called the Integrated Operational Nuclear Detection System (IONDS).

Manned Military Spacecraft

USAF MANNED ORBITAL LABORATORY

Manned reconnaissance satellites were under study by the U.S. Air Force and the Department of Defense before the first reconnaissance satellites were being developed.

Initial plans were for two separate military manned reconnaissance projects. First was the small manned orbital vehicle capable of hypersonic reentry, the second a multi-crew orbiting surveillance laboratory that would use modified Gemini flight hardware. The orbiter-glider called the Dyna Soar and later designated X-20 was to be launched with the heavy-lift Titan booster being built for the USAF. Because its lift capacity exceeded the Atlas, Titan was also adopted by NASA for its Gemini capsule launcher. Combined, the small Dyna Soar reentry glider and the larger Manned Orbital Development System (MODS) offered versatile space surveillance platforms. These two vehicles would have provided the first crew capability to diagnose and repair equipment problems, and provide surveillance target selection. In a congressional hearing in 1962, Air Force Lieutenant General James Ferguson described the potentially vital role of man-in-space operations:

> Man has certain qualitative capabilities which machines cannot duplicate. He is unique in his ability to make on-the-spot judgments. He can discriminate and select from alternatives which have not been anticipated. He is adaptive to rapidly changing situations. Thus, by including man in military space systems, we significantly increase the flexibility of the systems, as well as increase the probability of mission success (Ferguson 1962).

Part of NASA's charter was to establish cooperative agreements with the Department of Defense and the Armed Forces in its primary role developing and directing civil space exploration programs. There was also an expectation that NASA would benefit from its collaboration with the DoD and the Armed Forces by sharing equipment and expertise. Evidence of this informal but welcome agreement is seen in NASA's Gemini program, which was integrated into the USAF manned orbital projects in 1962 and 1963. Proposals were made to fly Air Force astronauts on the Gemini capsules in preparation for the military surveillance program. Initial proposals were to have NASA astronauts help in training the USAF crews, followed by USAF crews operating both NASA and military experiments. The shared equipment and training was dubbed Blue Gemini because of the Air Force colors. While this was attractive to NASA managers because it brought in increased funding, Air Force involvement in the civilian Gemini space program raised concerns about responsibilities and roles of the two disparate agencies and their missions. In addition, the USAF leadership felt its contribution to the NASA Gemini program could jeopardize funding for the fledgling Dyna Soar program.

Both the Dyna Soar and MODS began development in the early 1960s, and both underwent significant concept redefinitions during their evolution, ranging from a one-man reentry test vehicle for the X-20, to a twelve-man orbital reconnaissance satellite using Gemini transfer and cargo craft to support the MODS. In addition, both had progressed enough for astronaut selection to have started in both programs, beginning with the Air Force Dyna Soar/X-20 program. The group of seven Air Force astronauts in the initial X-20 selection included Neil Armstrong, who also flew in the NASA-Air Force X-15 project, along with two other NASA test pilots.

Dyna Soar was a less ambitious and less costly project than the Manned Orbital Development System, and began earlier. As a result, it reached the prototype stage ahead of MODS in 1963. But even with an added deorbit stage and expanded capabilities, the X-20 Dyna Soar lacked substantial justification for its manned mission. In spite of the Dyna Soar's utility for crew transportation in space, it suffered from the same flaw that plagued all of the Air Force's manned programs: their projects were costly duplications of NASA's manned exploration programs. The shortcoming can be found in the Air Force's penchant for manned operations, from its nuclear bomber fleet to its participation in the space shuttle program. These manned programs came at the expense of automated weapons. As a result, the USAF manned space program efforts have suffered a string of cancellations including the Dyna Soar project, terminated in 1963, and the Manned Orbiting Laboratory, which was cancelled in 1969. The Air Force's participation in the space shuttle program ended after the Challenger accident in 1986, when launch costs and safety requirements forced the Air Force to return to expendable launch vehicles for their reconnaissance satellites.

Termination of the X-20 was consistent with the direction and needs of the USAF at the time, although it may have served as a prototype for the space shuttle had the Orbiter not been scaled up to serve USAF reconnaissance program demands. Nevertheless, the termination of the Dyna Soar in December 1963 was made in the same announcement from Defense Secretary McNamara as the approval of the MODS, renamed the Manned Orbiting Laboratory (MOL). Dyna Soar's legacy did not end with its cancellation, however. Dyna Soar was the first spacecraft that would have employed pulse code modulation, used in nearly every subsequent spacecraft communications system. X-20 project hardware also included the "minimum reaction space tool" wrench that was used in Apollo, Skylab, and even the space shuttle EVA programs (Pealer 1995a). The X-20 Dyna Soar vehicle would also have been the first to use metallic thermal heat protection for hypersonic flight. While the Mercury capsule also used a Renee and titanium shell, it was more closely related to a ballistic vehicle than the X-20 or space shuttle gliders.

In 1964, the USAF Manned Orbiting Laboratory began to take on a more military character to justify it overtaking the X-20 and its expense above and beyond the NASA Gemini project. Operational mission definitions for the Manned Orbital Laboratory came primarily from Defense Secretary McNamara's and NASA Administrator James Webb's decisions that the two programs be developed and operated separately. For the MOL program to survive, it would have to weather several rounds of cost reduction. This translated into the MOL inheriting its basic components and design features from existing or emerging space program hardware. Fortunately, the available launchers and military surveillance hardware were much more mature than just a few years earlier. Ultimately, MOL inherited its primary hardware from both the military's Titan III launcher and NASA's Gemini flight capsule. The laboratory module was to be specified during contractor studies that followed.

Design of the Manned Orbiting Laboratory began as a basic cylinder with a diameter the same width as the Titan's 10 ft payload shroud. For the length, a minimum volume for two crew members was approximated using a 10 ft diameter cylinder with a length of 10 ft. The combined vehicle that included a converted Gemini capsule, the laboratory unit, and shroud, had an estimated mass of 9,300 kg (20,500 lb) which was well under the 25,000 lb payload capacity of the Titan III-C (Peebles 1987).

Figure 5.8. Artist's sketch of the USAF Manned Orbiting Laboratory with the Gemini B crew capsule separating for deorbit. Source: USAF.

Goals defined for the MOL were almost exclusively military, focusing on the development of technology for improving military space capabilities, and the manned assembly and servicing of large space structures. Experiments planned for the laboratory included ballistic missile early warning; ballistic missile defense; satellite detection, inspection, and repair; reconnaissance and surveillance; and nuclear test detection (Pealer 1995a). Early estimates of the program costs for six MOL vehicles, twelve launchers, and operations, were approximately $1.5–$2 billion in 1964 dollars (today more than $10–$13 billion).

Early program projections were for test flights to begin in 1967 or 1968, which constrained the completion of the design and the vehicle integration to three years—four at most. The rushed schedule made it necessary to borrow from Gemini and Apollo technology, and it pushed the Air Force to use a program office within the USAF to coordinate, monitor, and manage the work by the contractors and subcontractors. This unusual management style was effective, but only because the program was terminated early. This agency-level approach was not attempted again on any U.S. manned space flight programs.

One of the first and fundamentally most important decisions in the structural design of the MOL was how the Gemini capsule was attached to the laboratory. The method for attaching the Gemini to the MOL also had to solve the problem of transferring the crew between the MOL laboratory unit and the capsule. Four main methods for attachment and transfer were considered, each with its own significant risk. The first and simplest was a passive attachment of the Gemini capsule to the laboratory. The crew would transfer between capsule and laboratory by EVA at deployment, and again for deorbit (and if needed, for an emergency return). This was the most hazardous of the four options because of the vulnerability of the astronauts during the EVAs. The second was an inflatable transfer tunnel that would also expose the astronauts to the hazards of space, but to a lesser extent, and would require major modification to both the laboratory and Gemini structure. Third was a hinged capsule with matching hatches that could be latched and opened after deployment in orbit. This method also required structural modifications in both the laboratory and Gemini vehicles. The fourth was a heat shield hatch mechanism that allowed the crew to transfer between the capsule and laboratory directly through the center of Gemini's heat shield. This concept was more protective for the crew but compromised the integrity of the Gemini reentry face and heat shield. Since this design represented the lowest risk for the crew and needed only a few minor vehicle structural modifications it was chosen for the capsule-laboratory attachment and integrated into the abort separation mechanisms. Modifications on the Gemini heat shield were tested to verify the integrity of the vehicle and shield on the only MOL hardware launch (Pealer 1995a).

The next development phase for the MOL was determination of the basic functions and the supporting systems for the vehicle. Since electrical power represented one of the largest mass elements of the orbital lab, an effort was made to specify this system early. Fuel cells were an immature technology, but appeared promising as a power source for missions under thirty days. Solar photovoltaic systems were more efficient in mass and cost beyond that, but had more liability in spacecraft dynamics and volume than fuel cells. The decision to use fuel cells for the MOL was based on the expectation of the thirty-day mission length, at least initially, and the anticipated success in developing the new, although difficult, fuel cell technology. Another important decision was the MOL lab atmospheric composition. While pure oxygen at 5 psi could be furnished and controlled with simple systems, the safety hazards and the associated physiological problems of operating in a pure oxygen environment for extended periods suggested a two-gas atmosphere. Nitrogen was inexpensive and well-researched, but could also induce the bends if the astronauts went from a mixed-gas environment to lower-pressure pure oxygen used in space suits. Helium, although not used on manned spacecraft before, was the most attractive alternative and chosen as the secondary gas for the manned laboratory module (LM) atmosphere (Pealer 1995a). The decision on a dual-gas atmosphere was made following the pure oxygen Apollo 1 cabin fire that killed three astronauts in their preliminary countdown test.

Contracts and Studies

Contracts were awarded through 1964 and 1965 for MOL studies and hardware development projects. These included the launch vehicle, the manned laboratory, the modified Gemini B crew vehicle, and the laboratory, capsule, and launcher systems. Naturally, the same contractor supplying the Gemini vehicle for NASA's manned program was chosen as the primary contractor for the Gemini B—the McDonnell Aircraft Corporation. The Martin Company, which built the Titan launchers, was contracted to build and integrate the larger Titan III-M needed for placing the MOL in polar orbit. Since the mass of the completed MOL and Gemini B was now over the orbital lift capacity of the Titan III-C, it was necessary to increase the thrust and/or thrust duration of its first two stages. This was accomplished with the addition of two solid rocket motor segments on the five already used for each of the two solid rocket motors (SRMs) on the Titan III-C, along with an increase in the SRM thrust. Man-rating requirements for the larger seven-segment motors built by the United Technology Center would later lead to the development of the larger solid rocket boosters used on the space shuttle. Titan's second stage, called stage 1 (SRMs were called stage 0), consisted of two Aerojet LR87-AJ-9 liquid-fuel engines. Thrust duration was extended by lengthening the fuel and oxidizer tanks to increase the burn time from 150 to 170 seconds. These and other improvements increased the Titan III-C LEO polar orbit payload capacity from 11,340 kg to 14,515 kg (25,000 lb to 32,000 lb) (Pealer 1995a).

The primary contract for the MOL's lab module was awarded to the Douglas Aircraft Company, with subcontracts let by Douglas for five major elements of the MOL, including (Pealer 1995b):

* Environmental control and life support (Hamilton Standard)
* Attitude and translational control (Honeywell)
* Communications (Collins Radio)
* Fuel cells (TRW Systems)
* Data management (Sperry Rand)

Approval at the Top

Debate on the MOL program's utility and cost ended in 1965 when President Johnson formally approved the project. Implied in the approval was the assumption that the project could support manned operation requirements for the DoD as well as provide manned orbital experience relevant to the NASA mission, at least temporarily. Within a year of its approval, many of the structural details of the MOL lab were finalized. In the year that followed, the first two MOL astronaut selections were announced, and the property adjacent to the Vandenberg launch site was purchased to build the Space Launch Complex 6 (SLC-6) specifically for MOL launches.

At this stage, the MOL laboratory consisted of the laboratory module (LM) and the mission module (MM), which was attached to the Gemini B crew module that would bring the astronauts back after their completed mission. The laboratory module had a diameter of 2.8 m (9.2 ft) and length of 5.8 m (19 ft), with an unpressurized forward compartment 2.4 m (8 ft) in length used for life support and power equipment and instruments. An attached pressurized laboratory compartment was 3.4 m (11 ft) in length. The mission module, measuring 11.2 m (36 ft, 11 in) in length, carried equipment and experiments for specific missions and served as the structural attachment to the booster until separation (Pealer 1995a).

The MOL's pressurized compartment consisted of two cylindrical shells: a smaller, inner pressure vessel, composed of aluminum alloy, and a slightly larger protective outer aluminum alloy shell covered with an aerodynamic shroud for the ascent through the atmosphere. The two concentric cylinders were separated by a vacuum but contained multiple insulation layers. The outer shell was also designed to help reduce radiation exposure and micrometeoroid hazards. Initial plans for the MOL thermal control system called for circulating cooling fluid through the Gemini capsule radiators to reduce costs, but this was later changed to outboard radiators placed on the MOL laboratory.

Guidance, navigation, and control for the MOL were patterned after the Gemini Inertial Guidance System and the Titan III-M booster guidance system, called the Booster Inertial Guidance System (Pealer 1995a). Dual guidance systems were not just a prudent redundancy, but a requirement for man-rating the vehicle. A total of three major propulsion systems were to be used on the MOL, including the Reaction Control System for attitude control, the Retrograde Propulsion System for deorbit and orbit changes, and the Pad Abort System for launch and ascent abort contingencies (Pealer 1995b).

Like NASA's Gemini capsule, the modified Gemini B needed crew escape for launch and ascent emergency aborts. The original Gemini's ejection seats were replaced with a launch escape rocket tower similar to the Mercury and Apollo capsule abort systems. Mass and dynamics considerations reintroduced the dual ejection seat abort configuration, along with additional assemblies to provide release and protection from the Pad Abort System blast (Pealer 1995b).

Victim of the Vietnam War

Agreements as well as disputes among the DoD, the USAF, and NASA planners over the Manned Orbiting Laboratory design served to mold its mission profile and catalyze interagency participation. What began as an experimental manned platform demonstrating the effectiveness of manned reconnaissance during the Cold War became an operational program for that same end in order to justify its expense. Forced separation of the MOL program from NASA's Apollo Advanced Applications Program avoided duplication of efforts, but obligated NASA to contribute some Gemini hardware and astronaut training to the MOL program.

Program delays and budget overruns in 1967 stretched the first manned flights to 1970, then into 1971, and doubled the project costs to more than $3 billion by 1968 (Pealer 1996). Efforts to complete the MOL hardware were progressing successfully, including construction of the Vandenberg launch complex and completion of elements of the Titan M and MOL systems, although the budget continued to bloat. Just as much a problem for the MOL program was the success of the unmanned reconnaissance satellites that uncovered important details of the Soviet missile installations and provided the basis for strategic arms limitation verification. What proved to be the final blow to the MOL project was the federal funding drain by the Vietnam War. Reductions in allocated funds in 1969 stretched the project over a longer period, in turn, inflating projected costs even more.

Cancellation came quickly, and as a surprise to most involved with the program. The announcement was made on June 10, 1969, by Defense Secretary McNamara after it was realized that the cancellation would save $1.5 billion out of the $3 billion estimated cost. The fourteen astronauts remaining in the program were interviewed for possible transfer into the NASA astronaut corps. Of the fourteen MOL astronauts, eight were accepted and integrated at various levels in NASA's space flight programs.

Even though the MOL project hardware had been shared, little use was made of the equipment. One exception was the Titan III-M's booster design that was used to improve the Titan IV and subsequent models. Titan's SRMs were also used as a preliminary design for the larger space shuttle solid rocket boosters, which have now evolved for use as five-segment solid motor boosters for NASA's two new Constellation rockets: the Ares I crew launcher and the Ares V cargo launcher.

MOL's Single Flight

An important flight test of the MOL hardware was launched before the program's cancellation, though it had no influence on the decision to end the project. NASA supplied a refurbished Gemini capsule for the test that had previously been launched on a Gemini vehicle test flight in January 1965. The modified Gemini (2R) included a heat shield with a 0.66 m (25.8 in) diameter hatchway to evaluate the effects of reentry on the hatch seal and vehicle's structural integrity. A mockup of the MOL LM and MM attached to the Gemini capsule was constructed from a 10.4 m (34 ft) long, 3.1 m (10 ft) diameter Titan II first-stage tank. Nine experiments and three satellites were placed onboard the test structure, along with a Titan transtage booster to put the vehicle and satellites into orbit (Pealer 1995b). The November 3, 1966, MOL launch used a Titan III-C for the multi-satellite launch to boost the Gemini 2R on a thirty-three-minute suborbital flight. Recovery of the Gemini 2R capsule was made near Ascension Island in the South Atlantic by the USS *La Salle*.

Figure 5.9. MOL test launch November 3, 1966, from Cape Canaveral, with a reused Gemini capsule and a prototype station hull structure. Source: USAF.

MOL Astronauts

Of the seventeen MOL astronauts that were selected by the USAF in three groups, fourteen were considered for NASA assignments, and seven flew on space shuttle missions. Those included Karol Bobko, Robert Crippen, Gordon Fullerton, Henry Hartsfield, Robert Overmyer, Donald Peterson, and Richard Truly.

- MOL Group 1 selection—November 1965
  - Michael J. Adams (Air Force)
  - Albert H. Crews Jr. (Air Force)
  - John L. Finley (Navy)
  - Richard E. Lawyer (Air Force)
  - Lachlan Macleay (Air Force)
  - Francis G. Neubeck (Air Force)
  - James M. Taylor (Air Force)
  - Richard H. Truly (Navy)
- MOL Group 2 selection—June 1966
  - Karol J. Bobko (Air Force)
  - Robert L. Crippen (Navy)
  - Charles G. Fullerton (Air Force)
  - Henry W. Hartsfield, Jr. (Air Force)
  - Robert F. Overmyer (Marine Corps)
- MOL Group 3 selection—June 1967
  - James A. Abrahamson (Air Force)
  - Robert T. Herres (Air Force)
  - Robert H. Lawrence, Jr. (Air Force)
  - Donald H. Peterson (Air Force)

ALMAZ—SOVIET MANNED RECONNAISSANCE LABORATORY

Soviet space reconnaissance programs were developed and carried out to keep pace with the U.S. programs, often in spite of the burdensome Soviet bureaucracy. The Soviet Almaz ("diamond") orbital laboratory was a program response to the USAF MOL project, and initiated the same year the approval of the MOL was announced by President Johnson. While the Dyna Soar and MOL projects were both canceled, the Soviets successfully launched and operated several Almaz manned military stations and later unmanned derivatives.

In order to save money and to advance both programs concurrently, the Soviet Almaz project was developed along with the Salyut civil space station, but in competing design groups—a common practice in the USSR. By combining the military and civil stations in name, the Soviets could also obscure the military objectives of Almaz. Hence, the Almaz missions were not launched under its designation but with the name Salyut. Both the civil and military versions incorporated similar station structures and the same Proton heavy-lift booster, but their missions and the respective equipment differed dramatically.

The military Almaz and the civil DOS Salyut orbital stations had the same three-cylinder structure, although with the attached solar arrays on opposite ends of the vehicle: aft on the Almaz and forward on the DOS. Both versions of the Salyut station used the same Soyuz vehicle to transfer crews to and from the stations. Interior differences in the two Soviet station models were far more extensive than on the exterior. The greatest difference between the two was a large telescope and camera installed roughly in the center of the Almaz. The 10 m (33 ft) focal length Cassegrain telescope, optics, and camera extended from the floor almost to the ceiling of the Almaz (Peebles 1987). Pointing of the orbital telescope was accomplished by pointing the entire station, making station stability one of the biggest design challenges. An antisatellite cannon was also a part of the Almaz design, but was fired only once—unmanned.

Design of the Almaz, also known as the Orbital Piloted Station (OPS), had several striking similarities to the USAF MOL station, including the three-man crew vehicle attached to the station at launch and a heat shield pass-through hatch design between

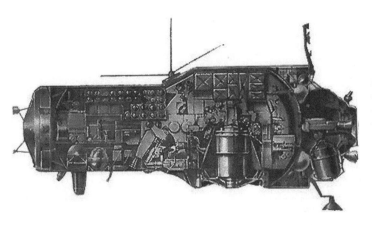

Figure 5.10. Cutaway sketch of the Almaz station structure and primary systems, including the dominating camera-telescope installed near the vehicle bottom-center. Source: NASA NIX.

the Almaz and the crew vehicle called the VA (Vozvrashemui Apparat, or return vehicle). The VA was developed as part of a combined crew and cargo vehicle called TKS (Transportniy Korabl Snabzheniya, or transport supply ship). The VA crew vehicle was attached to a resupply vehicle called the FGB (Funktsional'no Gruzovoi Blok, or functional cargo block) that was similar in function and shape to the Apollo command module, and with a launch-abort-escape assembly attached.

As useful as the TKS could have been, it was launched only three times to the Salyut stations, but none of them were manned. Launched under the Cosmos designation to obscure the mission type, the TKS was used to support the Salyut 6 (Cosmos 1267) and Salyut 7 (Cosmos 1443 and 1686) with civil and military experiments onboard (Portree 1995). The TKS did, however, serve as an important prototype for later Soviet space station laboratory vehicles. TKS's most versatile element was the FGB cargo structure that became the design blueprint for seven space station modules and one large military satellite. Construction of four of five Mir research modules were based on the FGB design, as well as the Zaria FGB that is currently part of the International Space Station. The FGB also served as the primary block for the Polyus "battle station" military satellite launched on the first Energia booster flight on May 15, 1987.

More recently, a second FGB constructed by the Russians as backup for the ISS Zaria is undergoing development as a replacement vehicle for the original, which has already passed its design life. FGB-2, known as the multipurpose laboratory module, is expected to serve as a Russian research module instead of storage—Zaria's current role—with an anticipated launch in 2011.

ALMAZ MISSIONS

Seven Soviet Salyut space stations were launched into orbit in the 1970s and early 1980s as a mix of civil and military laboratories. Although the early Salyut missions were plagued with problems, only one of the Salyut stations did not reach operational orbit. DOS civil stations included the Salyut 1, 4, 6, and 7, while the Almaz military stations were Salyut 2, 3, and 5.

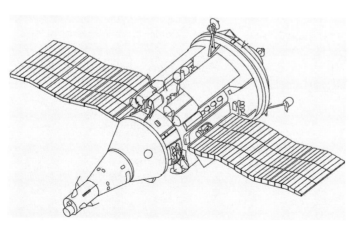

Figure 5.11. Sketch of the solar-powered TKS transport vehicle consisting of the VA crew module (dual-conical structure on the left) and the FGB cargo transport structure to the right of the VA capsule reentry shield. Source: Portree.

Salyut 2 (Almaz 101, OPS-1)

Salyut 2 was the only Soviet station that failed to reach stable orbit long enough for crews to dock. Launched on a Proton on April 4, 1973, Salyut 2 was put into a successful orbit by the third stage of the Proton, but penetration by shrapnel from the upper-stage booster damaged the pressurization system and the vehicle's control functions. Loss of attitude and orbit control was followed by complete power loss several days later. Salyut 2 reentered the atmosphere on May 28, 1973.

Salyut 3 (Almaz 101.2, OPS-2)

Salyut 3 was the second successful Salyut station placed in orbit, and the first successful military space laboratory. Launched on June 25, 1974, Salyut 2 had two scheduled crew flights, but like Salyut 1, had only one successful crew visit. The Soyuz 14 three-man crew completed a fifteen-day expedition onboard that involved station systems operations, Agat-1 camera targeting and filming, and film canister ejection from the aft airlock mechanism for deorbiting the film canisters in similar fashion to the American Corona satellites.

Onboard systems on the Salyut 3 reportedly included the first gyrodyne attitude control system, the first water recycling system, and an antiaircraft cannon that was fired only after the end of the manned missions, one day before the station was deorbited on January 24, 1975.

Salyut 5 (Almaz 103, OPS-3)

Salyut 5 was the second and the last of the Soviet manned military space stations, although the unmanned Almaz stations continued as synthetic aperture radar satellites. Almaz 103, launched on June 22, 1976, had three crew visits in 1976 and 1977, two of which were successful (Soyuz 21 and 24). A fourth planned mission on Soyuz 25 was also scheduled but later canceled when the station orbit decayed, leading to an early deorbit on August 8, 1977.

OPS-4

A fourth manned Almaz military orbital station was canceled when the entire Almaz program was scrubbed in 1978 because of its high cost and marginal productivity. The remaining Almaz and TKS vehicles were redirected to other programs that focused on unmanned reconnaissance satellites.

UNMANNED ALMAZ

The large Almaz structure made a convenient platform for a large radar surveillance payload consisting of several large antennas, sophisticated electronics, a large electrical power system, and a precise attitude control system. The conversion to an orbital synthetic aperture radar (SAR) imaging system used for high-resolution reconnaissance was arguably more successful than the Almaz manned stations.

Almaz-T

The Almaz-T was the first SAR laboratory built from the Almaz structure, and the first Almaz vehicle to be designated Almaz. The launch on October 29, 1986, from Baikonor was a failure due to the unsuccessful separation of the first and second stages of the Proton launcher.

Cosmos-1870

The second unmanned Almaz SAR satellite was the Cosmos-1870 launched on July 25, 1987, into a 72°, 260 km (161 mi) circular orbit. The spacecraft functioned for two years, providing radar imagery of the Earth's surface with a resolution as small as 25 meters (82 ft). Cosmos-1870 was deorbited on July 30, 1989.

Almaz-1

The third Almaz SAR spacecraft was launched on March 31, 1991, not as Cosmos, but as Almaz-1. Even though the launch of Almaz-1 was successful, the complete deployment of the antennas and solar panels was not. Almaz reportedly generated SAR image data for eighteen months before its deorbit on October 17, 1992.

Almaz-2

The last of the Almaz SAR spacecraft known as Almaz-1V was planned for launch in the mid-1990s but was canceled due to the collapse of the Soviet Union and the loss of funding.

References

Angelo, Joseph A. Jr., and David Buden. 1985. *Space Nuclear Power*. Malabar, FL: Orbit Books.

Banther, Chris. 2004. "A Look into the History of American Satellite Navigation." *Quest* 11, no. 3.

Bennett, Gary L. 1989. "A Look at the Soviet Space Nuclear Power Program." In *IEEE Conversion Technologies and Space Nuclear Reactor Systems*. Vol. 2. Proceedings of the 24th Intersociety Energy Conversion Engineering Conference.

Day, Dwayne A. 2007. "Tinker, Tailor, Satellite, Spy." *Space Review*, October. Retrieved from: http://www.thespacereview.com/article/989/1

DoD (Department of Defense). N.d. Archives. DoDLink. http://www.defenselink.mil/multimedia/multimedia.aspx

FAS (Federation of American Scientists). 2000. "Improved Crystal." Retrieved from: http://www.fas.org/spp/military/program/imint/kh-12.htm (accessed 6/12/08).

Ferguson, James. U.S. Congress, House, Department of Defense *Appropriations for 1963: Hearings before a subcommittee of the Committee on Appropriations*, 87th Cong., 2nd sess., 1962, pt. 2: 477–88.

Goodwin, Robert. 2003. *Dyna-Soar: Hypersonic Strategic Weapons System*. Burlington, ON: Apogee Books

JPL (Jet Propulsion Laboratory). N.d. JPL Mission and Spacecraft Library—Corona. Retrieved from: http://samadhi.jpl.nasa.gov/msl/Programs/corona.html

Nicks, Oran W. 1985. *Far Travelers: The Exploring Machines*. NASA SP-480. Washington, DC: NASA Scientific and Technical Information Branch.

NSA (National Security Agency). N.d. "GRAB II Elint Satellite Exhibit." Retrieved from: http://www.nsa.gov/MUSEUM/museu00027.cfm (accessed: 6/16/2008).

Pealer, Donald. 1995a. "Manned Orbiting Laboratory (MOL), Part 1." *Quest: The Magazine of Spaceflight* 4, no. 3, Fall.

———. 1995b. "Manned Orbiting Laboratory (MOL), Part 2." *Quest: The Magazine of Spaceflight* 4, no. 4, Winter.

———. 1996. "Manned Orbiting Laboratory (MOL), Part 3." *Quest: The Magazine of Spaceflight* 5, no. 2.

Peebles, Curtis. 1987. *Guardians: Strategic Reconnaissance Satellites*. Novato, CA: Presidio Press, CA.

Portree, David S. 1995. *Mir Hardware Heritage*. NASA RP-1357. Houston, TX: Johnson Space Center, Information Services Division, March.

Richelson, Jeffrey T. 1989. *The U.S. Intelligence Community*. New York: Ballinger Publications.

Stokely, C. L., and E. G. Stansbery. 2008. "Identification of a Debris Cloud from the Nuclear Powered SNAPSHOT Satellite with Haystack Radar Measurements." *Advances in Space Research* 41, no. 7.

USAF AFLink Photo Archive. Retrieved from: http://www.af.mil/photos/

NRO Image Archive. Retrieved from: http://www.nro.gov/corona/imagery.html

NASA GRIN (Great Images in NASA). Retrieved from: http://grin.hq.nasa.gov/

NASA NIX (NASA Image Exchange). Retrieved from: http://nix.nasa.gov/

RussianSpaceWeb. N.d. Retrieved from: http://www.russianspaceweb.com/

CHAPTER 6

Mercury

Preparing for Manned Space Flight

Much of the hardware developed for the manned Mercury space flights was not born of fresh ideas, but was adopted from the advances in high-performance military aircraft. It was high-altitude aircraft powered by high-performance jet and rocket engines that fostered the development of technologies that were used for many of the components and systems in early manned space vehicles. One of the most significant leaps in flight technology was the first supersonic aircraft, the X-1. Stable flight at speeds beyond Mach 1 was not possible in conventional propeller-driven aircraft, which was demonstrated in a number of World War II fighter aircraft accidents. Even more critical were the design requirements for vehicles reaching space and returning at speeds well beyond Mach 1. Yet the progression from high-altitude aircraft flight to space flight had to be taken incrementally. Supersonic flight had to be conquered before hypersonic flight (flight beyond Mach 5) could be attempted. And hypersonic flight would have to be mastered before controlled reentry from Earth-orbit reaching speeds of Mach 25 could begin. Nevertheless, the myriad hardware challenges that engineers faced in the incremental progression to space flight were tackled methodically with the help of an experimental aircraft series born in the National Advisory Committee for Aeronautics (NACA) that began with the X-1.

One of the greatest advances made in the experimental supersonic X-1 project was the rocket propulsion system. While the X-1 aircraft neither reached above the atmosphere nor reached speeds that would allow it to approach those heights, the liquid-fuel rocket motor served as an important prototype for later rocket-powered X-series experimental aircraft. Improvements in the X-1's XLR-11 rocket engine powered the first American aircraft to reach space, the X-15. Not only did the X-15 set impressive altitude and speed records in its 199 flights spanning nine years, it also provided an important research platform for developing the hardware for the Mercury manned space capsule.

MAN-IN-SPACE-SOONEST

The X-15 hypersonic flight project was in its early development stages when the Soviet Union launched Sputnik on October 4, 1957, which signaled the beginning of both the space race and the space age. America's response to the first two Soviet Sputnik flights was a launch attempt of its first satellite three months later. The embarrassing Vanguard failure overshadowed several space projects already underway, but also served to stimulate funding for both military and civil programs. The classified USAF manned orbital surveillance laboratory project under development was hastened by these events, and by the agreement by the Department of Defense, the U.S. Air Force, and the National Advisory Committee on Aeronautics (NACA) to collaborate on a manned orbiting satellite program. In early 1958, the USAF proposal to place a manned laboratory and surveillance satellite in orbit was redirected to the Man-in-Space-Soonest project, which had the advantage of being a simpler project that could be launched earlier than the manned orbiting laboratory.

NACA's expanded role in the planning and development of the manned satellite project, and its anticipated conversion to the National Aeronautics and Space Administration (NASA) later in the year, swayed the decision to place the manned program under civilian control. Although the Air Force would continue to pursue their manned orbital laboratory, the first manned satellite project would become the responsibility of NACA. On October 1, 1958, the transformation of NACA into NASA was complete, including the transfer of NACA's personnel, laboratories, and research facilities. The newly organized NASA also had its first marching orders to develop the manned satellite program. A separate panel known as the Space Task Group (STG) was soon created to help guide development of the new Mercury manned satellite project. NASA, the new STG panel, and the Mercury project also had the collaborative support of the Department of Defense's new Advanced Research Projects Agency (ARPA, today known as DARPA) created just two years earlier.

Top priority in the new project was given to the capsule design and the booster selection, and by December 1958, specifications and contracts for both had been issued. McDonnell Aircraft Corporation was selected as the contractor responsible for developing and supplying the Mercury capsule, while the Atlas ICBM built by Convair was chosen as the contractor to develop and furnish the man-rated launcher. But tests of the Atlas design were not complete, and Mercury's flight hardware testing would have to begin within a year. This left Redstone as the only launcher available to carry out the preliminary Mercury capsule tests. By coincidence, Redstone was chosen for the initial Mercury flights at the same time contracts for future Atlas launchers were being completed. In early 1959, the announcements were made of the official name of the Mercury program, and NASA's first astronaut selection.

In retrospect, NASA and the project Mercury managers had performed near-miracles in developing the hardware for the first U.S. manned flights in space, organizing the almost-flawless program operations, and selecting and training the astronauts, and all without critical failures. The Mercury project took just over four and a half years to complete, from approval to the last manned flight. NASA's Mercury program was not only the first to fly American astronauts in space, the project help build an efficient and effective partnership of engineers, contractors, agency directors, and project planners that continued through the Gemini and Apollo programs. Mercury's legacy also includes the development of specifications for man-rating boosters, capsules, and launch facilities, as well as the creation of the infrastructure for manned space flight programs that includes mission and crew training facilities, aeromedical research facilities, communications installations, flight control facilities, test and assembly facilities, and launch facilities, and all in less than five years. Mercury was not just America's first manned space flight program, it furnished the Gemini and Apollo programs with the personnel, expertise, and program infrastructure to succeed in their mission objectives.

MERCURY OBJECTIVES

The Mercury project objectives were established in 1958 to articulate the basic requirements for proving humans could fly in space safely and reliably. Three program objectives were adopted when the Mercury program was approved, which were:

1. To place a manned spacecraft in orbital flight around the Earth
2. To investigate man's performance capabilities and his ability to function in the environment of space
3. To recover the man and the spacecraft safely (Grimwood 1963)

To help expedite the program and to enhance the safety of the flight and operational crews, guidelines were detailed in order to attain the program objectives. These included:

1. Existing technology and off-the-shelf equipment should be used wherever practical.
2. The simplest and most reliable approach to system design would be followed.
3. An existing launch vehicle would be employed to place the spacecraft into orbit.
4. A progressive and logical test program would be conducted.

During the design and development phases of the Mercury program, a thorough and consistent approach was established for project testing, including test flights and mission flights. Since cost was a concern throughout the life of the program, already-available hardware that could be used for the project was to be integrated whenever possible. Input for improvements in the vehicle design, fabrication, operations, and flight procedures was provided by the Mercury astronauts, and very often adopted. General program requirements for Mercury were established as the following:

1. The spacecraft must be fitted with a reliable launch-escape system to separate the spacecraft and its crew from the launch vehicle in case of impending failure.
2. The pilot must be given the capability of manually controlling spacecraft attitude.
3. The spacecraft must carry a retrorocket system capable of reliably providing the necessary impulse to bring the spacecraft out of orbit.

4. A zero-lift body utilizing drag braking would be used for reentry.
5. The spacecraft design must satisfy the requirements for a water landing. (Grimwood 1963)

Fundamental design requirements for the completed Mercury capsule and the launchers were covered in the program's preliminary objectives, although not expressly. Implied but unstated in these objectives were the concepts of redundancy, and sound, safe reliability alluded to in an early Mercury program summary: "Should individual components or even entire systems fail, some means would exist either to complete the mission safely or to conduct a successful mission abort so that crew safety would be maintained" (NASA 1963).

The Mercury capsule and launch hardware underwent extensive planning, design review, and qualification testing, from launch abort errors, to reentry anomalies, to life-support system failures, in order to reduce potential critical failures to an absolute minimum. Safe and reliable flight operations in the Mercury program were not simply policy statements, but details that were spelled out in the development documents; in the specifications; in the operations manuals; and in the preparation, training, and flight procedures. Emphasis on safety in the Mercury program is contained in its operational record; there were no crew injuries or fatalities during the entire Mercury program.

Basic Hardware

The Mercury project had a daunting list of hardware and equipment that had to be developed for the program, which included new test facilities for flight, launch, and training, as well as new command and control facilities, new crew and support training operations and facilities, new communication facilities, and new fabrication facilities for NASA and its contractors. With less than three years between project assignment and the first manned flights, almost every major component, system, and vehicle had to be designed and built concurrently. For brevity's sake, this section will cover only the capsule and launch hardware.

CAPSULE STRUCTURE

The Mercury manned flight vehicle design began with the basic capsule structure, followed by the safety escape system, then the capsule systems. But before the capsule structure could take shape, the capsule profile had to be determined for a safe and stable reentry. That difficult problem was solved earlier in the work carried out by NACA engineer H. Julian Adams in the early 1950s at the Ames Research Center. Rather than the needle-nosed shape that appeared more intuitive as an optimum shape for hypersonic reentry, a blunt nose could dissipate far more heat energy at hypersonic speeds from the resulting shock wave than a pointed nose could. The broad, slightly curved heat shield would also increase the ablation and insulation surface area, adding to its effectiveness. One of the complications with the blunt-nose design which did not affect simple ballistic weapons reentry was that controllable, stable hypersonic flight required some aerodynamic lift. The simplest solution used by both American and Soviet designers to generate lift on these early capsules was to offset the center of gravity from the center of aerodynamic pressure.

One of the family of shapes considered for the Mercury capsule was the lifting body designs tested in supersonic flight during the early 1950s. Since hypersonic flight generates extreme heat in orbital reentry, the design would have to accommodate high temperatures over a large surface area. And while the lifting body provided substantial lift, surface heating during reentry was calculated to be more than the lifting body could survive. Atmospheric reentry would expose a reentry vehicle to heat exceeding 1,927°C (3,500°F) for orbital flights, depending on the reentry angle, and less for suborbital missions.

During reentry, a vehicle undergoes significant deceleration and heat loading that varies with the orbital/suborbital altitude, the entry angle, and the ballistic coefficient. Higher deceleration loading and higher temperatures are encountered in higher orbital or suborbital altitudes, higher descent angles (a more vertical entry), and with a higher ballistic coefficient (mass divided by the surface cross section). Conversely, a shallower entry angle and/or a lower ballistic coefficient would have lower maximum surface temperatures but a longer reentry heating phase. A more shallow entry makes thermal control of the vehicle interior more difficult and trajectory calculations more challenging.

To help develop and coordinate the Mercury manned program, NASA management created the Space Task Group in 1958. The STG team directed by Maxime Faget, who was also responsible for many of the design details on the Gemini and Apollo capsules, and the space shuttle orbiter, worked closely with the USAF and ARPA design staff to take the preliminary concept of the blunt-nosed Mercury capsule to production. With some revisions in the first months, the truncated cone shape that offered minimal heating, maximum flight stability, and practical parachute and recovery placement was incorporated into the Mercury capsule design. In hindsight, the hurried Mercury capsule design was an unqualified success. Evidence of the Mercury capsule's functional design is seen in the use of the same basic design for the Gemini and Apollo capsules, and even for today's Orion crew flight vehicle.

Mercury's structure was made up of three titanium sections in a semimonocoque design, meaning that the skin provides part of the structural strength. The three sections made up the primary structural assembly as follows:

1. Afterbody—the small cylindrical top that housed the reentry parachutes and recovery components, and, for later vehicles, allowed the astronaut emergency egress.
2. Midbody—the main conical structure consisted of a dual shell. The inner shell provided the primary structural strength, while the outer shell added to the structural integrity and furnished thermal control with its beryllium and thin Rene (nickel alloy) shingles.
3. Forebody—the reentry face was comprised of three shells, the inner one a pressure bulkhead for the cabin, the second a heat shield support, and the third an outer ablation shield composed of glass fiber and high temperature resin.

The exterior finish of the spacecraft body was coated with a dull black finish for maximum heat rejection (high emissivity). A retrorocket package and attachments were placed on the outside of the ablation shield for deorbit, then released before reentry.

During the initial design of the Mercury capsule, two reentry heat shields were considered. The simplest concept was a beryllium slab placed near the reentry face to

Table 6.1 Mercury Capsule Specs

Construction	Titanium shell with beryllium and nickel alloy outer layers
Height	3.51 m (11.5 ft, 28 ft including Launch Escape System tower)
Diameter	2.0 m (6.5 ft)
Interior volume	1.7m³ (60 ft³)
Launch weight	1,950 kg (4,300 lb – MA-6)
Orbit weight	1,360 kg (3,000 lb – MA-6)

simply absorb heat. Although rejected as a final solution, the beryllium heat shield was tested on suborbital flights. The second heat shield concept was a heat rejection ablative shield that was used for both suborbital and orbital manned missions because of its high heat rejection capability. Less heating on suborbital trajectories made the simpler beryllium heat sink an attractive alternative because of its ability to release the absorbed heat quickly at splashdown, but the design was not adopted because of the uncertainties in the heat flow. Before the ablation shield was adopted, however, proof of the heating effects from higher-speed orbital flight had to be tested on a booster capable of placing the Mercury capsule in orbit. The booster selected was the first converted Atlas ICBM used in the Mercury program, the Big Joe. The ablation heat shield would prove more difficult to design, but because of its greater versatility and

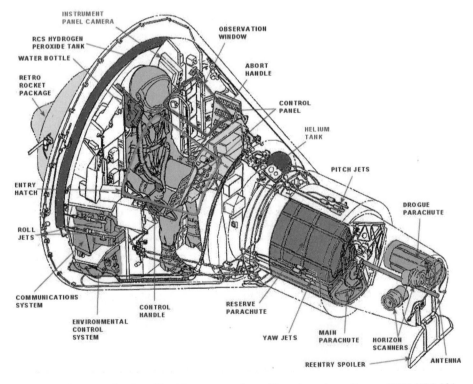

Figure 6.1. Cutaway drawing of the Mercury capsule and its major systems. Source: NASA SEDR-104.

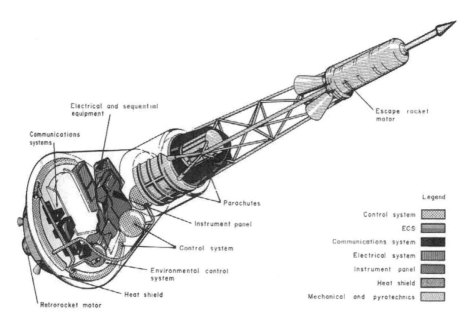

Figure 6.2. Mercury capsule cutaway and ejection tower mechanism attached at the recovery section. Source: NASA SEDR-104.

effectiveness, and its lower mass, it was used for both the suborbital and orbital flights. Mercury's basic reentry heat shield design was also adopted for the Gemini, Apollo, and the new Crew Exploration Vehicle Orion capsules.

NASA completed its initial specifications for the Mercury capsule in January 1959 and awarded McDonnell Aircraft the primary contract for its development and construction. Although a number of major decisions were yet to be resolved, an initial award was made for twelve spacecraft capsules. Vehicle development and completion was the responsibility of McDonnell even though NASA provided assistance for the project after the award by offering McDonnell the use of some of its facilities and personnel. The award was expanded to a final count of twenty production spacecraft. Of the twenty spacecraft capsules built, fifteen were launched, two were destroyed during test flights, and one was lost at sea. Gus Grissom's MR-4 capsule was recovered from the Atlantic Ocean thirty-eight years later with funding from the Discovery Channel and was displayed around the United States before being placed on permanent exhibit in the Kansas Cosmosphere and Space Center, in Hutchinson, Kansas.

CAPSULE INTERIOR

The Mercury capsule design was an exercise in integrating available high-performance aircraft components with advanced technology, while minimizing mass and cost, and maximizing redundancy. The result was a compact cabin layout that offered a single, small, contoured seat for the pilot, and an instrument console for capsule operations. The cockpit layout provided the pilot-astronaut access to the flight and operation controls for manual flight, and for communications, navigation, and system monitoring. A manual hand controller was placed to the astronaut's right with several switch panels on his left. An emergency and survival kit was placed to the astronaut's upper left, while water supplies and food were placed on the left (see figure 6.1 for cabin layout).

ELECTRICAL POWER

Electrical power for the Mercury capsule was supplied by primary and backup batteries since the missions were short in duration. The battery-supplied main electrical buses were maintained at 24 Vdc and divided into two main groups: the high priority circuits for critical operations, and low priority circuits for normal operations. Total primary and backup power was supplied by the following:

- Three main batteries—3,000 WHr (Watt-hours) each
- Two standby/backup batteries—3,000 WHr each
- One isolated battery—1,500 WHr

Alternating current was supplied to the AC loads by inverters feeding off of the DC battery buses. AC was generated for two purposes: one was to isolate the DC buses from the noise of the capsule fan motors; the second was to supply the AC avionics/electronics in the Automatic Stabilization and Control System (ASCS). Two primary and one standby AC buses carried the 115 Vac, 400 Hz power. The AC bus circuitry was the only power buses not fused since each had inverter overload protection (McDonnell Aircraft, 1961).

COMMUNICATIONS

Communications on the Mercury spacecraft were divided into several subsystems. Four different frequency bands were used for the spacecraft communications system's uplinks and downlinks, with over a dozen specific frequencies for the data and audio channels.

Communications functions for the Mercury capsule included:

- Capsule audio communications between astronauts and ground controllers
  - UHF and HF systems used for audio communications
    - Employed separate microphones in astronaut's helmet
    - UHF was active for the entire mission—from launch to recovery
    - HF backup communications were available from launch to reentry
- Biotelemetry and spacecraft telemetry to ground stations
  - Capsule systems and astronaut telemetry data totaling ninety-seven separate inputs were transmitted through a UHF-FM telemetry communications link
- Command signals from ground control
  - Secure, encoded signal transmissions were relayed to the capsule command receiver through a UHF-FM receiver circuit and decoder
- Recovery signals from the capsule
  - Communications between the capsule and the recovery ships/vehicles were handled with UHF and HF links

- S-band and C-band beacon signals were also used for descent and splashdown tracking calculations
- Radar tracking from ground and/or recovery vehicles
  - Ground tracking was facilitated with Mercury's onboard transponders that used both C-band and S-band beacon systems

Communications between the pilot-astronaut and ground stations were linked through NASA's newly created worldwide ground station network of fifteen sites that augmented the two existing ground stations. Together, these made up NASA's new Spaceflight Tracking and Data Network (STDN). These communications sites extended the coverage of telemetry, command, and audio links deemed necessary for the safety of the Mercury astronauts. A number of the ground stations also integrated C-band and S-band radar units for capsule tracking and identification. UHF and HF band communications were installed at the seventeen ground sites, although not all sites were capable of carrying audio, telemetry, and command data. Each of the sites' antennas had pointing and tracking capability to follow the orbiting Mercury capsule as it passed overhead.

STDN ground stations for the Mercury orbital flights included (McDonnell Aircraft, 1961):

- Cape Canaveral, Florida
- Grand Bahamas
- Grand Turk Island
- Bermuda Island
- Grand Canary Island
- Kano, Nigeria
- Zanzibar Island
- Muchea, Australia
- Woomera, Australia
- Canton Island
- Kauai Island, Hawaii
- Point Arguello, California
- Guumas, Mexico
- White Sands, New Mexico
- Corpus Christi, Texas
- Eglin AFB, Florida

In addition, there were two ship-based communications links in the Indian Ocean and the Atlantic Ocean that were maintained for the orbital flights. The Atlantic station was also a continuous communications link for the suborbital flights during the descent, splashdown, and recovery period.

ENVIRONMENTAL CONTROL AND LIFE SUPPORT (ECLS)

Many of the life-support systems for the Mercury capsule were first-of-a-kind, although a number of components were derived from life-support systems used in the hypersonic X-15 spacecraft. The Mercury capsule design included a duplicate life-support system circulating oxygen through both the cabin and the astronaut's space suit. Pressure for both breathing loops was 5.5 psi pure oxygen, with carbon dioxide removal and thermal control included in the circulation loops. Separate ECLS components provided a redundant environment in case of a suit failure or malfunction for up to two days in orbit.

Mercury's water supply was used for drinking, and for cooling the astronaut, the cabin, and the electronics. The excess waste water could either be dumped overboard, or stored in waste tanks. Water used for the capsule's active cooling system was sprayed

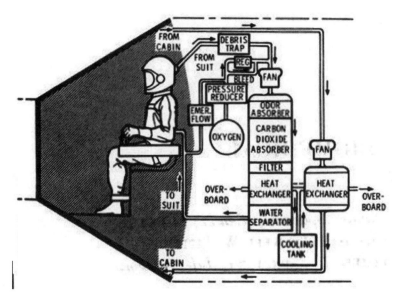

Figure 6.3. Schematic sketch of the major components of the environmental and life support system on the Mercury capsule. Note the separate atmosphere circulation loops for the suit and cabin environment. Source: NASA SEDR-104.

onto a heat exchanger/evaporator that vented the steam overboard. Although heating was needed for some sections/components, the Mercury capsule had a net heat gain from solar input, and from the onboard electronics and the astronaut's body heat.

Pure oxygen was supplied to the two life-support circulation loops by two 7,500 psi pressurized tanks that maintained a 5.5 psi interior pressure. Capsule pressurization functions also vented the cabin and suit environment to the outside during the descent. A vent door was ejected at approximately 21,000 ft for venting the capsule and suit to the atmosphere at altitudes below 18,000 ft. Suit pressurization and circulation was maintained after splashdown to provide cooling for the astronaut.

Carbon dioxide was removed from the cabin and suit lines as it circulated through the oxygen circuit by the suit compressor/fan. The circulated air that contained O_2, H_2O, and CO_2 passed through an activated charcoal filter, then a lithium hydroxide canister to remove CO_2, then a heat exchanger. Odors and trace gases were stripped out by the activated charcoal, while carbon dioxide was chemically removed from the circulating gas by combining with the lithium, displacing the hydroxide (OH). This would, in turn, combine with the oxygen to make water, all of which remained in the lithium hydroxide canister. The purified oxygen flow then passed through an exit filter to remove particles, then to the heat exchanger for water vapor and heat removal (see figure 6.3 for circulation pathway).

Thermal control of the cabin and suit was automated, although a number of environmental control functions could be manually controlled, or controlled from NASA's command center through the command communications system. The capsule's two heat exchangers cooled the circulated gas, and in the process, condensed the water vapor in similar fashion to an air conditioner. Heat exchanger water condensate was collected by a wicking water separator, then fed into a condensate (waste) storage tank and vented overboard, along with the heat exchanger cooling water (McDonnell Aircraft, 1961).

GUIDANCE, NAVIGATION, AND CONTROL

Mercury's guidance, navigation, and control system was called the Stabilization Control System (SCS). Automatic and manual components of the SCS operated outboard thrusters adapted from the X-15 design and powered by hydrogen peroxide propellant. Thrust operation was commanded by ground control, or by an automated sequencing control from onboard computations, or by manual control input from the pilot-astronaut. Capsule attitude control and navigation functions were used for all flight segments, from booster separation to reentry. The primary functional systems within the Stabilization Control System included the following:

- Automatic Stabilization and Control System (ASCS)
- Manual Rate Stabilization and Control System (RSCS)
- Reaction Control System (RCS)

Automatic Stabilization and Control System (ASCS)

Capsule attitude control was integrated into both automated function sequencing and manual control with two separate systems. Mercury's automated attitude control system, the ASCS, provided stability, orientation, and rotation functions for orbital flight, and during the deorbit sequence. Sensor input for the ASCS calculations and capsule attitude corrections came primarily from the horizon sensors and gyroscope platforms. Primary coordinate reference came from the capsule's two perpendicular horizon sensors located in the antenna faring, and from the attitude gyros. Subsystem components on the automatic stabilization system included the directional gyro, the vertical gyros, three-axis rate gyros (yaw, pitch, roll), the amplifier calibrator unit, and the 0.05-g accelerometer switch.

Manual Rate Stabilization and Control System (RSCS)

Critical capsule attitude/orientation controls could be taken over manually during nearly all phases of flight. Manual operations were, in general, selected by the astronaut with panel-mounted mechanical and electrical switches. The independent manual stabilization control system was included in the Mercury capsule design to provide backup and emergency flight control for the pilot-astronaut. The two ASCS and RSCS control systems provided the same functions, differing primarily in the thruster control valves.

Reaction Control System (RCS)

Attitude control for the capsule was modeled after the X-15 reaction control system, but with several differences. The reaction control system was part of the ASCS system, providing yaw, pitch, and roll control for both stabilization and rotation. Control of thruster operations was linked to the automatic control functions of the ASCS, and to manual control from the pilot's manual flight hand controller.

RCS thrusters were powered by hydrogen peroxide (H_2O_2) that spontaneously decomposed into water and oxygen, but at room temperature and at pressures near one atmosphere, the decomposition rate is slow. For usable thrust in the RCS units, a catalyst bed was placed in the fluid stream entering the thruster to accelerate decomposition heating and increase thrust.

ESCAPE AND JETTISON BOOSTER

Astronaut safety drove many of the early design features of the Mercury flight hardware, just as it has for each of NASA's crew vehicles since. The launch and ascent phase is one of the two most extreme flight environments and requires nearly perfect operations to ensure crew survival. The same requirements apply to the extreme conditions encountered during reentry.

Ensuring crew survival of a number of possible launch accidents was looked at carefully; this led to the general conclusion that a launch abort system could offer a reliable method for crew escape and survival for most contingencies. From that assumption, the Mercury capsule escape system was created by one of America's premier but unheralded spacecraft engineers, Maxime Faget. His escape system proved to be simple, reliable, effective, and inexpensive. A similar launch escape system design was adapted for use on the Apollo Command Module and NASA's new Orion Crew Exploration Vehicle.

Included in the Mercury capsule emergency escape tower system was the capsule separation rocket and a jettison rocket attached to the same frame, but at the top of the tower. The smaller, single jettison rocket motor was used to pull the tower away from the capsule before reaching orbit. It was also to be used in an abort sequence before parachute sequencing for emergency capsule descent and landing. A sketch of the escape and jettison components is shown below.

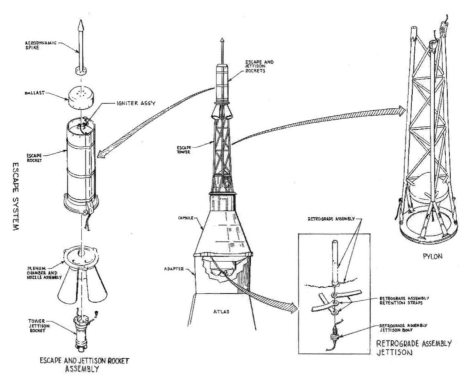

Figure 6.4. Escape and jettison rocket component layout used to pull the astronauts and Mercury capsule from the top stage of the launch vehicle. Source: NASA SEDR-104.

POSIGRADE BOOSTER

After launch, the Mercury capsule would separate from the launch vehicle and capsule adapter and enter a preprogrammed orbit using a solid-propellant posigrade rocket. The posigrade booster unit consisted of three small rocket motors, although only one of the rockets was needed for separation. Redundancy was also supplied in a dual-igniter assembly in each of the three motors.

RETROGRADE BOOSTER

The Mercury capsule's retrograde booster was a deorbit rocket package used to slow the capsule enough for the orbit perigee to dip far enough into the atmosphere for atmospheric drag to start the reentry phase. Mercury's three retrograde rockets were housed in the same container that housed the three posigrade motors. The entire booster package was held down with three metal straps anchored to the capsule bottom with explosive bolts. The booster unit was released sixty seconds from the retrograde burn and ejected from the capsule by coil springs (McDonnell Aircraft, 1961).

HARDWARE TESTING

Tests on the Mercury hardware were comprehensive and directed mainly toward the capsule structure, the ablation shield, and the capsule systems. These tests ran parallel with the Redstone and Atlas booster tests to save time. But before the first manned flight, all of the Mercury flight hardware had to be tested and man-rated in a rigorous testing and qualification process that included suborbital and orbital launches, and without significant glitches. Escape tower tests, capsule separation and reentry tests, capsule recovery tests, as well as subsystem and component tests, all had to be completed before the first manned flight could take place in 1961.

Another of the major challenges in the Mercury program was man-rating the two launch vehicles selected for the program—the Redstone and the Atlas. To save time, the first capsule and escape tower tests were flown with a smaller and much less expensive booster to finalize the basic capsule design. The small, inexpensive solid-fuel rocket booster dubbed Little Joe was designed specifically for the first capsule testing with an altitude capability similar to the Redstone. Award of the contract for seven Little Joe boosters was made to North American Aviation, and, within a year, the first launch of the Little Joe booster began a successful series of inexpensive ascent, separation, reentry, and splashdown tests for the early Mercury capsules. These early dummy vehicles used for aerodynamic and load tests were called boilerplate because the original capsule mockup was fabricated at a shipyard. The boilerplate designation was used, sometimes inappropriately, for a number of later manned-capsule prototype flights.

Launch Vehicles

The escalating Cold War between the United States and the USSR in the 1950s advanced ballistic missile programs quickly, and, by sheer coincidence, under the direction of the two space pioneers Wernher von Braun and Sergei Korolev. The Soviets began a large-scale, unified program to develop and deploy long-range missiles, while the U.S. efforts were slowed by the competitive and uncoordinated efforts of the three branches of the Armed Forces. In 1947, the Army Air Force was separated into the U.S. Air Force and the U.S. Army, which divided responsibilities for missile types, as well as diluted funding for the U.S. missile efforts.

Nonetheless, the three military branches contributed directly or indirectly to the first space exploration programs, as well as the first manned space flights. The Army Ordnance division was reorganized to include a ballistic missile group comprised of American engineers and German rocket teams brought to Fort Bliss under Operation Paperclip after the war. Army Ordnance, which later became the Army Ballistic Missile Agency, set out to develop large and small battlefield missiles, taking advantage of the Germans' expertise in advanced missile technology. A smaller-scale version of the V-2 called the Hermes was the first in a series of missile designs leading to a more powerful booster and advanced guidance systems that were incorporated in the Redstone missile design.

General Electric was contracted to oversee the launch of captured V-2 launches at White Sands, then to develop longer-range missiles based on the newly acquired technologies. Although their Hermes vehicles were never put into production, the research efforts were responsible for the intermediate-range ballistic missile named Major, and later renamed Redstone for the new missile development facility in Alabama. Chrysler Corporation was awarded the primary contract for Redstone, along with the North American Aviation (NAA) rocket group that built the engines. After initial fabrication in 1952, the first Redstone test flight took place in only one year due primarily to the advances from the Hermes project. Redstone's engine, although appearing similar to the V-2 motor, was a new design based on North American Rocket Division's (which later became Rocketdyne) XLR-43 engine, which had a rich progeny. NAA/Rocketdyne's XLR-43 Navaho engine became a blueprint for the basic design of the Atlas, Thor, Jupiter, and Saturn 1 and Saturn IB booster engines.

Redstone's design quickly matured, and in 1956 the booster was launched on a reentry test flight 600 miles in altitude and 3,300 miles down range. By the time the Mercury program was approved in 1958, the Redstone had proven itself as a reliable missile for Army battlefield units, and the space launch vehicle for America's first satellite. But the Redstone was limited in its payload capacity and could only push the Mercury capsule through a suborbital trajectory.

What was needed for Mercury orbital flights was a much larger booster—either tested or near completion—chosen to be U.S. Air Force Atlas ICBM. NASA's order for the already tested Redstone launcher was supplemented with orders for the converted Atlas ICBM. A contract was also given to North American Aviation in December 1958 for the first capsule booster dubbed, Little Joe. A distant relative called Big Joe was an ablation shield test launch on an Atlas booster made in September of the following year.

Table 6.2. Little Joe Specs

Thrust	1,044 kN (235,000 lb)
Length	15.2 m (50 ft)
Diameter	2.03 m (6.7 ft)
Weight	12,700 kg (28,000 lb)
Fuel	Solid (8 motor cluster)
Burn Time	37 s

LITTLE JOE

The Little Joe program was conceived in 1958 and first flown in early 1959 to investigate the Mercury capsule flight dynamics and aerodynamic characteristics. It would also be used to check the capsule parachute and drogue parachute operations. Because of the booster's low cost and utility, Little Joe missions were extended to include evaluation of the physiological effects of suborbital flight on primates (Grimwood 1963).

The basic vehicle design drawn up by NASA engineers consisted of a cylindrical structure 15 m long with a 2 m diameter. The dummy Mercury capsule was attached

Figure 6.5. Little Joe launch vehicle in preparation for launch at the Wallops Island, Virginia, launch site facility. Source: NASA GRIN.

to the top with pyrotechnic separation bolts. No active guidance was used for the eight-motor, solid-fuel first stage, although fins were attached for aerodynamic stability at higher speeds. Little Joe's eight-motor first stage was made up of four modified Sergeant solid rocket motors (named Castor and Pollux, depending on the modification) in a simple radial configuration, with four smaller Recruit motors arranged inside the Sergeant mix. Total thrust generated was designed to accelerate the capsule to the approximate speed and altitude provided by the Redstone and Atlas boosters near capsule separation. Electrical power was supplied by onboard batteries.

REDSTONE

The Army Ballistic Missile Agency's Redstone was a single-stage, liquid-fuel booster that used liquid oxygen and alcohol-water propellants, with a hydrogen peroxide–driven turbopump. Even though much of the man-rated Redstone redesign was designed and initially fabricated by NASA at the Huntsville facility, the Chrysler Corporation and North American Aviation were responsible for the final production. Redstone's 78,000 lb thrust and 150 sec duration were capable of boosting a 1,293 kg (2,850 lb) payload 185 km (115 mi) downrange. Although the last flight of the Mercury Redstone was July 1961, the original Redstone intermediate-range ballistic missile ended production in 1964 and was replaced with the Pershing missile.

In preparation for Redstone's integration with the Mercury capsule, NASA ordered eight Redstone missiles from the ABMA in January of 1958. Preparations for use as the Mercury launcher required a number of alterations and improvements for higher performance and improved reliability, and because of the need to space-qualify America's first manned space launch vehicle.

A major difference between the Redstone ballistic missile and the Mercury Redstone booster was the simplification of many of the systems. An increase in reliability and a decrease in complexity also proved to be beneficial to the Mercury project, as shown in the success record of its flights. A number of improvements in the engine and propulsion system performance would also be needed to bolster the heavier rocket and its payload. Some of those 600 alterations included:

- Structure
  - Lengthened 6 ft to accommodate larger tanks in order to provide an additional 20 seconds of thrust
  - Total weight increased to 66,000 lb
- Engine
  - Increased thrust from 75,000 to 78,000 lb
  - Hydrogen peroxide turbopump improvements
  - Antifire squelch components added
  - Reliability and stability improvements
- Instrumentation
  - Control sensing unit added to signal errors or malfunctions
  - Telemetry added to provide readings on attitude, vibration, acceleration, temperatures, pressures, thrust level, etc.

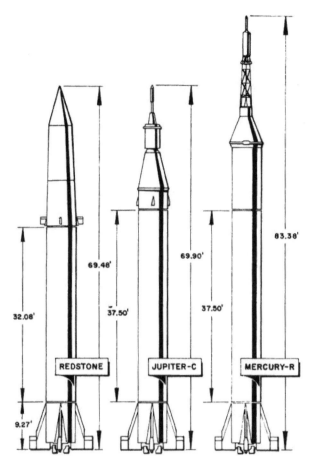

Figure 6.6. Comparison sketch of the Redstone missile, Jupiter-C Redstone, and Mercury-Redstone. Source: NASA MSFC.

- Control
  - A simpler, more reliable unit was incorporated to increase stability and reduce drift
- Abort
  - Added instrumentation and control components in order to identify problems in thrust levels, engine vibration, electrical failure, and other significant abnormalities
  - Control functions added to operate engine shutdown, booster separation, and escape rocket activation, either from an automated sequencer or from manual commands from the pilot-astronaut (Cassidy et al. 1963)

Table 6.3. Mercury-Redstone Specs

Length	25.3 m (83 ft)
Diameter	1.78 m (70 in)
Engine	Rocketdyne A-7 (XLR-43)
Thrust	35,380 kg_f (78,000 lb_f)
Propellants	LOX - oxidizer
	Ethyl alcohol + water - fuel

ATLAS

The Atlas ICBM, designed and built by the Convair Division of General Dynamics, which is now Lockheed-Martin, introduced a number of innovations that made it practical both as a nuclear missile launcher and as a civil space booster. One of its most unique features was the propellant tanks constructed from thin stainless steel sheet. When pressurized, the thin tanks also served as the missile's primary structure. The balloon-tank monocoque design made an extremely lightweight booster with the largest payload-to-structure weight ratio of any previous launcher.

Atlas was also the first booster with digital control and telemetry data systems, and was one of the first heavy-lift launchers to use an independent inertial guidance system (models E and F). Although not the first, Atlas also employed gimbaled engine thrust guidance for its primary engines. The two booster engines and the main sustainer engines were gimbaled for ascent guidance, while the two outboard vernier engines were used for the final guidance during the coast phase, which improved its targeting accuracy as an ICBM dramatically.

Table 6.4. Atlas D Specs

Diameter	3.05 m (10 ft–16 ft at the base)
Length	23.12 m (75.8 ft–85.5 ft in ICBM configuration)
Weight	117,930 kg (260,000 lb) maximum at launch
Engines	• 2 Rocketdyne LR105-NA strap-on boosters—69,850 kg_f (154,000 lb_f) thrust • 1 Rocketdyne LR89-NA-3 sustainer—25,855 kg_f (57,000 lb_f) thrust • 2 small vernier rockets for attitude correction—454 kg_f (1,000 lb_f) thrust
Engine thrust at launch	163,300 kg_f (360,000 lb_f)
Propellants:	
Fuel	RP-1 (kerosene)
Oxidizer	LOX

Mercury Astronauts

NASA's Mercury program presented many unique demands on the new agency, from major hardware design, to the management of contractors, and the collaboration with federal agencies in coordinating a program on a scale never attempted before. Challenges also arose in establishing, for the first time, the standards and operations that were critical to safe manned space flight. At the center of these decisions were the criteria for selecting and training the first Americans to fly in space, the Mercury astronauts.

Mercury project planning began within the NACA organization nearly a year before the creation of NASA. At the same time, the Department of Defense's newly created Advanced Research Projects Agency was working with the Air Force to organize the Man-

In-Space-Soonest project. Another proposal being considered in early 1958 originating in the Army Ballistic Missile Agency was Project Adam, which would loft an astronaut into a suborbital trajectory on a Redstone booster. Although this proposal had little backing, the initial flights may have been accomplished earlier than NACA's Mercury project or the USAF's Man-In-Space-Soonest. A third proposal to ARPA came from the Navy and outlined a suborbital mission incorporating a two-stage booster to push an inflatable cylinder through a suborbital arc, and later expand to return as a glider. The Navy's Manned Earth Reconnaissance (MER) vehicles were intended to make reentry more benign than the ballistic reentry procedures of the other three proposals, though the technology was purely conceptual (Swenson, Grimwood, and Alexander 1966).

Under the influence of the Eisenhower administration and Congress, a consensus arose that the manned space flight programs, like the space exploration initiatives, should be placed under civilian management. As the NASA agency approached its inception in the latter part of 1958, the decision was made to pursue manned space flight in the Mercury program, which effectively killed the three competing military proposals. The National Space Act that established NASA on October 1, 1958 was quickly followed by the formation of the Space Task Group to help direct the many details of the Mercury project, including the selection and training of the astronaut flight crews. In November, the STG outlined preliminary procedures for astronaut selection, while the aeromedical staff attached to the STG outlined astronaut selection criteria with assistance from the Armed Services and the aerospace industry. An announcement was then made on December 22, 1958, for research astronaut-candidates "with minimum starting salary range of $8,330 to $12,770 (GS-12 to GS-15) depending upon qualifications" (Swenson, Grimwood, and Alexander 1966). The announcement further described the prospective duties and expectations of the astronauts:

> Although the entire satellite operation will be possible, in the early phases, without the presence of man, the astronaut will play an important role during the flight. He will contribute by monitoring the cabin environment and by making necessary adjustments. He will have continuous displays of his position and attitude and other instrument readings, and will have the capability of operating the reaction controls, and of initiating the descent from orbit. He will contribute to the operation of the communications system. In addition, the astronaut will make research observations that cannot be made by instruments; these include physiological, astronomical and meteorological observations (Swenson, Grimwood, and Alexander 1966).

An effort to include civilian candidates for the space program was quashed by the administration, leaving only military pilots with test pilot experience as candidates. In January 1959, the minimum selection criteria were established for the Mercury astronauts, which specified the following:

1. Age—less than 40
2. Height—less than 5 feet, 11 inches
3. Excellent physical condition
4. Bachelor's degree or equivalent
5. Graduate of test pilot school
6. 1,500 hours total flying time
7. Qualified jet pilot

Eisenhower's decision to use only military pilots for the first astronaut selection greatly simplified the process, and by February 1959, the Pentagon had identified 110 qualified applicants from a pool of 508 pilots. The 110 candidates were sent to Washington, D.C., for interviews in three groups. Following initial interviews in early March, thirty-two men were chosen for testing and evaluation at the Lovelace Aeromedical Clinic in Albuquerque, New Mexico, and then sent on to Wright Patterson in Dayton, Ohio, for further evaluation. The expectation was that the top six candidates would enter training for the Mercury program, but by the end of March, medical, psychological, and skill tests had only pared the group to eighteen. The Space Task Group had to resort to technical merits for the final choice, but even then could not reach the favored six. After further deliberation, NASA announced the seven astronaut finalists to the world in April of 1959 as the Mercury Seven. The seven astronauts were Scott Carpenter, Gordon Cooper, John Glenn, Gus Grissom, Wally Schirra, Alan Shepard, and Deke Slayton.

All but one of the Mercury Seven astronauts flew in the Mercury program. Deke Slayton, who was disqualified from flying in the Mercury program because of a heart murmur, did fly on the Apollo-Soyuz Test Project in 1973. Of the six that did fly in the Mercury program, two flew only once: John Glenn and Scott Carpenter. John Glenn did, however, fly on the space shuttle after his retirement. Three Mercury astronauts went on to fly in the Gemini program—Gus Grissom, Gordon Cooper, and Wally Schirra. Alan Shepard was the only Mercury astronaut to land on the Moon on Apollo 14, and did so after being medically disqualified, then requalified after surgery on his inner ear. Wally Schirra was the only Mercury astronaut to fly in three different space programs—Mercury (Sigma 7), Gemini 6, and Apollo 7.

MERCURY 13

Thirteen female candidates were recommended to NASA in 1961 as suitable candidates for the Mercury program. Assisting in the effort was Dr. Randolph Lovelace, the same doctor

Figure 6.7. Photo of the Mercury 7 astronauts at the Redstone facility that would later become the Marshall Space Flight Center, in Huntsville, Alabama. Wernher von Braun is shown at the center. Source: NASA MSFC.

that performed the initial astronaut physiological and psychological evaluations at his Lovelace Aeromedical Clinic. Although the female candidates passed the same qualification tests and examinations as the seven Mercury astronauts, a lack of support from NASA and the lack of access to high-performance aircraft flight training killed the effort. It would be more than twenty years before the first American female astronaut would fly in space, coincidentally, twenty years after the first Soviet female flew in space. The astronaut was Sally Ride, selected in the first space shuttle program astronaut class and launched on STS-7 in 1983.

The first female group recommended for the Mercury astronaut program were:

- Myrtle Cagle
- Jerrie Cobb
- Jan Dietrich
- Marion Dietrich
- Wally Funk

- Jane Hart
- Jean Hixson
- Gene Nora Jessen
- Irene Leverton

- Sara Ratley
- B. Steadman
- Jerri Truhill
- Rhea Woltman

MILITARY ASTRONAUT SELECTION

Several military astronaut selections were held before NASA's Mercury astronaut selection in 1959, although none resulted in putting military crews in space. The first and least documented was the July 1958 selection for the Man-In-Space-Soonest (MISS) program. Selected were Neil Armstrong, Bill Bridgeman, Scott Crossfield, Iven Kincheloe, John McKay, Robert Rushworth, Joe Walker, Alvin White, and Robert White. Most notable was Neil Armstrong, the mission commander on Apollo 11 and the first to walk on the Moon. Armstrong, in addition to Crossfield, McKay, Rushworth, Walker, and Robert White, also served as pilots in the USAF-NASA X-15 program.

The Air Force Dyna Soar program included two astronaut selections, the first in April 1961, with Neil Armstrong, Bill Dana, Henry Gordon, Pete Knight, Russell Rogers, Milt Thompson, and James Wood. The Dyna Soar astronauts were announced in September 1962, with the addition of Albert Crews.

The Air Force Manned Orbital Laboratory project also involved a selection process for USAF astronaut crews, the first in November of 1965. The original MOL group included Michael Adams, Albert Crews, John Finley, Richard Lawyer, Lachlan Macleay, Francis Neubeck, James Taylor, and Richard Truly. Richard Truly was the only MOL astronaut from this group to enter the NASA astronaut program. A second USAF MOL selection was held in June of 1966 that included Karol Bobko, Robert Crippen, Gordon Fullerton, Henry Hartsfield, and Robert Overmyer. All five of the second MOL astronaut group transferred to NASA and flew on the space shuttle as Commander/Pilot Astronauts. The third and final MOL astronaut selection was in June of 1967 and included James Abrahamson, Robert Herres, Robert Lawrence, and Donald Peterson.

POST-MERCURY ASTRONAUTS

Project Gemini and the coming Apollo lunar missions prompted the next astronaut selection in 1962, often referred to as the Next Nine. NASA astronauts selected in September 1962 were Neil Armstrong (transferred from MISS to MOL, then to NASA), Frank Borman, Charles Conrad, Jim Lovell, Jim McDivitt, Elliott See, Tom Stafford, Ed White, and John Young. All of these astronauts flew in the Gemini program with the exception of Elliot See, who was killed in a training flight accident. All of the rest flew in the Apollo program except for Ed White, who died in the Apollo 1 capsule fire. While six of the astronauts would make a trip to the Moon (Borman, Lovell, Stafford, Young, Armstrong, and Conrad), Lovell and Young made it to the Moon twice, albeit only landing once. Young went on to command the space shuttle on its first flight (STS-1), and again on STS-9.

NASA's third group of fourteen astronauts was announced in October 1963 and included Buzz Aldrin, William Anders, Charles Bassett, Alan Bean, Eugene Cernan, Roger Chaffee, Michael Collins, Walter Cunningham, Donn Eisele, Theodore Freeman, Richard Gordon, Russell Schweickart, David Scott, and Clifton Williams. All but three of the astronauts in this group flew in the Apollo program. Roger Chaffee was killed in the Apollo 1 fire, while Theodore Freeman and Clifton Williams died in aircraft accidents before reaching space.

NASA's need for scientists on the future Apollo missions and the Skylab station influenced the makeup of NASA's fourth astronaut group, which included scientists Owen Garriott, Edward Gibson, Duane Graveline, Joseph Kerwin, Curt Michel, and Harrison Schmitt. Geologist Harrison Schmidt was the only scientist to walk on the Moon, while Garriott, Gibson, and Kerwin all flew on Skylab (Garriot also flew on STS-9). Neither Graveline nor Michel flew in space.

NASA's fifth group of astronauts was the largest of the Apollo-era astronaut classes, and all but five were destined for Apollo flights. The fifth group, announced in April 1966, included Vance Brand, John Bull, Gerald Carr, Charles Duke, Joseph Engle, Ronald Evans, Edward Givens, Fred Haise, James Irwin, Don Lind, Jack Lousma, Thomas Mattingly, Bruce McCandless, Edgar Mitchell, William Pogue, Stuart Roosa, John Swigert, Paul Weitz, and Alfred Worden. Of these, Bruce McCandless and Don Lind were slated for flight on the canceled Apollo missions 18, 19, and 20. Joesph Engle was bumped from Apollo 17, while John Bull left NASA before flying in space. Edward Givens was killed in an automobile accident after astronaut training. In addition to Apollo flights, Paul Weitz, Jack Lousma, and William Pogue flew aboard Skylab. Engle and Haise commanded flights on space shuttle Enterprise test flight missions, while Engle, Mattingly, McCandless, and Lind would fly on STS missions.

Astronaut group six, announced in October 1967, was selected primarily as backup crews for the final Apollo missions and the later Skylab missions, although seven of the eleven flew as mission specialists on STS flights. The sixth astronaut class included Joseph Allen, Philip Chapman, Anthony England, Karl Henize, Donald Holmquest, William Lenoir, John Llewellyn, Story Musgrave, Brian O'Leary, Robert Parker, and William Thornton. All but Chapman, Holmquest, Llewellyn, and O'Leary flew on STS missions.

NASA's seventh astronaut selection was actually a transfer of the USAF MOL astronauts who later flew on NASA's early STS flights. The seventh group, announced in August 1969, included the seven astronauts Karol Bobko, Robert Crippen, Gordon Fullerton, Henry Hartsfield, Robert Overmyer, Donald Peterson, and Richard Truly.

NASA's eighth astronaut class, announced in January 1978, was the largest of any selection, at least through the year 2009. The 1978 class of thirty-five candidates, known affectionately as "Thirty-Five New Guys," consisted of two astronaut classifications that would operate NASA's new National Space Transportation System—the space shuttle. In previous selections, astronauts were all astronaut-pilots, but with space shuttle payloads and operations at a new level of complexity, scientists and specialists had to be included on all but the early test missions. Hence, NASA added mission specialist astronaut to augment the revised pilot-commander astronaut classification. Astronaut group eight—the first space shuttle class—included commander/pilots Daniel Brandenstein, Michael Coats, Richard Covey, John Creighton, Robert "Hoot" Gibson, Frederick Gregory, Frederick Hauck, Jon McBride, Francis "Dick" Scobee, Brewster Shaw, Loren Shriver, David Walker, and Donald Williams. Mission specialist astronauts included: Guion Bluford, James Buchli, John Fabian, Anna Fisher, Dale Gardner, David Griggs, Terry Hart, Steven Hawley, Jeffrey Hoffman, Shannon Lucid, Ronald McNair, Richard Mullane, Steven Nagel, George Nelson, Ellison Onizuka, Judith Resnik, Sally Ride, Rhea Seddon, Robert Stewart, Kathryn Sullivan, Norman Thagard, and James van Hoften.

NASA's first astronaut class selected for space shuttle missions produced an impressive list of accomplishments and records. A few of these included Sally Ride, who was the first American female to fly in space on the STS-7 mission in 1983. She was also a mission specialist on Kathleen Sullivan's first female spacewalk during STS-41G in 1984. Shannon Lucid logged the longest duration flight for both American men and women on the Mir space station, until later missions eclipsed her record on the International Space Station.

RECENT ASTRONAUT SELECTIONS

NASA has scheduled astronaut selections approximately every two years since the first STS class in 1978. Selection criteria for the two astronaut categories has remained much the same as the first STS selection, although two new categories of astronaut have been added. Payload specialist and astronaut educator are now included in the selection procedures, the last being added in the selection of 2004.

Commander/Pilot Astronaut

Pilot astronauts serve as both space shuttle commanders and pilots. During flight, the commander has onboard responsibility for the vehicle, crew, mission success, and safety of flight. The pilot assists the commander in controlling and operating the vehicle and may assist in the mission experiments and payloads and their operations.

Mission Specialist Astronaut

Mission specialist astronauts work with the commander and the pilot, with overall responsibility for coordinating operations of the shuttle systems, crew activity planning, consumable usage, and experiment/payload operations. Mission specialists are trained in the details of the orbiter onboard systems, as well as the operational characteristics, mission requirements and objectives, and supporting equipment/systems for each of the experiments conducted on their assigned missions. Mission specialists perform EVAs (extravehicular activity), operate the remote manipulator system, and are responsible for payloads and specific experiment operations.

Payload Specialist

Payload specialists are non-NASA astronauts and may include foreign nationals that are assigned specialized payload duties. These specialists participate with the shuttle crews and are assigned to specific mission payloads.

Astronaut Educator

The new classification of astronaut educator was introduced in the 2004 astronaut class. Astronaut educators train for and perform their duties just as the other astronauts, but in a program that is intended to inspire students to consider careers in math, science, engineering and technology, and ultimately the nation's space program.

QUALIFICATIONS

Basic education and experience requirements for mission specialist and commander/ pilot astronaut candidates are a bachelor's degree from an accredited institution in engineering, biological science, physical science, or mathematics. In addition to the qualifying degree, three years of related, progressively responsible professional experience are also required. An advanced degree is desirable and may be substituted for all or part of the experience requirement. A master's degree is equivalent to one year of work experience, while the doctoral degree is equivalent to three years of experience. U.S. citizenship is mandatory to be considered for selection as a U.S. astronaut.

Commander/Pilot

- B.S. in one of the following degrees, plus three years of related professional experience (degrees not qualifying include technology, nursing, social sciences, aviation-related)
  - Engineering
  - Biological science
  - Physical science
  - Mathematics
- 1,000 hr pilot-in-command (PIC) jet time
- Test pilot experience preferred
- Ability to pass NASA Class I physical
- Vision 20/50 correctable to 20/20 each eye (distance visual acuity)

- Height 62 to 75 inches
- Blood pressure 140/90 or lower (sitting)

Mission Specialist

The same as for pilot except for the following:

- Vision 20/150 correctable to 20/20 each eye (distance visual acuity)
- Fight time not required
- Ability to pass NASA Class II physical

SELECTION AND TRAINING

Astronaut applicants that are offered interviews spend approximately two weeks in the interview and examination period at the NASA Johnson Space Center near Houston, Texas. The examinations are used to establish the candidate's physical and psychological fitness for the program. Those selected as astronauts spend two years in general training on space shuttle and International Space Station systems, basic science, mathematics, geology, meteorology, guidance and navigation, oceanography, orbital dynamics, astronomy, physics, materials processing, and more. Training also includes parachute jumping, land and sea survival training, scuba diving, and space suit operations.

Flight training for commander/pilot astronauts includes fifteen hours per month on the NASA T-38 trainers. Mission specialists fly a minimum of four hours per month in the NASA training aircraft.

A few numbers:

- While there are no age restrictions for the program, astronaut candidates selected in the past have ranged between the ages of twenty-six and forty-six, with an average age of thirty-four.
- Approximately 400 fully qualified applicants are notified for a prescreening medical examination and references.
- Approximately 100 are invited for interviews.
- Approximately 20 are selected in each cycle.

Mercury Flights

Mercury hardware test flights began in 1959 with the suborbital launch of the inert (boiler plate) capsule to examine the vehicle structure, capsule reentry heating, parachute operation, and escape tower separation. Each of these preliminary test flights used the Little Joe booster launched from the Wallops Island test site in Virginia. Unmanned missions with the production capsules were launched on Little Joe and Redstone rockets beginning in 1960. Manned Mercury flights began with the Redstone launch of MR-3 on May 5, 1961.

LITTLE JOE PRELIMINARY TEST FLIGHTS

The Little Joe program was brought to life in 1959 to investigate the Mercury capsule thermal and aerodynamic characteristics, as well as the capsule's parachute operations. The low cost and the utility of the small booster expanded the Little Joe flight program to include primate experiments.

Table 6.5. Little Joe flights

Launches	8
Failures	2
First launch	August 21, 1959
Last launch	April 28, 1961

Table 6.6. Launch Chronology

LJ-1	*21 Aug 1959*	Preignition failure
LJ-6	*4 Oct 1959*	Successful 5 min boilerplate launch
LJ-1A	*4 Nov 1959*	Almost successful relaunch of LJ-1 (8 min duration)
LJ-2	*4 Dec 1959*	Successful launch of Sam (rhesus monkey) (11 min)
LJ-1B	*21 Jan 1960*	Successful launch of Miss Sam (8 min)
LJ-5	*8 Nov 1960*	Unsuccessful test of Launch Escape System, destroyed capsule #3 (2 min, 22 s)
LJ-5A	*18 Mar 1961*	Successful relaunch of LES test (5 min)
LJ-5B	*28 Apr 1961*	Successful LES test using the same Mercury capsule (#14) (5 min)

UNMANNED MERCURY FLIGHTS

Mercury-Redstone 1

Mercury-Redstone 1 (MR-1) was the first launch attempt for the Mercury-Redstone combination on November 21, 1960. The unmanned suborbital launch attempt was a true comedy of errors, beginning with an accidental abort only one second after lift-off. The MR rocket rose about four inches, then settled back onto the launch pad without tipping over. The launch escape system jettison rockets then fired, launching the launch escape system but not the capsule 4,000 feet into the air. The Mercury capsule was then released, but remained on the Redstone, still sitting on the pad. Three seconds later, the spacecraft ejected its radio canister and deployed its drogue, main, and reserve parachutes, which dropped to the ground. Although the Redstone was damaged beyond repair, the Mercury capsule (capsule #2) was refurbished and launched a second time on the MR-1A flight.

Mercury-Redstone 2

Mercury-Redstone 2 (MR-2) launched January 31, 1961, was a successful qualification flight for the following manned Mercury-Redstone mission. Mercury spacecraft number 5 carried Ham the chimp on a 16 min 39 sec suborbital flight. A higher speed than planned and a 14.7 g reentry was accompanied by a 130 mi overshoot and a six-hour delay in Ham's recovery. Ham, in his own way, elected not to fly again.

Mercury-Atlas 1

Mercury-Atlas 1 (MA-1) was launched on July 29, 1960, in the first flight test of the Mercury capsule on an Atlas booster, but without a launch escape system attached. Thermal loading and dynamical tests of the dummy spacecraft were unsuccessful because of the failure of the adapter ring structure that mated the capsule to the booster upper stage. Due to the failure, a range-safety destruct command was transmitted to the Atlas booster approximately one minute after launch. Although the Atlas was completely destroyed, the severely damaged spacecraft was recovered and reassembled for failure analysis.

Mercury-Atlas 2

The successful retest of the failed MA-1 capsule reentry heating test and the operational test of the converted Atlas was launched on February 21, 1961, for an eighteen-minute suborbital flight. A reinforced adapter ring on capsule number 6 proved effective in correcting the problem encountered on MA-1.

Mercury-Atlas 3

MA-3, launched from Cape Canaveral on April 25, 1961, was a single-orbit flight of capsule number 8 and the Atlas booster. However, the Atlas launcher was commanded to self-destruct forty-three seconds into the flight because of guidance errors. The escape system did operate properly and the capsule was recovered, refurbished, and reused on the subsequent MA-4 test flight.

Mercury-Atlas 4

MA-4 was a relaunch of the single-orbit MA-3 mission on September 13, 1961. The flight of 1 hr 49 min was successful in proving the integrity of the spacecraft structure, the ablation shield, and the afterbody shingles used to protect the capsule from orbital reentry. The reused Mercury capsule number 8 was identified as 8A for this flight.

Mercury-Atlas 5

The last of the Mercury-Atlas test and qualification flights was MA-5, launched on November 29, 1961, with capsule number 9. The three-orbit mission carried the chimpanzee Enos in the man-rating checkout for the first manned orbital mission, MA-6. The flight was considered successful, although a control system malfunction terminated the flight after only two orbits.

MANNED MERCURY FLIGHTS

A successful man-rating of the Mercury capsule and the Redstone booster with MR-2 in January 1961 set the stage for launching the first astronaut into space. Had the Atlas booster completed testing before the Redstone, the first missions would have been orbital flights instead of suborbital. All manned flights were launched from the Air Force Eastern Test Range (ETR) at Cape Canaveral, Florida.

Mercury-Redstone 3

America's first astronaut in space was Alan Shepard, launched on May 5, 1961, in Mercury capsule number 10, named *Freedom 7*. The suborbital flight lasted 15 min 22 sec, reaching an altitude of 187.4 km (116.5 mi) and landing 487 km (303 mi) downrange. Posigrade rockets on the capsule were fired for separation from the Redstone approximately 10 sec after the engine shutdown of the Redstone 2 min 20 sec after liftoff. Following rotation of the capsule for reentry, Mercury's three retrorockets were fired near apogee, approximately 5 min 15 sec into the flight. Alan Shepard experienced weightlessness for roughly 5 min, encountering a maximum of 6 g's during final stage boost and just under 12 g's during reentry (Grimwood 1963). Shepard's inaugural space flight mission was considered a complete success.

Objectives for the MR-3 flight included:

1. Familiarize man with a brief but complete space flight experience including the lift-off, powered flight, weightless flight, reentry, and landing phases of the flight
2. Evaluate man's ability to perform as a functional unit during space flight by
   (a) Demonstrating manual control of spacecraft attitude before, during, and after retrofire
   (b) Use of voice communications during flight
3. Study man's physiological reactions during space flight
4. Recover the astronaut and spacecraft

Mercury-Redstone 4

NASA's second and last suborbital manned mission was the MR-4 flight launched on July 21, 1961, with Virgil (Gus) Grissom at the controls of *Liberty Bell 7* (capsule number 11). Maximum altitude for the flight was 190 km (118 mi), with a downrange distance of 486 km (302 mi) at splashdown. As with Shepard's MR-3 flight, the weightless period lasted approximately five minutes of the fifteen-minute flight. Neither of the first two astronauts in suborbital missions experienced any ill effects from their flights, including space adaptation syndrome.

Figure 6.8. MR-3 launch preparations for America's first astronaut in space at Cape Canaveral, Florida. Source: NASA KSC.

Grissom's capsule was lost during the recovery operations when the explosive side hatch activated before helicopter recovery. Grissom exited the capsule immediately and was retrieved after swimming in the water three to four minutes. With this second successful suborbital flight, the Space Task Group felt there was nothing further to be gained from the suborbital phase of Project Mercury, and the remaining suborbital Redstone flights were canceled.

Primary objectives for the MR-4 flight included (Grimwood 1963):

1. Familiarize man with a brief but complete space flight experience including lift-off, powered flight, weightlessness, atmospheric reentry, and landing phases of the flight
2. Evaluate man's ability to perform as a functional unit during space flight by
   (a) Demonstrating manual control of spacecraft during weightless periods
   (b) Using the spacecraft window and periscope for attitude reference and recognition of ground check points
3. Study man's physiological reactions during space flights
4. Qualify the explosively actuated side egress hatch

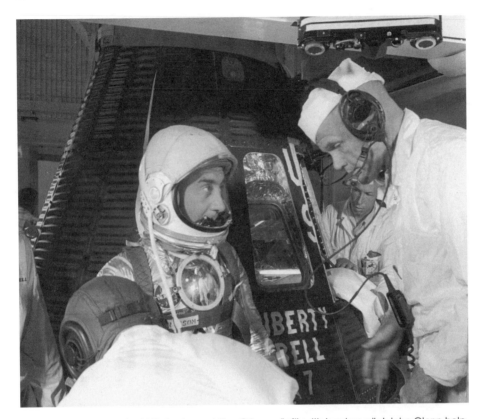

Figure 6.9. Photograph of MR-4 astronaut Gus Grissom (left) with backup pilot John Glenn helping in flight preparations. Source: NASA JSC.

Mercury-Atlas 6

America's first manned orbital flight used the larger Atlas booster since the velocity needed to reach orbit was 28,400 km/h (17,600 mi/h) and the Redstone could supply a ΔV of only 8,300 km/h (5,100 mi/h) for the Mercury capsule. Although NASA hoped for orbital missions for the first astronaut flights, the MA flights had to wait for the preliminary qualification flights of the Mercury capsule and Atlas booster.

MA-6 was launched on February 20, 1962, with John Glenn piloting capsule number 13, named *Friendship 7*. MA-6 was the first manned orbital mission for the United States that covered three-orbits, lasting 4 hr 55 min. Glenn was originally scheduled for a suborbital flight following Grissom's MR-4 mission, but because German Titov was launched on a one-day mission by the Soviets in August, further Mercury-Redstone suborbital flights were considered unnecessary.

John Glenn encountered control jet malfunctions and retrorocket circuitry problems during his orbital flight but maneuvered the spacecraft with the manual fly-by-wire controls to a successful splashdown. He suffered no physiological or psychological effects from the 4.5 hr weightlessness period, although there was a final complication to the mission. Concern about a loose heat shield and possible retro pack reentry damage, and the possibility of the loss of the capsule and astronaut, proved to be unfounded. Early deployment of the drogue chute was used to counter the severe oscillations of the capsule during reentry and the loss of reaction control fuel.

Primary mission objectives and most of the secondary objectives were met and the flight was considered a complete success. Primary objectives included (Grimwood 1963):

1. Evaluate the performance of man-spacecraft system in a three-orbit mission
2. Evaluate the effects of space flight on the astronaut
3. Obtain the astronaut's evaluation of the operational suitability of the spacecraft and supporting systems for manned space flight

Mercury-Atlas 7

America's second manned orbital mission, MA-7, was launched on May 24, 1962, with backup astronaut Scott Carpenter in the *Aurora 7* spacecraft (number 18). Deke Slayton was slated for the mission but was removed from flight rotation because of

Figure 6.10. Liftoff of the first Mercury manned orbital flight with John Glenn onboard from the Launch Complex 14 pad. John Glenn's MA-6 flight on February 20, 1962, in the *Friendship 7* Mercury capsule covered three orbits with a flight duration just under 5 hours. Source: NASA KSC.

a heart murmur detected during a flight-clearance medical. The three-orbit mission lasted for 4 hr 56 min and followed the same primary mission as MA-6, which was to evaluate spacecraft modifications and network communications. A critical malfunction occurred when the pitch horizon scanner circuitry failed, leading to excessive attitude-control fuel consumption. Control thruster operations were corrected for the following MA-8 flight.

A delay in the scheduled deorbit led to a splashdown some 400 km (250 mi) from predicted impact, leaving Carpenter and his capsule in the water for three hours before recovery.

Primary mission objectives for the MA-7 mission were (Grimwood 1963):

1. Evaluate the performance of man-spacecraft system in a three-pass orbital mission
2. Evaluate the effects of space flight on the astronaut
3. Obtain the astronaut's opinions on the operational suitability of the spacecraft systems
4. Evaluate the performance of spacecraft systems replaced or modified as a result of previous missions
5. Exercise and evaluate further the performance of the new Mercury Worldwide Network

Figure 6.11. Photograph of Scott Carpenter suited up in preparation for transport to the MA-7 vehicle. Source: NASA MSFC.

Mercury-Atlas 8

Mercury-Atlas 8 (MA-8, spacecraft number 16), designated *Sigma 7*, was launched from Cape Canaveral on October 3, 1962, with astronaut Wally Schirra at the controls for a scheduled six-orbit engineering test flight. Modifications were made to the thruster system because of the excessive fuel burns on previous missions, in addition to alterations of the capsule antenna to aid in ground communications. Experiments onboard MA-8 included photographic studies with a Hasselblad hand-held camera fitted with six filters, test samples placed on the capsule exterior to measure the reentry heating effects on ablation materials, and emulsion radiation monitors placed in the pilot-astronaut's seat to measure galactic cosmic rays.

Schirra's 9 hr 13 min mission was considered nearly flawless with only a few minor problems, one being elevated suit temperatures on the flight. MA-8 ended with the first Mercury capsule splashdown in the Pacific, and off-course by only six kilometers from the planned splashdown.

Primary mission objectives for MA-8 were (Grimwood 1963):

1. Evaluate the performance of the man-spacecraft system in a six-pass orbital mission
2. Evaluate the effects of an extended orbital space flight on the astronaut and compare this analysis with those of previous missions and astronaut-simulator programs
3. Obtain additional astronaut evaluation of the operational suitability of the spacecraft and support systems for manned orbital flights
4. Evaluate the performance of spacecraft systems replaced or modified as a result of previous three-pass orbital missions
5. Evaluate the performance of and exercise further the Mercury Worldwide Network and mission support forces and establish their suitability for extended manned orbital flight

Figure 6.12. Navy divers preparing the recovery of the MA-8 capsule and its pilot using an inflated flotation device. Source: NASA JSC.

Mercury-Atlas 9

NASA's final Mercury mission was the MA-9 flight launched on May 15, 1963, with astronaut Gordon Cooper piloting the *Faith 7* spacecraft (capsule number 20). MA-9 was the longest Mercury mission, with a duration of more than thirty-four hours covering twenty-two orbits. Eleven experiments onboard included visual acquisition and perception tests of ground targets that were illuminated specifically for the missions. A flashing beacon was also released in orbital flight to evaluate future Gemini docking exercises. An experimental balloon device was activated in order to measure the weak aerodynamic drag in orbit but failed to deploy. Photographic experiments during most of the mission included the identification of zodiacal light along the ecliptic on the Earth's night side. Radiation experiments similar to but more extensive than those on MA-8 were conducted with passive dosimeters worn on and near the astronaut, along with Geiger-Muller tubes attached to the retro pack used to measure residual radiation from atmospheric atomic weapons tests.

    In an otherwise uneventful mission, a series of malfunctions began on Cooper's nineteenth orbit when the 0.05 g reentry light came on erroneously, followed on the twenty-first orbit by a short-circuit in the main inverter bus that disabled the ASCS automated guidance system. Rising CO_2 in both the cabin and his suit during the last

Figure 6.13. Atlas launcher and the MA-9 Faith 7 capsule are shown being prepared for launch from pad 14 at Cape Canaveral on May 15, 1963. Source: NASA KSC.

orbit preceded a busy but successful manual deorbit, assisted from the ground by the backup pilot John Glenn. The modified manual reentry went smoothly with a splashdown just seven kilometers from predicted impact.

Primary objectives for the mission were:

1. Evaluate the effects on the astronaut of approximately one day in orbital flight
2. Verify that man can function for an extended period in space as a primary operating system of the spacecraft
3. Evaluate in a manned one-day mission the combined performance of the astronaut and a Mercury spacecraft specifically modified for the mission

Mercury-Atlas 10

The Mercury program came to an end with Cooper's MA-9 flight, although NASA considered an additional Mercury flight with Alan Shepard piloting a three-day mission. MA-10 would have filled the gap between the end of the Mercury flights and the beginning of the Gemini flights scheduled more than a year later. Adding an extended-duration mission would have also provided important space flight experience for the Gemini program preparations. The decision was made to cancel the proposed MA-10 mission as the launch date of MA-9 approached in mid-May, if MA-9 was successful. The MA-10 capsule (15B) was stored at the Cape Canaveral Air Force Station until it was placed on display at the Ames Research Center in Mountain View, California, then later transferred to the National Air and Space Museum in Washington, D.C. (Grimwood 1963).

MERCURY ADVANCED PROGRAMS

The manned Mercury program was not only NASA's first major space project, it was also responsible for ushering in a host of revolutionary new technologies. The Mercury program was also successful in teaching industry and agency managers how to plan, coordinate, and execute large, complex projects using new methods and new technologies. During the program, and more so as it came to a close, a number of proposals were presented to NASA and the Space Task Group relating to capsule improvements and modifications that ranged from modest to the ambitious.

Mercury's simple one-man capsule was not capable of orbital maneuvering, which meant that it lacked rendezvous capability and the capability for any complex reentry or landing operations. In addition, the capsule's one-day mission duration severely restricted its research utility for studying the impact of weightlessness on the astronaut-pilots. Late in 1959, well before the first flights, capsule manufacturer McDonnell Aircraft Corporation submitted several proposals to modify the vehicle that would allow functional control for manual touchdown and for maneuvering in orbit. Additions were also proposed for a self-contained guidance system, as well as for expanding the life-support and power capacity to allow mission durations to reach fourteen days. Several other concept proposals suggested that Mercury-style vehicles could also be developed for manned

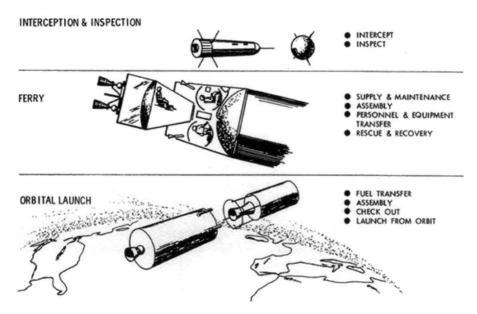

Figure 6.14. Concept sketch of several of the STG's preliminary plans for Mercury follow-on projects. Source: NASA SP-4002.

reconnaissance missions, and for high-speed, lunar-orbit reentry (Grimwood 1963).

In 1960, NASA's Space Task Group was convinced that advanced Mercury capsule studies would be worthwhile and began writing specifications for an increased-lift capsule with an onboard digital electronic navigation system. Design specifications were adopted for study within a few months and became the foundation of the Mercury Mark II proposal that, in turn, became the foundation for the Gemini capsule design. The Titan booster that was proposed for lifting a heavier advanced Mercury capsule into orbit became the Gemini's booster. More by research than by planning, what emerged from the advanced Mercury capsule project was the Gemini-Titan hardware that served as the bridge between the Mercury and Apollo programs.

In 1961, the same year as the first Mercury manned flight, a small one-man space station was proposed for further study that was to employ a Mercury manned capsule and an Atlas-Agena booster. The small cylindrical space laboratory was to have accommodations for a single astronaut for a fourteen-day mission duration. But because the NASA managers were increasingly focused on the Gemini and Apollo projects, the proposed Mercury orbital mini-station was rejected (Grimwood 1963).

References

Cassidy, J. L., R. I. Johnson, J. C. Leveye, and F. E. Miller, eds. 1964. *The Mercury-Redstone Project*. NASA TMX-53107. NASA, Marshall Space Flight Center.

Grimwood, James M. 1963. *Project Mercury: A Chronology*. NASA SP-4001. Houston, TX: NASA, Office of Scientific and Technical Information, Historical Branch, Manned Spacecraft Center.

Grimwood, J. M., B. C. Hacker, and P. J. Vorzimmer. 1968. *Project Gemini—Technology and Operations: A Chronology*. SP-4002. Washington, DC: NASA, Scientific and Technical Information Office.

McDonnell Aircraft. 1961. *Project Mercury Familiarization Manual*. NASA SEDR 104, NASA CR 555570.

NASA. 1963. *Mercury Project Summary: Including Results of the Fourth Manned Orbital Flight*. NASA SP-45. Washington, DC: NASA, Office of Scientific and Technical Information, October.

Swenson, Loyd S. Jr., James M. Grimwood, and Charles C. Alexander. 1966. *This New Ocean: A History of Project Mercury*. NASA SP-4201. Washington DC: NASA, Scientific and Technical Information Division, Office of Technology Utilization.

NASA GRIN (Great Images In NASA). Retrieved from: http://grin.hq.nasa.gov/

NASA MIX (Marshall Space Flight Center Image Exchange). Retrieved from: http://mix.msfc.nasa.gov/

NASA KSC Media Gallery. Retrieved from: http://mediaarchive.ksc.nasa.gov/

NASA JSC Image Collection. Retrieved from: http://images.jsc.nasa.gov/

CHAPTER 7

Gemini

NASA was in preparation for John Glenn's first orbital Mercury flight in 1961 at the same time its Space Task Group was being relocated from Langley, Virginia, to the new facility in Houston, Texas. The move coincided with the Space Task Group's name change to the new facility, which was named the Manned Space Flight Center. At the same time came the announcement that NASA had undertaken a manned space project that was yet to be named. Project names had been circulating through NASA and the contractors such as Advanced Mercury, the Mercury Mark II, and the two-man Mercury, but it would take three more months before the release of the official program name. The dual-pilot capsule configuration inspired the name Gemini for the new capsule and program, after the zodiacal constellation Gemini, The Twins. The Mercury program heritage retained a subtle presence in the Gemini name, since the Gemini astronomical symbol is "II," which was part of the designation for the project known earlier as Mercury Mark II.

Project Gemini was not created as an evolutionary step beyond the Mercury program, but as an important transition between the simple, single-crew Mercury hardware and the much more demanding Apollo boosters and lunar mission hardware. Important decisions faced NASA's management going forward with the Apollo program, which included specifications for the heavy-lift launchers, design concepts for the lunar crew vehicles, and the flight method to get the astronauts to the Moon and back. The flight mode proved to be the bottleneck that held up the crew vehicle design, and the specifications for the heavy-lift boosters. Yet even more decisions faced the NASA directors on the role and configuration of the Gemini transition program in order to perfect the systems and operational skills needed to carry Apollo astronauts on an unspecified flight mode to the Moon and back.

Gemini's objectives were crafted in general terms to prepare Apollo crews for the difficult rendezvous and docking essentials, and with mission durations equivalent to the longest Apollo missions. Fortunately, Gemini could be completed using technology and hardware that was already available, or under development, and without significant impact from the choice of a specific lunar flight mode. As 1961 came to a close, the Gemini program's approval placed the project on high priority, while NASA managers struggled with the Apollo flight-mode choice and the Mercury program headed toward completion the following year.

Concept Development

Little time was available for NASA and its contractors to develop and complete the Gemini program in preparation for the much larger Apollo program. This made the Mercury Mark II, also called Mark II, an attractive proposal because of its shorter development time and its lower cost compared to a completely new capsule design. Many elements of the Mercury capsule design were considered useful for the new project for reducing costs and saving time, but there were pitfalls in using first-generation hardware. Taking cautious advantage of the Mercury template meant that improvements in the new Gemini capsule design would be focused on the worst of Mercury's design flaws.

One of the major shortcomings of the Mercury hardware was the launch escape system. Its launch escape tower was not only heavy, it required manual and automated electromechanical operations. Even though the LES had a perfect record on manned Mercury missions, it was prone to failure because of its complexity. NASA and STG managers felt that a better alternative for crew aborts during launch and ascent were pilot ejection seats, but using ejection seats had far-reaching consequences, including the choice of booster. The Atlas ICBM used in the Mercury program had an explosive mixture of RP-1 (refined kerosene) and liquid oxygen that required a more powerful ejection seat rocket than was available for pulling the crew away from an explosion. The logical solution was to use the Air Force's new Titan heavy-lift booster. While not yet ready, Titan had more than twice the payload-to-orbit capacity of the Atlas, and its storable propellants were less explosive than the Atlas propellants.

Another weakness in the Mercury capsule design was the disjointed collection of system, subsystems, and components. To simplify Gemini's systems design, improve capsule preparations, and shorten the troubleshooting process involved with wringing out the inevitable problems, the advanced Mercury capsule design for the first time incorporated modular components and subsystems, as well as digital electronics.

PROGRAM AUTHORIZATION

In the last few months of 1961, a number of decisions critical to the Gemini project had yet to be made. For one, the program had yet to be authorized, or even named. The decision had yet to be made on the practice docking module and its booster as well. In addition, a method for landing the capsule had not been agreed on, although one concept being considered employed a paraglider to land the capsule on the ground instead of a splashdown in the ocean. A paraglider had much more directional control than a parachute, but feasibility studies were all that had been completed at the time. Paraglider landings were also questionable since many details of the untried system had yet to be evaluated.

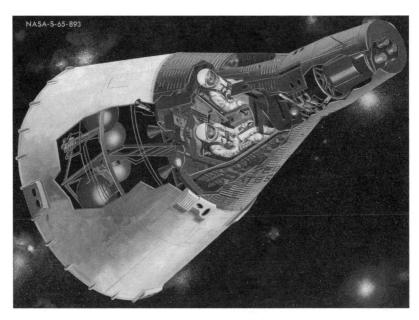

Figure 7.1. Cutaway of the Gemini capsule in orbit with the launch vehicle adapter skirt attached that would enclose the consumable storage section shown at the left. Source: NASA GRIN.

As 1961 came to a close, the contractors were working on final hardware proposals in expectation of being selected for the Mercury Mark II project. Prospective contractors included:

- Spacecraft—Mercury Mark II (2-man capsule) McDonnell Company
- Launcher—Titan II ICBM (man-rated) Martin Company
- Paraglider—North American Aviation Corporation
- Rendezvous vehicle—Agena McDonnell Company
- Agena launcher—Atlas (coordinated with USAF)

In early December, the Mercury Mark II program, along with the anticipated hardware development and the flight schedule, was approved by NASA management and Department of Defense officials. Atlas-Agena was chosen as the booster and docking vehicle, with the Titan II specified as the Mark II launcher. The newly-formed Manned Spacecraft Center project guidelines recommended that the first flights begin in 1963, and that rendezvous flights begin in 1964. This schedule left less than two years for the completion of the capsule, launcher, and support equipment (Grimwood, Hacker, and Vorzimmer 1968).

GEMINI GOALS

Like other manned programs, goals for the Gemini project were listed as simple objectives. Those included:

- To subject man and equipment to space flight up to two weeks in duration
- To rendezvous and dock with orbiting vehicles and to maneuver the docked combination by using the target vehicle's propulsion system
- To perfect methods of entering the atmosphere and landing at a preselected point on land (Grimwood, Hacker, and Vorzimmer 1968)

NASA completed the Gemini program's initial goals and a host of later goals and objectives, with the exception of the paraglider landing concept, which was canceled in 1964.

SPACECRAFT STRUCTURE

NASA's Gemini program had its own legion of obstacles to overcome in preparation for crew capsule launch and flight operations. Included were the testing and acceptance of the Titan launcher; conversion of the Agena target vehicle and its Atlas booster for manned flight; development of capsule recovery equipment; completion of expanded communications facilities; and most important, design and completion of the new capsule. As with the Mercury project, NASA's centers that were inherited from NACA were given program assignments related to their specialized functions. Along with the center facilities were the engineering offices and a number of distinguished spacecraft designers, including Jim Chamberlin, a Canadian who headed the Space Task Group's engineering division.

At the heart of the Gemini project was the renamed Mark II capsule, which was designed and constructed in three separate sections. The complete Gemini capsule could support a two-man crew in space for up to two weeks and protect them during the reentry and landing. Gemini's three sections were the crew capsule, or reentry module; the reentry control system section; and the supply and service adapter module. The crew capsule/reentry module in turn had three primary sections (see figure 7.2). These were the cabin (center) section, the rendezvous and recovery section (top), and the reentry control system section (bottom).

Crew Capsule

The Gemini capsule consisted of a truncated cone-shaped, welded titanium frame much like the Mercury capsule, but scaled up for a crew of two. The crew cabin was a double-walled pressurized unit, with dual-hinged crew hatch doors that could be opened manually for EVA, and with pyro activation for emergency escape. Each hatch contained a triple-pane Vycor (96% silica) window with antireflection and UV-reduction coatings (Hacker and Grimwood 1977).

Insulation shingles made of a nickel-alloy called Rene surrounded the primary structure for thermal protection. A thin inner layer of gold provided more thermal protection because of its low IR heat transmission. An ablation shield was placed on the widest-diameter bottom section, as it was on both the Mercury capsule and the Apollo Command Module. The ablative layer provided protection for reentry heating by boiling off, or ablating, the fiberglass and resin surface. Gemini's heat shield was composed of a silicone elastomer that filled a phenolic-impregnated fiberglass honeycomb.

Rendezvous and Recovery Section

Located at the top of the crew capsule was a detachable recovery section that carried parachutes, recovery gear, rendezvous equipment, and rendezvous radar.

Reentry Control System (RCS) Section

Gemini's reentry control system included the attitude control subsystems located between the cabin and the rendezvous and recovery sections (see figure below). This cylindrical titanium section housed the fuel (hydrazine) and oxidizer (nitrogen tetroxide) tanks, thrusters, and control plumbing used for reentry. Beryllium shingles surrounded the RCS section for thermal protection.

Supply and Service Adapter Module

Gemini's supply and service adapter was made up of two separable components—the equipment segment and a retrograde segment. The equipment segment housed the major components of Gemini's electrical power system, the maneuvering propulsion system that included eight thrusters for orbital operations and rendezvous, and six orbital attitude and maneuvering system (OAMS) thrusters. It also housed the equipment cooling system and the primary oxygen supply. Gold foil covered the aft segment to control temperature extremes and to reduce solar heating.

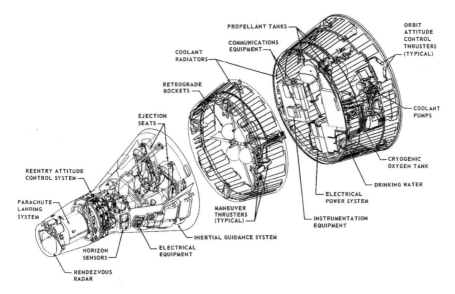

Figure 7.2. Cutaway of the major systems located in the Gemini capsule and its adapter components. Source: NASA SEDR 300.

The retrograde segment contained four solid-fuel retrograde rockets for capsule deorbit, as well as components for the equipment cooling system. This segment was jettisoned before atmospheric reentry.

SPACECRAFT SYSTEMS

Guidance and Control System

Gemini mission objectives were much more extensive than for the Mercury missions, involving objectives that were shaped primarily by the rendezvous and docking exercises needed in preparation for the Apollo missions. Gemini, the first manned spacecraft that could change its orbit, was outfitted with orbital maneuvering thrusters, attitude control thrusters, and a controllable-lift device to improve the capsule's landing precision. Improvements in capsule control functions incorporated an automated onboard navigation and guidance system as part of the digital computer functions—components needed for controlling the vehicle in addition to reporting its health. This capability was significant in several ways, including the first time that digital controls were used on a manned space vehicle. From Gemini's digital data system came a more sophisticated digital computer required for the Apollo missions to the Moon. These early developments led to the explosive growth of microelectronics, and powerful, portable digital computers within a little more than a decade.

Functions of the Gemini guidance and control system included orbital rendezvous guidance, reentry guidance, onboard navigation, backup launch guidance, and attitude reference. A brief list of the important elements of Gemini's guidance and control system that were precursors of the Apollo and space shuttle vehicles included the following:

Inertial and Spatial Reference Gemini's inertial guidance system was used to compute trajectories and control the spacecraft orbital operations from ascent through reentry, and was also used as backup for the launch vehicle guidance during ascent. At the heart of the unit was the inertial measuring unit (IMU) that contained three rate gyros and four gimbaled accelerometers for motion (rate) information. Position information with respect to the Earth, or spatial reference, was furnished by two horizon scanners that supplied data on the limb (edge) of the Earth contrasted by the cold background sky. The capsule was also equipped with rate gyros for accurate rotational motion measurements. Information from the rate gyros, horizon scanners, IMUs, and the rendezvous radar was combined in the onboard computer to calculate navigation parameters and any necessary corrections.

Gemini Computer Gemini's digital computer was the first to control a manned spacecraft autonomously. The Gemini prototype computer was a small, solid-state processor that had an even smaller 4,096-word programmable core memory. Digital, erasable memory consisted of tiny ferromagnetic rings wound with three sets of wires: two wound in opposite directions to write either a zero or one, and the third to read the setting. Computer technology in this age and at this scale was less than reliable, yet the requirements for the computer controls were much more demanding than for ground-based operations. Since mission operations exceeded the memory size of the 4,096 32-bit words, an external mass-memory source was added to the system. Operational commands were recorded on a tape drive loaded with three separate identical instruction sets that were loaded into the active memory. Each of the instructions fed from the tape drive was tested for integrity using a polling circuit. A similar technique was later used for the space shuttle computers that were also designed by IBM, and suffered from the same memory limitations. Gemini's computer also allowed the crew to enter manual instructions via dual ten-button keypads, called the manual data insertion unit (MDIU), which included dual data readout displays.

The Gemini computer operated in six mission modes:

- Prelaunch
- Ascent backup
- Insertion
- Catch-up
- Rendezvous
- Reentry

Time Reference Gemini's computer program operations were directed by several clock and time signals generated by an event timer that drove the outputs of an eight-day, twenty-four-hour clock. The event timer and the computer were started when the Gemini launch vehicle lifted 1.5 inches off the launch pad. This provided two time references; time reference since liftoff (later called mission elapsed time), and time to retrofire, although the retrofire time had to be selected by manual input commands (Tomayko 1988).

Flight Control System Gemini's on-orbit flight operations consisted of attitude and orbit maneuvers executed by the guidance and control system. Position and motion information was calculated by the computer's program instructions, then fed into the attitude control and maneuvering electronics (ACME) for command signals to the orbit and attitude control system thrusters, or to the reentry control system thrusters. Attitude operations were selected from seven flight modes that included:

- Horizon Scan
- Rate Command
- Direct
- Pulse
- Platform
- Reentry Rate Command
- Reentry

Orbit Attitude and Maneuvering System (OAMS) Gemini was designed with orbital maneuvering capability for rendezvous and docking exercises, a system that the Mercury capsule lacked. Attitude control for Gemini was integrated with the orbital functions into the orbit attitude and maneuvering system. In addition to the control functions, the OAMS operated sixteen thrusters powered with the hypergolic propellants monomethyl hydrazine (fuel) and nitrogen tetroxide (oxidizer). Eight of the thrusters were low-thrust vernier engines located around the capsule for attitude control, rated at 11 N (25 lb_f). Six thrusters rated at 444 N (100 lb_f) were used for orbital maneuvers, as were the two slightly lower-thrust forward engines rated at 387 N (85 lb_f).

Reentry Control System (RCS) Four solid-fuel motors were placed in the Gemini reentry control section located between the reentry crew capsule and the supply and service adapter module. The retrograde rocket package was also designed to separate the capsule from the launcher during an emergency ascent abort. Deorbit was possible with only three of the four titanium-case motors operational. The four motors were fired in a 5.5 sec ignition sequence for the deorbit, or simultaneously if triggered by an ascent abort command.

Rendezvous Radar The Gemini capsule was designed with a transponder-based radar system for rendezvous activities with the Agena target vehicle. The long-range rendezvous radar package was located in the top recovery section with dual polarized antennas to receive both transponder ID and reflected pulse data. The rendezvous radar also contained a command channel that allowed Gemini astronauts to control the Agena vehicle. Data readout from the rendezvous radar included target vehicle range, range-rate (velocity), and relative angle.

FUEL CELLS

Electrical power for the Mercury capsule was supplied by batteries that have an inherently low energy density. While batteries were adequate for powering Mercury's short-duration missions, they were unsuitable for Gemini's longer flights. Fuel cells and solar panels were considered the most likely candidates for powering the Gemini capsule, with fuel cells winning out over solar arrays for their weight savings, system simplicity, and the water generated in their operation, which was the biggest advantage. In early 1962, the decision was made to find a contractor to develop Gemini's fuel cell with the

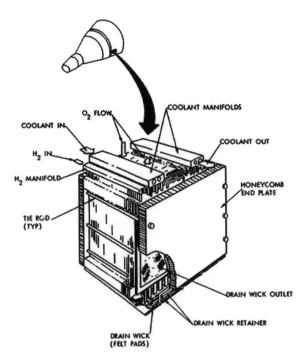

Figure 7.3. Cutaway of the Gemini fuel cell components showing the manifolds for circulating oxygen and hydrogen reactants, and the Freon coolant loop. Source: NASA SP-4203.

intention of incorporating the advances made in the Gemini program fuel cells into the Apollo capsule power supply.

Fuel cell development for Gemini was not simple, nor without its problems. Nevertheless, the fuel cells were a much better compromise than the heavier batteries, or the solar arrays that required a secondary battery power source while the capsule was in Earth's shadow. General Electric won the contract to develop and deliver fuel cells for the Gemini spacecraft, rated for 490 amps at 24 Vdc using hydrogen and oxygen reactants. Because of the difficulty in fuel cell development, however, the first two manned Gemini flights used an all-battery electrical supply from six silver-zinc batteries rated for 400 Amp-hours at 24 Vdc each. Gemini V through Gemini XII were outfitted with dual GE fuel cell units, with additional power supplied by backup silver-zinc batteries, along with two silver zinc squib batteries for pyrotechnic operations (Grimwood, Hacker, and Vorzimmer 1968).

EJECTION SEATS

Safety concerns multiplied as NASA expanded manned space flight programs and technologies. And as the systems became more complex and the boosters grew in size and thrust, the increased hazards demanded improvements in the capsule systems and their reliability for the entire collection of Gemini hardware. One of the best examples of the safety improvements on Gemini was the adoption of the Air Force's new Titan booster, which offered much greater payload lift using less explosive propellants. This gave designers the flexibility to use the lighter, simpler, safer ejection seats for the Gemini capsule.

The final design of the Gemini capsule was outfitted with dual ejection seats beneath dual hatches and hatch doors. Gemini's dual hatch and door arrangement allowed capsule clearance for crew ejection, crew ingress and egress, as well as on-orbit EVA operations. Ejection seats selected for Gemini were similar in design to the rocket-catapult (ROCAT) aircraft ejection systems that housed landing parachutes and flotation hardware for water landings. Although not practical for the Apollo capsule, ejection seats were adopted for the high-risk initial space shuttle flights. STS 1–4 missions were shakedown test flights using the Columbia orbiter that carried a crew of two with individual ejection seats.

LIFE SUPPORT AND ENVIRONMENTAL CONTROL

Gemini's crew life-support system, called the environmental control system (ECS), consisted of five subsystems. These included three oxygen supply loops, a water- and waste-management system, and an active coolant loop. Gemini's ECS design was configured for mission durations of up to two weeks, with shorter missions carrying consumable masses in proportion to the mission duration.

Oxygen Supply System

Oxygen was supplied to the Gemini crew in the cabin as well as in the pressure suits to support the capsule's EVA capability, in similar fashion to Mercury. Three oxygen supply sources were furnished for three types of spacecraft operations: Primary, Secondary and Egress.

- Primary
  - Stored as cryogenic fluid
  - Supplied to both the suit and cabin during normal orbital operations
- Secondary
  - Stored as gas in pressure tanks
  - Used as backup for the O_2 primary system, but became the primary source when the adapter supply (primary supply located in the adapter) was jettisoned for deorbit
- Egress (EVA)
  - Stored as gas in pressurized tanks
  - Used as O_2 supply for suits during EVA and for emergency ejection below 70,000 ft

Oxygen was supplied in two separate loops for redundancy in case of cabin or suit depressurization.

- Cabin loop
  - Supplied cabin environment O_2 for normal orbital operations (backup loop)
- Suit loop
  - Supplied space suit O_2 for most mission operations (primary loop)

Atmosphere Revitalization

- CO_2 was removed from the cabin and suit loops by lithium hydroxide canisters
- Particulates in the cabin and suit oxygen loops were removed by fine-mesh filters
- Odor and toxic vapors were removed by activated charcoal filters

Water- and Waste-Management System

Water for the crew was supplied by fuel cells and stored in tanks used to collect and dispense water.

- Water disposal was through external water dump line or water evaporator (coolant loop)
- Solid (fecal) waste was stored in bags

Coolant System

Cooling for the electronics and the crew on Gemini required a system capable of three times the heat load of the Mercury capsule and ten times the duration. A dual-redundant, active, liquid cooling loop system was designed with a pump and circulation subsystem for each loop, as well as heat exchangers and dual space radiators placed on the capsule's aft panels. During ground operations and during peak heat loads, the system was supplemented with a water boiler/evaporator for additional cooling.

- A cooling loop fluid called MSC-198 was developed by Monsanto with a freezing point below -100°F.
- Heat loads from fuel cells, batteries, and electronics were conducted through cold plates and transferred to the cooling loop.
- Radiator surface area in the retro section panel and equipment section panel was 15.3 m² (165 ft²).

TITAN II LAUNCHER

Titan, like Atlas and the Redstone, was a military ballistic missile adapted for space launch duties in the Cold War era. The first Titan launcher, Titan I, was built by the Martin Company for the USAF as a backup for the Atlas ICBM, but with a higher payload capacity. Titan was the insurance that at least one of the USAF's ICBMs would be available for strategic deployment in the early 1960s. However, Titan I had several design weaknesses, one of which was the LOX and RP-1 propellants. Liquid oxygen requires cryogenic storage and cannot be stored in the missile's propellant tanks for an extended period. Hence, LOX must be loaded before launch, allowing only a limited window for the missile's availability. A slower response time for missiles with cryogenic propellants was just one more reason why the first ICBM designs were retired early. Although the Titan I did conform to the contract specifications and the expectations of the USAF, it was soon replaced by the Titan II, also built by Martin.

The two-stage Titan II was an improved version of its predecessor, but with storable propellants consisting of hydrazine and unsymmetrical dimethyl hydrazine (UDMH), and nitrogen tetroxide (NTO). Although highly toxic, the storable hypergolic propellants offered the rapid response needed for its intended ICBM role. The Titan booster, which came into service as an ICBM in 1963 also proved useful as a medium-lift launcher for the Air Force reconnaissance satellites, for NASA's Gemini manned and exploration missions, and for NOAA's weather satellite launches. Later Titan III and IV models included strap-on solid rocket boosters that more than doubled the payload lift

Table 7.1. Gemini Capsule Specs

Height	5.67 m (18.6 ft)
Diameter	3.05 m (10 ft)
Volume	2.55 m³ (390 ft³)
Weight (total)	3,851 kg (8,490 lb)
Reentry module	1,983 kg (4,372 lb)
Retrograde module	591 kg (1,303 lb)
Equipment module	1,277 kg (2,815 lb)

of the improved Titan II core which, until retirement in 2005, was America's highest-payload-capacity expendable launch vehicle.

Titan's first stage was powered by two Aerojet LR-87-7 engines with gimbaled steering. Total thrust for the first stage was 1,900 kN (430,000 lb$_f$). The Titan II second stage was powered by a single Aerojet LR-91-7 engine that generated 445 kN (100,000 lb$_f$) thrust and also burned hydrazine and nitrogen tetroxide. Payload capacity to low Earth orbit for the two-stage booster was 3,750 kg (8,250 lb). Production for the Titan II encountered several delays due to "pogo" fluid oscillations in the fuel lines, and from combustion instabilities in the first-stage engine. Approval of the Gemini program was delayed even further because of the more critical qualification specifications, and the reorganization of the design group, which no longer included the program's lead designer, James Chamberlin (Hacker and Grimwood 1977).

AGENA TARGET VEHICLE

Gemini program objectives were straightforward in their explicit support for the Apollo program. Operations and hardware focus was on rendezvous and docking procedures that were essential for the lunar-orbit rendezvous (LOR) lunar mission, and for connecting the command and service module with the lunar module on the trip to the Moon and back. To simulate Apollo's dual spacecraft operations, Gemini flights required two vehicles, as well as two launchers, although little time was available to develop either. Hence, Gemini planners needed flight hardware that was already developed, or at least in the development process. The dual-pilot Gemini capsule was in the final stages of design while the Titan II launcher was undergoing man-rating. The last major hardware item needed was a rendezvous vehicle, and that already existed as a military satellite upper-stage booster called Agena, already adapted for launch on the Atlas.

Agena was developed as an upper-stage satellite booster for the early military surveillance satellites that included the Corona/Discover, the MIDAS, and the ELINT.

Table 7.2. Titan II Specs

Stages	2	
1st stage	Engines	2 LR87-AJ-5
	Thrust	1,900 kN (430,000 lb$_f$)
	Burn time	156 seconds
	Fuel	Hydrazine (50%) + UDMH (50%)
	Oxidizer	Nitrogen tetroxide (N$_2$O$_4$)
2nd Stage	Engine	1 LR91-AJ-5
	Thrust	445 kN (100,000 lb$_f$)
	Burn time	180 seconds
	Fuel	Hydrazine
	Oxidizer	Nitrogen tetroxide
Payload (LEO)	3,750 kg (8,250 lb)	

Agena A was first launched successfully in 1960, followed by a redesign to improve engine performance and to double its fuel capacity. Agena A became the Agena B series, then the C model that was proposed but never built. The most successful in the series was the Agena D upper stage. At the program's conclusion in the 1970s, Lockheed had delivered a total of 365 Agena vehicles for DoD and NASA missions. Just as impressive were the variety of launchers that carried the Agena, including the Thor (predecessor of the Delta), Atlas, and Titan converted missiles.

One of the most important improvements on the Agena D spacecraft was the capability for main engine restart. That gave the Agena target vehicle the ability to change orbit for rendezvous and docking exercises, and for changing the orbit altitude of the docked Gemini. It also provided the necessary propulsion for Gemini and Agena to deorbit. Agena proved so versatile that it was used as an upper stage booster on NASA's Lunar Ranger, Lunar Orbiter, Mariner interplanetary missions, and on several other space exploration programs.

Agena was made available to NASA planners by virtue of the cooperative agreements between NASA, the Department of Defense, the military branches, and the intelligence agencies. Agena serves as a good example of the procurement agreements made between NASA and the USAF Space Systems Division (SSD) during and after the Cold War space race. Similar efforts preceding Agena included the U.S. Army Redstone IRBM launcher, the USAF Atlas ICBM, and the USAF Titan II used for the Mercury and Gemini projects.

NASA's Gemini-Agena vehicle, designated the Gemini Agena Target Vehicle (GATV), was built originally for the USAF by Lockheed but underwent numerous

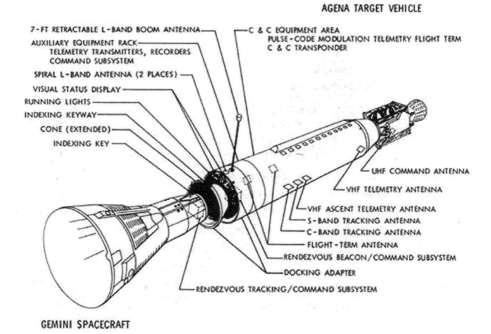

Figure 7.4. Sketch of the external component layout of the Agena specific to the Gemini target vehicle configuration. Included are the external communications, docking, and rendezvous components, and a docked Gemini capsule. Source: NASA SP-4203.

modifications to accommodate the docking unit, communications gear, a rendezvous radar transponder, and additional propulsion systems. Increased propulsion capability for the vehicle consisted of a restartable main engine for orbital operations and an attitude control system, both of which were controlled by an autonomous GN&C system. Communications control channels were added for flight commands from both ground and Gemini astronauts. Identification, docking beacon lights, and a status panel were also added to the front section to aid Gemini astronauts in docking operations.

Gemini's Agena primary propulsion was supplied by a gimbaled 70,280 N (15,800 lb$_f$) thrust, multiple-start, Bell XLR-81 liquid-fuel engine. Augmenting the main engine was a secondary propulsion system consisting of four thrusters—two 890 N (200 lb$_f$) and two 71 N (16 lb$_f$) engines—for small velocity adjustments. Secondary engine propellants were UDMH and MON (mixed oxides of nitrogen). Attitude control of the Gemini-Agena was powered by compressed nitrogen gas thrusters (Grimwood, Hacker, and Vorzimmer 1968).

Control commands for the Gemini-Agena were routed through the UHF command link, while spacecraft telemetry was transferred to ground with the VHF downlink. Onboard transmitters used for tracking operated in the C-band and S-band, in addition to L-band tracking radar. Power was supplied to the Agena vehicle by twelve onboard batteries; AC power needed for avionics was generated by three-phase, 400 Hz inverters, fed by the DC battery buses. A number of science experiments and telemetry instruments were also added to the Gemini-Agena vehicles (Grimwood, Hacker, and Vorzimmer 1968).

Table 7.3. Agena Specs

Length	7.56 m (24.8 ft)
Diameter	1.52 m (5 ft)
Weight	7,160 kg (15,800 lb)
Propulsion: Engine	Bell XLR-81 BA-9
Thrust	71.1 kN (16,000 lb$_f$)
Propellants	Inhibited red fuming nitric acid + UDMH

Gemini Missions

Gemini flight tests began with two unmanned qualification flights for the Titan II launcher and functional Gemini capsule designated Gemini I and Gemini II. Before that, individual launcher and capsule tests had to be completed that were delayed by a number of problems with the Titan II, with Agena's redesign, and with the Gemini capsule. The delay pushed back the projected schedule for the initial capsule flights from 1963 to early 1964. Resolution of the production problems and operational testing complications led to the first successful launch of the Gemini-Titan pair on April 8, 1964.

Figure 7.5. Agena Test Vehicle photographed before docking with the Gemini-8 capsule. The docking collar is on the right, the main engine on the left. Source: NASA NIX.

UNMANNED

GT-I (GLV-1)

Launch of GT-1 on April 8, 1964, was a successful test of the orbital sequencing on the Titan launcher and the Gemini capsule. The three-orbit test completed the planned objectives, followed by spacecraft reentry four days later without a ground-initiated deorbit command.

GT-II (GLV-2)

Gemini-Titan's second launch was a suborbital test of the capsule's reentry heat protection equipment on a trajectory that produced a maximum heating rate reentry. Instrument telemetry confirmed the capsule's structural integrity, and completed the man-rating for the Titan II launcher.

MANNED

Gemini III (GT-3)

The inaugural flight of Gemini was a three-orbit, 4.9-hour mission launched on March 23, 1965, with Virgil (Gus) Grissom (command pilot) and John Young (pilot) at the controls of Gemini III. This was the first American multiple-crew space mission, just five months after the Soviet's Voskhod three-man flight. The Gemini III mission was an important milestone in America's space technology overtaking the Soviet Union's, though neither side was aware of the shift until years later. Since this was the first manned Gemini flight, the mission objectives were primarily evaluation of the systems and basic flight operations, which included the following:

- Evaluate the overall Gemini capsule design
- Verify communications with the worldwide tracking network
- Evaluate the orbit attitude and maneuvering system (OAMS)
- Evaluate reentry control, landing, and recovery
- Verify spacecraft systems (Hacker and Grimwood 1977)

Primary objectives for Grissom and Young's GT-3 mission were almost all satisfied, with the exception of the controlled reentry, which was only partially achieved, since the reentry angle of attack was lower than expected. Predicted splashdown was 84 km off, and recovery was delayed for about thirty minutes.

The GT-3 spacecraft was named "*Molly Brown*" by Gus Grissom, as an ironical reference to the character in the stage play *The Unsinkable Molly Brown*—and his not-so-unsinkable Mercury capsule. NASA administrators did not share Grissom's sense of humor and ended the practice of crews naming their capsule for the rest of the Gemini program. After Gemini III, the capsules were assigned Roman numerals. The informal custom that began with the Mercury astronauts' naming their capsules was restarted during the Apollo lunar missions when the crews were allowed to choose the name of their command and service modules and lunar modules. The GT-3 flight was punctuated with a little more humor when John Young began eating a smuggled corned beef sandwich aboard, but had to stop when crumbs began floating in the cockpit. The GT-3 capsule is on display at the Grissom Memorial in Spring Mill State Park, near Grissom's hometown of Mitchell, Indiana.

Gemini IV (GT-4)

The second manned Gemini-Titan mission, designated GT-4, was launched on June 3, 1965, for a four-day, sixty-two-orbit flight. GT-4 is best known for astronaut Ed White's first American spacewalk. Command pilot James McDivitt remained in the capsule in his pressurized suit, while pilot Ed White conducted his twenty-two-minute EVA that included the first ever use of an independent propulsion system. The pressurized nitrogen thruster system called the handheld self-maneuvering unit was one of the EVA highlights that ended when he was ordered back into the capsule by commander McDivitt to avoid depleting oxygen needed for the remaining mission.

Objectives of the second manned Gemini mission were to evaluate work procedures and work schedules while in orbit and the effectivity of flight planning for an extended space flight. Secondary objectives included demonstrating extravehicular activity in space, conducting station-keeping and rendezvous maneuvers, evaluating spacecraft systems, demonstrating the capability to make significant in-plane and out-of-plane maneuvers, using the maneuvering system as a backup reentry system, and conducting eleven experiments. Most of these were completed, although several important operations were not, or were only partially satisfied. A rendezvous demonstration with the Titan second stage was not successful because of the flawed technique used for catching up with the booster. Forward thrust (thrusters on the Gemini firing aft to increase vehicle's forward speed) moved the capsule to a slightly higher orbit, and actually slowed the capsule relative to the booster. Later missions used rendezvous operations that placed the Gemini in a lower orbit, then used aft thrust to slow the vehicle, which in turn moved the capsule forward relative to the Agena target vehicle.

Automated reentry was not possible because of onboard computer problems in the last half of the mission. This forced the crew to manually execute the deorbit, which resulted in a high-g ballistic reentry. Splashdown was approximately 80 km short of the intended impact, although recovery was not delayed significantly. The Gemini IV capsule is on display at the National Air and Space Museum in Washington, D.C.

Gemini V (GT-5)

The primary purpose of the Gemini V mission, launched August 21, 1965, was to double the previous Gemini mission length with an eight-day flight. This was equivalent to the flight duration of a trans-lunar mission to the Moon and back, and also eclipsed the longest Soviet space mission at the time. Gemini V's crew members, Gordon Cooper (command pilot) and Charles Conrad (pilot), completed the seven-day, twenty-two-hour, fifty-five-minute checkout flight of the capsule systems and the new fuel cells that were used for the first time on a manned flight. System evaluation covered the new guidance and navigation system to be used on future Agena rendezvous missions. Fuel cell tests were considered successful, although there was an oxygen supply problem that curtailed the tests. A separate radar device used to evaluate rendezvous equipment and operations was released from the Gemini capsule as a sub-satellite, but fuel cell power problems on Gemini forced cancellation of the experiment (Hacker and Grimwood 1977).

- Primary objectives
  - Demonstrate a long-duration orbital mission
  - Evaluate the effects of long periods of weightlessness on the crew
  - Test rendezvous capabilities and maneuvers using a rendezvous radar evaluation pod
- Secondary objectives
  - Demonstration of all phases of guidance and control systems to support rendezvous and controlled reentry guidance
  - Evaluate the fuel cell power system and rendezvous radar
  - Test the capability of either pilot to maneuver the spacecraft in orbit to close proximity with another object
  - Conduct seventeen experiments

Although an OAMS thruster stopped working during the fifth day in orbit, reentry went as scheduled. A programming error sent from ground brought the capsule down about 130 km short of its intended landing. The Gemini V capsule is on display at NASA's Johnson Space Center.

Gemini VII (GT-7)

The fourth manned flight in the Gemini-Titan series was GT-7, launched on December 4, 1965, with astronauts Frank Borman (command pilot) and James Lovell (pilot) aboard. The fourteen-day mission (13 days, 18 hr, 35 min) was to follow the Gemini VI flight into orbit, but the Agena launched on October 25 as a rendezvous vehicle for the Gemini VI capsule ended with a propulsion failure six minutes and sixteen seconds into the flight. Because of the importance of the rendezvous mission, the decision was made to substitute the Gemini VII capsule for the Agena target vehicle on the Gemini VI mission. Gemini VI, which was then called Gemini VIA, was launched eleven days after Gemini VII (Hacker and Grimwood 1977).

Objectives for the GT-7 mission included:

- Successful demonstration of a two-week flight
- Perform station keeping with the Gemini launch vehicle's second stage
- Evaluate the Gemini capsule's shirt-sleeve environment and the lightweight pressure suit
- Act as a rendezvous target for Gemini VI
- Demonstrate controlled reentry close to the targeted landing point
- Perform three scientific, four technological, four spacecraft, and eight medical experiments

Gemini VII had achieved the manned mission endurance record even before launch of Gemini VIA. The twenty experiments on Gemini VII were the most of any prior mission, many of which were rendezvous and station-keeping exercises with the Gemini IVA capsule. Fuel cells were flown on the capsule, although backup batteries were used for primary power on this and the previous mission.

Figure 7.6. View of the Gemini VII capsule from Gemini VI-A. Source: NASA JSC.

Although problems arose with the OAMS thrusters and the fuel cells during the mission, deorbit and splashdown were uneventful and went as planned, with the landing only 10 km from the planned point. The Gemini VII capsule is on display at the Smithsonian National Air and Space Museum annex at Washington Dulles International Airport, Virginia.

Gemini VI-A (GT-6)

Gemini VI was delayed in its launch because of the Agena Target Vehicle launch failure. GT-VI was placed in orbit on December 15, 1965, for a one-day mission, with the designation VI-A. The crew of Walter Schirra (command pilot) and Thomas Stafford (pilot) participated in America's first space rendezvous with Borman and Lovell on Gemini VII. Rendezvous exercises included station-keeping for more than five hours, at distances that ranged from 0.3 m to 90 m (1 to 295 ft) (Hacker and Grimwood 1977).

With the Agena target vehicle unavailable, the primary objective of the Gemini VI-A mission was switched to a rendezvous exercise with the Gemini VII capsule.

Secondary objectives included:

- Perform closed-loop rendezvous in the fourth orbit
- Station keeping with Gemini VII
- Evaluate reentry guidance capability
- Conduct visibility tests for rendezvous using Gemini VII as target
- Perform three experiments

Gemini VI-A's one-day mission ended successfully, with a nominal deorbit and reentry, and a splashdown within 18 km of the programmed splashdown. The Gemini VI-A capsule is on long-term loan from the Smithsonian Institution, which includes appearances at several sites throughout the United States.

Gemini VIII (GT-8)

Gemini-Titan VIII was launched on March 16, 1966, just one hour and forty-one minutes after the Agena target vehicle was placed into orbit for rendezvous and docking exercises. The ten-hour, forty-one-minute Gemini flight was commanded by Neil Armstrong and piloted by David Scott. This was the first crew ever to dock with another vehicle in space.

Primary and secondary mission objectives for this flight, which included rendezvous and docking tests with the Agena target vehicle, EVA, and ten experiments, were scrubbed after the docked Gemini and Agena began tumbling twenty-seven minutes after the initial docking. One of Gemini's OAMS thrusters was stuck open, making manual control of the capsule and docked Agena ineffective. Since telemetry from the capsule was not available to ground controllers because the spacecraft was between ground stations, Armstrong uncoupled the Gemini capsule from the Agena to reduce the rotations. But because of the unrecognized OAMS thruster failure, separating the vehicles increased the rotation rates. As the Gemini capsule was reaching structural limits and blackout conditions for the crew, Armstrong and Scott deactivated the OAMS system, followed by activation of the Reentry Control System (RCS) to stabilize the capsule. Having depleted most of the RCS propellants, the crew was forced to deorbit for an immediate landing because of Gemini safety rules. The Gemini VIII mission ended after just seven orbits. Since the Gemini capsule landed in the Pacific instead of the Atlantic Ocean, recovery was switched to an alternate ship, which, by sheer luck, was close enough to the splashdown to have the crew on the recovery ship deck just three hours after splashdown (Hacker and Grimwood 1977).

Armstrong was recognized for his calm, analytical recovery from a potentially fatal space flight emergency with the prime assignment as commander on the first Apollo lunar landing. The GT-8 capsule is on display at the Neil Armstrong Air and Space Museum in Wapakoneta, Ohio.

Gemini IX-A (GT-9A)

GT-9 was the seventh manned flight and the third rendezvous mission in the Gemini series, and arguably the most jinxed of any of NASA's manned missions. GT-9A's three-day twenty-one-hour mission was flown by backup astronauts Thomas Stafford (command pilot) and Eugene Cernan (pilot), who replaced the prime crew of Elliott See and Charles Bassett, who were killed in their training aircraft at the McDonnell aircraft hangar in St. Louis on February 28, 1966. The original launch, scheduled for May 7, 1965, was postponed until June 3 when the Agena target vehicle exploded two minutes into the launch. A substitute target vehicle, called the Augmented Target Docking Adapter (ATDA) was readied, then launched on June 1. Because of delays in the Agena program, the ATDA was introduced as a simpler target vehicle with neither main engine nor propellant tanks. Its lower cost and use of already-tested hardware made it attractive enough to fabricate in case of delays in the Agena, and timely enough to be available for Gemini IX on short notice.

GT-9A's fortunes did not improve with its launch, however. After reaching orbit, the crew succeeded in a rendezvous with the ADTA on its third orbit. As they approached, the crew confirmed visually what ground telemetry had already indicated—the ATDA's shroud was still attached to the target vehicle. It was clear that GT-9A would not be able to dock with what Stafford described as the "angry alligator." And with the docking port blocked, the crew could only execute close-proximity rendezvous maneuvers and an EVA exercise before returning to Earth (Hacker and Grimwood 1977).

While the three modified rendezvous exercises went as planned, Cernan's EVA did not. Delayed until day three, the EVA was primarily a checkout of the new astronaut maneuvering unit (AMU), the predecessor to the manned maneuvering unit (MMU) used in the space shuttle program. Cernan found that movement was difficult, and that activity on the capsule was difficult because of a lack of handholds. Forced-air cooling of his space suit was also inadequate in removing perspiration and water vapor resulting from strenuous exercise that fogged up Cernan's visor.

However troubled, Gemini IXA was not without its rewards. In addition to the success of the difficult rendezvous procedures, a number of important hardware improvements were introduced. Because of the limited thermal capacity of the Gemini suits, the follow-up Apollo space suits incorporated liquid water-cooling as a garment underlayer for more effective heat transfer. Handholds and footholds were also placed on the Apollo and Skylab vehicles to facilitate spacewalks that proved difficult on the Gemini EVAs. In addition, the simple and somewhat ineffective astronaut maneuvering unit was improved over the years and later used to increase mobility on space shuttle EVAs. NASA's MMU was the first to allow astronauts to distance themselves from their spacecraft without a tether line.

- Initial mission objectives for the GT-9A flight were:
  - Demonstrate rendezvous techniques and docking with a target vehicle to simulate maneuvers to be carried out on future Apollo missions
  - Perform a spacewalk to test the astronaut maneuvering unit
  - Demonstrate Gemini's precision landing capability
- Scientific objectives were:
  - Obtain zodiacal light and airglow horizon photographs
  - Complete two micrometeorite studies, as well as one medical and two technological experiments

Deorbit, reentry, and splashdown were three events in the GT-9A mission that went as planned. In fact, the precision reentry landed the capsule and crew

Figure 7.7. Photograph of the "angry alligator" Augmented Target Vehicle taken from the GT-9A capsule. Source: NASA JSC.

just 700 meters from the landing point. But because the rest of the mission was so riddled with problems, a commission was empanelled to evaluate future Gemini missions to ensure their objectives would support the Apollo development program.

Gemini X (GT-10)

NASA's eighth manned mission in the Gemini series was launched on July 18, 1966, with crew members John Young (command pilot) and Michael Collins (pilot). Mission objectives for GT-10 focused on further testing of Gemini's docking and rendezvous capability with the GATV target vehicle, in addition to EVA activities and Gemini systems tests.

Both the Agena GATV and the Gemini capsule were launched successfully, with the Gemini X launched one hour and forty minutes after the GATV into a lower catch-up orbit for a planned rendezvous during its fourth orbit. But an orbit plane adjustment to bring Gemini to the same orbit plane as the Agena consumed nearly 60% of the propellants, resulting in the cancellation or scaling-back of a number of the objectives. Docking with the Agena target vehicle was successful, which satisfied the primary mission objective. And for the first time, the docked Agena propulsion system was used for an orbit altitude change, taking both vehicles to an apogee of 763 km.

Two EVA exercises were completed on the three-day mission (2 days, 22 hr 47 min), the first being a standup EVA that included UV telescope observations and photographs from the cabin. The second EVA followed an undocking with the Gemini X GATV, then a rendezvous with the nearby Gemini 8 GATV spacecraft. Collins retrieved a micrometeoroid experiment package from the Gemini Agena 8 target vehicle on his tethered but difficult spacewalk between spacecraft, which that lasted about thirty-nine minutes.

- Primary objective:
  - Rendezvous and dock with Gemini Agena target vehicle
- Secondary objectives:
  - Rendezvous and dock in fourth revolution
  - Rendezvous with Gemini Agena target vehicle GATV-8 using the Agena propulsion systems
  - Conduct two EVAs
  - Practice docking
  - Perform fourteen experiments
  - Perform system evaluation on bending-mode tests, docked maneuvers, static discharge, post-docked Agena maneuvers, reentry guidance
  - Park the Gemini Agena target vehicle in a 352 km (190.3 nautical mile) orbit

Deorbit, reentry, and splashdown were executed as planned, with a landing less than 6 km from the planned impact point. Crew and capsule were recovered by the USS *Guadalcanal* on July 21, 1966. The Gemini X capsule is currently on display at the Kansas Cosmosphere and Space Center, Hutchinson, Kansas.

Gemini XI (GT-11)

GT-11, the ninth Gemini manned mission, was launched on September 12, 1966, with an experienced crew that included Charles Conrad (command pilot) and Richard Gordon (pilot). The complex three-day mission assignment included tethered operation of the Gemini and Agena vehicles, as well as an automated reentry. While difficult, the mission was successful in establishing a number of firsts and records, which included the first one-orbit rendezvous between two spacecraft (the GATV was launched 1 hr 40 min before the Gemini), the rotation of two spacecraft about a common center-of-mass connected by a tether, and a manned orbital altitude record of 1,374 km, which was accomplished with the docked Agena propulsion system. That altitude record still stands for manned orbital missions, although the nonorbital Apollo lunar missions far exceeded the GT-11 altitude record.

One of the most unique and possibly far-reaching experiments on a manned flight was the tethered spacecraft rotation exercise that began with the tether attached to both spacecraft by Gordon on his first EVA. On the following day—the last day of the mission—Gemini was undocked from the Agena GATV and backed away from the GATV until the tether was taut. Gemini's thrusters were then fired for a right-angle push to start the two vehicles rotating about the center of mass, generating modest artificial gravity in the two vehicles. Testing was done at higher rotation rates, and, while constant tension was difficult to maintain on the tether, oscillations between the two vehicles were damped successfully, all of which proved the concept of rotating tethered vehicles and artificial gravity workable (Hacker and Grimwood 1977).

A single, primary objective for the GT-11 on its two-day, twenty-two-hour, seventeen-minute mission was to rendezvous and dock with the Gemini Agena target vehicle.

Secondary objectives included:

- Practice docking
- Perform three EVAs
- Conduct eleven experiments
- Maneuver while docked (high apogee excursion)
- Conduct tethered vehicle test
- Demonstrate automatic reentry
- Park GATV-10 in a 352 km orbit

Following the two-orbit tether experiment, the Gemini capsule was prepared for deorbit and reentry using an automated system for the first time. Conrad and Gordon's GT-11 mission was as close to a complete success as any of the Gemini missions, including its deorbit and reentry, which brought the capsule and crew to within 4.5 km of the intended splashdown. The GT-11 capsule is on display at the California Science Center in Los Angeles, California.

Gemini XII (GT-12)

NASA's Gemini bridge program was coming to a close in 1966 as the hardware test flights for the Apollo project were just beginning. Since the first Apollo manned flight was scheduled in early 1967, the final Gemini docking and rendezvous mission was of prime importance. Also critical to the preparations for Apollo missions was a demonstration that EVA activities could be done efficiently and with minimal difficulty. To aid in the EVA training, NASA introduced underwater spacewalk simulation at the Johnson Space Center for the tenth and last Gemini manned mission. This training technique carried over to all of NASA's subsequent spacewalk missions, including EVA operations on the International Space Station.

GT-12, NASA's last Gemini mission, was launched on November 11, 1966. The planned four-day mission piloted by James Lovell (command pilot) and Edwin (Buzz) Aldrin (pilot) launched to orbit one hour and thirty-six minutes after the Agena target vehicle. Lovell and Aldrin's successful three-EVA mission included the longest spacewalk in the program and two additional standup EVAs. A manual docking with the Agena confirmed the crew's ability to dock after an automated guidance system failure. Following the second EVA, and, after undocking from the GATV, Gemini XII astronauts executed a second tethered experiment with the assistance of gravity gradient stabilization and controlled artificial gravity in a four-hour exercise. A planned altitude boost using the Agena propulsion system was one of the few cancelled objectives during a productive mission that closed out the Gemini program. The three-day, twenty-two-hour, thirty-four-minute flight included a successful automated reentry and a splashdown less than 5 km from the planned splashdown point. The GT-12 capsule is now on display at the Adler Planetarium, Chicago, Illinois.

- Primary objectives for the GT-12 mission included:
  - Rendezvous and docking with the GATV
  - Execute three EVAs
- Secondary objectives included:
  - Tethered vehicle operation
  - Perform fourteen experiments
  - Rendezvous and dock in the third orbit
  - Demonstrate automatic reentry
  - Perform docked maneuvers
  - Practice docking
  - Conduct system tests
  - Park Gemini Agena target vehicle in a 556 km (300 mi) orbit

Although the Gemini project came to a close with the return of the Gemini XII crew and capsule, the engineering experience, hardware technology, and seasoned crews went on to serve as an important foundation for the Apollo program.

MILITARY GEMINI

Success in the Gemini program went far beyond its program accomplishments, and beyond the hardware and operational legacy that was transferred to the Apollo project. Among a long list of other important innovations introduced in the Gemini program was the capsule design. Starting simply as an enlarged Mercury vehicle, structural improvements in the capsule design allowed a more versatile platform, while improvements in the systems advanced the necessary technology that enabled the Apollo command module to reach the Moon. And in a period when NASA and the USAF were looking beyond the Apollo lunar program, Gemini's innovations sparked interest in more advanced vehicles for both civil and military applications.

Blue Gemini

Cancellation of the USAF X-20 Dyna Soar project in 1961 removed the manned mission capability from the DoD's strategic military operations plans. But as the Gemini program progressed toward the first hardware tests in 1962–1963, the USAF planners became more interested in its capability. Rather than develop a new manned vehicle specifically for reconnaissance, the Air Force forwarded a proposal to incorporate the Gemini launcher and capsule in a variety of orbital military programs. An obvious plus was that the USAF heavy-lift Titan III boosters could launch the two-man military Gemini into polar orbits for more effective global coverage, and offer more secure reentry paths. Titan III could also boost heavier payloads into orbit that could be accommodated in modified single-passenger Gemini capsules. Informally known as Blue Gemini, the USAF proposals started with the simple concept of adopting NASA's Gemini hardware for initial training and evaluation. In a relatively short period, the concept progressed to developing highly modified Gemini hardware and sophisticated instrumentation to replace the canceled Dyna Soar lifting body orbital and reentry reconnaissance vehicle. In support of the project, two Air Force astronaut selections were held to begin training exercises for future Blue Gemini military pilots to fly alongside NASA's Gemini astronauts. But Gemini was firmly established as a civil space project, forcing the USAF planners to begin looking at a more radical approach for their future manned reconnaissance missions. On the drawing boards this time was an orbital space station called the manned orbital laboratory (MOL).

Big Gemini

As Gemini was coming to a close in 1966, USAF planners were negotiating the construction of a scaled-up Gemini spacecraft for the crew module on their manned orbital laboratory. After refurbishment, the second capsule flown (GT-2) had been turned over to the USAF for test flights, although by 1968, the MOL project appeared to be destined for cancellation. To take advantage of the advances in its design, Gemini contractor McDonnell Aircraft Company proposed a large manned vehicle for use in cargo and crew transport for either military or civil space stations. McDonnell's Big Gemini was to have a crew capacity of nine to twelve, along with a cargo capacity of two and a half tons. Large boosters for the vehicle would have been selected from those available at the time, including the Titan III, the Saturn IC, and Saturn IV-B, as well as the Saturn IB (Hacker and Grimwood 1977).

To make the proposal more attractive in terms of cost, McDonnell revived the concept from the early Gemini project proposals that employed a large parafoil for

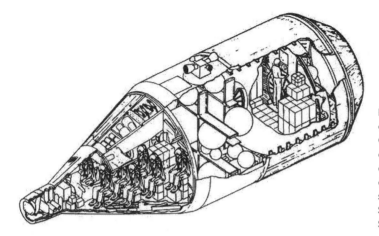

Figure 7.8. Cutaway of the Big Gemini concept crew vehicle (left) docked with a concept space station at right. Source: NASA SP-4011.

surface landings. A reusable capsule that had the capability to bring flight crews back to the launch site or to convenient landing strips was attractive because of its versatility and its much lower recovery costs. The one carryover directly from the Apollo and Mercury projects for Big Gemini would have been the launch escape tower.

A larger Gemini crew and cargo vehicle had other advantages. It would not interfere with the Apollo spacecraft and mission schedule, and it would have twice the capacity of the Apollo command module, but with the same capsule weight. In addition, NASA's Apollo Applications Program (AAP), which evolved into the Skylab space station, needed a crew vehicle, as did the USAF manned orbiting laboratory. But Big Gemini could not weather the budget crises that terminated the MOL project. Budget woes also had trimmed the AAP, which limited the crew flight vehicles to the Apollo Saturn rockets. In the end, Gemini vehicle derivatives never matured to the design completion and fabrication stage, although a prototype of the USAF's MOL was launched in 1969.

References

Grimwood, J. M., B. C. Hacker, and P. J. Vorzimmer. 1968. *Project Gemini—Technology and Operations: A Chronology*.SP-4002 Washington, DC: NASA, Scientific and Technical Information Office.

Hacker, Barton C., and James M. Grimwood. 1977. *On the Shoulders of Titans: A History of Project Gemini*. NASA SP-4203. Washington, DC: NASA Scientific and Technical Information Division, Office of Technology Utilization.

NASA. 1965. *Project Gemini Familiarization Manual*. SEDR 300.

Newkirk, Roland W., Ivan D. Ertel, and Courtney G. Brooks. 1977. *Skylab: A Chronology*. NASA SP-4011. Washington, DC: NASA Scientific and Technical Information Office.

Tomayko, James E. 1988. *Computers in Spaceflight: The NASA Experience*. NASA Contractor Report 182505, Wichita, KS: Wichita State University.

NASA GRIN (Great Images in NASA). Retrieved from: http://grin.hq.nasa.gov/

NASA NIX (NASA Image Exchange). Retrieved from http://nix.nasa.gov/

NASA. JSC Image Collection. Retrieved from: http://images.jsc.nasa.gov/

CHAPTER 8

Apollo

The first practical plans to put humans on the Moon came from the two space pioneers and adversaries, Wernher von Braun and Sergei Korolev. Initiatives for manned lunar missions first surfaced in the military rocket projects that were also directed by von Braun and Korolev, not because of the natural fit for lunar projects, but because both worked as missile designers and hoped to reach the Moon with their military boosters. In fact America's first plans to reach the Moon appeared on the Air Force and Army drawing boards well before the creation of NASA in 1958 and well before Kennedy's announcement of Apollo in 1961. Both the U.S. Army's Project Horizon and the Air Force's Lunex (Lunar Expedition) programs involved manned lunar outposts supported by crew and cargo vehicles that called for heavy-lift boosters far more powerful than the ICBMs being developed at the time. And while the Lunex and Horizon projects did not reach the hardware fabrication stage, studies of the launcher and flight hardware requirements and the need for extensive lunar research, as well as crew selection and training recommendations, proved useful for the later Apollo program plans.

A far-sighted but completely unrelated report was published in 1957 by the American Rocket Society that recommended funding lunar missions—specifically a manned lunar project—to begin in 1959. Also outlined in the report was the recommendation to establish a national organization to manage civil space projects and a manned lunar project, along with a timetable that was less than a year off the actual Apollo flight schedule (Compton 1989). Even earlier, the British Interplanetary Society, organized in 1933, began publishing articles in its popular journal on manned lunar missions.

Presidential Initiative

The creation of NASA on October 1, 1958, by the Congressional National Aeronautics and Space Act ushered in a number of space initiatives and program proposals, some already underway. These included the manned Mercury program and the embryonic manned lunar program that had yet to be named. Even with a name and a preliminary proposal, Apollo could not be positioned in the program planning process in 1960, since it had not yet been approved. Nevertheless, Apollo was considered important enough to become a key component of NASA's future space exploration programs. Since Apollo was an extraordinarily ambitious project in its goals, the budget was expected to be as just as lofty. Apollo's approval would therefore take more than the agency's blessing or congressional approval: it would require the popular support of the entire nation.

Kennedy's defeat of Nixon in the 1960 presidential election had minimal influence on the direction of U.S. space policy until the launch of the first Russian cosmonaut into orbit. Yuri Gagarin's historic flight in April 1961 ended the faint participation of the White House in the space race, much the same as the launch of the second Sputnik changed Eisenhower's reluctance to launch a satellite. Now under increasing pressure because of Gagarin's first manned flight, and perceived as losing ground in the Soviet missile buildup, Kennedy used his political acuity to create his own space race that could be won. A manned lunar program that was already named Apollo was researched by the vice president's office, then announced and promoted as a renewal of the American spirit, and even more important, a project that could reestablish America's military and national supremacy.

Mercury manned flights were being readied for launch in 1961, but were not part of a spectacular program that could surpass the Soviet space firsts that were quickly mounting. In a memo to Vice President Lyndon Johnson in April 1961, Kennedy asked for a review of space exploration projects that could reinvigorate America after falling behind the Soviets. A number of options were available, with the two most promising projects being a manned lunar landing project and a manned research space station. Johnson responded with a green light for the Apollo program already in progress. Gagarin's flight on April 12, 1961, was followed by Shepard's launch on May 4 and then by President Kennedy's congressional proclamation of a manned lunar program on May 25, 1961:

> I believe that this nation should commit itself to achieving the goal, before this decade is out, of landing a man on the moon and returning him safely to the earth. No single space project in this period will be more impressive

THE WHITE HOUSE
WASHINGTON

April 20, 1961

MEMORANDUM FOR

VICE PRESIDENT

In accordance with our conversation I would like for you as Chairman of the Space Council to be in charge of making an overall survey of where we stand in space.

1. Do we have a chance of beating the Soviets by putting a laboratory in space, or by a trip around the moon, or by a rocket to land on the moon, or by a rocket to go to the moon and back with a man. Is there any other space program which promises dramatic results in which we could win?

2. How much additional would it cost?

3. Are we working 24 hours a day on existing programs. If not, why not? If not, will you make recommendations to me as to how work can be speeded up.

4. In building large boosters should we put out emphasis on nuclear, chemical or liquid fuel, or a combination of these three?

5. Are we making maximum effort? Are we achieving necessary results?

I have asked Jim Webb, Dr. Weisner, Secretary McNamara and other responsible officials to cooperate with you fully. I would appreciate a report on this at the earliest possible moment.

to mankind, or more important for the long-range exploration of space; and none will be so difficult or expensive to accomplish.

PROGRAM APPROVAL

Congress was swayed sufficiently by Kennedy's address to increase funding almost immediately for NASA and its Apollo program, and for several other space projects. NASA's future was also buoyed by the public's interest in the new "Moon" project. Added to that were the successful Mercury launches that bolstered America's confidence in regaining the lead in space. A confluence of positive events and the approval of his new space project, as well as strong public support, gave President Kennedy the national backing needed to overtake the Soviets in the race to the Moon.

Apollo project approval in mid-1961 brought coherence to the program, but also burdened NASA with tremendous challenges as well. With Project Mercury underway and Gemini in the development phase, NASA's efforts did not have to change direction, since Apollo was already in progress. Yet dramatic changes in the scale of NASA's manned space program would be needed, considering the technologies that had to be developed and the experience needed to accomplish the Apollo program goals outlined by President Kennedy.

Apollo hardware efforts had a head start with the heavy-lift boosters being developed for the military. Booster engines for the earliest Saturn launchers emerged from Rocketdyne's Navaho engine, while the much-larger F-1 engines slated for the Nova and Saturn V had been turned over to NASA by the DoD and the USAF in 1959. Perceived as unnecessary by the DoD, the Saturn C5 proposed by von Braun and the F-1 engine were adopted by NASA for the Apollo program soon after. Apollo crew flight hardware also had the advantage of the Mercury and Gemini programs but was slowed by NASA's indecision on the method to reach the Moon.

Major Decisions

Three major decisions faced NASA managers, engineers, and contractors after Apollo's approval. The first was the basic design of the crew's flight and reentry vehicle. Design proposals coming from aerospace contractors had a wide range of possibilities—from an enlarged Mercury capsule to a cocoon-like lifting body. Just as challenging was the Apollo launch vehicle that would have to transport a crew of three to the Moon and back, with a stay on the lunar surface for up to one week. Most important was the choice of a flight method to get the hardware and crew to the Moon and back. Apollo's flight mode was especially important because it dictated the size of the booster, the configuration of the flight vehicle, and the choice of most major hardware options. Although the mass of the vehicle needed to support the crew for several days on the lunar surface was fairly well established, the hardware needed for the entire mission and the size and cost of the booster depended heavily on the flight method chosen. Apollo flight hardware awaited the decision on the transfer orbit and the flight vehicle combinations based on three possibilities:

1. Earth-Orbit Rendezvous (EOR): multiple launchers would place the Apollo lunar mission hardware into Earth orbit for docking followed by departure to the Moon.
   a. Advantages
      i. Required only small, low-cost boosters
      ii. Faster development time for small boosters
   b. Disadvantages
      i. Any delay or failure in any of the launches to Earth orbit would mean a delay or cancellation of the entire mission. This option presented a greater risk than the sum of advantages it offered.
2. Lunar-Orbit Rendezvous (LOR): launch with a single vehicle into Earth orbit, then boost the crew vehicles to the Moon for lunar orbit. To save propulsion mass, the vehicles would be divided into an orbiter and a lander at the Moon for lunar operations. This flight mode was ultimately selected for Apollo lunar missions.
   a. Advantages
      i. Launched with a single, large booster
      ii. Would result in the lowest combined mass for the flight vehicle and launch vehicle
   b. Disadvantages
      i. Booster design would be costly and could take five years or more
3. Direct Ascent: launch of the flight vehicle on a single booster would be placed in Earth orbit, then boosted to the Moon for a direct lunar landing. The direct ascent mode would bring the astronauts back without the need for a lunar orbiter.
   a. Advantages
      i. Would result in a simpler crew flight vehicle
   b. Disadvantages
      i. Enormous launcher needed
      ii. Much longer development time and cost for the booster than option 1 or 2

LAUNCH VEHICLE DECISION

The Apollo project's quick approval was a positive for the program's progress, but the end-of-the-decade deadline declared by President Kennedy limited the time available for development and testing. The first manned lunar missions would have to launch before 1970, yet decisions on the booster, the crew vehicles, and the flight mode had yet to be made. Because the booster design was completely dependent on the flight mode, NASA managers began in earnest to find the flight option that could satisfy the minimum payload mass, which, in turn, would determine much of the booster design.

In the meantime, NASA's Saturn heavy-lift launcher concepts were being prepared to help identify the most attractive booster for the various flight modes. Nearly all of the early Saturn series launcher designs employed Rocketdyne's F-1 engine under development for the Air Force. Even so, the possible combinations of booster stage sizes and the number of engines were almost limitless, exemplified by a separate booster concept named Nova. Like Saturn, Nova was a generic name given to a variety of heavy-lift boosters, though Novas were significantly larger than the Saturn C-5/V. Nova booster concepts were attractive because of their enormous lift capacity, but only

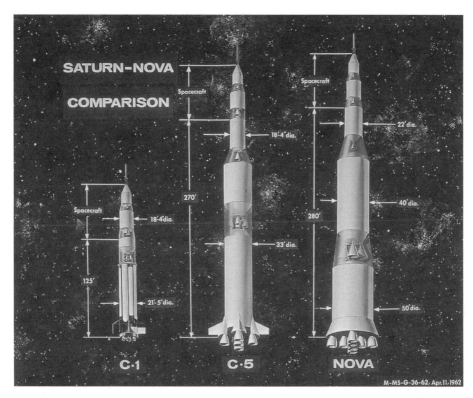

Figure 8.1. Scale comparison of the early Saturn 1 and Saturn V designs conceived in 1962 (designated C-1 and C-5 at the time), with the heavier-lift Nova booster on the right. Source: NASA MIX.

suitable for the direct-ascent mode. Even with its proposed use of the F-1 engines and the potential of fewer engines and/or smaller tanks, Nova's versatility, especially the engine and propellants for the upper stages, created another problem. Its larger capacity and enormous size would increase the time needed for the design and integration work and likely extend development time well beyond what would be required for the Saturn C-5/V. Ultimately, the Saturn C-5/V design won out over the smaller and larger breeds.

The Saturn C-5 was itself one variation of earlier Saturn concept classes named Saturn A, B, and C. Each of the classes had size designations ranging from 1 to 5. Within this collection, the largest was the Saturn C-5, the smallest, the Saturn A-1. Initial efforts by NASA and the STG were directed at developing the Saturn C-1 and C-4 until the early months of 1962. Then, on June 22, 1962, one of the most important meetings of the Apollo program took place in which the lunar orbit rendezvous was chosen for the flight mode and the Saturn C-5 was selected as the lunar mission booster (Ertel and Morse 1969). Although the complete vehicle designs were not established by the choice of the flight mode, the basic vehicle specifications were.

First in the booster development process came the first-stage engine design that was powered by the F-1. By 1962, both the J-2 and F-1 engines had been specified and were already under a development contract with Rocketdyne. Preceding both the J-2 and F-1 was the H-1, which served as a prototype for a number of the components, an engine that itself arose from previous ballistic missile engine designs.

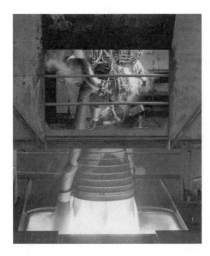

Figure 8.2. Test firing of the H-1 engine. Source: NASA NIX.

Table 8.1. H-1 Specs

Thrust (sea level)	889 kN (200,000 lb$_f$)
Specific impulse	263 s
Propellants	LOX & RP-1 (refined kerosene)
Engine weight (dry)	998 kg (2,100 lb)
Area ratio	8:1
Chamber pressure	633 psia (4.4 MPa)
Used on:	First stage on Saturn 1 and Saturn 1B
Manufacturer	Rocketdyne

Figure 8.3. Test firing of the J-2 engine at the Stennis facility. Source: NASA NIX.

Table 8.2. J-2 Specs

Thrust (sea level)	889 kN (200,000 lb$_f$) initial SA-504 and subsequent 1.023 MN (230,000 lb$_f$)
Specific impulse	418 s
Propellants	LOX & LH2
Engine weight (dry)	3,480 lb (1,580 kg)
Area ratio	27.5:1
Used on:	Second stage Saturn 1B, second stage Saturn V, third stage Saturn V,
Manufacturer	Rocketdyne

Figure 8.4. F-1 engine checkout at NASA's Marshall facility. Source: NASA MIX.

Table 8.3. F-1 Specs

Thrust (sea level)	6.7 MN (1.5 Mlb$_f$) initial 6.77MN (1.522 Mlb$_f$) SA-504 and subsequent
Specific impulse	260 s (263 s on SA-504 and subsequent)
Propellants	LOX & RP-1
Engine weight (dry)	8,353 kg (18,416 lb)
Area ratio	16:1
Used on:	First stage Saturn V (5)
Manufacturer	Rocketdyne

Rocketdyne moved quickly to develop the H-1 for the first-stage propulsion on the Saturn-1 and Saturn-IB, since those vehicles were needed for the early Apollo booster test and validation flights. For the Saturn IB's second stage, a more powerful and efficient J-2 engine was progressing more slowly because of its cryogenic oxygen and hydrogen propellants. Development was also complicated by the requirement to be restartable—a first for a large liquid-fuel engine. The J-2 engine's liquid oxygen (LOX) and liquid hydrogen (LH2) propellants had a higher I_{sp} efficiency, but since low-density hydrogen demanded a much greater tank volume than other fuels, the J-2 was best suited for Saturn's upper stages. Rocketdyne's restartable, high-efficiency J-2 engine became the most versatile of the Saturn engines, powering the second stage of the Saturn IB (S-IVB), and the second stage (S-II) and third stage (S-IVB) of the Saturn V. The remarkable J-2 engine would also serve as the prototype for the current J-2X being readied for NASA's Ares boosters for use in the Constellation program.

Saturn's Origin

In 1958, the Army Ballistic Missile Agency proposed a multistage, clustered-engine booster for the Defense Department that was based on the Jupiter missile. The proposal's primary proponent was none other than the director of the ABMA's missile design, Dr. Wernher von Braun. Jupiter was a variant of von Braun's Redstone, and codeveloped for the Army and Navy. After the Navy's cancellation, the missile was put to use as an Army IRBM. Von Braun's proposal of the heavy-lift booster, called the Super-Jupiter, included eight clustered engines from the Jupiter to provide a total of 6.67 meganewtons (1.5 M lb_f) of thrust on the first stage. The proposal was timely since it was introduced just before the Sputnik launch, the creation of ARPA, and the creation of NASA. Von Braun's Super-Jupiter project was soon approved and funded, but with limited resources. To reduce costs, Super-Jupiter was recast in its basic design, using borrowed tanks from the Redstone and Jupiter missiles and eight enhanced H-1 engines, and renamed the Juno V, which later became the Saturn I. The Jupiter missile continued in its role as an IRBM, with a second assignment as NASA's first interplanetary launcher, the Juno II. Its predecessor, Juno I, was a converted Redstone missile that launched the first Explorer satellite.

ABMA's Juno V was prepared as the first heavy-lift launcher for use by the USAF, the Navy, the DoD, and NASA. NASA also saw the need for a much larger booster for deep-space and lunar missions with a projected thrust of 26.6 MN (6 M lb_f) or more. By 1959, the Juno V and Nova rockets had become the focus of NASA's future heavy-lift booster plans, while lighter payloads were expected to be launched on the converted Atlas and Titan ICBMs. But in the same year, ARPA realized that the heavy-lift Juno V (now called the Saturn after ABMA's lead) and NASA's proposed Nova boosters were too large and too expensive for future DoD needs. Atlas and Titan ICBMs were thought to have sufficient payload capacity for their spacecraft and warheads. For this and a host of other reasons, the decision was made to separate civil and military space projects, which resulted in the transfer of the Saturn program to NASA, along with the ABMA and its resources, including the Jet Propulsion Lab (Bilstein 1996).

Figure 8.5. Photograph of the Saturn I launcher (right), the Jupiter IRBM (rear left), and the Redstone IRBM (left front) during assembly at the Marshall Space Flight Center fabrication facility. Source: NASA MIX.

SATURN I

Saturn I's first stage consisted of eight Jupiter engines in two sets of four. The four outboard were gimbaled for directional control, while the four engines mounted inboard were fixed to the bulkhead for stability. The eight first-stage H-1 engines were fed from the central Juno II oxidizer tank that provided much of the structural strength for the first stage. The single Jupiter/Juno II tank was surrounded by a circular arrangement of eight Redstone/Juno I tanks, with the four LOX tanks painted white and the four RP-1 tanks black. All nine tanks were extended four meters for increased propellant capacity to increase thrust duration.

Table 8.4. Saturn I Specs

	First stage (S-1)	*Second stage (S-IV)*
Mass (wet)	432,681 kg (196,673 lb)	50.576 kg (22,989 lb)
Engines	Eight H-1	Six RL-10
Propellants	LOX & RP-1	LOX & LH2
Thrust	7.1 MN (1.6 Mlb$_f$)	400 kN
I$_{sp}$	288 s	410 s
Burn time	150 s	482 s
Payload to LEO	9,000 kg (4,090 lb)	

Table 8.5. RL-10 Specs

Thrust (sea level)	66.7 kN(15,000 lb$_f$)
Specific impulse	433 s
Propellant	LOX & LH2
Engine weight (dry)	928 lb
Area ratio	40:1
Used on:	S-IV, Centaur, upper stages on Atlas and Titan
Manufacturer	Pratt & Whitney

The second stage of the Saturn I, designated S-IV, housed six RL-10 engines that were later replaced by five of the more powerful J-2 liquid oxygen + liquid hydrogen engines. A third stage was proposed for the Saturn I, but was never used since the second and third stages were replaced by the more powerful S-IVB second stage. Saturn's third stage was not discarded, but was transformed into the ubiquitous Centaur rocket later used as an upper stage for the Atlas and Titan launchers, and as a booster for a number of satellites launched from the space shuttle orbiters before the Challenger accident. Centaur continues to fly today as an upper-stage booster on the heavy-lift Atlas V.

Ten test launches of the Saturn I took place from Cape Canaveral, Florida, as Apollo hardware development missions, which ran from October 1961 to July 1965. The launch designation for these early Apollo hardware test flights were SA-1 to SA-10 (SA for Saturn-Apollo), with the first four flights simply first-stage booster tests carrying a dummy second stage. Saturn I's fifth launch was the first to carry an operational S-IV second stage, followed by the

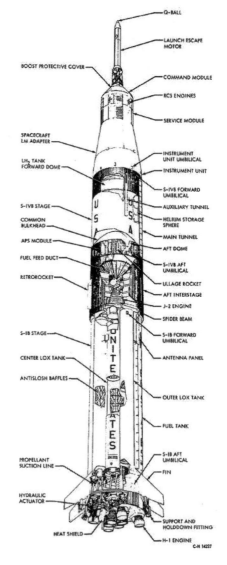

Figure 8.6. Cutaway diagram of the Saturn IB booster. Source: NASA TM-X-70137.

Table 8.6. Saturn 1B Specs

	First stage (S-1)	Second stage (S-IVB)
Mass (wet)	454,500 kg (1,000,000 lb)	115,000 kg (253,000 lb)
Engines	Eight H-1	One J-1
Propellants	LOX & RP-1	LOX & LH2
Thrust	7.58 MN (1.6 Mlb$_f$)	1,001 kN (225,000 lb$_f$)
Burn time	2.5 min	7 min
I$_{sp}$	288 s	421 s
Manufacturer	Douglas Aircraft Company	
Payload to LEO	18,180 kg (40,000 lb)	

sixth and seventh SA launches that also carried a S-IV second stage. The three final Saturn I launches included the S-IV second stage used to boost command and service module dummy payloads that carried Pegasus micrometeoroid research satellites. The first eight Saturn I first-stage boosters were built at the Marshall Space Flight Center by von Braun's Redstone engineers and technicians, many of whom were transferred from the Army Ballistic Missile Agency. Subsequent Saturn I and Saturn IB first-stage boosters were manufactured by the Chrysler Corporation. Saturn S-IV and S-IVB upper stages were built by Douglas Aircraft.

SATURN IB

NASA needed a larger booster than the Saturn I to reach Earth orbit in order to test the combined Apollo Command and Service Module. By 1962, the enhanced booster had been approved, concurrent with the LOR flight mode decision. The projected flight date of 1965 for the new man-rated Saturn IB was quickly approaching, though the H-1 and J-2 engine designs were already in progress and the Saturn I's S-IVB was moving from development to production. Help also came from von Braun's successful lobbying for the program, and from the reduction in costs for Apollo hardware testing that the Saturn IB would provide.

The more powerful Saturn IB's first stage was almost identical to Saturn I, with the exception of the uprated H-1 engines that generated 912 kN (205,000 lb$_f$) thrust instead of the Saturn I's 890 kN (200,000 lb$_f$). But one of the greatest differences was the Saturn IB's second stage which carried an improved S-IVB also being developed as the upper stage for the Saturn V. Saturn's S-IVB booster incorporated the new, smaller, and more efficient J-2 engine that had an initial thrust rating equivalent to the H-1 at 890 kN. Saturn IB also incorporated digital systems that allowed automated checkout and launch cycling that greatly improved reliability while reducing processing time (Bilstein 1996).

Fourteen of the Saturn IB launchers were built for the Apollo program, with its maiden flight on February 26, 1966. The first four of the fourteen launched were for

Apollo hardware flight tests, the fifth carrying the first Apollo astronauts into orbit on Apollo 7, designated AS-205 (AS for Apollo-Saturn). Four of the Saturn IBs left over from the cancelation of the final missions of Apollo were used for three Skylab manned flights—SL-2, SL-3, and SL-4—and the Apollo-Soyuz project (ASTP) launched on July 15, 1975. A Skylab rescue mission planned for one of the Saturn IB boosters, designated AS-209, was never flown and is on display at the Kennedy Space Center's Visitor's Center's Rocket Garden.

SATURN V

Saturn V was by far the largest launch vehicle built for its time, and the tallest of any rocket flown. Only the Soviet Energia booster, developed in the late 1980s, had a comparable lift and payload capacity. Energia, however, had a short life with only two production launches, one of them with their space shuttle named Buran. Saturn V's lift was about the same as the total LEO payload capacity of the Space Transportation System (space shuttle), both of which were roughly 108,860 kg (240,000 lb).

Throughout 1960 and 1961, a variety of booster and engine combinations were being considered for the Apollo lunar missions, powered by combinations of the F-1, the H-1, and the J-2 engines. Saturn C-2, for example, consisted of an eight-engine S-1

Figure 8.7. Saturn V first-stage being delivered to the Vehicle Assembly Building at the Kennedy Space Center in preparation for stacking on the mobile launcher platform. Source: NASA KSC.

Table 8.7. S-IC Specs

Height	42 m (138 ft)
Diameter	10 m (33 ft)
Mass	2,178,000 kg (4,792,000 lb)
Engines	5 F-1 engines
Thrust	33,400 kN (7,500,000 lb$_f$)
Burn time	150 s
Propellant	LOX + RP-1
Manufacturer	Boeing Company

first stage, like Saturn I, along with an S-II second stage. The limited-thrust C-2 could be used for Earth-orbit-rendezvous (EOR) flight mode, although other booster options were available for EOR. The four-launches needed for the Saturn C-2 EOR mission mode were soon replaced with the C-4 booster design that required only two EOR launches. C-4 would use four F-1 first-stage engines, along with two upper stages, but could not carry the lunar module (LM) and the command and service module (CSM) to lunar orbit. Then in January 1962, NASA managers made a number of decisions on the Apollo hardware and flight modes that fixed the planning and development process.

1. The lunar-orbit-rendezvous flight mode would be used for the manned lunar missions.
2. The heavy-lift Saturn C-5 would be used for the LOR manned missions to the Moon. Saturn C-5 would use five F-1 engines on the first stage, designated S-IC; with five J-2 engines on the second stage, called S-II; and a single-engine J-2 third stage called S-IVB. The designation Saturn C-5 was contracted to Saturn V in 1963.
3. An improved Saturn C-1 called Saturn IB would be used for Apollo hardware development and initial manned flight testing.

S-IC First Stage

Saturn V's enormous scale posed several serious design hurdles, especially the first stage, which used the most powerful engines ever built. Adding to the challenge was the design, fabrication, and testing of two new engines, the F-1 and J-2, that had no parallel in scale and complexity. Propellants for the Saturn V's first stage were the same as those used on the Saturn I and IB launchers, which provided a higher-density mix than LOX and LH$_2$, and furnished reasonable performance. Five F-1 engines, each with 6.67 MN (1.5 M lb$_f$) thrust, totaled 33.4 MN (7.5 M lb$_f$), which was beyond the capabilities of any added guidance thrusters. To accomplish first-stage directional control, the four outer F-1 engines were gimbaled under hydraulic control with the center engine fixed to the thrust structure, which was also the largest and heaviest piece on the S-IC.

Tanks on the S-IC presented welding problems unmatched on any other booster because of their size and the flight loads. Perfect welding in both of the 10 m diameter tanks was needed for man-rating the vehicle. Tank suspension within the main

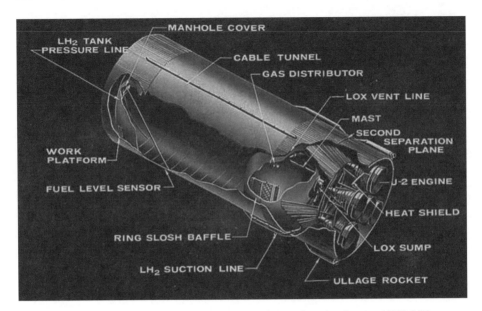

Figure 8.8. Cutaway diagram of the Saturn S-II second-stage booster. Source: NASA MIX.

structure required extensive testing as well. Work on the S-IC was carried out at the NASA-Marshall Space Flight Center (MSFC) facilities where both NASA and the contractor participated in the development, tooling, and fabrication. Completion work on the S-IC was officially transferred to Boeing in 1962, with MSFC engineers still participating. Development and testing continued for five years before the first launch of the Saturn V on November 7, 1967 as Apollo 4. Both Apollo 4 and Apollo 8 had S-IC boosters built by MSFC, with the remaining boosters manufactured by Boeing (Bilstein 1996). Of the fifteen Saturn V launches, all but three were Apollo missions. Skylab Orbital Workshop was the last launch of the Saturn V and S-IC on May 14, 1973. The two remaining Saturn V boosters planned for two of the three canceled Apollo missions are on display at the Johnson Space Center and the Kennedy Space Center.

Table 8.8. S-II Specs

Height	24.9 m (82 ft)
Diameter	10 m (33 ft)
Mass	471,400 kg (1,037,000 lb)
Engines	Five J-2 engines
Thrust	5.040 MN (1,125,000 lb$_f$)
Burn time	360 s
Fuel	LOX + LH2
Manufacturer	North American Aviation

S-II Second Stage

Saturn V's S-II second stage was designed expressly to boost the Apollo modules into Earth orbit after the S-IC first-stage burnout. Like the S-IC, the second-stage S-II was not used on either the Saturn I or the IB. In fact, a separate launcher, designated Saturn II, was proposed that was based on the S-II and S-IVB, serving as the first and second stages. Saturn II was to replace the Saturn IB as a more effective and efficient expendable launcher, but never left the drawing board.

The S-II second stage incorporated five of the new yet-to-be-tested J-2 engines fueled by LOX and LH_2 propellants. The propellant combination was needed for the larger lunar payload, since the LH_2 and LOX mix had a higher I_{sp} and generated more thrust per unit fuel mass than any other propellant combination. Total thrust for the S-II booster stage was 4.45 MN (1 Mlb_f). As with the S-IC first stage, the four outer engines were gimbaled, while the center engine was fixed on its mount to the thrust structure.

Several unique features of the S-II were integrated into the propellant tank design. The LH_2 fuel was stored in six cylindrical tanks, while the LOX was stored in a single ellipsoidal tank. Also unique was the use of a common bulkhead between the bottom of the LH_2 tank and the top of the LOX tank. This saved nearly three meters in length and several tons in mass by avoiding the use of an intertank connecting structure. Since the cryogenic liquid hydrogen was much colder than the liquid oxygen, tank design and welding of the aluminum alloy pieces were critical, as was the uniform application of external insulation. Spraying the insulation on the LH_2 tank, followed by careful trimming, was used instead of applying prefabricated sheets of insulation. The foam spray technique was later used on the space shuttle's external tank, which contained the same propellants (Bilstein 1996).

Nineteen of the S-II stages were built by North American Aviation for preliminary tests, qualification flights, and planned Apollo missions. The first four were static test articles, with two destroyed on the stand and one not completed. The remainder were used on the Apollo flights, with the exception of the last three planned for the canceled Apollo missions to the Moon. The last two are on display as part of the Saturn V exhibits at the Johnson and Kennedy Space Centers. The remaining S-II was used as the upper stage for the Skylab launch.

S-IVB Third Stage

The only Saturn V booster stage used on other launchers was the third stage S-IVB, used as the Saturn IB's second stage, although there were some minor differences between the two. Since the Saturn IB's S-IVB second stage was not intended to be restarted as the Saturn V's was, the helium pressurization storage on the Saturn IB was smaller. The larger-diameter S-II stage on the Saturn V also required a larger flare on the interstage structure between the S-IVB and the S-II.

The origin of the S-IV booster came from the design of the fourth stage of the C-4 launcher (hence the designation S-IV), which later evolved into the Saturn I's second stage. In its transformation, the S-IV was the first booster stage of the Saturn vehicles to be built by a contractor—Douglas Aircraft. The first incarnation of the S-IV on the

Saturn I was powered by six RL-10 engines since the J-2 was not completed for the preliminary design. The S-IV's six RL-10 engine cluster was replaced by a single J-2 on the S-IVB model.

Development and extensive testing of the booster and its engine, one of the hall-marks of von Braun and the Saturn program, was nearly finished by 1964 for the first complete S-IVB that was launched on a Saturn IB. Additional testing and qualification flights readied the booster for its inaugural flight on the Saturn V's first launch on Apollo 4 in November 1967. None of the S-IVB or the Saturn V boosters had a failure during launch or in flight.

S-IVB Flights

Saturn V's S-IVB third stage was a multifunction booster that first took the Apollo hardware to orbit, and was then restarted several orbits later to place the flight vehicle into a trans-lunar trajectory to the Moon, a boost called trans-lunar injection, or TLI. Following burnout and separation post-TLI, the S-IVB would have followed alongside the Apollo CSM and LM to the Moon, which could pose a significant collision hazard. Prior to Apollo 13, the S-IVB tanks were vented or reignited after separation to alter the boosters' trajectory to reduce that risk of collision. From Apollo 13 onward, the S-IVB was placed in a lunar impact trajectory after separation from the Apollo CSM flight vehicle. The high-energy S-IVB impact served as a probe of the lunar interior by generating shock waves for the seismic instruments placed on the surface during previous Apollo missions.

Undoubtedly the most intriguing booster flights of any on record occurred on the Apollo 12 mission that began with the post-TLI separation boost of the S-IVB. A burn that lasted slightly too long placed the third stage in an Earth orbit instead of a heliocentric orbit. After thirty years, gravitational perturbations returned the S-IVB to a distant Earth orbit, where it was thought to be a near-Earth asteroid, identified as J002E2. But a titanium dioxide spectral signature from the paint on the booster proved that it was instead the lost Apollo 12 S-IVB stage. The upper-stage booster returned to heliocentric orbit two years later in a seemingly chaotic trajectory. To the contrary, the Apollo 12 S-IVB orbits are an excellent example of the quasi-stable orbital paths available for boosting spacecraft on interplanetary missions with far less energy than needed for the traditional Keplerian (elliptical) transfer orbits.

Skylab Orbital Workshop

Apollo was the largest, most ambitious, most challenging, and most spectacular of any of NASA's space programs. And, with all their uncertainty, the Apollo manned lunar missions were selected as the most effective means to gain advantage over the Soviets after their impressive accomplishments during the early space race. Apollo was also considered an expedient means of advancing missile technology and weapons development, how-ever indirectly. Yet, the choice of manned lunar missions was not simple. Several other projects were considered as President Kennedy and Vice President Johnson searched for a space project that could inspire public support, while gaining technological superior-ity. Considered the next in line after Apollo was a manned orbital space station. Even

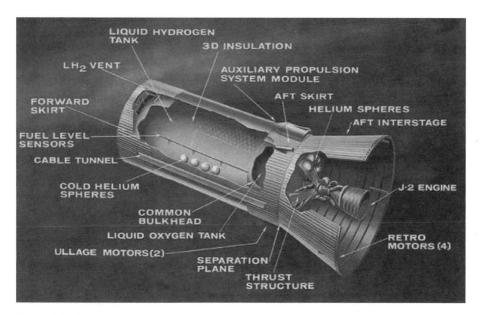

Figure 8.9. Cutaway of the Saturn S-IVB third-stage booster with its single J-2 engine. Source: NASA MIX.

though the space station project lost out to the lunar program, long-term manned orbital research had scientific and exploration utility that was appealing to many.

Early in the 1960s, NASA and Congress became increasingly interested in employing the Apollo hardware and its technology for other programs. The first of these conversion programs was called Apollo X, a program aimed at extending lunar missions and capabilities using both standard and modified Apollo hardware. As the Apollo flight tests progressed and Gemini was coming to an end, Apollo X was renamed the Apollo Application Project (AAP). With time, the AAP centered on less ambitious, less costly projects, while still using Apollo hardware. As the manned lunar landing missions began in late 1969, the Apollo Applications Program was transformed into the Skylab orbital space station project.

Initial project plans for the modest station were to launch the Saturn V with its S-IVB third stage as a conventional booster stage, then construct the orbital laboratory from the empty structure. The entire S-IVB would be outfitted with laboratory equipment and crew facilities during subsequent manned missions. This "wet workshop" concept was later considered impractical and replaced with the "dry workshop" design in which the Skylab was launched as an S-IVB converted into the orbital laboratory before launch. Manned missions to and from Skylab could then be flown using the Apollo command and service module vehicles that were modified for shorter flights than on lunar missions.

Three complete Apollo mission hardware sets were left over from the program following the cancelation of the Apollo missions 18, 19, and 20. Of the three sets, one was used for the Skylab project, which was also the last Saturn V mission. The two other flight-ready Saturn V boosters, as well as other flight hardware, were placed on exhibit at NASA's Johnson Space Center and the Kennedy Space Center.

Table 8.9. S-IVB Specs

Height	17.8 m (58.4 ft)
Diameter	6.6 m (21.7 ft)
Mass	119,100 kg (262,000 lb)
Engines	Single J-2 engine
Thrust	1.01 MN (225,000 lb$_f$)
Burn time	480 s (2 burns)
Fuel	LOX + LH2
Manufacturer	Douglas Aircraft Company

INSTRUMENT UNIT

Saturn V flight operation commands and deployment functions on the Apollo vehicles were controlled during ascent and checkout by computer instructions from the Saturn V Instrument Unit (IU) attached to the third stage. After initial deployment, the CSM computer systems progressively took over command of the vehicle and its mission. The IU was responsible for vehicle checkout on the launch pad, for flight navigation during the vehicle ascent and trans-lunar injection, for guidance instructions on the Saturn V's three stages, for sequence initiation and vehicle telemetry, and for booster and vehicle communications and telemetry. Subcontractor for the Saturn V Instrument Unit was International Business Machines (IBM).

One of the last sequence operations, yet one of the most important guidance and navigation functions on the Saturn V IU, was stabilization of the LM while still attached to the S-IVB. Following the CSM separation from the booster, the CSM was rotated 180° (transposed), then returned to dock with the LM still inside the S-IVB. The docked CSM and LM backed away from the S-IVB, again rotated 180°, then reestablished itself on the TLI trajectory to the Moon. The last task of the IU was to guide the S-IVB away from the CSM while both vehicles were en route to the Moon. The S-IVB was then guided to either a collision with the Moon or placed in a safe heliocentric orbit.

SATURN'S LEGACY

Saturn booster flights came to an end with the launch of Skylab, but its legacy continued for decades, and has even contributed to NASA's future plans for the next generation boosters and missions to the Moon. The heavy-lift launcher that is under development is not just close to the Saturn V in height, weight, and payload, but is also named Ares V in honor to the Saturn V. Ares V also houses engines similar to those carried on the second and third stages of the Saturn V. And, like the Saturn IB and Saturn V, the Ares 1 and V have a single LOX + LH$_2$ engine booster, with an engine similar in many ways to the Saturn's J-2. In fact, the upper-stage engines on the Ares 1 and V are improved versions of the J-2, called J-2X.

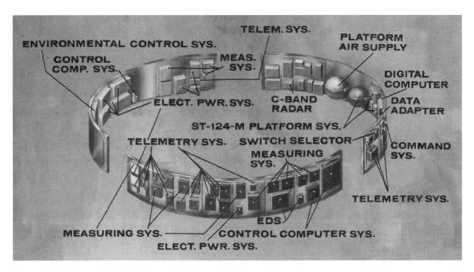

Figure 8.10. Layout of the Saturn V Instrument Unit and its major components that were powered by batteries located at several locations on the ring. A comparable unit with similar components was installed on the Saturn IB booster, as well as on the Skylab Orbital Workshop. Source: NASA MIX.

Saturn V's digital controls on the Instrument Unit were successful in managing complex booster and multiple-flight vehicle operations during deployment and initial vehicle operations, enough so that similar navigation and control operations units were placed on the Skylab, and recently, on the Ares boosters for their early flight operations. Even the engine placement, thrust vectoring, and cooling on the Saturn IB and Saturn V are being used for the design of the Ares launchers.

Many of the problems related to building the enormous Saturn V and making it fly contributed to later booster technology, not the least of which was the largest booster engine ever built, the F-1. Combustion-instability problems that were solved by NASA and Rocketdyne allowed the design of some of the most advanced rocket engines since, including the space shuttle main engines. Technology advances developed for the Saturn rockets, especially the Saturn V, introduced fabrication techniques that took years to perfect. Examples include the exacting welding and the fabrication techniques used on the propellant tanks that have since been employed in a host of applications. Saturn, like Apollo, generated so many spin-offs in aerospace industries, medicine—even improvements in the home—that their numbers have reached the tens of thousands. The venerable, forty-year-old Saturn V has not just changed the direction of space exploration since its first flight, it remains the most recognized booster of all time.

Apollo Flight Spacecraft

Following President Kennedy's declaration to commit the United States to a manned mission to the Moon by decade's end, planning for the program's structure, the Apollo

Table 8.10. Apollo Chronology

First launch: November 9, 1967
Last launch: May 14, 1973
Number launched: 13
Vehicle success rate: 100%
Launch mass: 2,965 metric tons

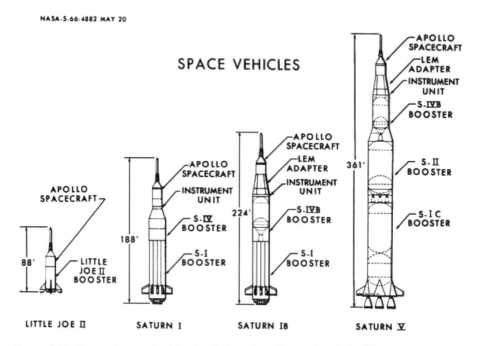

Figure 8.11. Comparison scale of the Apollo launchers. Source: NASA SP-4205.

manned vehicles, and Apollo boosters became the primary assignments of numerous NASA study groups, councils, and panels. The new vision for the Apollo Program that was being organized in 1961 consisted of three phases. The first was an orbital station, the second a circumlunar project, with both providing validation for the third and most important phase, the Apollo lunar landings (Brooks, Grimwood, and Swenson 1979.).

NASA's Space Task Group and their lead engineer, Max Faget, began drafting the basic layout for three flight vehicles. A trio of spacecraft represented a logical solution to the flight hardware needed for any of the three lunar mission modes; the EOR, the LOR, and the direct ascent. Experience gained from the Mercury design proved that a blunt-nosed return capsule would be suitable for the lunar missions, but a second vehicle would have to be added for the life-support and power requirements on the longer-duration Apollo missions. A third vehicle would also be needed for lunar surface operations, although that was far less certain in concept and function, since the flight mode had yet to be determined. But, until the flight mode was chosen, neither the launch vehicles nor the flight vehicles could be specified in any detail.

Table 8.11

	First stage (S-1C)	Second stage (S-II)	Third stage (S-IVB)
Mass (dry)	130,570 kg (287,860 lb)	36,395 kg (80,237 lb)	11,380 kg (25,088 lb)
Engines	Five Rocketdyne single-chamber F-1 engines	Five Rocketdyne J-2 engines	One gimbaled Rocketdyne J-2 engine
Propellants	LOX, RP-1	LOX, LH_2	LOX, LH_2
Propellant mass	2,149,500 kg (4,738,840 lb)	451,650 kg (995,720 lb)	106,940 kg (235,760 lb)
Thrust	33,851 kN (sea level—uprated) (7,610,000 lb_f)	5,116 kN (vac–uprated) (1,150,000 lb_f)	1,023 kN (vac–uprated) (229,980 lb_f)
Burn time	162 s (outboard), center engine: 135.5 s	Outboard: 390 s Center: 296.5 s	Initial 145 s Translunar: 345 s
Power	Two 28 VDC batteries	Four 28 VDC batteries	One 56 VDC and three 28 VDC batteries
Payload LEO	118,000 kg		
Trans-lunar trajectory	47,000 kg		

By the end of 1961, NASA managers arrived at several important decisions. In December 1961, North American Aviation was selected as the primary contractor for the crew flight module, a blunt-nose, truncated cone structure, similar in shape to the Mercury and Gemini capsules, but larger. Early in 1962 the decision was made to use an ablative coating for the entire crew vehicle instead of thermal shingles used on Mercury and Gemini. Although heavier, the coating added thermal insulation for the higher-energy reentry, and for radiation protection in space. The crucial decision in the same year to use the LOR flight mode to the Moon settled the questions about the choice of Apollo hardware. The Saturn V would be used as the primary launcher, and the three Apollo flight vehicles would be configured as a reentry capsule called the command module; a lunar orbiter, which would be the combined command module and service module; and a separate lunar lander module called the lunar module. North American designers accelerated their efforts on the crew command module and the service module. NASA managers quickly drew up specifications for the lunar module for prospective contractors. After bid submissions in September, the announcement was made that Grumman would be the primary contractor for the lunar excursion module (LEM), which was later shortened to LM (Brooks, Grimwood, and Swenson 1979.).

Command Module

Although a number of design features carried over from Mercury and Gemini to the Apollo command module (CM), most of the capsule's character was derived from design lessons learned and from the development of newer technology. Structural differences were significant, especially the Apollo CM frame structure, which was made of aluminum alloys instead of titanium. The conical shape of the Apollo CM structure was simpler, with a greater divergent angle and wider reentry face to accommodate higher reentry speeds. The Apollo CM also had four times the habitation volume of the Mercury capsule.

Notable similarities between the CM and earlier manned capsules included a launch escape system similar to Mercury, parachute landings to a water recovery, and a fuel cell power system first flown on Gemini. Also, like Gemini, the command module used onboard propellants for active guidance and attitude control during reentry. Propellants for the CM were also similar to Gemini; nitrogen tetroxide as the oxidizer, but with unsymmetrical dimethyl hydrazine (UDMH) instead of monomethyl hydrazine (MMH) as the fuel, since UDMH offered better performance.

Major components in the CM included the crew cabin, crew seating, control and instrument panel, optical and electronic guidance systems, communications systems, environmental control/life-support system, batteries, a heat shield, a reaction control system, a forward docking hatch, a side hatch, five windows, and the parachute recovery system.

CM STRUCTURE

The Apollo command module was constructed of an aluminum honeycomb sandwich bonded between two different sheet aluminum alloy skin panels. A heat shield

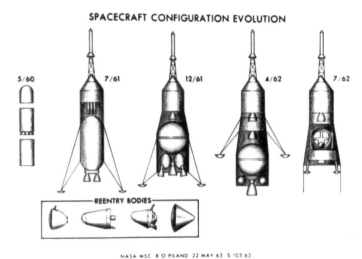

Figure 8.12. Command and service module design evolution from 1960 to 1962. Source: NASA SP-4206.

layer surrounded the inner aluminum sandwich structure, which served as the crew compartment. The outer heat shield consisted of an inner aluminum alloy sheet bonded to stainless steel honeycomb panels that were filled with a glass-phenolic ablation material. Five windows were integrated into the structure for crew observations, primarily for docking and navigation.

The CM's aft (bottom) base consisted of a blunt-nosed heat shield made in similar fashion, with brazed stainless steel honeycomb panels that were filled with a phenolic epoxy resin ablative material, but with an outer film of aluminized polyethylene terephthalate (PET) held in place with an outer metal ring and fairings (Brooks, Grimwood, and Swenson 1979).

Two hatch assemblies were located in the command module, one at the top for transfer to and from the lunar module, and a side hatch for ingress, egress, and EVA. The side access hatch was manually operated with handle, linkage, and latch mechanisms. A manual vent valve located on the hatch was used to bring the cabin to ambient pressure for opening the side hatch. An observation window was placed roughly in the center of the hatch.

The CM forward access hatch was placed at the top of the cabin with a pressurized tunnel and docking mechanism. This hatch provided a pressure seal and thermal insulation for the CM, as well as access to the lunar module when docked. A pressure equalization valve was used for hatchway opening between the CM and the docked LM.

The Apollo CM was designed with an impact attenuation system to reduce the landing shock on water, or on land if needed. The main structure included crushable ribs that were located on the outer section of the aft (bottom) inner sidewall surrounding the capsule. Since the parachute system brought the capsule to a landing with a 27° off-vertical angle, the crushable ribs were the first to be impacted. A set of shock struts were also included, and placed beneath each astronaut's couch seat assembly.

Table 8.12. Command Module Specs

Crew	3
Crew cabin volume	6.17 m³ (409 ft³)
Length	3.47 m (11.38 ft)
Diameter	3.90 m (12.80 ft)
Mass	5,806 kg (12,800 lb)
Heat shield mass	848 kg (1,870 lb)
RCS thrust	12 × 446 N (12 × 100 lb$_f$)
RCS propellants	N_2O_4/UDMH
Electrical system batteries	20.0 kW·h, 1000 A·h
L/D ratio hypersonic	0.3
Cabin atmosphere	Pure oxygen at 344.8 mbar (5 psi)
Max acceleration/deceleration	6 g (reentry)
Manufacturer	North American Aviation

CM SYSTEMS

Electrical Power System (EPS)

Power for the CM was generated by the three fuel cells located in the service module, with backup and postseparation power supplied by five silver-zinc batteries located in the lower equipment bay of the CM. Three 40 amp-hr batteries were employed as primary power sources for emergency power loss in the fuel cells, for peak load periods, for EPS controls and logic, and for CM operations after SM separation, which

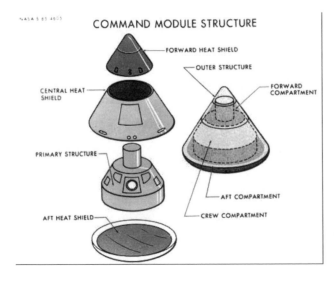

Figure 8.13. Exploded view of the early Apollo command module concept. Source: NASA SP-4205.

included reentry, landing, and recovery. Two smaller 0.75 amp-hr silver-zinc batteries were used for pyrotechnic operations and were separate from the main power buses. Main bus voltage was maintained at 28 Vdc (NASA 1969).

Two inverters were connected to the two main buses to provide 400Hz, 115 Vac power, primarily for avionics equipment. The three-phase inverters were load-shared with the primary dc buses (A and B) for either fuel cell or battery power. Other CM EPS circuitry was included for battery charge/discharge and power distribution throughout the command and service module's six bus types:

- Flight and postlanding buses
- Flight bus
- Nonessential bus
- Battery relay bus
- Pyro bus
- SM jettison controllers

Guidance, Navigation and Control

Apollo's command and service module's (CSM) primary guidance, navigation, and control system (PGNCS) measured position, attitude, and velocity data; compared the data with the computed trajectory; then commanded the primary booster and attitude thrusters for any necessary corrections. The PGNCS system contained three major subsystems:

- Inertial subsystem (ISS)
  - Inertial measurement unit (IMU) for continual inertial reference
  - Coupling data unit (interface for GNC subsystem data)
- Computer subsystem
  - Command module computer
  - Keyboard and display modules (2)
- Optics subsystem
  - Crew optical alignment sight (COAS)
  - Scanning telescope
  - Sextant (NASA 1969)

The CSM guidance and navigation system was aligned on the launch pad for the ascent phase in order to accurately measure the trajectory and compare it to predetermined values. Pertinent data was displayed for the crew and flight controllers showing critical parameters affecting abort decisions. Following ascent, the IMU and command module computer (CMC) were used to measure position and velocity changes. When not needed for sensing velocity changes, the IMU and CMC were placed in standby to conserve power. After reactivation, the CMC was updated with position and velocity data with the optical devices. Reentry guidance was supplied with the IMU inertial data fed into the CMC for command data to drive the reaction control system.

The command module's CMC contained both erasable and fixed memory for program operations. Navigation tables, trajectory parameters, programs, and constants were stored in the permanent memory, while computations and commands were transferred to/from the erasable memory. Commands and corrections were calculated from the input data, then transferred to the appropriate subsystems through the input-output (I/O) section. Apollo's command module computer was comprised of the following devices/sections:

- Timer
- Sequence generator
- Central processor
- Memory
- Priority control
- Input-output
- Power

Reaction Control System (RCS)

Reaction control on the command module was used primarily for the capsule's reentry trajectory control. During reentry guidance, the capsule was oriented so that the center of mass was offset from the center of pressure to create a positive lift component in order to follow the complex, predetermined trajectory. When lift was not needed, the capsule was rotated slowly to average the trajectory error and improve splashdown accuracy.

Twelve 420 lb_f thrusters were placed near the top of the capsule for pitch-down control and near the bottom for pitch-up, roll, and yaw control. Propellants were nitrogen tetroxide and UDMH.

Life Support

Life support for the Apollo command module crews was provided by the environmental control system (ECS) for the entire mission, from launch to landing. The three major ECS subsystems were the spacecraft atmosphere control, water management, and thermal control.

- Spacecraft atmosphere control
  - Pressure regulation of oxygen for the crew cabin (5 psi pure O_2)
    - Supplied by LOX stored in CM and SM
  - CO_2 removal by LiOH canisters
  - Trace contaminants and odors were removed with activated charcoal filters
- Water management
  - Potable water was supplied by the three SM fuel cells
  - Stored in potable water tank
    - Heated and cooled water were provided for drinking and food preparation
  - Second tank used for wastewater storage
- Thermal control was managed by two active cooling loops (primary and secondary)
  - Water-glycol cooling loop subsystem
    - Heat transferred from cabin to SM radiators using a heat exchanger
    - Wastewater evaporators used to remove excess heat

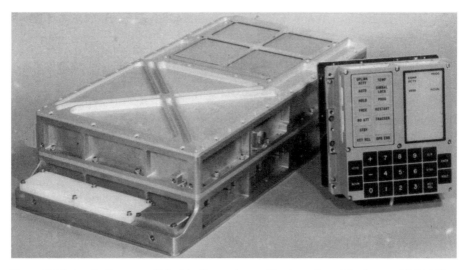

Figure 8.14. Apollo Command Module Computer and keyboard module. Source: NASA NIX.

- Pressure suit oxygen flow (3.5 psi O_2) was used to cool crew
  - Heat transferred to heat exchanger
- Post-landing ventilation subsystem

Other functions in the Apollo CM life-support system included:

- Waste management
  - Liquids were stored in waste tank or dumped overboard
  - Storage bags were used for solids and some liquids
  - Excess liquids were vented overboard
- Galley
  - Chilled drinking water
  - Hot water
  - Food
  - Food warmer
- Hygiene
  - Oral
  - Towels and wipes
- Medical kit

Communications

Communications between the Apollo spacecraft and the Manned Space Flight Network (MSFN) were enabled by the ground stations first built for the Mercury missions. Voice communications, telemetry, command, and identity data were transferred through the CSM and LM communications subsystems. Range and range rate (velocity) data for the CSM and for the CSM-LM rendezvous and docking maneuvers were generated by the various communications links and transponders on the CSM, LM, and the MSFN ground stations. Recovery communications signals were also active on the CM and its recovery units during reentry and splashdown.

Communications channels included:

- Voice communications
- Telemetry downlink (vehicle and biotelemetry)
- Tracking and ranging
- Command uplink
- Video downlink

An X-band radar used for LM rendezvous and docking data was part of the CM communications equipment, along with a longer-range VHF ranging channel.

Communications bands included:

- X-band
- S-band
- VHF
- UHF

Earth Landing System (ELS)

Three main parachutes were used for CM landing on the ocean. Pilot chutes were ejected using pyrotechnic charges, beginning with a mortar that pulled out two drogue chutes, in turn extracting the three main parachutes.

LAUNCH ESCAPE SYSTEM (LES)

Apollo's CM launch escape system consisted of rocket boosters to extract the crew and command module from the Saturn upper stage in emergency abort situations on the launch pad or during ascent. A second solid rocket motor was used to jettison the tower after the Saturn V's second-stage ignition. The 10.2 m long LES weighed 4,170 kg (9,200 lb) with a boost thrust of 689 kN (155,000 lb_f). A similar launch escape system is being integrated into the Orion crew exploration vehicle design.

Service Module (SM)

Apollo's service module was a life-support and booster complement to the command module that provided the crew and their capsule with the necessary consumables, electrical power, and propulsion, from launch to atmospheric entry at mission's end. The SM was a cylindrical structure approximately 7.5 m long, with the same diameter as the CM's widest section at the forward heat shield. The cylinder-shaped structure contained fore and aft bulkheads, with a connection faring at the forward bulkhead for attachments with the CM. Umbilical connections, power distribution, separation control, and plumbing connections linked the forward section of the SM with the forward section of the CM.

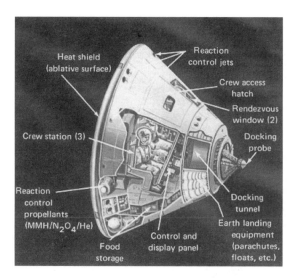

Figure 8.15. Apollo Command Module major component layout. Source: NASA NIX.

Attitude control thrusters were placed at four quadrant points, each with four thrusters called a quad unit. Each quad unit had separate fuel and oxidizer tanks, along with a propulsion unit that was located in one of the SM's six pie-shaped radial sectors. The sixteen 446 N (100 lb$_f$) thrusters were powered by NTO and MMH propellants. The main propulsion unit was the single high-thrust engine with a 92 kN (20,680 lb$_f$) rating, powered with NTO and UDMH propellants. Thrust vector control of the SM main engine was accomplished with two electromechanical actuators driven by the guidance system signals. The restartable main engine was used for lunar orbit injection (LOI) after arrival from Earth, and for lunar orbit exit (trans-Earth injection), as well as for mid-course corrections. Propellant capacity was 18,400 kg (40,570 lb), which afforded the CSM a total ΔV of approximately 2,800 m/s (SID-66).

Three fuel cells provided electrical power for the CM, SM, and LM during the portion of the mission that the modules were connected to the SM. Exceptions were

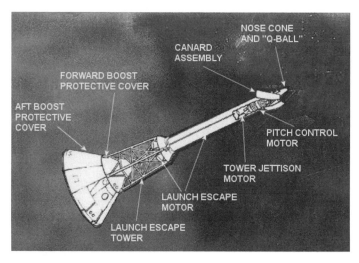

Figure 8.16. Sketch of the CM Launch Escape System component layout. Source: NASA JSC.

SATURN V APOLLO

SERVICE
MODULE

APPROX. DRY WT:
            11,000 LBS.
MISSION WT:
            50,000 LBS.
PROPULSION:
• 1 Restartable Engine
  Thrust: 22,000 lbs.
• 16 Reaction Control
  Motors
  Thrust: 100 lbs. Each

154"

22.5'

MSFC 68-MS-G 1336 E

Figure 8.17. Sketch and specifications of the Apollo Service Module. Source: NASA MIX.

during atmospheric reentry where CM batteries took over until splashdown and recovery. The LM also carried batteries that were used after separation from the CM en route to lunar landing and back. Reactants for the fuel cells were LOX and LH_2, which were stored in four cryogenic tanks in the SM: two for LOX and two for LH_2. AC current was generated by power inverters, driven by the fuel cell dc power sources supplying the electrical power for the CSM. The service module EPS was also used for recharging the CM's five zinc-oxide batteries as needed.

Lunar Module (LM)

Last on NASA's Apollo hardware development list was the lunar module that would carry astronauts from lunar orbit to the surface, support the astronaut's stay on the Moon, then return the astronauts for a rendezvous and docking with the orbiting CSM. Actual design specifications did not appear until late 1963 and early 1964 when the Saturn V and CSM specifications were complete, or nearly so. The lunar module was given a target mass and size that would support the mission and would fit in the Saturn V's third stage (S-IVB) shroud. A two-stage LM configuration provided the lowest mass for the lunar landing with a return to the CSM. Hence, systems and equipment were first specified, then designed in an iterative process to maximize the lunar module's payload mass while minimizing the LM's total mass.

In addition to the challenges of the new system designs, the limited allowed mass, and the changing configuration specifications, were the separate propulsion systems for descent and ascent stages. Both engines had to start after a cold soak in space of 4–7 days, while the descent engine had to be thrust-vectored and able to be throttled, which was a first. Bell Aerosystems had been assigned the ascent engine design, and Rocketdyne, the descent engine. For simplicity, both had to be powered by UDMH and NTO. Similar to the flight module and booster development programs, both

Table 8.13. Service Module Specs

Length	7.56 m (24.8 ft)
Diameter	3.90 m (12.8 ft)
Mass	24,523 kg (54,064 lb)
RCS thrusters	16 × 446 N (100 lb$_f$)
RCS propellants	NTO + MMH
Service propulsion (SPS) engine mass	3,000 kg (6,614 lb)
SPS engine thrust	92 kN (20,682 lb$_f$)
ΔV total	2,800 m/s
SPS engine propellants	NTO + UDMH
Electrical power	Three fuel cells, each rated at 1.4 kW max, 30-volts
AC power	Three inverters supplying 115 Vac, 400 Hz, three phase

engine projects encountered inevitable development delays, primarily from combustion instabilities. In an unusual switch of subcontractors in the middle of a project, an alternate design of Rocketdyne's descent engine was chosen and completed by the Space Technology Laboratories (Brooks, Grimwood, and Swenson 1979.).

A training device was developed for the Apollo LM pilots that mimicked the dynamics of the lunar module during descent and landing. The ungainly lunar landing research vehicle (LLRV) had a center-thrust turbofan engine to reduce the thruster reaction forces to simulate a 1/6th g lunar environment, along with an ejection seat for the single crew member. While training flights proved useful for the lunar module pilot astronauts' skills, the LLRV's stability proved elusive. Two of the LLRVs were built, as well as three later-model lunar lander training vehicles (LLTV). One of the LLRVs crashed during Neal Armstrong's training flight, while two of the three LLTVs crashed. Ejection seats adapted to both trainers saved all three pilots. In stark contrast, the lunar lander pilots in the secret Soviet manned lunar program were relegated to using helicopters with engine-out (auto rotation) landings for their lunar landing simulations.

DESCENT STAGE

The lunar module, called the lunar excursion module in the early program, was divided into two sections: the unpressurized descent module for landing, and the pressurized ascent module for liftoff from the lunar surface for rendezvous and docking with the CSM. Deorbit and descent functions required much more fuel and greater thrust than the ascent stage, since the descent stage carried all of the LM components and propellants, and was then left on the surface. Lunar surface power units, surface experiments including the Apollo lunar surface experiment package (ALSEP), and on missions 15–17, the lunar rover, were also a part of the descent stage. The descent stage structure was composed of aluminum alloy skin protected by micrometeoroid shielding and insulation layers. This was segmented into four quadrants surrounding the descent engine and four landing-gear assemblies. Each of the four landing gear attached to the descent module frame included shock struts and footpads with crushable materials instead of hydraulic damping.

ASCENT STAGE

In addition to ascent propulsion, the ascent stage doubled as the control center and the crew quarters while on the Moon's surface. The pressurized cabin provided life support for the two astronauts when not on EVA. Life support on the ascent section also stored consumable oxygen and water for the space suits used on EVA, and the lithium hydroxide canisters for both the cabin and the suits. Heat was removed from the space suit, the cabin, the electronics, and the batteries by dual glycol loops connected with internal water loops, which transferred heat to dual sublimators (evaporative coolers). The ascent stage also housed the docking port for crew and equipment transfer to and from the CSM. Additional equipment on the ascent stage included navigation radar and attitude thrusters for guidance during landing, and for ascent, rendezvous, and docking with the CSM.

The Apollo LM differed from all other manned spacecraft in several ways. First, the LM was operated only in space, which made aerodynamic shaping unnecessary. Second, the LM was exposed to only light loads during its operation: generally the weightlessness of orbit, or the 1/6th g gravity of the Moon. Because of the reduced gravity, seating for the LM needed only to be a simple restraint system for flight operations, which reduced the weight compared to a more conventional couch or chair seating. Micrometeoroid shielding and thermal insulation were placed on the outside of the thin aluminum alloy shell of the ascent stage. Layered insulation was separated from the skin, and outside that was the thin, outer micrometeoroid bumper shield made of sheet aluminum.

Three windows were installed in the ascent stage forward section: two on the lower wall for landing, and one overhead for docking. The twin-pane windows were fabricated to resist micrometeoroid damage and coated to reduce IR and UV light in the cabin.

LM DERIVED VEHICLES

A variety of proposals for lunar vehicles based on the LM design appeared in the later phases of the Apollo program motivated primarily by the possibility that the program would be expanded, though no vehicles were ever approved or built. Design of the LM was versatile enough to appear in a wide variety of lunar vehicles proposals that ranged from orbiters to cargo landers. Several of those proposed by the LM manufacturer, Grumman Aircraft, are listed here.

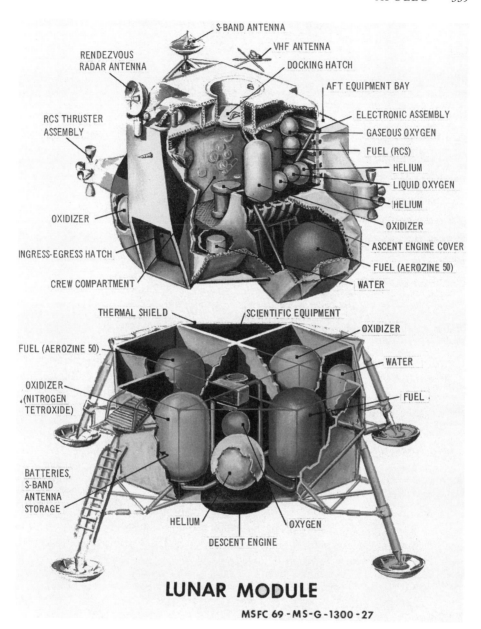

S-BAND ANTENNA

VHF ANTENNA

RENDEZVOUS
RADAR ANTENNA

DOCKING HATCH

AFT EQUIPMENT BAY

ELECTRONIC ASSEMBLY

RCS THRUSTER
ASSEMBLY

GASEOUS OXYGEN

FUEL (RCS)

HELIUM

LIQUID OXYGEN

HELIUM

OXIDIZER

OXIDIZER

ASCENT ENGINE COVER

INGRESS-EGRESS HATCH

FUEL (AEROZINE 50)

CREW COMPARTMENT

WATER

THERMAL SHIELD

SCIENTIFIC EQUIPMENT

OXIDIZER

FUEL (AEROZINE 50)

WATER

OXIDIZER
(NITROGEN
TETROXIDE)

FUEL

BATTERIES,
S-BAND
ANTENNA
STORAGE

HELIUM

OXYGEN

DESCENT ENGINE

LUNAR MODULE

MSFC 69 - MS-G -1300 - 27

Figure 8.18. Exploded view of the Apollo Lunar Module descent stage and ascent stage, along with major system and component layout. Source: NASA MIX.

Table 8.14. LM—Ascent-Stage Specs

Height	3.76 m (12.33 ft)
Diameter	4.2 m (13.8 ft)
Mass	4,670 kg (10,300 lb)
Empty weight	2,045 kg (4,500 lb)
RCS Thrusters	16 × 446 N (100 lb$_f$)
RCS Propellants	NTO + UDMH
Ascent engine thrust	15.6 kN (3,500 lb$_f$)
ΔV (ascent)	2,220 m/s
SPS main engine propellants	NTO + UDMH
Electrical power	Two 296 A-hr silver-zinc batteries, 28 VDC
AC power	115 Vac, 400 Hz, three phase from inverters
Cabin pressure	3.8 ± 0.2 psi

Extended LM

Improvements in the design and experience gained from the Apollo lunar missions led to the proposal to increase the payload at the lunar surface to 454 kg (1,000 lb), and extend the lunar surface exploration to three and a half days without adding propellant or vehicle structure mass.

Lunar Reconnaissance Module (LRM)

An orbital version of the LM was proposed for lunar surface reconnaissance and exploration by replacing the ascent stage with an instrument package, and using the descent stage as a lunar orbit insertion booster, requiring significant modifications. The LRM would undock with the CSM in lunar orbit, then support an orbital mission with as many as two crew members for up to fourteen days. The LRM would then rendezvous with the CSM for return to Earth.

Table 8.15. LM—Descent-Stage Specs

Height	3.2m (10.5 ft)
Diameter	4.2 m (13.8 ft)
Mass	10,334 kg (22,783 lb)
Empty weight	1,860 kg (4,100 lb)
Descent engine thrust	45.0 kN (10,125 lb$_f$)
ΔV (descent)	2,470 m/s
SPS engine propellants	NTO + UDMH
Electrical power	Four 400 A-hr silver-zinc batteries, 28 VDC

LM Taxi

A lunar module almost identical to the LM was proposed for crew transport to the Moon, following an LM shelter vehicle landing that would supply cargo and equipment for a stay of up to fourteen days. A radioisotope thermoelectric generator would supplement the power for both the LM taxi and shelter vehicles.

LM Truck

An unmanned lunar lander proposal included a carrier for landing cargo on the lunar surface or delivering a surface rover for transporting cargo or crews on extended missions. The LM Truck would be capable of transporting a 2,404 kg (5,300 lb), 2.55 m^3 (90 ft^3) science package to the lunar surface, or provisions for a fourteen-day stay for a crew of two.

LM Shelter (LMS)

A modified LM without ascent capability would be landed on the lunar surface with up to a sixty-day standby period for later use as a surface habitation module for up to fourteen days. The LM taxi would be used in combination with the LMS to transport the crew from the CSM to the shelter, then back to the CSM at mission's end.

Lunar Payload Module

Another version of the modified LM would replace the ascent-stage propulsion and part of the descent stage with cargo and equipment that could be used for extended crew missions using the LM taxi.

It is interesting to note that NASA's new Constellation program lunar module called Altair (also known as the lunar surface access module, or LSAM) may include several configurations that are similar to those proposed by Grumann, including the LM shelter and the lunar payload module.

EXTRAVEHICULAR MOBILITY UNIT (EMU)

Apollo missions employed two types of space suits, both called the A7L. The first was a space suit with fewer layers and an umbilical attachment for EVA outside of the CSM that used oxygen circulation for cooling the astronaut. The second, a more durable, self-contained extravehicular mobility unit, was used for spacewalks on the lunar surface with more and heavier layers that included an inner liquid cooling garment (LCG) in contact with the astronaut's skin. On the outside of the inner cooling garment was the pressure garment assembly (PGA) that held in the pressurized atmosphere of pure oxygen. The PGA was a part of both A7L suit types that included gloves, and a helmet and visor assembly.

The outer layers of the lunar surface EMU comprised the integrated thermal micrometeoroid garment (ITMG) for protection from the lunar environment including micrometeoroids, radiation, vacuum, and penetration or tearing by sharp objects. Also unique to the lunar surface EMU suit was the self-contained portable, rechargeable, environmental control and life-support system called the portable life-support system (PLSS). This rigid backpack was designed to supply the suit oxygen, remove carbon dioxide, circulate cooling water, furnish electrical power, provide communications, and house the EMU controls. Other components placed in the lunar EMU were a basic medical kit, an EMU repair kit, drinking water, and a waste management system.

Heat was removed from the lunar surface EMU by a sublimator that was fed by water circulating through the inner liquid cooling garment. The sublimator was cooled by circulating water past a porous block, allowing the water to seep toward the open face exposed to the cold vacuum of space. Outward flowing coolant water would freeze before reaching the face, exposing only ice in the outer layers to space vacuum. Ice formation blocked the direct flow of water out of the porous sublimator, but allowed sublimation to cool the outer region of the porous block and the coolant water circulating deeper in the sublimator. The same cooling method is used on the current space shuttle and ISS EVA space suits.

LUNAR ROVER

The Apollo lunar roving vehicle (LRV), also called the lunar rover, was designed to extend the range of the astronaut's reach on the lunar surface since their walking range was roughly 1 km. Boeing's LRV was built in less than eighteen months for the J-class missions, which were Apollo 15, 16, and 17. The two-person, four-wheeled rover was powered by batteries, limiting its life to the duration of the mission.

Each of the LRV's open mesh wire wheels was driven by a ¼ hp electric motor and gear reduction system, with an independent suspension assembly. Each wheel was covered across the top with a molded fender to reduce spreading the powderlike,

Table 8.16. Apollo A7L Space Suit Specs

Lunar surface life support duration	4 hr Apollo 11–14 8 hr Apollo 15–17
Weight	111 kg (245 lb)
Electrical power	One silver-zinc battery with 25.7 amp-hour capacity
CO_2 removal	LiOH canister
Odor, toxic gas removal	Activated charcoal filter
Cooling	Circulated cool water • Inner liquid cooling garment • Sublimation effect evaporative cooler

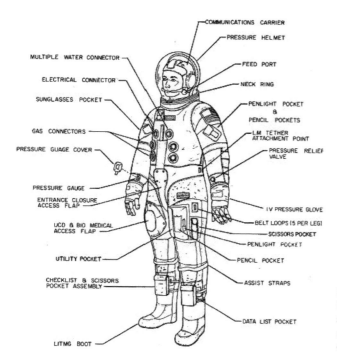

COMMUNICATIONS CARRIER
PRESSURE HELMET
MULTIPLE WATER CONNECTOR
FEED PORT
ELECTRICAL CONNECTOR
NECK RING
SUNGLASSES POCKET
PENLIGHT POCKET
&
PENCIL POCKETS
GAS CONNECTORS
LM TETHER
ATTACHMENT POINT
PRESSURE GUAGE COVER
PRESSURE RELIEF
VALVE
PRESSURE GAUGE
ENTRANCE CLOSURE
ACCESS FLAP
IV PRESSURE GLOVE
UCD & BIO MEDICAL
ACCESS FLAP
BELT LOOPS (5 PER LEG)
SCISSORS POCKET
PENLIGHT POCKET
UTILITY POCKET
PENCIL POCKET
CHECKLIST & SCISSORS
POCKET ASSEMBLY
ASSIST STRAPS
DATA LIST POCKET
LITMG BOOT

Figure 8.19. Apollo A7L lunar surface space suit showing exterior component layout. Source: NASA GRIN.

electrostatically charged lunar surface soil. Mechanical brakes were attached to each wheel and operated by two methods. One was a cable system, the other was a power disconnect on the drive motors. Simultaneous steering of both front and rear wheels was operated by dual drive motors and directed by the center-position hand controller. Both speed and steering were commanded by the hand controller, which when pushed forward translated into speed commands. The controller input was also used for steering and braking which was regulated by the aft movement of the controller. Full-aft detent was used for the parking brake.

Power for the LRV was supplied by two 115 A-hr silver-zinc batteries located in the forward chassis. A switching and monitor system was connected to the electrical distribution system interconnected with the caution and warning system. To aid navigation through unfamiliar terrain and provide accurate position and directional information on survey, sample, and crater sites, as well as manage the LRV's solar exposure to prevent overheating, navigational instruments were included on the control and display console. These included an attitude indicator for pitch and roll information, a heading indicator, a distance indicator showing accumulated travel (odometer), and a bearing indicator. Power and temperature instruments were also placed on the console as part of the electrical monitoring system.

The LRV chassis was an aluminum alloy assembly consisting of three chassis sections with independent suspension mounts for each wheel. The two side-by-side seats located on the center chassis were made of aluminum alloy tubing and webbed nylon fabric. Equipment storage was provided in the seat backs, forward section, and

Figure 8.20. Photograph of the deployed Apollo 15 Lunar Rover. Source: NASA JSC.

aft section. The LRV could accommodate as much as 635 kg (1,400 lb) of equipment (weight in Earth's gravity), which was twice its own weight. Attached equipment included a color camera as well as low-gain and high-gain antennas. Ascent videos for the Apollo 16 and 17 missions were recorded by the parked LRVs and transmitted through the LM-CSM communications link.

Table 8.17. LRV Specs

Mass	210 kg (463 lb–1,540 lb w/max. payload)
Length	3.5 m (11.5 ft)
Wheelbase	2.3 m (7.54 ft)
Power supply	Two silver-zinc 36 Vdc batteries 115 A-hr capacity
Surface speed	14 km/h (8.7 mph) (design) 17 km/h (10.6 mph) (Apollo 16)
Maximum endurance	78 hr
Maximum cumulative range	92 km (57 mi)

Apollo Missions

Early in the Apollo project, NASA managers classified the progressively more advanced mission types alphabetically; a convention that remained throughout the program. In 1968, NASA extended the A–G list with longer mission types that were designated H, I, and J. A summary of the mission classification types is shown in table 8.18.

TEST FLIGHTS

Tests flights of the Apollo hardware started with the early tests of the launch escape system pulling dummy command modules, followed by the launch of mock flight modules on Little Joe II to evaluate structures and adverse flight dynamics. Following these were booster-spacecraft compatibility test launches using the Saturn I booster for suborbital launches from Cape Kennedy (Cape Canaveral today). Next in approximate sequence were the orbital tests of the Saturn IB boosters, then the command and service module flight hardware tests. Following the qualification of the Saturn IB and CSM modules the first manned launch was made on Apollo 7. Following the final stages of testing, the Saturn V launcher was used to take the astronauts and flight hardware to the Moon and back on Apollo 8. An all-up test of the complete hardware including the LM in Earth orbit followed on Apollo 9. A complete test of all systems with the exception of an actual lunar landing was flown to the Moon and back on Apollo 10—the final dress rehearsal for Apollo 11. This completed the testing and qualification for the Apollo lunar landing hardware, and the procedures for man's first steps on another world.

Table 8.18. Apollo Mission Type Designation

Designation	Mission Type	Apollo Mission
A	Unmanned tests of the launch vehicles and the Command Module	
B	Unmanned tests of the LM	Apollo 5
C	Manned Earth-orbit tests of the command module	Apollo 7
D	LM/CSM tests in Earth orbit	Apollo 9
E	Tests in high Earth orbit	None flown
F	Lunar orbit tests	Apollo 10
G	Lunar landing	Apollo 11
H	Pinpoint landings at more diverse sites	Apollo 12, 13, 14
I	Lunar orbit-only science flights	None flown
J	Longer-duration missions, LM design changes, addition of the Lunar Rover	Apollo 15, 16, 17

Launch facilities for the Apollo missions were located at Cape Canaveral, renamed Cape Kennedy from 1963–1974, thereafter reverting to Cape Canaveral. NASA's Kennedy Space Center retained the Kennedy name.

Convention for the Apollo mission identification was based on the launch vehicle and sequence number beginning with the Saturn 1. The mission sequence identification is as follows:

- Saturn 1 flights, SA-1 to SA-10
- Saturn IB flights, AS-201 to AS-205
- Saturn V flights, AS-501 to AS-512

Unmanned Tests

PRELIMINARY

- Pad Abort
  - November 7, 1963, White Sands, New Mexico (PA-1)
  - Launcher: LES solid rocket booster
  - Objectives: Test of the LES to perform pad abort before launch
  - Results: Successful
  - Spacecraft: Boiler plate CM capsule 6 (BP 6)
- Little Joe II
  - May 13, 1964, White Sands, NM (A-001)
    - Launcher: Little Joe II
    - Objectives: Transonic abort test to simulate a Saturn V abort
    - Results: Successful
    - Spacecraft: CM BP 12
  - December 8, 1964, White Sands, NM (A-002)
    - Launcher: Little Joe II
    - Objectives: Max-Q abort test of launch escape system and Earth landing system test
    - Results: Successful
    - Spacecraft: CM BP 23
  - May 19, 1965, White Sands, NM (A-003)
    - Launcher: Little Joe II
    - Objectives: High-altitude LES test of canard and LES orientation
    - Results: Booster guidance malfunction produced premature low-altitude abort, although the landing system was successful
    - Spacecraft: CM BP 22
  - June 29, 1965, White Sands, NM (PA-2)
    - Launcher: Little Joe II
    - Objectives: Pad abort test of LES to qualify it for launch, including test of flight control canards, boost protective cover, dual reefed drogue chutes
    - Results: Successful
    - Spacecraft: CM BP 23A

- January 20, 1966, White Sands, NM (A-004)
  - Launcher: Little Joe II
  - Objectives: Last abort test of spacecraft
  - Results: Successfully qualified LES for manned flights
  - Spacecraft: SC 002

Saturn Tests

SATURN I

- SA-5
  - January 29, 1964, Cape Kennedy, FL
  - Objectives: Launch vehicle development test of Saturn I and S-IV second stage
  - Results: Successful
  - Spacecraft: CM BP 13
- SA-6
  - May 28, 1964, Cape Kennedy, FL
  - Objectives: Further qualification of Saturn I, and test of the spacecraft compatibility with launcher (launch to orbit phase)
  - Results: Successful
  - Spacecraft: CM BP 13
- SA-7
  - September 18, 1964, Cape Kennedy, FL
  - Objectives: Successful test of the LES on Saturn I
  - Spacecraft: CM BP 15
- SA-9
  - February 16, 1965, Cape Kennedy, FL
  - Objectives: Successful launch of first Pegasus micrometeoroid detection satellite carried in CSM+LM shroud; boilerplate CSM ejected into separate orbit by LES
  - Spacecraft: CSM BP 16
- SA-8
  - May 25, 1965, Cape Kennedy, FL
  - Objectives: Launch of second Pegasus micrometeoroid detection satellite carried in CSM+LM shroud; boilerplate CSM ejected into separate orbit by LES
  - Results: Successful
  - Spacecraft: CSM BP 26
- SA-10
  - July 30, 1965, Cape Kennedy, FL
  - Objectives: Final launch of Saturn I, and third Pegasus micrometeoroid detection satellite launch carried in CSM+LM shroud; boilerplate CSM ejected into separate orbit by LES
  - Results: Successful
  - Spacecraft: CSM BP 9A

SATURN IB

- AS-201
  - February 26, 1966, Cape Kennedy, FL
  - Objectives: First launch of Saturn IB and CSM; test of CM reentry from orbit, test of CM and SM propulsion
  - Results: Successful
  - Spacecraft: CSM 9
- AS-203
  - July 5, 1966, Cape Kennedy, FL
  - Objectives: Test flight of S-IV second stage systems and engine restart; test of IU
  - Results: Successful
- AS-202
  - August 25, 1966, Cape Kennedy, FL
  - Objectives: Second launch of CSM, test of CM reentry from orbit
  - Results: Successful
  - Spacecraft: CSM 11
- Apollo 5 (AS-204): First launch of lunar module (LM)
  - January 22, 1968, Cape Kennedy, FL
  - Objectives: First space flight of LM; test of ascent and descent engines, and lunar landing abort test
  - Results: Successful
  - Spacecraft: LM-1

SATURN V

- Apollo 4 (AS-501): First launch of Apollo-Saturn V
  - November 9, 1967, Cape Kennedy, FL
  - Launcher: Saturn V (SA-501)
  - Objectives: First launch of complete Saturn V and included a CSM; test of CM reentry from lunar mission after S-IVB boost to simulated TLI; CM encountered reentry conditions of approximately 40,200 km/h (25,000 mi/h), 2,760°C (5,000°F)
  - Results: Successful
  - Spacecraft: CSM 17
- Apollo 6 (AS-502): Second launch of Apollo-Saturn V
  - April 4, 1968, Cape Kennedy, FL
  - Launcher: Saturn V (SA-502)
  - Objectives: Second launch of Saturn V was for manned flight qualification; test of Service Module engine propulsion; launch vehicle problems on first and second stages, and SM boost redirected mission objectives
  - Results: Partially successful
  - Spacecraft: CSM 20

Manned Missions

APOLLO 1 (AS-204): LAUNCH PREFLIGHT TEST

- January 27, 1967, Cape Kennedy, FL
- Objectives: Countdown test was made in the command module atop an unfueled Saturn IB on Launch Pad 34 in preparation for their scheduled first manned flight in February.
- Results: The deaths of astronauts Grissom, Chaffe, and White in the CM capsule fire delayed the Apollo program for more than a year and a half while safety evaluations were formulated and redesigns completed.

Problems on the preflight test for the first manned Apollo mission multiplied throughout the afternoon during the AS-204 launch countdown, when a "capsule fire" radio transmission came from one of the astronauts in the cockpit. Within seventeen seconds, communications and telemetry ended. The CM cabin and crew were quickly engulfed in fire from a spark in the pure oxygen environment of more than one atmosphere. Several major design flaws conspired to prevent the astronauts' escape from the fatal capsule fire. First, the over-pressure from the cabin fire prevented the inward-opening door from being opened either from the inside or the outside. Second, a thirty-second delay for the hatch release mechanism was too long to allow the astronauts an exit from the flash fire, even if the hatch opened outward. Third, the astronauts' space suits did not have a fireproof layer needed for basic fire safety. And fourth, the intense fire kept rescuers from opening the cabin hatch for approximately five minutes.

The AS-204 countdown test would be later designated Apollo 1 in the astronauts' honor. When launched on January 22, 1968, AS-204 was reassigned the designation Apollo 5. No missions or flights were ever designated Apollo 2 or 3.

APOLLO 7 (AS-205): FIRST MANNED APOLLO LAUNCH IN THE CSM

- October 11, 1968, Cape Kennedy, FL
- Crew
  - Walter Schirra, commander
  - Don Eisele, CM pilot
  - Walter Cunningham, LM pilot
- Objective: Orbital test of manned CSM systems and operations
- Results: Successful
- Spacecraft: CSM 12

The Saturn 1-B mission included the flight lunar vehicles, although the lunar module was a dummy (boiler plate) vehicle. Following Earth-orbit entry, and CSM and S-IVB tests, CSM separation and transposition maneuver and simulated docking were executed successfully, along with CM and SM separation, deorbit, reentry, and CM landing.

Numerous firsts were made on the first manned Apollo flight, including:

- First U.S. three-man mission
- Third mission for Schirra (Mercury 5, Gemini VI), and first for Eisele and Cunningham
- First flight of Block II Apollo CSM spacecraft
- First flight operation of the Apollo space suits
- First flight with full crew support equipment
- First live national TV from space during a manned space flight
- Apollo 7 spent more time in space than all Soviet space flights combined at that time
- Mission duration: ten days twenty hours

APOLLO 8 (AS-503): FIRST MANNED TEST OF SATURN V AND THE FIRST MANNED FLIGHT TO THE MOON

- December 21, 1968, Cape Kennedy, FL
- Launcher: Saturn V (SA-503)
- Crew
  - Frank Borman, commander
  - James Lovell, CM pilot
  - William Anders, LM pilot
- Objectives: Operational test of the command module systems, including communications, tracking, and life-support in cislunar space and lunar orbit; evaluation of crew performance on a lunar orbiting mission. Lunar module was a dummy vehicle used for mass balance.
- Results: All systems operated within allowable parameters and all objectives achieved
- Spacecraft: CSM 13
- TLI burn: 319 sec (S-IVB)
- Lunar orbit entry burn: 247 sec (SM)

Apollo 8 was first launched to Earth-parking orbit in preparation for checkout and trans-lunar injection burn (TLI). Apollo 8 astronauts entered lunar orbit sixty-nine hours and eight minutes after launch. Photographic and visual lunar exploration from orbit was completed during the mission's ten orbits around the Moon before returning. Experiments and observation objectives included photographs of the lunar far-side and near-side surface to provide information on topography and landmarks for future Apollo landings. The only anomaly was crew fatigue from interrupted sleep that canceled some of the later photography tasks. A televised broadcast was made from Apollo 8 on Christmas Eve of all three astronauts reading from the book of Genesis.

Mission firsts included:

- First manned Saturn V flight
- First humans to leave Earth's gravity
- First humans to reach the Moon
- First pictures of Earth from deep space
- World speed record for a manned vehicle at 38,938 km/h (24,200 mi/h) during reentry
- First live TV coverage of the lunar surface

Other mission details:

- Mission duration was 147 hours 59 minutes 49 seconds
- This was the third mission for Lovell (Gemini VII, Gemini XII), the second for Borman (Gemini VII), and the first for Anders

APOLLO 9 (AS-504): FIRST COMPLETE MANNED TEST OF LUNAR HARDWARE IN EARTH ORBIT

- March 3, 1969, Cape Kennedy, FL
- Launcher: Saturn V (SA-504)
- Crew
  - James McDivitt, commander
  - David Scott, CM pilot
  - Russell Schweickart, LM pilot
- Objectives: Test all aspects of the lunar module in Earth orbit, including operation of the LM as an independent, self-sufficient spacecraft and execution of docking and rendezvous maneuvers
- Results: Successful
- Spacecraft: CSM 14 (*Gumdrop*), LM-3 (*Spider*)

The Apollo 9 launch was the first Saturn V+CSM+LM in full lunar mission configuration, which carried the largest payload ever placed in orbit. After orbit insertion, the CSM separated from the S-IVB, then transposed (rotated 180°) to dock with the LM three hours after launch. One hour later, the S-IVB and CSM-LM were separated. A second S-IVB burn for sixty-two seconds raised the vehicle's apogee to 3,050 km. During checkout, the SM's service propulsion system (SPS) was fired five times to change orbit in preparation for rendezvous maneuvers, and to test flight dynamics. Deorbit, reentry, and landing went as scheduled.

Apollo 9 mission firsts included:

- First mission to use names for spacecraft after the practice was suspended in the Gemini program, following Gus Grissom's Gemini naming of the capsule the "*Molly Brown*" in oblique reference to his Mercury capsule sinking in the Atlantic

- First test of LM flight hardware in space
- First test of portable life-support system in space

Other mission details:

- Rendezvous and docking were executed in orbit with as much as a 113-mile separation in space
- Total vehicle weight at liftoff: 2,901,681 kg (6,397,005 lb)
- Mass launched to low Earth orbit was 135,000 kg (297,000 lb)
- Mission duration: ten days (241 hr 53 sec)
- This was the second mission for McDivitt (Gemini IV) and Scott (Gemini VIII), and the first for Schweickart.

APOLLO 10 (AS-505): FIRST MANNED TEST OF LUNAR LANDER HARDWARE AT THE MOON

- May 18, 1969, Cape Kennedy, FL
- Launcher: Saturn V (SA-505)
- Crew
  - Thomas Stafford, commander
  - John Young, CM pilot
  - Eugene Cernan, LM pilot
- Objectives: Extensive checkout of all Apollo vehicle, flight systems, and communications equipment in lunar orbit, in preparation for Apollo 11; photographic surveys of the lunar surface, especially near the planned Apollo 11 landing site
- Results: Successful
- Spacecraft: CSM 106 (*Charlie Brown*), LM-4 (*Snoopy*)

This was the first manned test of lunar hardware at the Moon and the second manned mission with the complete lunar hardware. After a nominal launch to parking orbit, the S-IVB was restarted two and a half hours after launch for the trans-lunar injection boost. Lunar orbit insertion occurred seventy-two hours after launch in preparation for the LOR sequence that consisted of the complete lunar landing mission procedures, with the exception of the lunar landing. On the fifth day of the mission, Stafford and Cernan descended in the lunar module to an altitude of about 14,326 m (47,000 ft) from the Moon's surface, where two passes were made over the planned Apollo 11 landing site. Evaluation was made of the landing site for Apollo 11, as well as observations of the lighting conditions at the surface. Following descent and boost back to lunar orbit, the LM descent stage was jettisoned into a temporary lunar orbit. After rendezvous and docking with the CSM in lunar orbit, then separation, the ascent stage of the LM was boosted into heliocentric orbit. The LM and CSM rendezvous and redocking occurred eight hours after separation on 23 May, verifying the maneuver to be used on Apollo 11. A single mid-course correction was executed approximately three hours before CM-SM separation, just before reentry. Reentry and landing went as planned.

Apollo 10 mission firsts included:

- First color TV broadcast on an Apollo mission
- First complete mission hardware in lunar orbit
- Apollo 10 was the only Apollo mission to launch from Launch Complex 39B

Other mission details:

- All aspects of the Apollo 10 mission duplicated conditions of the lunar landing mission as closely as possible.
- The mission was the third space flight for Stafford (Gemini VI and IX) and Young (Gemini III and X), and the second for Cernan (Gemini IX).
- Maximum separation between the LM and the CSM during the rendezvous sequence was about 563 km (350 miles), providing an extensive checkout of the LM rendezvous radar, as well as the backup VHF ranging device aboard the CSM, being flown for the first time.
- Mission duration was approximately eight days (192 hr 3 min 23 sec).

APOLLO 11 (AS-506): FIRST MANNED LANDING ON THE MOON

- July 16, 1969, Cape Kennedy, FL
- Launcher: Saturn V (SA-506)
- Crew
  - Neil Armstrong, commander
  - Michael Collins, CM pilot
  - Edwin Aldrin, LM pilot
- Objectives: Perform a manned lunar landing and return to Earth
- Results: Successful
- Spacecraft: CSM 107 (*Columbia*), LM-5 (*Eagle*)
- TLI burn: 5 min, 48 sec (S-IVB)
- Mid-course correction burn: 3 sec (SM)
- Lunar orbit entry burn: 358 sec (SM)

Apollo 11 was the culmination of a decade of efforts by the United States in the space race with the Soviet Union, and more specifically the race to the Moon. The Soviet's long list of space firsts was eclipsed as the world watched Neil Armstrong take the first steps on the Moon on July 20, 1969.

Apollo 11's landing site was chosen to ensure as safe a landing as possible on a flat, unobstructed terrain. Sites with few craters and no significant hills or obstacles were reviewed within 10° of the Moon's equator. Limits on the latitude range came from the need for a free-return trajectory, meaning a simple return path in case of an abort condition near lunar orbit, and because booster energy is generally at a minimum for landing near the lunar equator. Site selection also required a minimum distance from the eastern limb of the Moon to allow time for landing data to be relayed back from

Earth (LM landing trajectory was from east to west). A maximum terrain slope of 2° was chosen to avoid complications during approach and landing (NASA 1969).

Although well prepared, the Apollo 11 landing proved to be the first major challenge on the mission. Eagle's thirty-second descent burn was followed by a descent trajectory that was controlled primarily by the onboard guidance computer until pitch-over, at which time flight control of the LM was handed over to the two astronauts. As the LM approached the surface, rendezvous radar returns flooded the guidance computer with data. The overload initiated a 1202 and then a 1201 executive overload alarm that immediately created concern in the cockpit and in the control room back at JSC. Although the error was quickly diagnosed in the control room, the error reappeared several times, in addition to several brief communications losses. A landing abort back to lunar orbit was under consideration, although never executed, since the guidance computer succeeded in managing the landing tasks by overriding less important instructions and errors. Because Neil Armstrong found a large crater and boulders in the landing path, he chose to extend the descent range until a suitable site was found. This extended the descent under thrust control to a point roughly 300 meters beyond the crater, then slowed for a controlled touchdown with only twenty seconds of fuel remaining (NASA 1969).

During a White House ceremony in 1969, the Apollo 11 astronauts were recognized for their daring and success, along with Stephen Bales, who was decorated for his contribution to the Apollo 11 mission success by quickly and correctly diagnosing the Apollo computer 2102 and 2101 alarm malfunction, and recommending that the landing continue.

During their twenty-one-hour stay on the Moon, astronauts Armstrong and Aldrin conducted the first-ever spacewalk on another celestial body with a five-hour EVA. The first task on Armstrong's lunar spacewalk was to collect surface samples in the event that an emergency required an immediate departure. After Aldrin joined

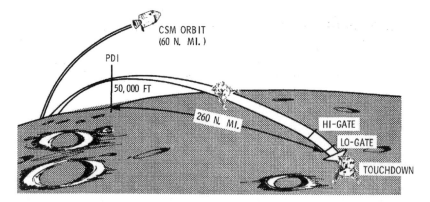

Figure 8.21. Lunar Module descent profile sketched from the PDI (Powered Descent Initiation) to touchdown. PDI is reached from the LM descent burn while in orbit, with its new transfer orbit periapsis (perilune) at the point corresponding to the PDI. At this point, a continuous burn is initiated for velocity reduction to touchdown. Final approach is from Hi-Gate to Lo-Gate, at which time the visual assessment of the landing site is made. The Lo-Gate to touchdown stage is flown manually by the astronauts with the assistance of readouts on altitude, vertical and horizontal speeds, and fuel remaining. Source: Apollo 11 Press Kit.

Armstrong nineteen minutes later, both began setting up scientific experiments from the Early Apollo Surface Experiments Package (EASEP), took photographs, and collected 21.5 kg (47 lb) of lunar samples. Their 250-meter excursion was followed by seven hours of rest and systems checkout before departure. The LM took off from the Moon on July 21 for a CSM rendezvous, with the trans-Earth injection on July 22. After a single mid-course correction, the astronauts returned to Earth on July 24.

Mission firsts for Apollo 11 included:

- First human landing on another celestial body
- First return of samples from another celestial body
- These first samples included basalts dated at 3.7 billion years old
- Moon rock/soil sample mass returned was 21.7 kg (47 lb)

Surface experiments and sampling included:

- Solar wind composition (SWC) experiment
- Panoramic photographs of the region near the landing site and the lunar horizon
- Close-up photographs of in-place lunar surface material
- Deployment of a laser-ranging retroreflector (LRR)
- Deployment of the passive seismic experiment package (PSEP)
- Collection of two core-tube samples of the lunar surface.
- The mission was the second space flight for Armstrong (Gemini VIII), Collins (Gemini X) and Aldrin (Gemini XII)
- Lunar ascent was 21 hr 36 min after the lunar landing
- Mission duration was approximately eight days (195 hr 18 min 35 sec)
- Landing site was in the Mare Tranquillitatis (Sea of Tranquility)

APOLLO 12 (AS-507): SECOND MANNED LUNAR LANDING

- November 14, 1969, Cape Kennedy, FL
- Launcher: Saturn V (SA-507)

Figure 8.22. Photo of Edwin Aldrin completing the deployment of the Early Apollo Surface Experiments Package (EASEP) on Apollo 11. Source: NASA JSC.

- Crew
  - Charles Conrad, commander
  - Richard Gordon, CM pilot
  - Alan Bean, LM pilot
- Objectives
  - Perform inspection, survey, and sampling in lunar mare area
  - Deploy the Apollo lunar surface experiment package (ALSEP)
  - Develop techniques for a point-landing capability
  - Extend capability to work in the lunar environment
  - Obtain photographs of candidate exploration sites
  - Apollo 12's secondary objective was to retrieve portions of the Surveyor III spacecraft, which had been exposed to the lunar environment since the unmanned spacecraft soft-landed on April 20, 1967
- Spacecraft: CSM 108 (*Yankee Clipper*), LM-6 (*Intrepid*)
- TLI burn: 5 min 45 sec (S-IVB)
- Lunar orbit entry burn: 6 min (SM)

Clouds and rain at the launch site were not considered flight hazards until lightning struck the Apollo vehicle and launcher twice—the first thirty-two seconds after liftoff, and again fifty-two seconds into the flight. The lightning surges temporarily interrupted power and communications on the CM. After switching to backup battery power, the electrical system was reset manually with no further incidents. Launch criteria were changed thereafter to prohibit launches with cumulonimbus in the launch pad vicinity, or when potential lightning or high electric-field hazards exist.

Apollo 12's landing site was chosen in the Sea of Storms, some 1,500 km west of the Apollo 11 Tranquility base, with nearly the same selection criteria as Apollo 11, which was dictated by exploration objectives, flight performance, and safety. The site offered a relatively diverse geology with several nearby dominant craters and the possibility of a pinpoint landing next to a Surveyor probe launched less than three years earlier. The landing site proved fruitful in Alan Bean's and Richard Conrad's exploration surveys during their two EVAs. A total of 34 kg (75 lb) of rock and soil samples were collected, in addition to the retrieval of the camera and several other parts of Surveyor III.

Apollo 12 astronauts Conrad and Bean deployed an improved experiment package on their first EVA, called the Apollo lunar surface experiment package (ALSEP), that was powered by the first radioisotope thermoelectric generator (RTG) taken to the Moon. The ALSEP's attached RTG was used for extended operations of experiments on this and subsequent missions.

Apollo 12 mission highlights:

- The crew remained in quarantine for twenty-one days after their return
- Mission duration: Approximately ten days (244 hr 36 min 24 sec)
- Lunar stay: 31 hr, 31 min
- Moon rock sample mass returned: 34.4 kg (75 lb)
- Landing site: Oceanus Procellarum (Sea of Storms)

Apollo 12's S-IVB booster was reignited in the first-ever maneuver intended to curve its trajectory away from the Apollo spacecraft as both headed toward the Moon. The burn was slightly longer than planned, placing the S-IVB in Earth orbit. After thirty years, the booster was recaptured by the Earth-Moon system, only to return to heliocentric orbit two years later in 2003.

For the first time, the LM was jettisoned toward the Moon, resulting in Intrepid's impact on the Moon at more than 8,047 km/h (5,000 mi/h), creating the first artificial Apollo moonquake. The event registered on the seismometers left on the Moon in the Apollo surface experiments. On later Apollo missions, controllers directed the S-IVBs to impact with the Moon, removing the booster from the lunar environment, and creating shock waves that could be picked up by the Apollo seismic instruments.

Surveyor III parts returned to Earth were reportedly found to contain Earth-borne bacteria (Streptococcus mitis) that would have survived over two years of extreme temperatures, radiation, and the vacuum of space. Later testing indicated that the samples were most likely contaminated after arrival.

The Apollo 12 backup crew inserted several *Playboy* pinup photos in the cuff-mounted EVA checklist worn by Conrad and Bean, copies of which can be found in the online Apollo Lunar Surface Journal files.

Public excitement over the Apollo 12 mission was far less than for Apollo 11 for several reasons, the most obvious being that Apollo 11 was a worldwide sensation of

Figure 8.23. Apollo 12 astronaut Pete Conrad inspecting the Surveyor III spacecraft in the Sea of Storms with the LM lander in the background. Source: NASA JSC.

the first humans to land on the Moon. Unfortunately for the program's public support, camera coverage of Apollo 12's surface activities ended by accident when the camera was pointed into the Sun. Neither NASA nor the public could see the crew's excitement about being on the Moon, which was not expressed by the Apollo 11 crew. Without the excitement of the live TV broadcasts, public interest quickly waned. Even before the launch, President Nixon had trimmed the Apollo program to end after the Apollo 20 mission. And, because of the declining public and congressional interest in Apollo and the shrinking budget, NASA canceled the Apollo 20 mission in January 1970. On September 2, 1970, Apollo 18 and 19 were also canceled.

APOLLO 13 (AS-508)

- April 11, 1970, Cape Kennedy, FL
- Launcher: Saturn V (SA-508)
- Objectives
  - Precision lunar landing in the Fra Mauro highlands
  - Surveying and sampling the Imbrium Basin
  - Deploying and activating the Apollo lunar surface experiments package
  - Further developing the capability to work in the lunar environment
  - Photographing exploration sites for future missions
- Spacecraft: CSM 109 (*Odyssey*), LM-7 (*Aquarius*)
- Crew
  - James Lovell, commander
  - John Swigert, CM pilot
  - Fred Haise, LM pilot

Apollo 11 and 12 landing sites had been chosen for their benign topography of few hills and craters, and a minimum of obstacles that could interfere with a safe touchdown. After suitable landing sites were identified, the most interesting geology was chosen for as much diversity in surface sampling as possible. The Sea of Tranquility offered broad, flat regions of the volcanic mare plains, but only scattered rubble from distant impact craters and little of the older highlands rock. Safety concerns in the first two Apollo surface landing sites did compromise geological diversity in the rock samples returned. Hence, the next flights targeted smaller landing sites that were still flat but within reach of rubble from the highlands. Although landings in rugged terrain were never considered, the landing accuracy demonstrated on Apollo 12 expanded the list of attractive and available exploration sites. Frau Muro highlands in the Sea of Rains was chosen for this Apollo mission because of a young nearby crater that added to the rubble field, with even more ejecta from the large Copernicus crater 360 km to the north.

Apollo 13's nearly flawless launch and uneventful early flight were violently interrupted fifty-six hours into the flight. An oxygen tank explosion in the service module's cryogenic oxygen supply had serious consequences, since it reduced the command module's available electrical power, drained critical breathing oxygen, and trimmed the astronauts' water supply. Just a few minutes after a televised broadcast from the CM, internal fans were turned on to stir SM oxygen tanks 1 and 2. The Accident Review Board concluded that the power lines to the oxygen tank 2 circulation fans had been damaged during preflight testing. When power was applied on the mission, the wires shorted, which set the Teflon insulation on fire. The fire spread within the tank, raising the pressure until oxygen tank 2 exploded, which also damaged oxygen tank 1.

Apollo 13's mission abort plan called for a minimum duration, minimum energy return. These constraints were satisfied only by a circumlunar trajectory, meaning continuing the rendezvous with the Moon before swinging back to Earth. Return time was just under six days, but even that exceeded the damaged CSM life-support capability. Fortunately, consumables and power from the lunar module, including oxygen, batteries, CO_2 removal canisters, and an independent propulsion system were available to bring the astronauts back on a stressful but successful return, reentry, and landing. Nonetheless, the success of the perilous aborted lunar flight far outweighed the hardware failures and lost lunar landing. After the crew's dramatic return, NASA spokesmen described the Apollo 13 mission as their most successful failure.

Apollo 13's more important mission details included:

- First aborted Apollo mission
- First impact of the S-IVB on the lunar surface
- Use of lunar module to provide emergency propulsion and life support after loss of several critical service module systems
- Fourth mission for Lovell (Gemini VI, XII, Apollo 8), first for Swigert and Haise
- Use of backup CM pilot for Thomas Mattingly (Apollo 16). Mattingly was removed 72 hr prior to scheduled launch due to exposure to the German measles.
- Mission duration was approximately six days (142 hr 54 min 41 sec)

APOLLO 14 (AS-509)

- January 31, 1971, Cape Kennedy, FL
- Launcher: Saturn V (SA-509)
- Objectives: Duplication of Apollo 13 mission objectives
- Spacecraft: CSM 111 (*Kitty Hawk*), LM-9 (*Antares*)
- Crew
  - Alan Shepard, commander
  - Stuart Roosa, CM pilot
  - Edgar Mitchell, LM pilot

The minor problems on the Apollo 14 mission were a relief to the crew and ground controllers following the near-disaster of the Apollo 13 flight. A forty-minute launch delay due to clouds and rain was the first launch delay for the Apollo lunar missions. After departing on the trans-lunar cruise, the crew was frustrated by a series of six attempts before a successful CSM-LM docking was completed. Lunar orbit and landing operations were uneventful, with the exception of a landing radar glitch that delayed radar acquisition of the surface until late in the landing sequence. A pinpoint landing within thirty meters of the selected site was successful, even though the landing placed the LM on an 8° tilt, which interrupted sleep on the overnight stay. On the first of two EVAs, a contingency sample was first collected, then both Shepard and Mitchell set out to deploy the ALSEP. One addition to the equipment was a hand cart, called the modular equipment transporter (MET), that allowed the astronauts to carry more samples, equipment, and tools.

Apollo 14's landing site was the same as that selected for the aborted Apollo 13, near the Frau Muro formation, and near a crater rim next to the Mare Imbrium (Sea of Rains). The rim of ejecta was expected to offer a convenient stratigraphic marker for sampling the Moon's interior to the depth of the crater, some 150 km. Shallow excavation of the site that was also near the smaller, more recent Cone Crater was expected to provide samples of the ancient bedrock below the regolith more than 4 billion years old.

Mitchell and Shepard toured the Cone Crater region on their EVAs for more than 9 hours. Their second excursion just reached the planned Cone Crater rim objective, although there was confusion about the rim feature. With the help of the MET, Shepard and Mitchell collected 42.9 kg of lunar samples. As their last EVA ended, Shepard executed a physics experiment and a little one-handed golf practice by hitting two smuggled golf balls "miles, and miles, and miles." Their total time on the lunar surface was 33.5 hours, with a total distance traversed of 3.3 km on their two EVAs. Return of the CM capsule to Earth, reentry, and landing were uneventful, with the CM splashdown within 1 km of its planned impact.

Figure 8.24. Apollo 14 ALSEP after deployment with the SNAP-27 RTG nuclear power supply in the foreground and the central station unit in the background. Source: NASA JSC.

- For the last time, the Apollo 14 crew was placed in quarantine after their return (twenty-one days from completion of the second EVA)
- Mission duration was approximately nine days (216 hr 1 min 58 sec)
- Moon rock sample mass returned totaled 42.9 kg
- Alan Shepard was the only Mercury astronaut to reach the Moon and the second to fly in the Apollo program
- Red stripes were placed on the arms, legs, and hood of Shepard's space suit to make him more identifiable in the returned photographs and films, a procedure that carried through to the space shuttle and International Space Station crews

APOLLO 15 (AS-510)

- July 26, 1971, Cape Kennedy, FL
- Launcher: Saturn V (SA-510)
- Objectives (primary)
  - Perform selenological inspection, survey, and sampling of materials and surface features in a preselected area of the Hadley Apennine region
  - Emplace and activate surface experiments
  - Evaluate the capability of the Apollo equipment to provide extended lunar surface stay time, increased extravehicular operations, and surface mobility
  - Conduct in-flight experiments and photographic tasks from lunar orbit
- Spacecraft: CSM 112 (*Endeavor*), LM-10 (*Falcon*)
- Crew
  - David Scott, commander
  - Al Worden, CM pilot
  - James Irwin, LM pilot

Apollo 15 was the first of three J-class missions, which included the battery-powered Lunar Rover and the improved space suit that contained a waist joint to allow the astronauts to sit. It was also the first mission with a landing and exploration site outside of the Apollo landing zone situated 5° either side of the lunar equator. Apollo 15's landing site, located at 26° latitude, required more orbital energy, but allowed access to a much more diverse terrain. A more complex mission and more challenging landing site were now permitted because of the demonstration of relatively precise landings by the crew and equipment on prior missions. With the new freedom at hand, geologists and flight planners selected Mare Imbrium (Sea of Rains), located in a small, flat region between tall crater rims in the Apennine Mountains, as the landing site. The site's diverse geology also included an unusual and intriguing riverlike valley called the Hadley Rille that appeared to be in a field of ancient rocks strewn from some of the earliest craters on the Moon.

Lunar deorbit, descent, and landing preparation were nominal, but a lack of distinctive features on the undulating terrain required mission controllers to call the touchdown for Scott and Irwin. After touchdown and a brief survey just outside the cabin, the crew decided to rest. For the first time in the program, the crew was able to sleep outside of their suits. The first EVA began with Scott and Irwin deploying the LRV, followed by a brief excursion before setting up the ALSEP experiments.

Problems with the drill design prevented Scott from completing several bore holes for experiments, resulting in just one bore hole, and that only partially completed. Three hours into their second EVA, the crew encountered a light-colored crystalline rock on a breccia outcrop that raised the level of excitement for almost everyone watching, especially the astronauts and the geologists. Knowing that the sample was anorthocite, which are the oldest surface rocks on the Moon, Scott and Irwin realized that they had found one of the most important samples yet. The sample was later called the Genesis Rock in reference to the ancient highlands material that solidified and formed the primordial lunar crust. Scott and Irwin's second EVA was also successful in collecting anorthocite samples, but smaller and less dramatic in their geological impact. The third EVA was devoted mostly to a survey of the Hadley Rille, a long, narrow, meandering valley that had the superficial appearance of an ancient riverbed. Photographs and samples were returned, and several more core samples were attempted, which again proved difficult.

Additional experiments for this mission were placed on the command module and attached to a new scientific instrument module (SIM) located in one of the service module's bays. During Scott and Irwin's nearly three-day lunar stay, Al Worden conducted experiments in the command module and in the service module SIM, including photography of the lunar surface and the lunar environment, radar and laser altimetry measurements of the lunar surface, and X-ray and gamma-ray spectroscopy to evaluate elemental abundance in the highlands and maria. Before leaving lunar orbit, Worden deployed a sub-satellite to measure interplanetary magnetic fields and the Earth's magnetic field near the Moon. The tiny moonlet also contained charged particle detectors and electronics for measuring the mass variations in the Moon called mass concentrations, or mascons.

Return to Earth, separation from the SM, and reentry of the Endeavor command module was nominal until parachute deployment. One of the three main parachutes failed after deployment from several damaged risers, although the system was designed so that only two of the main chutes were needed for landing. A hard but successful splashdown landing was followed by a swift recovery of the crew and capsule.

Apollo 15 astronauts were the first to use the lunar rover, although the decision to develop the LRV was made before the launch of Apollo 11 because of the rover's many advantages for the astronauts' surface exploration tasks. The LRV could not only carry the two crew, but could also transport tools, scientific equipment, communications gear, and lunar samples—a cargo capacity nearly three and a half times its own weight. The golf cart–sized lunar rover was developed and fabricated by Boeing in only eighteen months, at a cost of $40 million.

- Duration of lunar stay: 66 hr 54 min 53 sec
- EVA total of 18 hr 30 min, plus 33 min stand-up EVA
- Total distance traversed: 27.9 km (17 mi)
- Mission duration was approximately 12 days (295 hr 11 min 53 sec)
- Moon rock sample mass returned: 76.8 kg (169.3 lb)
- First mission to carry orbital sensors in the service module
- First launch of a sub-satellite into lunar orbit

APOLLO 16 (AS-511)

- April 16, 1972, Cape Kennedy, FL
- Launcher: Saturn V (SA-511)
- Primary objectives
  - Perform selenological inspection, survey, and sampling of materials and surface features in a preselected area of the Descartes region
  - Deploy and activate surface experiments
  - Conduct in-flight experiments and photographic tasks
- Detailed objectives
  - SM SIM orbital photography
  - Visual light flash experiments
  - Command module photography
  - Improved gas/water separator test
  - Body fluid balance analysis
  - Sub-satellite tracking for autonomous navigation
  - Improved fecal collection bag test
  - Skylab food package evaluation
  - Lunar rover vehicle evaluation
- Spacecraft: CSM 113 (*Casper*), LM-11 (*Orion*)
- Crew
  - John Young, commander
  - Thomas Mattingly, CM pilot
  - Charles Duke, LM pilot

NASA managers and scientists elevated the Apollo landing site challenges with each succeeding mission to reach more diverse and complex surface geologies, moving the landing site next to the highlands for the first and only time with the Apollo 16 mission. Samples and surveys from the previous missions suggested that the Descartes crater region would be useful for exploring cratered strata in the ancient crust, and distance Apollo 16 from the other landing sites. Separation would serve to diversify the lunar samples and help fill in some of the gaps in the theory of the Moon's formation. Thought originally to be volcanic in origin, the region chosen for this landing site was located between the two young craters named North Ray Crater and South Ray Crater, which proved to be highland bedrock, with breccia (broken) rock and bedrock fragments from both craters littering the surface.

Launch, Earth parking orbit, TLI, and lunar orbit insertion were executed without incident, although separation of the LM from the CSM was delayed to evaluate a service module SPS engine problem. Delay in the lunar descent and landing went as planned, but was complicated by the lack of shadows on the rolling surface. The first of three planned EVAs for Young and Duke began after an eight-hour rest period after landing. This was the first crew that slept uninterrupted on the Moon. The first EVA began with the deployment of the lunar rover, followed by deployment of the ALSEP experiment package. After the ALSEP central station placement, Young caught the heat flow experiment cable under his foot, which tore the cable from its connection on the instrument and rendered the experiment inoperative. Concerned but undaunted, the two astronauts continued on their EVA, including a lunar rover excursion through the South Ray Crater ejecta field. Site surveys were completed with 20 kg of samples returned, mostly highlands breccias.

While as successful as the first EVA, the second was slowed by damage to a rear fender that sprayed fine dust onto the astronauts and their equipment, which had some effect on the LRV equipment. The third and last EVA focused on the North Ray Crater field and on finding magma samples possible in large boulders. Several stops at these sites proved productive even with the excursion shortened because of the deorbit delay. Young and Duke had spent more than twenty hours in their EVAs, were on the Moon for three days, and collected nearly 95 kg of lunar samples.

Experimental equipment included a far-ultraviolet camera/spectroscope that was placed near the LM to study the Earth's atmosphere and radiation belt emissions during the lunar stay, making it the first lunar observatory. Although not the first one, a portable magnetometer was taken on the lunar rover excursions to measure the local fields in the lunar surface.

CM pilot Mattingly conducted a number of experiments placed in and on the service module, including observations with the scientific instrument module containing a gamma-ray spectrometer, an X-ray spectrometer, an alpha-particle spectrometer, a mass spectrometer, and a sub-satellite with three particles and fields experiments. The SM experiment suite also included a mapping camera, a panoramic camera, and a laser altimeter. During the trans-Earth coast, Mattingly retrieved film from the service module SIM camera and exposed a microbial experiment during his one-hour, twenty-three-minute EVA.

Apollo 16 firsts and records include:

- Commander John Young flew twice in the Gemini program, was the command module pilot on Apollo 10, the commander on Apollo 16, and commanded the first space shuttle flight, STS-1. Thomas Mattingly later flew on space shuttle missions STS-4 and STS-51-C.
- Second in the series of three science-oriented J class missions
- First use of the Moon as an astronomical observatory
- Mission duration: approximately 11 days (265 hr 51 min 5 sec)
- Moon rock sample mass returned: 94.7 kg (208.8 lb)

APOLLO 17 (AS-512): LAST APOLLO LUNAR MISSION

- December 7, 1972, Cape Kennedy, FL
- Launcher: Saturn V (SA-512)
- Primary objectives
  - Perform selenological inspection, survey, and sampling of materials and surface features in the Taurus Littrow region
  - Deploy and activate surface experiments
  - Conduct in-flight experiments and photographic tasks

- Detailed objectives
  - Obtain lunar surface photographs and altitude data from the service module SIM instruments
  - Obtain data on the visual light flash phenomenon
  - Obtain CM photographs of lunar surface features of scientific interest and photographs of low brightness astronomical and terrestrial sources
  - Record visual observations from lunar orbit of specific lunar surface features and processes
  - Obtain data on Apollo spacecraft-induced contamination
  - Obtain data on whole body metabolic gains or losses, together with associated endocrinological controls (food compatibility assessment)
  - Obtain data on the use of the protective pressure garment
- Spacecraft CSM 114 (*America*), LM-12 (*Challenger*)
- Crew
  - Eugene Cernan, commander
  - Ronald Evans, CM pilot
  - Harrison Schmitt, LM pilot

Apollo 17 was the last of the Apollo lunar missions, and the third of the J-class missions (Apollo 17 was also classified as the J-3 mission) that were characterized by heavier payloads, an added lunar surface rover, extended duration flights, and more scientific equipment and instruments. By taking advantage of the experience gained in the previous missions, Apollo 17 established the record for the longest duration with 12.6 days, the longest time on the lunar surface at 75 hours, and the longest total surface traverse distance of nearly 32 km.

Site selection for Apollo 17 was aimed at exploring and sampling the most diverse and unique geological formations accessible on the Moon. The site was in the southeast region of the Sea of Serenity (Mare Serenitatis) near a crater rim excavated by a large impact early in the formation of the lunar crust. The impact created uneven blocks of fractured layers, along with valleys within the mountainous rim. One of those valleys, the Taurus-Littrow valley, was located on a large lava flow that filled the Mare Serenitatis crater basin, surrounded by remnants from an ancient impact. The Taurus-Littrow landing site is a relatively flat region, surrounded with smaller craters that were attractive as sampling sites, and useful for landing cues. The lava flows responsible for forming the Mare Serenitatis also filled the lower valleys, producing localized fiery fountains that showered glass fragments in dark as well as orange-colored deposits discovered on the second Apollo 17 EVA. One of those responsible for the selection of the site, geologist Harrison Schmidt, was also a member of the Apollo 17 crew. Schmidt was originally assigned a seat on the Apollo 18 mission, but with its cancelation, was moved to the Apollo 17 mission because of his scientific expertise.

Launch of the Apollo 17 was delayed for 2 hours and 40 minutes by a sequencing delay, but subsequent launch and flight operations to the Moon were otherwise uneventful. TLI boost was lengthened to reduce the trans-lunar coast time to the Moon to compensate for the 2.7-hour delay. Impact of the S-IVB booster stage and the later LM impact on the lunar surface were recorded on the previous ALSEP instruments.

Lunar orbit, LM separation, and LM descent and landing were also executed as planned. Instead of a rest period, the crew began their first of three EVAs about 4 hours after landing. The very successful EVA that included the LRV and ALSEP deployment ended after 7 hours, 37 minutes, with a collection of about 14 kg (31 lb) of surface samples.

Apollo 17 carried the most comprehensive set of ALSEP instruments and experiments of any mission. Included was a portable traverse gravimeter that measured the gravity variations at the field sites, although a design flaw kept it from working properly. Also included were a surface electrical properties (SEP) experiment, and a neutron probe that identified the mixing depths of the top regolith after being placed in one of the boreholes. The SEP experiment used a transmitter placed near the ALSEP experiment package and a receiver and antenna placed on the lunar rover. Changes in the propagation of the radio waves along the surface were influenced by the conductivity of the surface soil and rock, providing an important measure of the physical makeup of the surface. Unfortunately, heat and dust affected the equipment to the point of providing little useful data. A heat flow experiment was also added to replace data from the experiment damaged on the previous mission. Added to the science instrument lineup on their ALSEP was a cosmic ray detector, while the solar wind collector, a corner-cube reflector for laser ranging of the Moon from Earth, and the passive seismometer were left off.

Cernan and Schmidt's second EVA began with a 6 km traverse using the LRV for survey and sampling at several sites, including a massif (large, displaced block of crust) to the south of their lander. Schmidt's discovery of light orange-colored soil on a later stop and his excitement at the find spread quickly to NASA geologists monitoring the mission at the Johnson Space Center. The significance of the find was enough to extend the stay at the site and shorten several others that followed. The clearly visible orange soil that ran along the rim of the Shorty Crater was thought to be part of an explosive outflow from a lunar volcano assumed to be responsible for the crater. Later analysis linked the unusual orange glass soil to a lower layer of volcanic glass created much earlier in the mare-forming period that was later spread by the impact that had formed the Shorty Crater. The second Apollo 17 EVA ended after a record seven hours and thirty-seven minutes with the collection of 34 kg (75 lb) of lunar surface samples.

Third and final in the Apollo 17 EVA series was an excursion of seven hours, fifteen minutes to nine stations, during which the astronauts collected 62 kg (137 lb) of samples. Like the other J-series final EVAs, the astronauts parked the lunar rover so that its camera could be directed by JSC controllers to cover the launch and departure of the LM ascent stage.

Mission totals for Apollo 17 were 110 kg (243 lb) in surface samples, a total traverse of 36 km on the LRV, a total EVA time of twenty-two hours and four minutes, and time on the surface totaling three days and three hours. Following the trans-Earth injection, Evans retrieved lunar sounder film and panoramic and mapping camera cassettes from the SM SIM bay during his EVA of one hour and seven minutes from the command module.

Firsts and records for the mission included:

- First geologist on lunar surface and the only scientist to walk on the Moon
- Longest LRV traversed on a single EVA

- Greatest lunar surface sample mass returned to Earth—110.5 kg
- Mission duration: 12.6 days (12 days 13 hours 51 minutes 59 seconds)
- Cernan previously flew in space aboard Gemini IX and Apollo 10, while Apollo 17 was the first flight into space for Evans and Schmitt.
- The landing site was on the southeastern rim of Mare Serenitatis in a dark deposit between massif units of the southwestern Montes Taurus
- Each of the Apollo LM descent stages carried a commemorative plaque dedicated to the exploration efforts of the astronauts and humankind. Apollo 17's Challenger plaque read: "Here Man completed his first explorations of the moon. December 1972 AD. May the spirit of peace in which we came be reflected in the lives of all mankind." Thus ended the most ambitious and fruitful exploration program to date—NASA's manned Apollo project.

THE APOLLO LEGACY

Apollo science contributed more to our understanding of the Moon and its formation, of the Earth-Moon system, and of terrestrial planet formation in general, than all previous research. But even more important, the Apollo program substantially advanced the state of technology for aviation, transportation, and nearly every technology industry. The often-quoted contribution of the Apollo program to the development of the miniaturized digital computer is truly an understatement, since microelectronic circuitry developed for the Apollo missions quickly evolved into a revolution in microelectronics employed in communications, radar, avionics, manufacturing automation, and, of course, the powerful digital computer. NASA's Apollo project also contributed to everyday technology found in automobiles, homes, medical equipment, and even food processing. The list of spinoffs from the Apollo program is long, but not limited to the material dimension. Apollo succeeded in its political purpose during the Cold War of galvanizing our nation in the triumph over the Soviet communists' attempt to establish superiority in space and advancing their military prowess using space-based weapons. And yet the same Apollo program was used to promote the peaceful cooperation between the U.S. and Soviet space programs in the Apollo-Soyuz Test Project (ASTP).

A monumental effort was required to carry the Apollo program through to a successful completion, from the broad industry participation in research and development to the contributions from all of NASA centers, in addition to the significant collaborative assistance from the Department of Defense and its military technology branches, including the extensive naval crew, vessel, aircraft, and communications divisions used for capsule recovery. The Apollo program was also a management triumph that created cooperative arrangements that succeeded in solving monumental problems associated with advanced technology and space flight operations on a scale never before attempted. NASA orchestrated the largest and most diverse program in history in taking men to the Moon in only eight years from the decision to commit funding until the touchdown of the first American crew on the Moon.

The Apollo Computer

The first major application of microelectronics was introduced by the missile industry to enable autonomous guidance and navigation without suffering the weight penalty of bulky onboard computers. In this era of the space age, the low-power transistor circuitry had shrunk enough to allow small multipurpose circuits to control many of the systems functions, communications tasks, and inertial navigation operations of larger ballistic missiles. The first digital computer systems for manned vehicles were developed for, and flown on, the Gemini missions because of the complex calculations required for the orbital positioning, and for the demanding rendezvous and docking maneuvers. Gemini's digital computers also permitted automated functions that included orbital maneuvers, deorbit, reentry, and descent, with much greater accuracy than for the Mercury flights.

Mercury spacecraft had no need for onboard computations since there was no capability for orbital changes, except for boost-to-orbit and deorbit burns. All of Mercury's trajectory calculations were made from ground-tracking and ground-based computers that relayed commands back to the capsule and crew. As advanced as the Gemini computer was for its time, the sophistication of the navigation and control systems was insufficient for the precision, complexity, and reliability required for the Apollo lunar flights. It was recognized early in the program that one of the important development efforts needed for the Apollo missions was a new and more powerful computer and navigation system that would be capable of complex, autonomous navigation and guidance to the Moon and back.

Three contractors were hired by NASA to develop the Apollo computers for the three primary Apollo vehicles: the Saturn V, the command module, and the lunar module. The most complex Apollo computer system was the command module's Apollo guidance computer (AGC), which was also used as the primary computer on the LM. Designed by the MIT Instrumentation Laboratory and built by the Raytheon Corporation, the AGC was responsible for the guidance and navigation operations, and the operation of the vehicle's control functions. Weight limitations on all of the Apollo vehicle components were critical, especially for the lunar lander, because of the tremendous booster energy required to reach the Moon, to slow it to enter lunar orbit, then descend to the surface, followed by a lunar surface launch for rendezvous back with the CSM, and finally a boost back to Earth. Electronic circuitry for the Apollo vehicles increased the density of the component configurations and the modular circuitry with a much smaller scale of logic units. The AGC was also the first to employ integrated circuitry. Resistor-transistor logic (RTL) was used to reduce the size of more than 5,000 gates so that they could fit into small modular units called integrated circuits that used wire-wrap connections encased in epoxy plastic for durability. Apollo's AGC became the prototype computer that rapidly flourished in myriad applications, ranging from inexpensive hand-held calculators to powerful manufacturing and business computers (Tomayko 1988).

The Apollo AGC was unusual in several ways, including its memory. Ferromagnetic core units, looking like tiny metal donuts, were wrapped with various wire combinations for random access rewritable memory, as well as for programmable memory. Core memory was reliable if manufactured correctly, and versatile enough to be used a decade later in the IBM AP-101 General Purpose Computer designed for the space shuttle orbiter. While durable and reliable, the bulky, power-hungry core ferromagnetic memory was replaced with the much lower-weight and much lower-power semiconductor modules in the space shuttle GPC computers in the early 1990s.

Software for the AGC was based on assembly language compiled into a higher-order language to save valuable memory and to speed operations. A 16-bit-word embodied the software and hardware architecture, along with a memory size of 32 k words of fixed memory (ROM) and 4 k of erasable memory (RAM) for the later Block II models. Cycle time for AGC operations was based on the internal 2 MHz oscillator clock (Tomayko 1988).

Backup flight control for the CSM was established through a ground-link command communications channel for the CSM, and a separate, simpler abort guidance system on the LM.

Apollo Advancements in Medicine

Extensive research on Apollo astronauts was conducted preflight, postflight, and during flight to determine the health of and physiological changes in the crews during their space flights. Microgravity and radiation exposure were some of the most serious concerns for NASA's medical teams. Since the lunar missions were the longest duration flights and extended beyond Earth's protective magnetic field, measuring the effects of the space environment on Apollo astronauts required sophisticated equipment and test procedures that had only begun in the Mercury and Gemini programs. Adding to NASA's concerns were possible toxin or even viral contamination of the astronauts by the lunar environment. Ultimately, contamination of the crew or the equipment was never encountered, but the screening and tests developed to identify exotic contagions and toxins led to more advanced medical testing and screening used today.

Digital signal processing developed by the Jet Propulsion Lab, used to computer-enhance lunar images, became the foundation of later body-scan imaging that is used for advanced diagnostics, known today as computer-aided tomography (CAT), and nuclear magnetic resonance imaging systems, now called magnetic resonance imaging (MRI) because of the public's discomfort with the word nuclear.

Ultrasound technology, developed by the aerospace industry for examining critical space components for irregular or imperfect surfaces/joints/welds, has been adapted for tissue analysis to determine the extent and depth of burns and aids in treatment of the trauma.

Biotelemetry, used in the manned missions to relay data on the astronaut's vital functions, has been adapted for use in hospitals and in home-care programs to monitor patients' health. Transmission of real-time data or near real-time data on a patient's specific functions allows quick response to potentially critical conditions, as well as low-cost diagnosis for clinics that cannot afford trained medical staff.

Improvements in targeting and tracking deep-space missions using image processors onboard autonomous spacecraft led to the development of techniques for aiding vision-impaired patients and patients suffering from macular degeneration.

Image processors tailored to the individual's vision anomalies have been used to improve image recognition by projecting remapped images onto the retina.

A thin, body-cooling layer used for temperature control for astronauts in their space suits was adapted as a cooling garment for patients with multiple sclerosis, cystic fibrosis, severe burns, and some forms of cancer. Smaller cooling units are also being used for limbs and smaller regions of the body that require regulated temperatures.

Blood analyzers, microbial detectors, specialized food processing, and a multitude of other advances spun off from the Apollo and other manned space programs have contributed much to our health care, medical treatments, medical diagnosis, and medical therapies. Long-duration exposure to microgravity has spawned numerous studies for remediating bone loss, a condition also found in those advanced in age, especially females. These and other medical research programs that were developed for solving the problems related to space flight are invaluable in understanding the aging process and related problems because of the extreme conditions associated with exposure to space that can accelerate deterioration of the bone, muscle, nerve tissue, and even immune system.

APOLLO-SOYUZ TEST PROJECT (ASTP)

Cooperative agreements between scientists in the United States and the Soviet Union during the Cold War were almost impossible, and space programs were no exception. Bilateral missile and weapons buildup and Stalin's cultivation of a hostile, militaristic Soviet government bred distrust between the two superpowers following World War II and continued until the collapse of the Soviet Union in 1991. In spite of the hostility, an early overture was made by President Kennedy in 1962 that would have joined the two manned lunar program; the offer was rebuffed by the Soviet Politburo almost immediately. Yet cooperative agreements in space research were made on a smaller scale. In 1962, NASA and the Soviet Union agreed to share magnetic field experiments on satellites and joint communications experiments using the U.S. Echo 2 satellite. A second agreement in 1965 led to the publication of joint research reviews in space biology and space medicine, followed by discussions of a possible manned vehicle rendezvous and docking mission by NASA administrator Thomas Paine and Soviet Academy of Sciences president M. V. Keldysh (NASA 1975). Continuing talks were held in 1970 and a working group was assigned the task of designing a docking unit that could connect the U.S. Apollo command and service module to the Soviet Soyuz vehicle. Additional agreements were made in 1972 between the United States and Soviets on shared meteorological data, sounding rocket data, and the exchange of some lunar samples. Most important, the breakthrough for the final decision to fly a combined Apollo-Soyuz mission, called the Apollo-Soyuz Test Project, came with the signing of the agreement for the mission by President Nixon and Soviet chairman Kosygin in May of 1972 (NASA 1975).

Primary objectives for the ASTP were simple, though an impressive number of experiments and secondary objectives were planned for and completed on the mission. An important element of the project was the docking adapter that furnished a

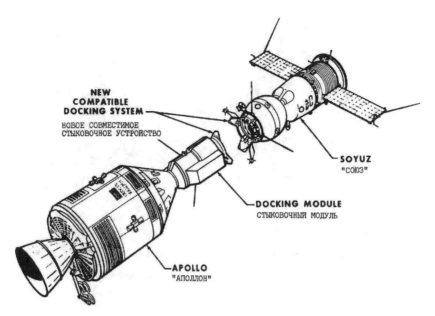

Figure 8.25. Sketch of the Apollo-Soyuz modules and the jointly developed Docking Module that was used to connect both flight vehicles with a pressurized access tunnel. The ASTP Docking Module was a precursor to the Shuttle-Mir docking module that was the prototype for the current ISS Pressurized Mating Adapter. Source: ASTP Press Kit.

mechanical connection between the two spacecraft docking ports, allowing crew passage while providing a positive pressure seal. Beyond the joint mission objectives were innumerable firsts that have since endured. The most important first was the joint agreement between Cold War adversaries to fly complex space hardware in complete cooperation. Apollo-Soyuz was the first international orbital laboratory, only surpassed in scope by the new International Space Station. Although the Soviet Mir space station was much larger, its international element comprised only experiments contributed by other nations.

ASTP Flight Mission

Both the Apollo CSM and Soyuz crew vehicles were launched on July 15, 1975, with the American crew onboard consisting of Thomas Stafford (mission commander), Vance Brand (CM pilot), and Deke Slayton (docking module pilot). The Soviets launched their crew, consisting of Alexi Leonov (mission commander) and Valeriy Kubasov (capsule engineer), onboard the Soyuz 19 vehicle about seven and a half hours before the Apollo CSM launch. A conventional lower-altitude orbit for the CSM chase vehicle brought it in proximity to the Soyuz 19 for docking on July 17. Prior to the docking, the Apollo CSM executed transposition maneuvers similar to the lunar missions in which the CSM pulled the LM from the S-IVB upper-stage booster. For the ASTP mission, the CSM extracted the docking mechanism, while the S-IVB

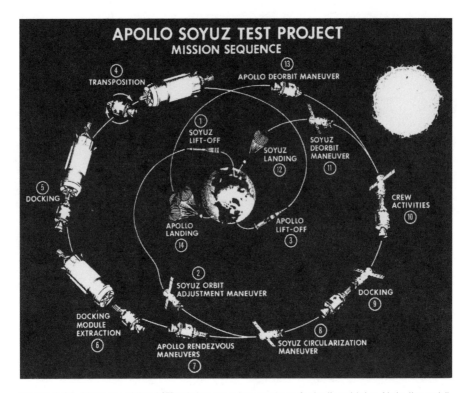

Figure 8.26. Diagram of the ASTP mission events sequence for both vehicles. Note the additional maneuvers used to extract the Docking Module from the Saturn IB upper stage booster (steps 4, 5, 6). Source: ASTP Press Kit.

was commanded to pull back. The Saturn IB's S-IVB second stage was then deorbited to burn up in the atmosphere. The CSM and attached docking module were rotated 180°, then proceeded to the Soyuz for rendezvous and docking.

Docking of the two spacecraft began with a mechanical latching of the CSM-mounted docking module with the Soyuz, then pressurization of the intercapsule tunnel, and finally the hatch opening for both crew capsules. During the forty-four hours docked, four crew transfers were made and a variety of internal and external experiments conducted on both spacecraft. After undocking on July 19, an artificial solar eclipse experiment was made for coronal observations of the Sun from Soyuz 19. Redocking on the same day, then a final undocking, also on that day, was followed by combined astrophysical, solar, and Earth observations. Two days later, the Soyuz deorbited, with Apollo 18 deorbiting five days after that. A Doppler experiment was performed on the jettisoned docking module before reentry.

ASTP Experiments

Twenty-seven experiments were placed on the Apollo CSM and the Soyuz 19 vehicles, which were conducted jointly and separately during the nine-day mission. The following table is a brief summary of those experiments.

Apollo-Soyuz Test Project Experiments

- Space sciences and astronomy experiments
  - Soft X-ray instrument to observe X-ray sources within and outside of our galaxy
  - Extreme ultraviolet survey instrument to examine our galaxy
  - Helium glow detector to observe the interstellar medium near our solar system
  - Artificial solar eclipse experiment to observe the solar corona
  - Crystal activation experiment to investigate the effects of particle radiation in Earth orbit on instrument noise levels of gamma-ray detectors
- Space sciences and Earth environment experiments
  - Ultraviolet absorption experiment to measure atomic constituents of the Earth's upper atmosphere
  - Stratospheric aerosol measurements to measure the stratosphere's aerosol content
  - Earth observations and photography experiments to study surface features on Earth
  - Doppler tracking equipment to measure mass distribution below the Earth's surface
  - Geodynamics experiments to measure mass distribution below the Earth's surface
- Life sciences experiments
  - Light flash experiment to measure the effects of particles upon the human retina
  - Zone forming fungi experiment to measure particle effect upon growing bacteria cells
  - Biostack experiment to measure particle effect upon seeds and eggs
  - Microbial exchange experiment
  - Cellular immune response experiment
  - Polymorphonuclear leukocyte response experiment
- Medical substances and materials processing applications experiments
  - Electrophoresis technology experiment
  - Electrophoresis experiment
  - Crystal growth experiment in which material is processed at ambient temperatures (NASA 1975)

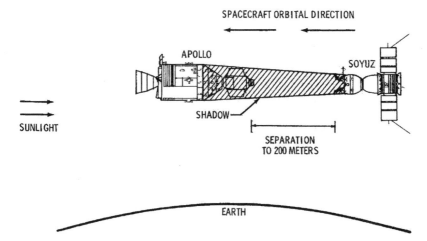

Figure 8.27. Sketch of the Apollo-Soyuz artificial eclipse experiment with the CSM providing the artificial eclipse disk for the co-orbital observation experiment housed on the Soyuz capsule. Source: ASTP Press Kit.

Table 8.19. Soyuz 19 Specs

Launch	Soyuz (R-7 derivative) 15 July 1975
Landing	21 July (96 orbits)
Commander	Alexi Leonov
Engineer	Valery Kubasov

Table 8.20. Apollo CSM Specs

Launch	Saturn IB 15 July 1975
Landing	24 July (148 orbits)
Commander	Thomas Stafford
Command Module Pilot	Vance Brand
Docking Module Pilot	Deke Slayton

SAFETY AND HUMAN FACTORS

In spite of more than 2,000 hours of crew training and preparation for the ASTP mission, and extensive flight training before becoming astronauts, as well as considerable space flight training in the astronaut program, a near-fatal pilot error occurred during the ASTP command module reentry. All three astronauts were responsible for overlapping duties in the reentry sequence checklist. Thomas Stafford, the mission commander, read the entry checklist items, while Vance Brand, the CM pilot, operated the switches. During the lower-atmosphere descent, the Earth landing system (ELS) switch was not activated, even though it was listed and a requirement since it deactivated the command module reaction control system (RCS) jets while the parachute sequence let the capsule down through splashdown. Without the ELS switches on, the drogue chute was not armed. As the crew passed through 30,000 feet, the apex cover and drogue chutes went undeployed. Realizing this, Brand activated the drogue chutes manually at about 23,000 feet. As the parachutes deployed, the capsule began to sway, which automatically activated the RCS jets. Gases from the RCS containing uncombusted nitrogen tetroxide and hydrazine entered the capsule from the cabin vent, flooding the cabin with toxic fumes. At splashdown, the crew was overcome with the fumes, coughing, nearly unconscious, and upside down in a capsule that was inverted in the water. Stafford was conscious enough to place the oxygen masks on the other two crew members and was able to right the capsule with the actuation device. Stafford then opened the vent valve which cleared the cabin of fumes (Ezell and Ezell 1978).

Although capsule recovery was normal, the crew was hospitalized for two weeks to recover from exposure to highly toxic nitrogen tetroxide and unsymmetrical dimethyl hydrazine. Deke Slayton was later diagnosed with a lesion in his lung which proved benign. The near-fatal results of a simple switch error in the flight sequence underscores the importance of extensive training and absolute discipline in the unforgiving environment of space flight. A simple error by a crew composed of the most highly qualified, highly trained professional pilots in the world provides a stark reminder that the imperfect human element is still a necessary component in operating the most advanced technology, and that human factors research is increasingly important in that increasingly complex world.

APOLLO 18, 19, 20

Contracts for the development, testing, and supply of the Apollo launch and flight vehicles to support the complete project included the CSM, the LM, the Saturn V, and the Saturn IB. The original contract for fifteen flight-qualified Saturn V vehicles permitted test and validation flight, plus ten manned lunar missions. However, as the program entered the first series of successful lunar landings in 1969, budget pressures and waning congressional and public interest forced the cancelation of the last three missions and the vehicle sets. Apollo flights 18, 19, and 20, the last three of the program, were not all canceled at once, but in a sequence beginning with Apollo 20 in January 1970, followed by Apollo 15 and Apollo 19 in September, 1970. The remaining missions were then renumbered 15 through 17, with landing sites and crews reshuffled to maximize scientific returns. After Apollo 13, the Tycho and Copernicus crater landing sites were dropped (NSSDC).

References

Bilstein, Roger E. 1996. *Stages to Saturn: A Technological History of the Apollo/Saturn Launch Vehicles*. NASA SP-4206. Washington, DC: NASA History Office.

Brooks, Courtney G., James M. Grimwood, and Loyd S. Swenson. 1979. *Chariots for Apollo: A History of Manned Lunar Spacecraft*. NASA SP 4205. Washington, DC: NASA Scientific and Technical Information Office.

Compton, William D. 1989. *Where No Man Has Gone Before: A History of Apollo Lunar Exploration Missions*. NASA SP-4214. Washington, DC: NASA Office of Management, Scientific and Technical Information Division.

Ertel, Ivan D., and Mary Louise Morse. 1969. *The Apollo Spacecraft: A Chronology*. Volume I. NASA SP-4009. Washington, DC: NASA Scientific and Technical Information Office.

Ezell, Edward Clinton, and Linda Neuman Ezell. 1978. *The Partnership: A History of the Apollo-Soyuz Test Project*. NASA SP-4209. Washington, DC: NASA Scientific and Technical Information Office.

NASA. 1969. Press Kit, Apollo 11 Lunar Landing. Washington, DC: NASA, July 6.

NASA. 1969. *Apollo Operations Handbook, Block II Spacecraft: Volume I, Spacecraft Description*. SID 66-1508. Washington, DC: NASA.

NASA. 1972. *Skylab Saturn 1B Flight Manual*. MAN-206, TM-X-70137. Huntsville, AL: Marshall Space Flight Center.

NASA. 1975. ASTP Press Kit, Apollo Soyuz Test Project. Washington, DC: NASA.

NSSDC (National Space Science Data Center). "Apollo 18 through 20: The Canceled Missions." Retrieved from: http://nssdc.gsfc.nasa.gov/planetary/lunar/apollo_18_20.html (accessed 8/24/2009)

Tomayko, James E. 1988. *Computers in Spaceflight: The NASA Experience*. NASA Contractor Report 182505. Wichita, KS: Wichita State University.

NASA MIX (Marshall Space Flight Center Image Exchange). Retrieved from: http://mix.msfc.nasa.gov/

NASA NIX (NASA Image Exchange). Retrieved from: http://nix.nasa.gov/

NASA JSC Image Collection. Retrieved from: http://images.jsc.nasa.gov/

NASA GRIN (Great Images In NASA). Retrieved from: http://grin.hq.nasa.gov/

CHAPTER 9

Space Stations

Early History

Widespread interest in space flight came to life with the spreading popularity of science fiction in the late 1800s from authors such as Jules Verne and his *Voyage to the Moon*, and *Around the Moon*, and H. G. Wells, *The Time Machine* and *War of the Worlds* from the 1890s. One of the earliest stories of a manned satellite appeared in the *Atlantic Monthly* in 1869 as a series written by Sir Edward Everett Hale on the accidental launch of space station called *The Brick Moon*.

With the arrival of the twentieth century came a turn from fictional to theoretical space flight outlined in a number of scientific publications including those by space pioneer Konstantin Tsiolkovsky. Building on the foundation of rocket propulsion theory and concept, Tsiolkovsky mapped out the first steps in space travel including the use of orbiting space stations as research tools and stepping stones to extend man's reach throughout the solar system. Tsiolkovsky's space station concepts introduced the first practical ideas in space station design that included crew accommodations, life-support systems, space suits, air locks, and space walks. His notional ring-shaped orbiting station was far from a complete blueprint, but it did weave together important subjects for his later works on space colonization. Within his 500 papers, manuscripts, and books were descriptions of the first staged launch vehicles, and the liquid-fuel rockets to launch orbiting stations and crews. More than imaginative, Tsiolkovsky's manned orbital station concepts pointed out the technological progression required to travel beyond Earth to the distant reaches of the solar system.

In 1928, a scientist and engineer commissioned in the Austrian Army named Hermann Noordung (a pen name for Herman Potocnik) published a book on space flight and rocketry with a insightful description of an orbital space station. Noordung's wheel-shaped, rotating station was constructed with modular components which allowed assembly in multiple launches like the construction technique used for the Mir space station and the International Space Station. Noordung examined earlier works of Goddard, Oberth, Hohmann, and others to determine the feasibility of a rocket large enough for the project and concluded correctly that high-energy fuels would be required for any vehicle to reach orbit. In spite of Tsiolokovsky's and Noordung's insight, it would be nearly three decades before boosters large enough to reach orbit were launched. And when the first space stations were launched, they used the same high-energy propellants discussed in the works of the early pioneers Tsiolkovsky and Noordung.

Noordung's rotating station outlined in his book *The Problem of Space Travel: The Rocket Motor* (1929) introduced a number of practical design concepts that included artificial gravity in the outer rim that housed the living quarters and laboratories. Centrifugal gravity was only one of Noordung's imaginative station design ideas, but an important consideration since rotational acceleration is so far the only solution to the problem of human adaptation to zero-gravity space flight. Yet for its utility, artificial gravity in space is still not practical since any mechanism capable of generating suitable artificial gravity would be enormous. A 1 g rotational acceleration at a rate that would not induce crew disorientation—a limit of about 1 rpm—requires a platform nearly 1 km in diameter.

Travel between the central hub and the outer ring of Noordung's station would use spiral staircases and elevators within the connecting spokes. Electrical power for the space station was to be supplied by the Sun heating a fluid that would drive onboard power generators with the expanding gas. Heat for the power system would come from a reflective mirror the diameter of the station focused on a heating device to concentrate the sunlight. The station's details were remarkable in their utility, including airlocks and space suits, three-dimensional reaction wheel attitude control, an artificial atmosphere, thermal controls, and attached observatories. Noordung also outlined atmospheric reentry for crew vehicles, and more complex maneuvers such as progressive aerodynamic braking employed on several Mars orbital missions decades later.

Noordung's 50 m diameter station was to be placed in a geosynchronous orbit, which made it an ideal platform for Earth-observation, celestial observations, and global communications. The orbiting platform predated Arthur Clark's pioneering concept of geosynchronous satellite orbits that eventually became the prevalent worldwide communication satellite location. Still, there were a number of serious practical problems related to the 36,300 km (22,500 mi) geostationary orbit. The two biggest problems were the huge boosters needed to reach that altitude and the intense particle radiation stored in the van Allen belts at this altitude, although Noordung was unaware of the latter since the radiation belt had not been measured until the Explorer 1 spacecraft was launched in early 1958.

Several other proposals for orbital stations came in the 1920s and 1930s from a Transylvanian scientist named Herman Oberth who worked with the Nazis on the V-2 missile program with Wernher von Braun. One of Oberth's unique but totally impractical concepts was a proposal for an asteroid to house a telescope, along with crew living quarters and research facilities. Proposals with more pragmatic design concepts appeared in the British Interplanetary Society journal *BIS* in 1948, and in the proceedings of the International Astronautical Union conference in 1951. A large rotating torrid and disk, described in the *BIS* paper by H. E. Ross and R. A. Smith, was similar to Noordung's station, including its solar mirror and solar heat-generated electric power. An astronomical observation platform and research base offered a 1 g

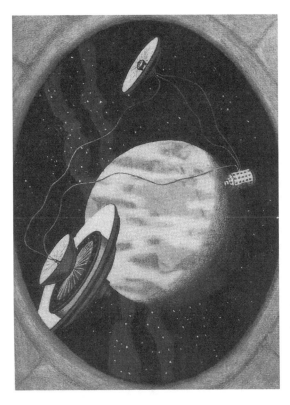

Figure 9.1. Illustration of Noordung's rotating space station (lower left) tethered to an observatory module (right) and a power workshop module (top) viewed through the window of a crew vehicle. Source: NASA SP-4026.

environment for the crew, along with dual telescopes that had their images corrected for the platform's rotation. The International Astronautical Union conference paper by H. H. Koelle was a more complete station concept that consisted of a rotating ring structure constructed of spheres and cylinders in a "necklace" arrangement with spokes extending between the central hub and the outer ring. Each sphere had individual functions with independent emergency life-support units for the crew. In addition to the thirty-six-sphere and cylinder ring, the 60 m station, weighing150 metric tons, included a solar concentrator for heating and electrical power. Koelle's paper was a more detailed plan than Noordung's and bore some of the same details as von Braun's rotating ring station that appeared soon after. Similarities in the two station concepts were not coincidental since Koelle and von Braun worked closely at the Army Ballistic Missile Agency in Huntsville, Alabama, in the 1950s (Massey 1960).

Perhaps the most striking space station concept appeared in a lavishly illustrated series of articles in *Collier's* magazine on space travel that began in 1952. Wernher von Braun's ring-shaped space station was detailed in several of the articles showing the station's modular construction, along with the launch and crew vehicles used to support the orbiting laboratory in its 1,600 km (1,000 mi) orbit. Like Koelle and Noordung, von Braun's design used a large reflective mirror-concentrator to provide heat for generating electrical power. The enormous rotating station consisted of a large, dual-spoke ring that was to be constructed of nylon and assembled in sections, first on Earth, then, after testing and disassembly, launched into orbit and reassembled

Figure 9.2. Artist Chelsey Bonestell's illustration of von Braun's space station orbiting 1,000 miles above Earth that appeared in the March 22, 1952, issue of *Collier's Magazine*. The rotating station's modular segments were to be assembled, first on Earth for testing, then launched with heavy-lift rockets similar to the one depicted on the left. The detailed drawing also includes crew ferry capsules and a space telescope. Source: Reproduced courtesy Bonestell LLC.

section by section. On orbit, the 83 m (250 ft) rotating space station would serve as a research station, a navigational aid, a meteorological station, a military platform, and a way-station for deep-space exploration. Located in the central hub were the support services, docking module, and laboratory facilities—all at zero-g (Massey 1960).

Wernher von Braun was a tireless advocate of manned space exploration, beginning with his work on the German V-2, then while designing missiles for the U.S. Army's Ballistic Missile Defense Agency in the 1950s, and ultimately, in his success fashioning the Apollo-Saturn program. Even though his orbiting station concept never reached the design stage, his efforts that put the first American in space and later put men on the Moon were a prelude to America's first space station. Von Braun's Saturn V launcher and much of the equipment from the Apollo program were used to build and launch the Skylab space station.

Early NASA

America's first manned spacecraft was not the orbital Mercury capsule, but the suborbital X-15 launched from beneath the wing of a B-52. The X-15 used rocket propulsion to climb above the atmosphere, requiring reaction jets to control vehicle orientation in space in addition to aerodynamic controls to glide to a conventional landing after returning from space. Because of the extensive applications of the X-15 project, a number of aerospace companies involved in the program and in several long-range missile projects proposed a variety of manned orbital satellites and orbital laboratories. Interest in manned satellites also came from the Department of Defense and the USAF, also underway at the National Advisory Council for Aeronautics (NACA), NASA's predecessor. Although the combined efforts were not convergent, the resulting project was. The Air Force's manned orbital reconnaissance laboratory proposal in 1957 was recast as the simpler Man-In-Space-Soonest to expedite the project after the launch of the Soviet Sputnik satellite in 1957. Soon after, NACA would be charged with the renamed manned space flight project called Mercury, and in October 1958, would itself be converted into the new National Aeronautics and Space Administration.

NACA's space technology group had been working on manned projects for several years before the transition from NACA to NASA in 1958 and inheriting the Mercury project. Following the personnel and facility transfer, a new group within NASA named the Space Task Group was assigned a number of proposals for manned space flight projects to review, in addition to its Mercury program. One proposal was to launch a two-man orbital station on the new Air Force Atlas ICBM.

Soon after the formation of the agency, NASA administrators began clarifying their objectives, which included a permanent manned orbiting laboratory, as well as more ambitious manned lunar outposts and lunar orbital laboratories. It was, however, military interests in a manned orbiting laboratory that led to the first concrete plans for a working orbital space station for both the United States and the Soviet Union.. It was inevitable that both countries, locked in an adversarial cold war, would develop the first manned orbital space stations for military purposes.

MANNED ORBITAL LABORATORY (MOL)

In theory, the early manned space stations presented numerous advantages over reconnaissance satellites for military planners. Orbital stations offered greater versatility since instruments, equipment, and even weapons could be added, removed, modified, or repaired by the flight crews. The crews could also be trained for specialized missions with rotating assignments to avoid the excessive stress of space flight. In fact, crews had already been selected for flight on the Air Force's Dyna Soar spacecraft in 1960. In 1963, the Manned Orbital Laboratory had been approved with its announcement coming at the same time the Dyna Soar project was canceled.

The MOL, also known in its early development as Dorian and KH-10, was the first U.S. manned orbiting laboratory program to reach the fabrication stage (Peebles 1987). An earlier Air Force project called the Manned Orbital Development System (MODS) was to be a four-man space laboratory that would incorporate an advanced "Blue" Gemini capsule for crew operations. Because the project appeared to duplicate similar NASA projects, and because of chronic federal budget woes, the USAF MODS was replaced by the MOL project and tied to NASA through the Gemini program.

NASA planners were moving quickly into the Apollo lunar project by 1963, but discussions on the Apollo X concept (Apollo X evolved into the Apollo Applications Program that would later become Skylab) introduced several attractive manned orbital and lunar projects because of the potentially low cost. While the USAF MOL project incorporated NASA's Gemini capsule, some in Congress felt it would be prudent to merge NASA's space station efforts with the USAF plans to build an orbiting laboratory (Brooks, Grimwood, and Swenson 1979). Ironically, Cold War fears of initiating a weapons race in space kept both projects separate.

The USAF Manned Orbiting Laboratory was designed to carry a crew of two for thirty-day missions using a modified Gemini for transporting the Air Force crews. The MOL was to be launched on the Titan III-C booster, later increased in capacity to the Titan III-M model. MOL's Gemini B capsule had the same dimensions as NASA's Gemini capsule, but with entirely new systems. The laboratory module and crew capsule launched together would avoid the inherent problems of rendezvous and docking. After the mission end, the Gemini B would be undocked for a deorbit and reentry and the station deorbited.

Seven of the Titan III boosters were ordered for the MOL project, although only one would launch laboratory hardware during the program. The only launch of MOL was on a test flight November 3, 1966 that included a MOL mockup made of a Titan II propellant tank and a recycled Gemini 2 spacecraft provided by NASA that was separated early for a suborbital reentry test. The MOL mockup continued to orbit and later released three satellites, while the Gemini 2 capsule was recovered near Ascension Island (Baker 1982).

Launches for the MOL space stations were to be from Cape Canaveral and from Vandenberg to accommodate both high inclination and polar orbits. Initial design plans for the military reconnaissance missions are still classified, but reportedly included large optics cameras and side-looking radar. Other mission types could have included interception and inspection of satellites, ocean/submarine surveillance, and space adaptation

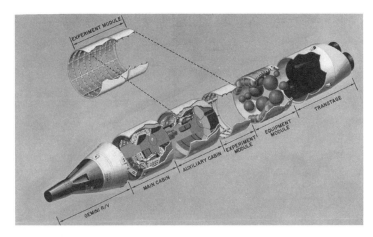

Figure 9.3.
Cutaway of the USAF Manned Orbital Laboratory with the docked Gemini 2 and access tunnel through the reentry shield (not shown) to the main cabin. Source: NASA-NIX.

studies. Power for the MOL laboratory was to be based on the new fuel cell technology developed for the Gemini and Apollo capsules, with either photovoltaic arrays or RTGs installed for extended missions. MOL's carbon dioxide removal system was going to be the first regenerable CO_2 removal device used on a manned spacecraft called a molecular sieve. Reaction thrusters for attitude control were to use nitrogen tetroxide and unsymmetrical dimethyl hydrazine as propellants (Massey 1960).

Three military astronaut selections were held in 1965, 1966, and 1967 for the MOL flight crews, some coming from the previously canceled X-20 Dyna Soar project, and some later joining the NASA astronaut corps to fly in the space shuttle program. The three MOL astronaut classes included the following military pilots:

- Group 1—November 1965
  - Michael Adams (Air Force)
  - Albert Crews, Jr. (Air Force)
  - John Finley (Navy)
  - Richard Lawyer (Air Force)
  - Lachlan Macleay (Air Force)
  - Francis Neubeck (Air Force)
  - James Taylor (Air Force)
  - Richard Truly (Navy)
- Group 2—June 1966
  - Karol Bobko (Air Force)
  - Robert Crippen (Navy)
  - Charles Fullerton (Air Force)
  - Henry Hartsfield, Jr. (Air Force)
  - Robert Overmyer (Marine Corps)
- Group 3—June 1967
  - James Abrahamson (Air Force)
  - Robert Herres (Air Force)
  - Robert Lawrence, Jr. (Air Force)
  - Donald Peterson (Air Force)

Following the cancelation of the MOL, a number of the program astronauts were accepted by NASA and later assigned to missions primarily on space shuttle classified missions. NASA-USAF military mission astronauts from the MOL program included Karol Bobko, Charles Fullerton, Henry Hartsfield, Donald Peterson, Richard Truly, Robert Crippen, and Robert Overmyer.

By 1969, Department of Defense officials had come to the realization that military reconnaissance satellites could be orbited in large numbers and at a much lower cost than manned missions. Even though the DoD's first surveillance satellites launched in 1961 were only modestly successful, second and third generation reconnaissance satellites proved their worth as rapid-response, detailed imaging systems that could reveal almost any Soviet

Figure 9.4. Launch of the mockup MOL laboratory and Gemini 2 capsule on a Titan III booster November 3, 1966. Source: USAF.

target. In reaction to the obvious disadvantages of the program, the Secretary of Defense canceled MOL on June 10, 1969. The cancelation was the end of the program, but not the end of its impact. Much of the hardware innovations from the Titan III-M booster was adopted or transferred to other Titan series rockets, including the later Titan IV. As a result of the MOL program, however, the USSR began development of their own manned military orbital station called Almaz, which flew as part of the Salyut space station series.

APOLLO-ERA STATIONS

NASA engineers, managers, and scientists considered manned orbital laboratories as one of the necessary steps in advancing space exploration beyond the pioneering but limited Mercury, Gemini and Apollo projects. A number of proposals emerged from within and outside of NASA that could make use of at least some space flight hardware already approved in order to minimize cost and shorten development time. A one-man station, for example, was proposed by Langley Research Center's Space Station Office in 1960 that would use a Mercury capsule and an inflatable structure for a short-term, inexpensive orbital laboratory. One study that begun in 1963 looked at extending Apollo orbital missions to 100 days using only Saturn IB boosters, and without the need for resupply. In the same year, a Lockheed study proposed the use of Saturn V boosters to orbit a station by 1968 with a twenty-four-man crew capacity, a five-year development and operation period, and a price tag of only $2.9 billion. Low-cost crew transfer vehicles could have included an expanded Apollo command module, or an unspecified lifting body design, either of which could provide crew emergency return. Conflicts in the proposed station's mission arose during efforts to merge the NASA station with the USAF's MOL project, which ended its consideration (Newkirk, Ertel, and Brooks 1977).

As the major Apollo hardware tests approached in the mid-1960s, more studies emerged on space station design utilizing the Apollo command and service module and the Saturn IB and Saturn V booster hardware. One such proposal came from the Marshall Space Flight Center in 1964 under the Apollo-X project designation that would combine an orbital module and a CSM, and be launched on a Saturn IB. Like the Apollo lunar module, the larger orbital module would be extracted by the CSM after booster separation and deployment but would remain attached for the duration of the mission. This variant of the Apollo-X projects proved important because it introduced a modest manned orbital laboratory using already-developed Apollo hardware, and ultimately the early prototype for the Skylab space station design.

Apollo-X/AES

Between the Apollo-X discussions and the Apollo Applications Program which would ultimately become the Skylab space station, was the Apollo Extension System (AES) which had the influential backing of Max Faget. AES was conceived to expand applications of the Apollo hardware and, at the same time, accelerate the development and versatility of the Apollo lunar landing vehicles. The proposal sparked a request for more complete studies within NASA and from outside contractors for an orbital laboratory in four different configurations that ranged from forty-five-day to five- to ten-year missions (Newkirk, Ertel, and Brooks, 1977).

Apollo Applications Program (AAP)

In mid-1965, NASA had delegated research of the AES project to several contractors and working groups with the objective of defining the most favorable configuration, while coordinating MOL efforts in the same direction. Early design of the AES was to use a spent S-IVB booster stage as the primary orbital laboratory. At this time, both NASA's Manned Orbital Research Laboratory (MORL) project and the Apollo Applications Program (AAP) were moving toward the same objective, but in different directions. MORL was coordinated by NASA to aid the USAF MOL project by instituting research goals on a modular space station, while the NASA AAP effort was to adapt the studies in progress on the AES project for a modular orbital station purely for research purposes. As 1965 came to a close, the Saturn S-IVB orbital workshop and telescope module became the focus for implementing an orbital laboratory design based on Apollo hardware. An ascent stage from the Apollo lunar module was identified as the best candidate for the astronomical telescope housing. What seems an isolated detail is actually the origin of the Skylab Apollo telescope mount design that survived a multitude of changes in the program's evolution

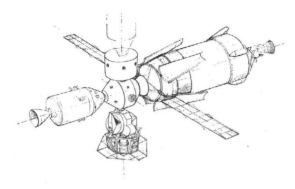

Figure 9.5. Sketch of the Apollo Extension System concept incorporating Apollo launch and flight hardware. The S-IVB upper-stage of a Saturn booster would be used as part of the station, along with one or more connecting nodes. Source: NASA SP-4011.

By 1966, NASA's unnamed orbital laboratory was taking shape piece by piece. The Saturn S-IVB upper-stage workshop was to have extensive research and laboratory facilities available, but, for the first time, its design would be adapted to the principal investigators' experiments instead of using standardized accommodations. Specifications were drawn up for experimental equipment on the S-IVB workshop, and for the modified LM telescope mount, and for the CSM experiments. Budgetary constraints paced progress on the new AAP orbital station, as did the requirement that the project could not interfere in any way with the Apollo lunar landing missions. Since the proposed Advanced Orbiting Solar Observatory was recently canceled, the newly-designated Apollo telescope mount would now be built from a duplicate LM ascent stage and carry a suite of solar telescopes to replace the canceled AOSO mission (Newkirk, Ertel, and Brooks 1977).

Wet Workshop

Concern mounted over the use of a spent Saturn IB second-stage S-IVB as an unequipped workshop called the Saturn IB Workshop—informally known as the wet workshop—as progress continued on the AAP orbital station. In question was the suitability of using the spent hydrogen tank as the research platform and living quarters for the astronauts. Alternatively, launching a completed orbital workshop within the S-IVB upper stage of a Saturn V began to gain acceptance for several reasons. First, the completed workshop would not have to be outfitted using equipment, construction, and assembly flights after reaching orbit. Second, the hydrogen tank would not be launched to orbit with explosive fuel residue. The Saturn V, or "dry" workshop, required a more expensive launcher, but was seen as saving costly resources in the end.

In August of 1966, the recommendation that an airlock module be included in the AAP design became reality since it was needed for crew EVA and CSM docking and was the logical choice for life-support systems. The most appropriate location for the airlock module was determined to be the entryway for the S-IVB workshop. By early 1967, the orbital workshop (OWS), the multiple docking adapter (MDA), the airlock module (AM), and the Apollo telescope mount (ATM) had been identified as the modular components making up the Apollo Application Project, but not yet specified except for the OWS—a spent S-IVB. A control moment gyro (CMG) system was also selected as the fine-pointing device needed for high accuracy on the station's telescopes and other instruments.

Dry Workshop

Debate over the choice of a wet or dry workshop configuration ended in 1968 when it was determined that the Saturn V (dry) Workshop would eliminate the hazards of outfitting a wet workshop and reduce project complexity as well as the number of launches. Launching the orbital workshop complete with laboratory facilities and equipment with major modules attached would establish an outfitted and tested laboratory in orbit with a single launch. With the switch of the Saturn IB workshop to the Saturn V workshop came a new designation called the AAP cluster which incorporated the integration of the entire assembly in one launch. Confirmation of the completed-workshop concept and its configuration approval came in mid-1969, along with a projected launch date of 1972, and a new name. In February 1970, the AAP orbital space station project was renamed Skylab—a contraction for laboratory in the sky (Newkirk, Ertel, and Brooks 1977).

NASA's Apollo Applications Program covered not only the Skylab project, but a broad range of other Apollo projects that included the actual Apollo landings, as well as post-Apollo missions involving extended lunar surface stays, a lunar orbit station, and a small lunar base. Cancelation of the post-Apollo programs due to budget limitations reduced the AAP package to just the Skylab project.

Skylab

NASA's Skylab orbital space station project was announced in 1970, along with details on the workshop hardware, the launch vehicles, and the design teams. The short three-year development and testing period culminated in the launch of the orbital workshop on May 14, 1973. But the success of the Skylab program was not limited to the clever engineering and existing Apollo hardware. Skylab's research and experimentation program proved the importance of the human element in space exploration, and established man's capability for productive work in the challenging, demanding weightlessness of space. In retrospect, the Skylab program more than satisfied its stated goals which were to:

- Prove that humans could live and work in space for extended periods
- Expand our knowledge of solar astronomy well beyond Earth-based observations

Skylab and its three crewed missions were expected to be the first in a series of space stations supported by NASA's new Space Transportation System—the space shuttle. But the same budget woes that shaped Skylab and cut the last three Apollo flights also killed the second Skylab vehicle. Since the first space shuttle flight was not made until two years after the loss of the Skylab station in 1979, NASA's manned exploration efforts were converted into short-duration missions on the space shuttle that were stretched out for nearly two decades. In 1998, the second American space station began a new legacy of long-duration orbital research with the launch of the International Space Station's (ISS's) first elements by the space shuttle. Ironically, the completion of the ISS signaled the end of the space shuttle which was designed with

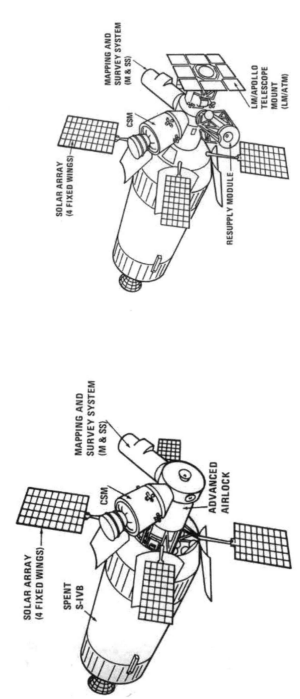

Figure 9.6. Sketch of the two configurations of the wet workshop concepts that preceded the final design of the Skylab station. The S-IVB upper stage of a Saturn IB would become the Orbital Workshop after evacuation of propellants, the attachment of modules, and the installation of the laboratory facilities. At a minimum, four additional launches were expected to be required to complete the orbiting laboratory. Source: NASA SP-4011.

the capacity to build the 400 metric ton space station. Following Skylab's replacement by the space shuttle, America's space station efforts have come full circle with the completion of the International Space Station and the retirement of the space shuttle.

LAUNCH AND DEPLOYMENT

The Skylab cluster was launched into orbit with the first two stages of a Saturn V left over from the three canceled Apollo lunar missions. Since the Skylab station was designed around the S-IVB third stage of the Saturn V, the deployment was nearly the same as it would be for a lunar mission, but with the Skylab station remaining in Earth orbit. Skylab's nominal launch and liftoff on May 14, 1973 was followed by anomalous telemetry soon after entering orbit. After just one orbit, the OWS interior temperature began rising steadily. It soon became apparent that the micrometeoroid shield had deployed early and damaged the photovoltaic system as it was torn away. The first crew launch to the station scheduled for the next day was canceled, and the SL-2 flight was put on hold until Skylab's serious problems were better understood and repairs could be planned.

Within days, the decision was made on the most practical solution to replace part of the missing protective shield and reverse the runaway temperatures in the workshop. The consensus of the mission managers and engineers was to deploy a lightweight extendable "parasol" that could be attached to the outer OWS wall from inside the workshop. The protective solar shield would shade the most exposed region of the OWS which was roughly the surface area of the solar panels. After critical tests, the parasol was readied, then launched with the SL-2 crew ten days after the orbital workshop. This technique required an internal spacewalk with the astronauts in their space suits because of the 52°C (126°F) OWS temperature. Soon after parasol deployment, temperatures began to drop, and within a day the astronauts began work inside the OWS in shirtsleeves.

A second problem remained after the Skylab launch accident unrelated to the runaway internal temperature in the workshop. Only one of the two solar panels survived the ascent and micrometeoroid shield damage, but was not completely deployed. Several days into the mission, the lack of solar power from the OWS resulted in malfunctioning secondary power batteries, which forced the first crew to perform a risky spacewalk to release the panel. Film from the earlier flyaround and astronauts simulating the panel release in the NASA underwater facility at the Johnson Space Center helped Skylab's first EVA crew plan how to cut the strap that bound the photolvoltaic panel and, after warming in the sunlight, deploy the panel.

SKYLAB STRUCTURE

Skylab's primary structure was constructed from an upper stage of a Saturn V with an outer shell and two inner tanks. The outer shell and bulkheads on the workshop supplied the structural frame to attach the other modules and the solar panels. The largest tank making up the S-IVB workshop structure served as the pressure vessel for the habitation and experiment rooms after considerable modification. The smaller

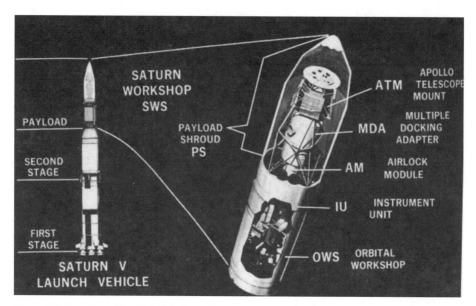

Figure 9.7. Illustration of the Skylab station within the Saturn V upper-stage shroud in launch configuration. Source: NASA SP-4011.

liquid oxygen tank was converted into a waste tank for the entire station. Attached to the forward bulkhead of the OWS was the airlock module that served as the structural connection for both the docking module and the Apollo telescope mount.

Skylab's overall length with a docked CSM module was 36.1 m (118.4 ft) with a diameter of 6.6 m (21.6 ft), or 30 m (98.4 ft) with solar panels extended. Loaded weight of the Skylab cluster was 86,210 kg (190,000 lb), which included equipment and consumables for all three planned crew missions. Habitable volume of the Skylab cluster was 361 m³ (12,750 ft³).

ORBITAL WORKSHOP

Attached to the Skylab workshop structure were two sets of solar arrays with secondary batteries housed inside both the OWS and the ATM. Also attached to the OWS exterior were external experiments and instruments, a combined debris and solar shield, and thermal radiators. In addition to the three other attached modules, the complete Skylab cluster included a converted Apollo CSM during manned missions used to transport astronauts to and from the station. Skylab's CSM vehicles were highly modified Apollo lunar mission hardware that needed only a fraction of the life-support required for the lunar missions. For even greater weight reduction, the three Apollo command modules had lighter reentry shielding because of the capsule's lower orbital entry speeds.

The orbital workshop was designed to provide a habitable environment for the crews, facilities and support for a wide variety of onboard experiments, and waste storage for the station's lifetime. Skylab's habitation section was divided into two levels. The lower level provided crew accommodations for sleeping, food preparation and

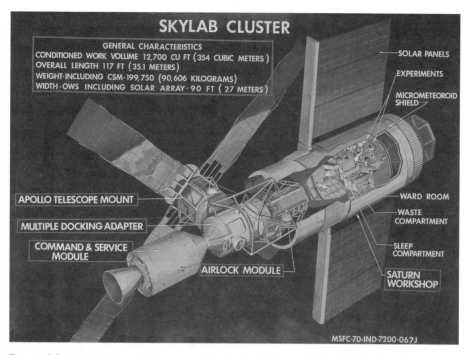

Figure 9.8. Skylab cluster shown complete with primary modules attached, and a docked Apollo Command and Service Module. Source: NASA MIX.

food consumption (the galley), personal hygiene, waste management, data management, exercise, and some experiment operations. The upper level housed water storage tanks, food freezers, bulk storage provisions, scientific airlocks, mobility and stability experiment equipment, and equipment for other experiments. Major elements that provided support for the OWS functions included:

- Waste tank—Waste storage was located in the converted liquid oxygen tank beneath the crew quarters for containment of liquid waste, solid waste, and trash throughout the station's life.
- Propulsion (attitude control)—Secondary attitude control employed compressed nitrogen gas to power six thrusters mounted in two areas of the aft end of the workshop. Cold gas thrusters augmented the three control moment gyros (CMGs) located in the Apollo Telescope Mount, and were primarily used to desaturate the CMGs.
- Solar arrays—Power for the OWS was supplied by two photovoltaic array wings mounted on each side of the workshop augmented by a second, independent array of four panels mounted on the Apollo telescope mount. A set of NiCad rechargable batteries supplied secondary power for both the ATM and OWS arrays.
- Micrometeoroid shield—A standoff shield was designed to be deployed some six inches from the workshop surface for protection from micrometeoroids and to reduce solar heating. The shield was deployed early due to unexpected vibrational loads which resulted in the complete destruction of the shield before reaching orbit.

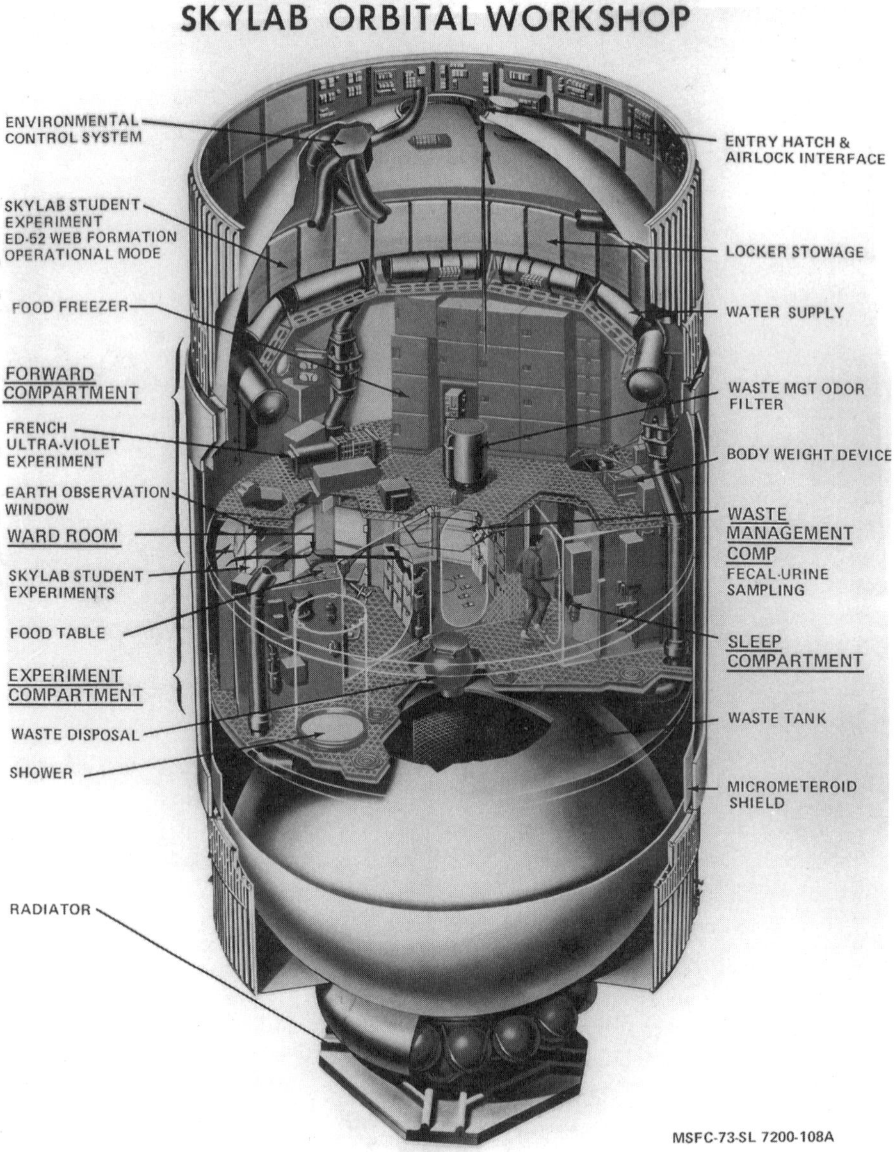

SKYLAB ORBITAL WORKSHOP

ENVIRONMENTAL CONTROL SYSTEM

SKYLAB STUDENT EXPERIMENT ED-52 WEB FORMATION OPERATIONAL MODE

FOOD FREEZER

FORWARD COMPARTMENT

FRENCH ULTRA-VIOLET EXPERIMENT

EARTH OBSERVATION WINDOW

WARD ROOM

SKYLAB STUDENT EXPERIMENTS

FOOD TABLE

EXPERIMENT COMPARTMENT

WASTE DISPOSAL

SHOWER

RADIATOR

ENTRY HATCH & AIRLOCK INTERFACE

LOCKER STOWAGE

WATER SUPPLY

WASTE MGT ODOR FILTER

BODY WEIGHT DEVICE

WASTE MANAGEMENT COMP
FECAL-URINE SAMPLING

SLEEP COMPARTMENT

WASTE TANK

MICROMETEROID SHIELD

MSFC-73-SL 7200-108A

Figure 9.9. Cutaway of the Skylab Orbital Workshop detailing the two levels or floors built into the converted S-IVB liquid hydrogen tank. The bottom section was the liquid oxygen tank shown here converted into the waste tank. Source: NASA MIX.

- Radiators—Thermal radiators were placed on the forward section of the OWS to remove internal heat from the main modules with the exception of the ATM which had its own thermal cooling system. An additional radiator was placed on the aft section of the OWS for the refrigeration system used for cooling samples and for cold storage.
- Instrument Unit—An attached guidance, sequencing, and deployment unit, called the instrument unit (IU), was built into the OWS, as it was on each S-IVB booster

of the Saturn V. The programmed IU systems provided the first seven hours of Skylab's operation before internal systems took over the space station functions.

A variety of minor assemblies were attached to the OWS that included astronaut mobility aids (handrails, tethers, and a central "fireman's pole"), as well as antennas and instruments.

AIRLOCK MODULE

Skylab's airlock module was a cantilever extension of Skylab's structural assembly connecting the forward modules—the MDA, ATM, and docked CSM—to the OWS. Life support functions for the entire station, including the OWS and the MDA, were contained in the airlock module because of its functional makeup, and its central location. The 79% O_2 and 21% N_2 atmosphere was maintained at 5 psi while crews were onboard. Spacewalks were conducted through an EVA hatch placed in the pressurization chamber of the airlock module to isolate the EVA astronauts from the rest of the station during the necessary depressurization and repressurization cycles. Major functions and components of Skylab's AM included:

- Storage for atmospheric gases (O_2, N_2) supplying the MDA, AM, and OWS
- Molecular sieve carbon dioxide removal system
- Pressurized passageway between the OWS and the MDA
- Control node for electrical power, environmental control, and communications for the Station
- Four optical observation windows

The AM structure consisted of a central pressurized tunnel, a structural frame used to attach the multiple docking adapter to the orbital workshop, a truss assembly to support the tunnel section, and gas supply containers. The AM also housed the deployment assembly for the Apollo telescope mount, and the fixed airlock shroud.

Electrical power, environmental control, and communications support provided by the airlock module for Skylab's habitable modules included the following:

- Eight rechargeable (secondary) batteries with individual charge-discharge regulator units
- Active thermal control system radiator
- Umbilical connections for EVA
- VHF systems for data and for command operations (NASA 1974a)

MULTIPLE DOCKING ADAPTER

Skylab's multiple docking adapter was a multifunction control center for the station and the ATM that also provided storage compartments including storage vaults for the telescope film. Built-in to the MDA were dual docking ports that could accommodate

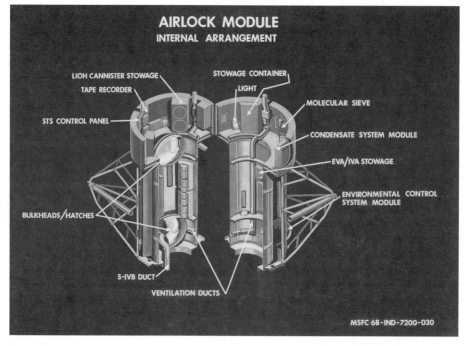

Figure 9.10. Sketch of the Skylab Airlock Module showing major components and their layout. Source: NASA MIX.

two crew Command and Service Modules simultaneously, although only the forward docking port was ever used. Instrument control and readout panels were located in the MDA for the Apollo telescope mount, and for the Earth resources experiment package (EREP), as well as for external experiments/instruments operations. The Apollo telescope mount control console was fitted with several video monitors and visual readouts for monitoring the telescopes and instrument output, and to aid in selecting targets and interpreting observations. In addition to the ATM display console, the multiple docking adapter contained controls for the ATM electrical power system, and the station's video, telemetry, and audio communication systems, and for the station's attitude control system.

Life support from the airlock module was augmented by MDA equipment that could circulate air through the command module for atmosphere revitalization. Heaters, lighting, guide rails, and foot restraints were placed in the MDA for crew operations.

APOLLO TELESCOPE MOUNT

Skylab's Apollo telescope mount was first conceived as a free-flying observation platform, then later planned as a docked observation module to be attached to the Skylab

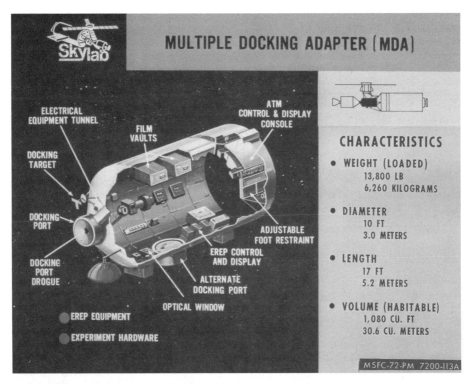

Figure 9.11. Cutaway view of the Skylab Multiple Docking Adapter and major components. Source: NASA MIX.

wet workshop. The ATM's final design was based on the Apollo lunar module ascent stage frame attached to the MDA for deployment on a swing-hinge. ATM's thermally controlled interior housed a number of instruments and telescopes for astronomical observations of the Sun, planets, stars, galaxies, and a variety of other celestial objects. Emphasis of these observations was on solar observations, including monitoring and recording the Sun's structure and behavior. Controls for the Skylab ATM allowed astronauts to manually operate telescopes in space for the first time, though optical devices were used for navigation on earlier Gemini and Apollo flights.

ATM's structure consisted of an outer truss assembly constructed from an Apollo LM ascent stage, and an inner instrument canister. The truss was attached to the upper OWS, along with a deployment mechanism that placed the observatory at right angles to the OWS-AM-MDA centerline after reaching orbit. Since this arrangement did not allow astronauts access to the ATM through a pressurized compartment, provisions were made for EVA spacewalks to retrieve and install film, and to attach and retrieve external experiments.

Along with its primary role as a solar observatory, the ATM module provided basic attitude control for the Skylab cluster by means of three control moment gyros located near the mount base. Each CMG was a double-gimbal-mounted 65.3 kg rotor powered

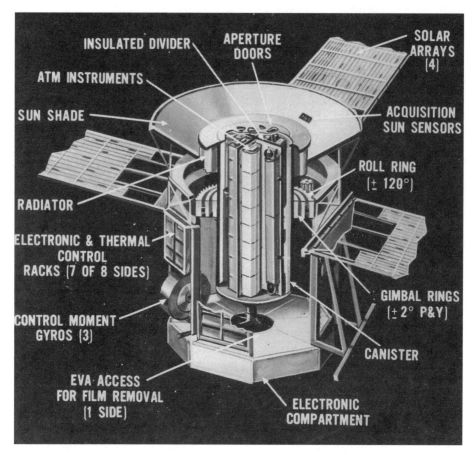

Figure 9.12. Cutaway view of the Apollo Telescope Mount showing the major component layout, including the gimbal assembly used for accurate pointing of the internal telescopes. Only one of the three Control Moment Gyros is depicted in the sketch. Note also the thermal control radiator. Source: NASA MIX.

by an electrical motor. Improved pointing accuracy in the instrument canister was added with a gimbal assembly and control unit called the pointing control system.

Four hinge-mounted solar arrays were used for primary power on the ATM. Secondary power was supplied by NiCad batteries charged by the ATM solar arrays—both independent of the OWS electrical power system. Like the OWS, the ATM electrical power system included charge-discharge circuitry and regulators, along with an electrical power distribution unit for the ATM systems and instruments. The ATM electrical power system was designed for power sharing with the Skylab as a whole, a fortunate detail that kept the station alive after its nearly fatal mishap at launch. With one array panel ripped off by the micrometeoroid shield and the other jammed in retracted position, the entire station relied on the ATM power supply for its early operations.

Figure 9.13. Checkout of the Instrument Unit at the Marshall Space Flight Center that was used on both the Saturn V and the Skylab OWS for initial deployment, navigation, power, and communications. Source: NASA MSFC.

ATM's temperature-sensitive instruments required an active cooling system that consisted of a water-ethanol fluid pumped through the external radiator. The ATM thermal control system regulated temperature inside the ATM canister to 12.8 ±2.8°C (55 ±5°F). More sensitive instruments added resistive heaters to maintain a maximum variation of ±1.4°C (±2.5°F).

ATM functions included digital control systems that allowed astronauts and ground controllers to initiate complex commands quickly and simply. Ground control of the ATM controls allowed instrument operations while the crew slept, between manned missions, and during emergencies.

INSTRUMENT UNIT

An instrument unit that controlled initial orbit and deployment operations was placed on the third stage of each Saturn V and on the Skylab OWS. The IU's navigation, communications, deployment operations, and initial power furnished control of the unmanned Skylab for a maximum duration of 7.5 hours. Skylab's S-IVB instrument unit also supplied sequencing commands for the deployment functions of the Apollo telescope mount and its solar arrays, for the workshop solar array deployment, and for initiation of the Skylab cluster attitude control system. In addition, the IU transferred digital command and telemetry data between the workshop and the ground stations prior to activation of the workshop communication systems.

COMMAND AND SERVICE MODULE

Three Apollo command and service modules were used for crew transport to and from Skylab with some modification to both the command module and the service module. The same CSM systems required for the Apollo lunar missions (propulsion, electrical power, environmental control and life support, communication, etc). were retained for the Skylab transport flights, with a reduction in some of the capabilities because of the shorter flight durations to the Station. Several modifications were also made to accommodate the standby role of the CSM during the Skylab missions. Instead of the Apollo fourteen-day flight duration, the Skylab CSM supported the crew with power and life-support consumables for the ascent to docking and descent to Earth phases. While docked during the three Skylab missions, the CSM was powered down and served in a standby role on the Station. As with the Apollo CSM lunar flights, the service module was separated from the command module before reentry and was not recovered.

SKYLAB SYSTEMS

Skylab hardware was designed to be compatible with the Apollo flight hardware since much of the Skylab equipment came from the Apollo inventory, and because the Skylab launch and flight vehicles were the same used for the Apollo lunar missions.

Figure 9.14. Photograph of a Skylab molecular sieve assembly that was used for removing heat, water, CO_2, and contaminant gases and odors from the pressurized modules. Source: NASA EP-107.

Electrical power voltages, docking equipment, data and voice communications, and thermal conditioning were designed for the Skylab station to match the CSM, Saturn V, and Saturn IB vehicles.

Electrical Power

Power for the Skylab orbital workshop and Apollo telescope mount were supplied by solar arrays, with secondary electrical supply from NiCad batteries during the thirty-six-minute eclipse period in the station's ninety-minute orbit. Independent power sources were also provided for the Skylab station during launch and its deployment, and during initial flight operations supplied by the instrument unit EPS. A separate fuel cell and battery power supplied in the command and service module, designated CSM EPS, that offered emergency power to the station, if needed.

Station EPS

- OWS PV (photovoltaic) power: 12,400 W design max—6,200 W single array
- OWS battery power: 3,700 W
- ATM PV power: 10,400 W max
- ATM battery power: 3,500 W (NASA 1974b)

The two station EPS solar array supplies were designed to be operated in parallel, although the right photovoltaic panel was torn from the OWS during launch. After regulation and conditioning, the single solar array electrical power was distributed throughout the OWS for the station loads, and for recharging the NiCad batteries. The OWS solar panel EPS also supplied power for the docked CMS to minimize the use of its fuel cell reactants.

Power Conditioning and Distribution The charging and conditioning functions for the Skylab EPS were provided by the ATM and the OWS power systems. The ATM and OWS circuits included charger battery regulator modules (CBRM) for the secondary power system recharge, and power conditioning groups (PCGs) for distributing the electrical power that was regulated at 28 Vdc (NASA 1974b).

IU EPS Skylab's instrument unit EPS, designed for a maximum use of 7.5 hours, was isolated from the other power systems with the exception of the monitor interface. The IU EPS unit consisted of four single-use silver-oxide batteries and an AC power supply that generated 3-phase 400 Hz at 26.6 Vac.

CSM EPS The Command and Service Module EPS consisted of two separate but shared supplies. Primary supply for the CSM was furnished by two fuel cells located in the SM, and battery power supplied for emergency and reentry placed in the CM. AC power for the CSM was generated by inverters connected to the fuel cell and battery power supply. The CSM's major components consisted of:

- 2 fuel cells (1,420 W max)
- 3 solid state inverters for three-phase, 400 Hz AC
- 3 entry/post landing batteries (40 amp-hr)

- 2 pyrotechnic batteries (40 amp-hr)
- 3 descent batteries (500 amp-hr) (Belew and Stuhlinger 1973)

Skylab Thermal Control

Active and passive systems were used on the Skylab space station to provide efficient thermal control throughout the station. Skylab's passive thermal systems were designed to limit heat flow through the interior to and from space, while the active systems were used primarily to dissipate heat from the cabin atmosphere and station hardware using a fluid loop circulated through the external space radiator.

Passive Thermal Control Passive thermal control for the station consisted primarily of the micrometeoroid-sun shield, surface coatings, and thermal insulation. Thermal coatings were used to either increase absorption with a black paint or decrease absorption and increase emission with a zinc oxide coating to reduce heat loading from the Sun. Thermal insulation was used extensively to reduce heat flow into, and heat loss from, the interior compartments which reduced the need for electrical power for active thermal systems.

A combined sun shield and micrometeoroid shield was installed on the exterior of the OWS to reduce solar heating and protect the structure from small space debris. The shield was placed next to the OWS shell for release at the deployment signal, then extended several centimeters outward in a stand-off position that would dissipate impact energy over a broader area on the hull than at the shield. Premature activation of the deployment mechanism was caused by unexpectedly large vibration loads as the booster and station passed through Max Q. Standoff deployment forced high-velocity, turbulent air around the shield, which tore off nearly all of the shield, taking with it one of the OWS solar panels. A replacement parasol was designed and fabricated before the first manned flight, and installed by the crew during their SL-2 mission to protect the OWS from the excessive solar heat.

Active Thermal Control Active cooling in the Skylab station consisted of electrical heaters for components and compartments exposed to low temperatures that could reach -121°C (-250°F), and cooling loops to collect and remove heat generated by electrical equipment and the crew cabin. Skylab's active thermal system employed a second cooling loop for refrigeration of food and biological samples. Internal heat was removed by circulating conductive fluid through the loops to the radiators on both the OWS and the ATM. A much smaller cooling effect came from the cryogenic oxygen warmer that supplied the crew with breathing oxygen.

- Heat input
  - Oxygen heat exchanger—heated cryogenic oxygen for crew respiration
  - Electric heaters—heated liquids and selected equipment to prevent cold exposure damage
- Heat removal—Heat was removed from the Skylab cluster by a system of heat collectors, heat transfer loops, and heat radiators. The major components of the active thermal system included the following:
  - Heat collection

- Condensing heat exchanger—heat from metabolism and respiration (water vapor) condensation was transferred to the cooling loop in similar fashion to an air conditioner
- Cold plates—heat was removed from larger electronic equipment by circulating coolant through conductive metal (cold) plates attached to the electronic units
- ATM control and display panel heat exchanger
- EVA/IVA suit cooling module—during EVAs, heat from the space suits was circulated through a separate loop connected through the umbilical lines attached to the suits
- Battery modules—heat was removed from the battery power conversion and charging modules through a cooling loop heat exchanger
- Heat transfer loops
  - Airlock module cooling loop
  - ATM cooling loop
- Heat radiators
  - Dual pass (bifilar) radiator was used as the primary space radiator for the OWS and the ATM
  - Refrigeration system radiator
  - Thermal capacitor—the cooling capacity was increased in the OWS coolant loop by adding dual paraffin-filled thermal capacitor units that partially melted and absorbed extra heat in the orbit phase exposed to sunlight, then resolidified and released heat in the eclipse phase of the orbit. The thermal capacitors not only allowed a smaller primary OWS radiator, but provided a narrower range of coolant line temperatures for EVA and IVA with its placement downstream of the OWS radiator.
- Refrigeration systems
  - OWS refrigeration subsystem—a refrigeration system loop was used for food storage, chilled potable water, and for freezing food and urine samples
    - Wardroom food freezer—low-temperature storage of food for a fifty-six-day supply (100 lb)
    - Food storage freezer—low-temperature storage of food for an eighty-four-day supply (150 lb)
    - Water chiller—drinking water supply included chilled water at 4°C

ENVIRONMENTAL AND LIFE-SUPPORT SYSTEM

Even though the Skylab environmental control and life support equipment inherited some of the Apollo CSM hardware, the station's larger size and longer duration missions required a completely new design for many of the subsystems and components. For the first time, a dual-gas atmosphere was used in a U.S. manned spacecraft cabin to reduce the pure oxygen hazards. It was also the first spacecraft to use a regenerable carbon dioxide scrubber to replace the heavier lithium hydroxide canisters. Storage for food and biological samples were added to the OWS for improving the astronauts' diet, and to store tissue, blood, urine, and fecal samples for research. Skylab's longer flight missions also required sleeping accommodations, an integrated waste management system, exercise facilities, as well as a galley and personal hygiene facilities that, for the first time, included a shower.

Crew life support on Skylab consisted of a number of subsystems and components that were grouped into seven systems as part of the environmental control system (ECS). These included the following (NASA 1974a):

- Atmosphere revitalization and control
- Thermal control
- Water and food management
- Waste management
- Personal hygiene
- Microbial control
- Medical support and monitoring

Atmospheric Revitalization and Control (ARC)

Atmospheric gas in the Skylab modules was supplied by pressurized tanks of oxygen and nitrogen filled before launch. Skylab's OWS, AM, and MDA pressurized modules were first pressurized with nitrogen gas before launch, then, as gas escaped from the modules during ascent and initial orbit, the atmosphere was replaced with oxygen until reaching preset limits. The reduced-pressure atmosphere in Skylab was maintained at 5 psi total, with a 3.6 psi oxygen partial pressure (72%) and the remaining gas pressure supplied by nitrogen (24–28%). Between the three crew missions, the Skylab module pressure was reduced to 0.5 psi to prevent gas loss due to leakage. OWS temperature was maintained at 15.6–32°C (60–90°F).

Humidity, heat, carbon dioxide, and odors were removed with two redundant molecular sieve assemblies which contained the following components:

- Two screen filters used for particle removal
- Two heat exchangers used for heat and moisture removal
- Two compressors used to force air through the heat exchangers
- Two molecular sieve beds for CO_2 removal
- One activated charcoal canister for odor and contaminant gas removal

Molecular Sieve

Skylab's molecular sieve was a regenerable carbon dioxide removal device that replaced the bulky and single-use lithium hydroxide canisters used on shorter-duration manned missions. The molecular sieve employed two absorbing beds of granulated amine that alternately absorbed then released water and carbon dioxide. The molecular sieve dual-bed system circulated cabin air through one bed that absorbed CO_2 and H_2O,

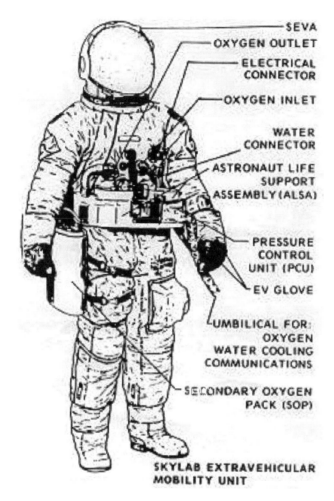

SEVA
OXYGEN OUTLET
ELECTRICAL
CONNECTOR
OXYGEN INLET
WATER
CONNECTOR
ASTRONAUT LIFE
SUPPORT
ASSEMBLY (ALSA)
PRESSURE
CONTROL
UNIT (PCU)
EV GLOVE
UMBILICAL FOR
OXYGEN
WATER COOLING
COMMUNICATIONS
SECONDARY OXYGEN
PACK (SOP)

SKYLAB EXTRAVEHICULAR
MOBILITY UNIT

Figure 9.15. Component layout of the Skylab Extravehicular Mobility Unit. Source: NASA EP-107.

as the second bed that was heated released the CO_2 and H_2O to the vacuum of space. Continuous CO_2 removal was maintained by switching the flow of cabin air and the space vacuum between the two beds. Cycle period for the dual bed molecular sieve was fifteen minutes. For complete regeneration of the amine beds, a twelve-hour bakeout cycle was completed at 204°C every twenty-eight days, which allowed complete water extraction.

Water Management

As with the other consumables, the Skylab water supply was launched with the station. Approximately 1,000 gal of water, sufficient to supply three crew members for 140 days, was supplied to the OWS and AM through three water-distribution networks. Stored potable water was provided principally for the following:

- Food reconstitution
- Drinking
- Crew hygiene
- Housekeeping
- Urine separator flushing
- OWS shower
- AM EVA/IVA cooling loop

Skylab's water supply was stored in ten 100 gal (272 kg, 600 lb) stainless steel tanks that were furnished with heating blankets to prevent freezing. Zero-gravity feed in each tank employed a bellows mechanism pressurized with nitrogen gas to maintain positive water pressure. An iodine compound was added to the water supply as a biocide, and for purification. Water for the Skylab galley was routed through a water chiller for drinking, and a water heater for food rehydration.

Food

Approximately 950 kg (2,100 lb) of food was launched on the OWS for the three planned crew missions. The five food types included:

- Dehydrated
- Intermediate moisture
- Thermostabilized
- Frozen
- Beverages

Waste Management

Waste management was a troublesome problem on each of NASA's short-duration manned programs, from Mercury to Apollo. As the mission duration expanded, so did the need for a more complete solution to the problems of waste accumulation, storage, and cleanup, especially for the urine and fecal waste. While the crew's waste products presented a biological hazard and their collection and storage were time-consuming, both urine and fecal samples had to be stored for later research on the physiological effects of space adaptation.

A solution to the waste problems in zero-gravity began with the design of the Skylab commode, called the waste compartment, which served as the core of the waste management system. The other major component of the waste system was the waste storage tank. The large oxygen tank on the converted S-IVB booster was dedicated to storing the waste generated by all three crews on Skylab for its orbital life. A pressure seal was used to prevent gases from leaking into the OWS while allowing the transfer of waste into the tank. Skylab's waste system also contained waste processors for solidifying fecal samples and trash containers. Refrigerating and freezing the waste samples allowed their return on the CSM for research back on Earth.

Extravehicular Mobility Unit (EMU)

The Skylab A7LB space suit (EMU) was used for both intervehicular and extravehicular operations (IVA and EVA) and provided life support, pressurization, and minor radiation and micrometeoroid protection. The A7LB was adopted from the Apollo program vehicle operations (not lunar surface missions) with atmosphere and cooling provided through an umbilical line. Skylab's A7LB suit was comprised of six primary components which included:

- Apollo pressure garment assembly—retained pressurized oxygen for respiration and ventilation
- Inner liquid cooling garment—a layer of cooling water tubes networked in a woven garment that was worn by the astronauts as an undergarment to remove body heat
- Pressure helmet
    - Skylab extravehicular visor assembly—attached to the pressure helmet to provide micrometeoroid, thermal, and UV protection
- EVA gloves
- Astronaut life support assembly—regulated and distributed water, and supplied oxygen and electrical power to the spacesuit through the life support umbilical
    - Pressure control unit—regulated oxygen pressure in the PGA
- Life support umbilical—space suit life-support umbilical lines 18.3 m (60 ft) in length were attached to the suit on one end and the AM consumable supply panel on the other

ATTITUDE AND POINTING CONTROL SYSTEM

Accurate station attitude control supplied by the control moment gyros and the thruster attitude control subsystem was needed to point Skylab's ATM telescopes and OWS experiments. A more stable and accurate pointing control unit called the experiment pointing control subsystem was added to the ATM instrument canister to improve the telescope guidance and pointing. A gimbal and control assembly inside the canister augmented the station's pointing system within about 10^{-5} radians (2 arc seconds). Skylab's complete three-axis attitude/pointing control unit, named the attitude and pointing control system (APCS), was used for three-axis attitude control, three-axis stabilization, and pointing equipment and experiments at specific targets. The APCS was operated manually by the astronauts in the MDA or from commands issued through ground controllers. The major subsystems of the APCS consisted of the following:

- Thruster attitude control subsystem (TACS)—provided initial control and maneuvering of Skylab following orbital insertion
- Control moment gyros subsystem (CMGS)—provided pointing control, maneuvering, and offset pointing of the station using three double-gimbaled CMGs. The

three primary control modes were solar inertial, Z-local vertical, and experiment pointing.

- Experiment pointing control subsystem (EPCS)—provided accurate pointing and stabilization of the ATM experiment package along with limited offset pointing from Sun center in the pitch, yaw, and roll axes. EPCS operation was separate from the CMGS/TACS operations, including several attitude sensors. (Belew and Stuhlinger 1973)

APCS Components

Skylab's attitude and pointing control system was the management system for maintaining or changing the station's orientation in orbit, for pointing the Skylab cluster, and for pointing the onboard telescopes and experiments. The attitude determination and attitude control components on the APCS included:

- CMGs—three CMG units provided accurate three-axis station attitude control. The CMG consisted of an induction motor-driven constant-speed wheel, and dual gimbal-supports for two degrees of freedom.
- Star tracker—provided star position inputs to the ATM digital computer for calculating the roll reference angles and the orbital plane errors
- Fine Sun sensor—provided accurate attitude information for the X and Y axes of the EPS
- Acquisition Sun sensor—provided attitude information for the X and Y axis control
- Experiment package caging and gimbal assembly—maintained attitude control of the experiment package during some operational phases
- Manual pointing controller—used to offset point the ATM instrument package or search for a desired star
- Experiment pointing electronics assembly—contained the electronics for amplifying, shaping, and mixing the output signals of the ATM experiment package sensors to obtain error signal for driving the EPCS actuators
- ATMDC/workshop computer interface unit—provided general purpose computing capabilities
- Memory load unit—provided ATMDC flight programs from the onboard tape recorder or from the ground (Belew and Stuhlinger 1973)

Thruster Attitude Control Subsystem (TACS) Components

Skylab's TACS was a three-axis attitude control system that used compressed (cold) gas thrusters for maintaining or changing attitude on the station and desaturating the CMGs. The TACS consisted of:

- Cold gas propellant (nitrogen)
- Twenty-two titanium propellant storage tanks located in the aft skirt section
- Two thruster modules of three thrusters each on the aft skirt of the station

- Four redundant valves for each thruster
- Power and Control Switching Assemblies

ATM Digital Computer (ATMDC)

Skylab's ATMDC was the first fully digital computer system on a manned space-craft. Although the digital computer was introduced in the Gemini program and greatly improved for the Apollo missions, the Skylab computer was the first to use digital systems to control the station that included complex navigation and CMG operations. Manual and ground commands allowed Skylab's redundant ATMDC systems to monitor and manipulate all onboard functions once control was turned over from the IU after orbit entry. ATMDC's processor had a 16,300 word memory of 16-bit words to control APCS functions and the station system operations. APCS control included orbital navigation and timing, operational mode control, CMG and TACS control, CMG momentum desaturation, and system redundancy management (NASA 1971).

Operation Modes

Skylab's ATMDC and APCS had six station operation modes that were mutually exclusive. Mode selection was made by internal, crew-initiated, or ground commands that included:

- Standby mode—no vehicle APSC control required
- Solar inertial mode—maintains Z-axis alignment with Sun within stabilization constraints
- Experiment pointing mode—used for telescope and experiment pointing and stabilization
- CMG nested attitude hold mode—used for inertial (fixed) reference
- TACS attitude hold mode—inertial fix maintained with TACS thrusters
- Z-local vertical (Z-LV) mode—continual horizontal alignment used for pointing (manned) or docking (unmanned) (NASA 1971)

COMMUNICATION SYSTEM

Skylab's communication system was used to transfer data, voice, and video between the station and ground and between modules and sections within the Skylab complex. NASA's Spaceflight Tracking and Data Network (STDN) was used for the ground station communications links for Skylab's CSM, AM, and ATM communication systems. Additional and separate communications channels were also installed on the Saturn V launch vehicle in the S1-C, S-II and SIV-B stages, as well as in the IU unit. Telemetry channels were included in each of the booster stages as well as the IU. In addition, command destruct communications uplink was installed on each booster stage. A nondestruct command channel was installed on the Skylab IU along with a C-band tracking transponder (Disher 1972).

Communications for the Skylab station could be routed through the command and service module for some data, but not all. CSM communications systems were nearly identical to the system used on the Apollo missions with some modifications, one being the lack of the Apollo CSM's large high-gain antenna. A second unique feature was that ground controllers could select video downlinks using the CSM FM transmitter.

A modified Gemini system was adapted for the Skylab Airlock Module communication system with four VHF transmitters to downlink real-time data and three tape recorders for recording systems, experiment data, and voice data. Command capability to the AM was through redundant UHF receiver decoders. The Apollo Telescope Mount communication system used modified Saturn hardware with two VHF transmitters for downlinking real-time and delayed-time data (Belew and Stuhlinger 1973). Two nonreplaceable recorders and related interface equipment were used to record and transfer experiment data.

During the manned phases of the mission, Skylab's communication system provided:

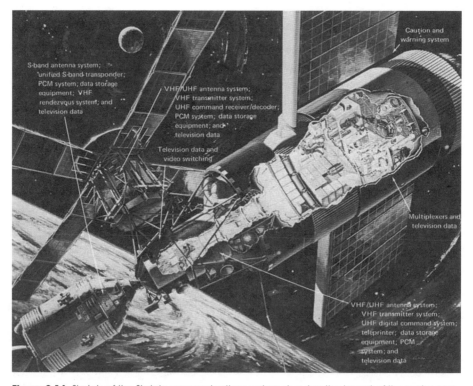

Figure 9.16. Sketch of the Skylab communications system showing the layout of the major components. Source: NASA SP-400.

Table 9.1. Skylab Uplinks (Disher 1972)

Module	System	Frequency (MHz)	Modulation
AM	UHF command	450	PSK/FM
AM	UHF teleprinter	450	FM
ATM	UHF command	450	FM
ATM	UHF command (redundant)	450	FM
CSM	VHF voice	259.7, 296.8	AM
CSM	USB up data	2106.4	PM
CSM	VHF ranging	295.8	AM

Table 9.2. Skylab Downlinks (Disher)

Module	System	Frequency (MHz)	Modulation
AM	VHF telemetry	230.4, 235.0	PCM/FM
AM	VHF voice	246.3	FM
ATM	VHF telemetry	231.9, 237.0	PCM/FM
CSM	USB voice, ranging, data	2287.5	PM
CSM	USB voice, TV, data	2272.5	FM
CSM	VHF voice	295.8	AM
CSM	VHF voice, ranging	259.7	AM

- Voice communication between Skylab crew members
- Voice communication between Skylab crew and Mission Control Center through the STDN
- Caution and warning alerts
- Range and rate information between the CSM and Skylab for rendezvous and docking
- Teleprinter messages from Mission Control Center to the crew
- Video data transmission through the STDN
- CSM S-band data, TV, and voice downlink
- CSM VHF voice downlink
- CSM C-band ranging
- OWS VHF and UHF command, data, TV, and voice channels
- ATM VHF and UHF command and data channels

SKYLAB MANNED MISSIONS

Skylab's first crew mission was scheduled to launch for rendezvous and docking one day after the station launch on May 14, 1973. Telemetry from the station in orbit indicated that serious problems with Skylab's systems required immediate attention which delayed the launch indefinitely. Because the perishable foods and temperature-sensitive experiments would likely be ruined after ten days in the elevated OWS temperatures which reached 52°C (126°F), a ten-day deadline was established for the docking and rescue mission. As a result, SL-2, the first scheduled crew mission, was launched on May 25.

A fly-around of Skylab and the video images that were downlinked showed the micrometeoroid shield had indeed torn loose except for several pieces that pinned the right solar panel in its stowed position. After careful preparation, an EVA was staged from the CSM into the OWS since the extreme temperatures prevented the astronauts from entering Skylab until the heat was reduced by the experimental sun shade. Further problems with the bound solar array on the OWS led to an EVA that used part of the backup sunshade mechanism and a tether to leverage the partially deployed solar panel into place. Skylab's remaining OWS array (panel 2) deployment was successful, and after checkout, roughly doubled the power for the orbital space station.

- SL-1 Skylab station launch
  - Launch: May 14, 1973, Saturn V
  - Crew: Unmanned
  - Duration: 3 yr, 2 mo (34,981 orbits)
- SL-2 First crew
  - Launch: May 25, 1973   Saturn IB
  - Crew: Charles Conrad (Commander), Paul Weitz (Pilot), Joseph Kerwin (Science Pilot)
  - Splashdown: June 22, 1973
  - Duration: 28 d (404 orbits)
  - EVAs: 0.67 hr, 3:30 hr, 1:44 hr
- SL-3 Second crew
  - Launch: July 28, 1973   Saturn IB
  - Crew: Alan Bean (Commander), Jack Lousma (Pilot), Owen Garriott (Science Pilot)
  - Splashdown: September 25, 1973
  - Duration: 59 d (858 orbits)
  - EVAs: 6:29 hr, 4:30 hr, 2:45 hr
- SL-4 Third and final crew
  - Launch: November 16, 1973   Saturn IB
  - Crew: Gerald Carr (Commander), William Pogue (Pilot), Edward Gibson (Science Pilot)
  - Splashdown: February 8, 1973
  - Duration: 84 d (1,214 orbits)
  - EVAs: 6:33 hr, 7:01 hr, 3:28 hr, 5:19 hr

SKYLAB EXPERIMENTS

During Skylab's development, emphasis shifted from applications research to scientific research. By the time specifications were completed and contracts prepared for the Skylab modules, systems, and instruments, the broad areas of research and numerous experiments were gathered into three major areas of research—astronomy, biology, and Earth observation—with numerous exceptions. The manned laboratory was ideal for observing at wavelengths absorbed by the Earth's atmosphere. It was also ideal for a broad view of the Earth's surface, like satellites, only Skylab had a much more sophisticated suite of instruments that could be operated and maintained, and even repaired, by trained astronaut-scientists. Skylab also offered the first opportunity for extensive medical research on humans and biological life exposed to microgravity for months at a time. Congressional leaders were fully aware of the potential advances in science and technology that the Skylab program offered and ensured that NASA would direct research on Skylab toward space science and medical-physiological programs.

In addition to the telescopes and instruments that belonged to the solar payload complement located in the ATM, Skylab carried an Earth resources experiment package (EREP) controlled from the MDA panel that consisted primarily of film cameras, spectrometers, radiometers, as well as numerous individual instruments and experiments. Skylab's EREP data provided the first systematic, comprehensive survey of the Earth from space. Skylab's EREP instruments included:

- Multispectral photographic camera system (six different camera systems)
- Earth terrain camera (high-resolution color film)
- Visible-infrared spectrometer (two-channel spectrometer)
- Multispectral optomechanical Scanner (Earth-surface feature observations)
- Passive microwave radiometer (measured emission/reflection brightness)
- Active scatterometer and radar altimeter (measured surface roughness and irregularities)
- UV airglow horizon photography (observed sunset/sunrise airglow for measuring ozone layer vertically) (Belew and Stuhlinger 1973)

Skylab's Solar Payload Complement was located primarily on the ATM and included:

- X-ray spectrographic telescope (X-ray images of the Sun from 0.2 to 6 nm)
- X-ray telescope (soft X-ray telescope and X-ray event analyzer)
- XUV spectrograph (solar X-ray and UV emissions—the only solar instrument located in the OWS)
- EUV spectroheliometer (spectral line data from specific atomic emissions from the Sun)
- White light coronagraph (high resolution photographs of the solar corona that blocked the Sun's central image)
- XUV spectroheliograph (solar X-ray and UV images)
- XUV spectrograph (high-resolution data in the X-ray and UV bands)
- Hydrogen-alpha telescopes (2) (solar emission images in the red H-alpha band)

Skylab laboratory facilities also included material science research conducted on metals, alloys, crystals, ceramic compounds, glasses, and plastics. In addition, extensive research was carried out in astronomy and astrophysics, and zero-g research in combustion, biology, physiology, and physics.

SKYLAB'S DEMISE

As the astronauts on the SL-4 mission were closing up Skylab for the last time, its systems were already showing their age. Two of the CMG attitude control units were losing their ability to point the station accurately, while several of the cooling loops were beginning to fail. Since the nine-year expected life of the Skylab complex would extend past the first space shuttle flights expected for the end of the 1970s, mission planners began putting together a rescue project that, among other things, would attach a booster to take the space station to a higher orbit so that repairs could be made on later space shuttle flights. Deorbit was projected to be in 1983, perhaps as late as 1984. This offered engineers and planners enough time for station rescue and possibly several refurbishment flights (Compton and Benson 1983).

By 1977, plans were being drawn up for a rescue project followed by reuse of Skylab in several stages that would begin in the early 1980s. First would be a propulsion module to boost the station's altitude for an additional five-year lifetime before atmospheric drag would deorbit the complex. Other proposals were made for a second Skylab using surplus hardware, and for launching two workshops on the remaining Saturn Vs in support of Apollo and Soyuz missions that were already scheduled for a linkup in 1975. Funding for these, however, never materialized.

As with many other space projects, costs became more important as the hardware reached the contract and fabrication stage. NASA's budget was near its lowest figure ever following Apollo, and complicating the project was the incompatibility of the Skylab and the space shuttle orbiter systems that made the operation, servicing, and support a serious challenge. Electrical power would be difficult to share between the two vehicles, the cabin atmospheres were a factor of three different in pressure, and reversed in composition. The transfer hatches on Skylab and the space shuttle were so dissimilar that a passageway adapter would have to be designed similar to the Apollo-Soyuz docking module.

Approval for funds to study the Skylab extension project came in 1978, the same year that the reality of the inevitable end of Skylab loomed. NOAA scientists had calculated a level in the Sun's coming peak activity that was far greater than projected by NASA. Controversy about unreliable data, and the failure to use solar astronomy data from NOAA, would become less important as 1979 approached. A lack of funds for adapting and reboosting Skylab delayed the first phase of the rescue plan which was to reboost the station in order to gain time on the reentry date. A rapidly expanding atmosphere increased the aerodynamic drag which shortened the time available for a reboost before the Skylab would fall from orbit, with or without NASA's participation. But there was more controversy that made Skylab's fate of concern to NASA managers. In 1978, the Soviet Union's high-powered radar satellite Cosmos 954 failed in orbit, forcing the reactor to reenter the atmosphere. The onboard nuclear reactor broke up on reentry, spreading its radioactive fuel and components over a 124,000 km² (48,000 mile²) area of the Great Slave Lake in the Northwest Territories on January 24, 1978. Skylab's approaching deorbit created concern worldwide about a similar mishap. Although no significant nuclear fuels were aboard Skylab, the potentially disastrous reentry from the massive satellite was real and a growing concern. If delayed further, Skylab could deorbit with little or no control over its reentry, nor over its impact point.

Communications had not been maintained with Skylab since its deactivation, hence control of the station systems had not been attempted for several years. But whether Skylab was to be rescued or deorbited, flight control of the station's attitude was needed. Attempts to align the solar arrays with the Sun to recharge the batteries in 1978 were successful, which continued to offer the option of deorbiting the station, or extending its onorbit life for later rescue. But without funds for a booster and the space shuttle several years from completion, the only alternative was the inevitable deorbit of Skylab. Implications of the Cosmos 954 reentry disaster forced NASA's hand in planning for an accurate, controlled reentry and possible contingencies.

Preparation for Skylab's deorbit intensified in early 1979 since preparations had to be made for controlled orbit decay while modeling the atmospheric changes that affected Skylab in its final orbits. NORAD radar and NASA tracking facilities targeted mid-June to early July for the critical altitude decay and reentry, and by June the reentry date converged on July 11. Careful control of the attitude to counter the reentry drag provided a reasonably accurate orbital altitude decay rate until reaching a higher-density region. At that boundary, the accuracy of the calculated trajectory to impact was improved considerably. Tumbling the station end over end created more accurate drag calculations and was programmed for the final orbit. As July 11th approached, worldwide press coverage waited for the spectacular burnup with the expectation of a splashdown between Africa and Australia. Skylab, however, was far more durable than planned, and as it passed over North America and the South Atlantic on its way to Africa, it had not broken up—even the solar panel remained attached. Decreased drag from the delayed breakup extended the flight path, and the remnants that were to expected to splash down in the Indian Ocean between Africa and Australia traveled several hundred kilometers east, some raining down on a remote region of Western Australia.

Russian Space Stations

SALYUT

The Soviet Union's race to the Moon officially but quietly came to an end when the manned lunar project was canceled in 1974, well after the last Apollo mission landed on the Moon in 1972. Some residual interest in lunar missions remained, but following the USSR's lunar program failure, the politicians, scientists, and engineers turned their efforts in a new direction. The Soviets continued their unmanned interplanetary exploration missions to Venus and Mars, while their manned missions were transformed into orbital space station projects.

America's military Manned Orbital Laboratory program that began in the mid-1960s was predictably countered by a Soviet manned satellite project. As much a competition with other rival designers as it was a directive from the Soviet leadership, spacecraft designer Vladimir Chelomei had already begun a manned military orbiting laboratory project called the Almaz ("diamond") when the announcement of a manned orbital laboratory was made by Secretary Leonid Brezhnev in 1969. Brezhnev and others proclaimed that the Soviet Union had little interest in reaching the Moon with their manned vehicles, which was not true. The statement that their intention all along was to develop orbiting space stations "...for the good of the people and for the good of science" (Siddiqi 2000) was only slightly more correct since three of the seven Salyut space stations were military missions. The decision was made to complete Chelomei's Almaz station for launch in 1970 before delays in the system and instrumentation development short-circuited the program. But delays and the political conflict between Chelomei and Dimitiy Ustinov, the man in charge of military defense and space programs, a competing civilian space station was chosen for the first launch.

The two rival station projects were approved under the name Salyut ("salute"), with Chelomei's Almaz military station incorporating the support systems from the civilian station under design bureau director Vasily Mishin. The civilian station design, designated DOS (Dolgovremennaya Orbitalnaya Stanziya), used a variation of the Almaz hull for the DOS structure. However, the two stations would not be the same, only similar. The Almaz contained a reinforced two-cylinder structure with the solar arrays and docking port at the aft end. The primary instrument on the military station was a large-aperture camera combined with a film ejection airlock that could deploy the exposed film canisters back to Earth in a reentry capsule. Defense weapons and a plethora of surveillance equipment were flown on the Soviet's three Almaz station missions.

The DOS orbital research laboratories were comprised of three cylindrical sections with the docking port at the forward end of the station, along with the solar panels, making them easy to distinguish from the Almaz vehicles. DOS-Salyut stations were outfitted with equipment and instruments for Earth observation, Earth environment studies, solar system exploration, astrophysics observations, and biological and physiological research. Like the Almaz, but on a smaller scale, an airlock was provided for ejecting small experiments and waste canisters. Military hardware was not excluded on the civilian DOS missions, just not emphasized, with the exception of several military expeditions on Salyut 7.

Together, the Almaz and DOS stations constituted the series of seven Salyut space stations launched by the Soviets. For increased safety and decreased weight, station crews were not launched with the station but on Soyuz ferry vehicles that were originally developed for the failed manned lunar project. The Soyuz ("union") transport ferry vehicle was adopted for the Salyut crew vehicle following its flight tests in the late 1960s that proved successful enough to be used for all of the Soviet and Russian space station crew flights, including the current missions to the International Space Station.

Launchers for the Salyut stations were the heavy-lift Proton boosters developed in the 1960s. The lighter Soyuz spacecraft were launched with the improved R-7 ICBM boosters that were also called Soyuz. All of the Salyut series stations as well as the Mir and research modules were launched on the Proton SL-13 from the Baikonor launch facility that is now part of Kazakhstan.

Salyut 1 (DOS 1)

- Launch: April 19, 1971
- Deorbited: October 11, 1971
- Crews: 1
- Occupied: 24 days
- In orbit: 175 days
- Orbital mass:18,400 kg

Construction   DOS versions of the Salyut series stations were composed of three cylindrical sections which consisted of:

- Airlock/transfer module—included an airlock and transfer module and docking adapter at the forward end of the station
- Working compartment—the center section which housed the galley, exercise equipment, control stations, food and water storage, toilet and hygiene station, and scientific equipment
- Service/engine module—based on the Soyuz service module and located at the aft end of the vehicle

A single docking port on the forward end of the station called the transfer module accommodated the Soyuz crew vehicle only. The single port arrangement limited the station's useful life to the stored consumables for the crews that could be launched on the station. Only several hundred kilograms of contingency equipment and supplies could be taken to or from the station on the Soyuz crew vehicle, making resupply impossible. For that reason, the first generation Salyuts with a single docking port had an operational life of approximately one year. Second-generation stations which included Salyuts 6 and 7 were constructed with a second docking port at the aft end permitting cargo craft to replenish the station during or even between the manned missions. This extra port allowed the operational lifetime of the Salyut station to stretch from one year to five years.

Crew missions   The first mission to Salyut 1, designated Soyuz 10 for the tenth flight of that generation of Soyuz, was launched successfully four days after the station was placed in orbit on April 19, 1971. Although the crew completed a hard docking, the mission had to be aborted because the hatch could not be opened. The second mission crew, designated Soyuz 11, completed the rendezvous and docking, followed by a successful twenty-four-day stay. Research projects on the flight included astronomical observations, biology and physiology experiments, and Earth observations that were first of a kind, although the last on Salyut 1. Tragically, the Soyuz 11 crew died during reentry from a pressure equalization valve opening prematurely and exhausting the cabin air to space. To save weight on the Soyuz flights, the cosmonauts were not given pressure suits to wear in case of emergencies, a policy that was changed for all of the following Russian manned space programs. All passengers on the Soyuz (as well as on all American space vehicles) now wear pressure suits for launch, deorbit, and reentry.

Figure 9.17. Artist's sketch of the Salyut first-generation station and Soyuz transfer vehicle. Source: NASA NIX.

Salyut 2 (Almaz 1)

- Launch: April 3, 1973
- Deorbited: May 28, 1973
- On Orbit: 54 days
- Orbital mass: 18,500 kg

Salyut 2 was the only unsuccessful station launch in the Salyut series. The mission began with an aborted launch in September 1972, which delayed the actual launch until April 3, 1973. A one-week delay in the initial crew flight to the orbiting station was needed because the station depressurized after being hit by debris from an explosion on the Proton upper stage during ascent. Soon after, the crew missions for the Salyut 2 were canceled when the station began tumbling and broke up. The station fragments reentered the atmosphere approximately three weeks later.

Cosmos 557 (DOS 2)

- Launch: May 11, 1973
- Deorbited: May 22, 1973
- On Orbit: 11 days

The unnamed launch of the second civilian DOS Salyut station a week after the failed Salyut 2 launch was an apparent attempt to beat Skylab into orbit, but without success. Controllers lost command of the vehicle, which began tumbling soon after launch due to the loss of onboard fuel. The launch failure became a Cosmos designation, which was commonly used to conceal failed missions.

Salyut 3 (Almaz 2)

- Launch: June 24, 1974
- Deorbited: January 24, 1975

- Crews: 1 - Soyuz 14 (Soyuz 15 could not dock)
- Occupied: 15 days
- On Orbit: 213 days
- Orbital mass: 18,500 kg

The first military Almaz station launched successfully into orbit was the Salyut 3 with more than a year's delay after the failed Salyut 2. Salyut 3 was outfitted with a number of new technologies that included a new electrically driven flywheel attitude control system called the gyrodyne that was improved over the years and incorporated into subsequent Soviet and Russian stations, including the Mir and the International Space Station. Also flown on Salyut 3 for the first time was a new thermal control system and a new water recycling facility. The first crew on Soyuz 14 succeeded in completing a shortened fifteen-day mission.

Salyut 4 (DOS 3)

- Launch: December 26, 1974
- Deorbited: February 3, 1977
- Crews: 2 - Soyuz 17, 18B
- Occupied: 92 days
- On orbit: 770 days
- Orbital mass 18,500 kg

Salyut 4 was an improved Salyut 1 with three solar arrays that could be rotated to better follow the Sun's position while on orbit, and a solar telescope that replaced the Almaz surveillance camera. More than four dozen pieces of equipment and experiments were placed onboard the station for the two crew flights. Soyuz 17, which was the first crew mission to Salyut 4, spent twenty-nine days onboard successfully operating the experiments except for the solar camera which was damaged before arrival. The Soyuz 18 mission which was scheduled for a two month mission experienced a second-stage abort

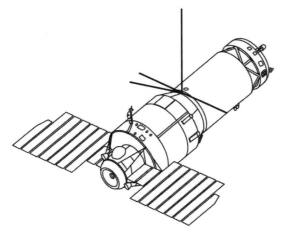

Figure 9.18. Sketch of the Almaz 2 (Salyut 3) station showing the two-cylinder structure with the single docking port at the aft (left) end. Source: Portree.

with a safe landing, although the crew was subjected to a 20-g deceleration during the descent. A follow-up Soyuz 18B mission was successfully launched nearly two months later for a successful sixty-three-day stay.

Salyut 5 (Almaz 3)

- Launch: June 22, 1976
- Deorbited: August 8, 1977
- Crews: 2 - Soyuz 21, 24
- Occupied: 67 days
- On orbit: 412 days
- Orbital mass: 19,000 kg

Salyut 5 was the Soviet's second and last military space station, and the last of the Soviet's first generation space stations. The station was structurally similar to Salyut 3 and included the Agat surveillance camera and the gyrodyne attitude control and pointing system. The first crew to reach Salyut 5 was on Soyuz 21 and executed a successful forty-nine-day observation and research mission until several events abruptly ended the mission, with the crew deorbiting and landing the following day. The following crew flight on the Soyuz 23 mission failed to complete the automated approach and docking, and returned the next day. The third crew mission aboard Soyuz 24 completed a shortened but successful eighteen-day stay, possibly to tie up loose ends left by the Soyuz 21 crew (Portree 1977).

Salyut 6 (DOS 5)

- Launch: September 29, 1977
- Deorbited: July 29, 1982
- Crews: 16
- Occupied: 683 days
- On orbit: 1,764 days
- Orbital mass: 19,000 kg

Salyut 6 was the first of the second-generation space stations built by the Soviets. The design was similar to the Salyut 1 and Salyut 4 DOS stations, but included a second docking port on the aft end of the station, in addition to numerous improvements in the systems and equipment. The aft port was designed to accommodate the new Progress cargo vehicle using the same docking mechanism as the Soyuz docked on the forward port. Resupply flights extended the station's operational lifetime by resupplying consumables and equipment that previously had to be launched with the first-generation stations. Visiting crews could also be transferred to and from the station using the second docking port.

During the five-year operational lifetime of the Salyut 6, five long-duration expeditions were carried out in conjunction with eleven short-duration flights for visiting crews from Warsaw Pact countries. A number of records were set during Salyut 6's operational life, including a 185-day crew mission. Two new transfer vehicles were introduced on this station that included the new Soyuz vehicle named Soyuz T. The second was the versatile TKS (Transportniy Korabl Snabzheniya) transport module. The TKS consisted of the FGB base module and the VA reentry module also called Merkur ("Mercury"). The first station mission for the TKS was designated Cosmos 1267, not because of a failure, but because of its secret mission. TKS, often referred to as the Star module, was originally developed for a crew and cargo carrier for the Almaz program, although it never carried a crew (Portree 1977). The Merkur reentry vehicle was not as practical as the FGB and was discontinued, but the FGB proved to be a very versatile prototype. FGB's basic design was used for the Kvant, Kvant-2, Kristall, Spektr, and Priroda research modules that were attached to the Mir complex. It also served as the base design for the Polus military satellite launched on the Energia booster in 1987. FGB's most recent incarnation is the Zarya FGB module that became the base element of the International Space Station with its launch in November 1998.

Salyut 7 (DOS 6)

• Launch: April 19, 1982
• Deorbited: February 7, 1991
• Crews: 10
• Occupied: 816 days
• On orbit: 3,216 days
• Orbital mass: 19,000 kg

The last of the Salyut series stations was Salyut 7, which had the same structure and many of the same features as the Salyut 6 but with a number of improvements. New instruments and equipment were always a part of each new station launched, but Salyut 7 had more than any previous. With the delay of the new third generation Mir space station, the decision was made to launch the backup station for Salyut 6 with some of the improvements scheduled for Mir.

Living quarters on Salyut 7 were designed with new windows that passed UV light from the Sun to help reduce microbiological buildup in the cabin. Surface coatings reduced fungal and bacterial buildup on the long-duration missions, while paint schemes improved crew adaptation to the microgravity environment. Electric stoves, a refrigerator, hot water for the galley, and improved exercise equipment were also added.

Externally, the three solar rotatable arrays now had adapters to mount secondary panels on each side of the array, making the Salyut 7 distinctly different in appearance from the other Salyut stations. Salyut 7 also had a wider forward docking port to accommodate the large TKS/Star resupply vehicles that were flown under the Cosmos series, and was almost as large as the Salyut vehicle.

Crew missions to the Salyut 7 station included six long-duration expeditions and four visiting flights. Resupply and crew flights employed the Soyuz, Progress, and TKS vehicles, totaling fifteen support vehicle flights in all. New vehicles used for Salyut 7 included the new Soyuz T and the TKS/Star cargo ships. The last of these was Cosmos 1686, a TKS/FGB module used to reboost, then aid in the controlled reentry and burnup of Salyut 7.

One other first in the Salyut 7 missions was the near-disastrous Soyuz T-10A launch which was the only pad abort of the Soyuz crew vehicles. A fuel spill was followed by a fire on the booster before launch, forcing ground controllers to initiate the abort command, but the signal cable had already burned through. Twenty seconds later, a backup radio command was given that successfully ejected the capsule and crew with the escape rockets. The brief flight ended in a successful but hard landing that was followed by a few rounds of vodka and a medical checkup (Newkirk 1990).

MIR

Rather than docking several Salyut vehicles together for their next-generation station, the Soviets decided to alter the docking arrangement for a more advanced Salyut that could accommodate five or six additional laboratory modules, each as large as the Salyut core. The multi-docking arrangement could also be used to berth the new Soviet space shuttle named Buran that was intended to help assemble and support the next two orbital space stations. TKS support vehicles were also expected to be used for cargo flights to and from the station, and ultimately did, but not as cargo craft.

Ambitious Soviet plans in the mid-and late-1980s incorporated their shuttle program, unofficially named Buran, which was launched on the huge Energia booster. In addition to Buran, their manned space program included a permanent space station, manned lunar missions, and even a manned Mars project. In the interim, the decision was made to construct a large, versatile space station complex using the existing technology from the Salyut, the Soyuz, and the TKS programs. Adding a multiple docking node to the station core module that could berth several research modules along with the already-developed cargo and crew transfer vehicles, offered the designers, planners, and Soviet leadership a rational path for continuing their successful space station program. The blueprint also prepared the Soviets for future manned missions beyond Earth orbit.

The new Salyut station complex renamed Mir (meaning peace or commune/village) was designed using the Salyut hull as the core module, with a multidocking adapter replacing the forward docking port. The new five-port adapter could berth the laboratory modules that used the TKS FGB template, as well as the Soyuz and Progress transfer vehicles. The new research modules had the same basic structures as the TKS/FGB, but each had an independent equipment configuration with distinctly different research objectives, and each was powered by a separate solar array.

The new Mir station core module was outfitted with six ports and used the fore and aft docking ports to accommodate the Soyuz and Progress vehicles, like Salyuts 6 and 7. The four radial ports on the fore section were designed to berth four laboratory modules on a semipermanent basis. The clever addition of a research module with a pressurized compartment and berthing/docking port on both front and back and berthed on the Mir aft port brought the research module count on the Mir station to five.

Mir was designed and launched without the same experiment equipment aboard that was the practice for the Salyuts since it was to be the station's habitation module. Some medical equipment and the control section remained on the Mir core, but the habitable volume was increased to make room for more crews on longer-duration missions. The new station design also extended the on-orbit lifetime to ten years which was expected to fulfill their needs for orbital research until the more-ambitious Mir 2 station could be completed.

Mir also included an important navigational and docking tool for the two fore and aft docking ports known as the Kurs ("course") automated radar docking system. Kurs could dock and undock transfer vehicles on the Mir core by radar-guided, computerized control of both the station and the modules, without reorienting the space station core as was required for the Salyuts using the older Igla system. Mir's docking unit also allowed docked modules on the forward node to be mechanically transferred to or from one of the four radial ports, saving time and fuel needed for more complex docking techniques.

The new Mir core module also included larger, more efficient gallium arsenide photovoltaic arrays that increased power, as well as a Strela seven-computer station computer system and a new Vozduyk carbon dioxide removal system to replace the less efficient single-use lithium hydroxide canisters. More exercise equipment was installed in the workroom, along with two separate, small rooms for crew privacy, each with a window.

Mir (DOS 7)

- Launch: February 20, 1986
- Deorbited: March 23, 2001
- Crews: 16
- Occupied: 4,594 days
- On orbit: 5,511 days
- Orbital mass: 20,400 kg (core), 124,340 kg (orbital complex)

Mir station upgrades from its Salyut predecessor included a new Soyuz crew transport vehicle called the Soyuz TM, and the improved Progress M cargo vehicle. Progress also served as a trash disposal vehicle when it was deorbited at the end of its stay, although the Progress M cargo spacecraft also had the capability to carry a small, ballistic reentry capsule named Raduga for returning research materials and films from the Mir.

Mir Systems

Like the Salyut series, the Mir core was launched as a complete, autonomous vehicle. The station base block, or core, included propulsion, navigation, electrical power, communications, and life-support systems, but without the experiment equipment and instruments installed in the Salyuts since Mir's function was principally a crew

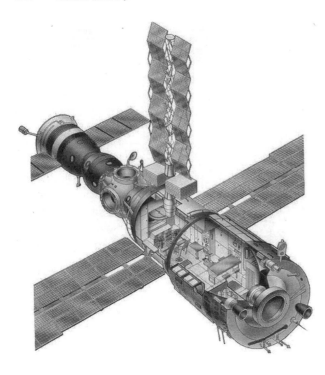

Figure 9.19. Cutaway sketch of the Mir core module with a Soyuz ferry vehicle docked. Source: NASA NIX.

support module. Most of the systems onboard Mir were inherited from the Salyut, but with many improvements and several significant advancements—some entirely new.

- Electrical power
  - Primary source—Three GaAs photovoltaic arrays
  - Bus voltage—28.5 Vdc
  - Total power available (complex)—26.8 kW @ beginning-of-life
  - Secondary power source—NiCd batteries (12 units in the Mir core)
    - Battery life—2–3 yrs at 50% depth-of-discharge
    - Battery capacity—720 Amp-hr for Mir core (1,440 Amp-hr for the complex)
  - Electrical connectors were installed on the forward docking port to allow power transfer between the Mir core and docked Soyuz TM vehicles
  - A floating ground was used for the Mir complex power distribution system
- Environmental control and life support
  - O_2 + N_2 atmosphere
    - 1,053–1,276 millibar pressure (~1 atmosphere)
    - Humidity 20–70%
    - O_2/N_2 mixture regulated at approximately 80% N_2, 20% O_2
- Max O_2 content 40%
- Max CO_2 content 3%
  - Oxygen supply —Water electrolysis ($H_2O \rightarrow O_2 + H_2$, with H_2 dumped overboard)

- Carbon dioxide removal—Regenerable amine beds (similar to Skylab's molecular sieve)
- Trace contaminant control—Thermally regenerable activated charcoal
- Temperature and humidity control—Condensing heat exchanger
- Water supply consisted of four loops
  - One for potable drinking and food preparation
    - Salts were added to potable water to improve taste
    - Potable water was treated with silver ion antimicrobial solution
  - Three recycle loops, each treated separately
  - Each purified by a multifiltration system
    - Hygiene water
      - Reclaimed for reuse as wash (hygiene) water
    - Urine distillate
      - Urine was first treated by the air evaporation process to produce distillate
      - Urine water was filtered, treated, and reclaimed for the production of oxygen in the electrolysis unit
    - Cabin (respiration) condensate
      - Reclaimed for use as drinking water and food preparation water
- Communications
  - Mir communications was designed for satellite relay through the Russian SDRN (Satellite Data Relay Network) communications link
    - Similar to NASA's TDRSS geostationary data relay satellite system
  - Communications was routed through Mir's aft-mounted parabolic antenna to the Altair (aka Luch) geosynchronous satellites
    - Altair/Luch satellites are the space elements of the Russian SDRN space communications network
- Propulsion
  - Both attitude control and orbit operations were integrated into the same guidance, navigation, and control system
  - Both operations used unsymmetrical dimethyl hydrazine (UDMH) and nitrogen tetroxide (NTO) propellants with crossover storage
  - Orbit entry and orbit maintenance for the Mir core used two aft-mounted 300 kg thrusters
    - When the Kvant search module was permanently berthed on the aft port, Mir thrusters were no longer usable
  - Orbital operations were then provided by Progress resupply ships
  - Propellant—UDMH (fuel) + NTO (oxidizer)
    - Stored on Mir base block
    - Resupplied by Progress cargo modules
    - Connectors on Mir's fore and aft docking ports allowed propellant and fluid transfer between Mir core, Soyuz TM, and Progress M
  - Attitude control (three-axis)
    - Provided by both thrusters (external) and momentum exchange gyrodynes (internal)

- 32 thrusters with 137 N thrust each
  - Two independent networks of 16 each
- Sofora thruster module
  - 4-jet thruster placed on 13 m (40 ft) boom on Kvant
  - Boom could be retracted
  - Thrusters offered increased roll control for the Mir complex
- Momentum exchange provided by gyrodynes (6 on Kvant, 6 on Kvant-2)
- Navigation
  - Attitude/navigation information was provided by IR Earth horizon scanning sensors, solar sensors, star sensors, Sun-presence sensors, automatic/manual star sextant, magnetometer, gyros, and linear accelerometers
  - Managed by onboard computers and ground-based commands
  - PINS platform-less inertial navigation system used for inertial reference
- Docking
  - Mir systems included two automated radar docking systems
  - Kurs ("course")—complete automated proximity navigation and docking, and undocking and departure operations
    - Installed on forward docking port and Soyuz TM (and later TMA) vehicles
    - S-band duplex link
    - Kurs held Mir stationary while docking vehicle executed all the maneuvers
  - Unmanned vessels docked under ground control
  - Igla ("needle")—automated approach with manual radar docking unit
    - Installed on rear docking port for Progress M cargo spacecraft
  - Kvant module housed both Kurs and Igla systems (Igla deactivated in later operations)
- Airlock and EVA
  - The Mir base block had no separate EVA airlock
    - EVA airlocks were installed on Kvant-2 and Krystall modules
  - Cosmonauts egressed from any of the forward docking/berthing ports
    - Each had a separate pressurized chamber to allow crew passage through docked modules

Mir Research Modules

The first Mir research module attached to the orbiting Mir core was the Kvant, which was berthed to the Mir aft port. This configuration permitted a total of five research modules instead of just four modules attached to the four forward radial ports. To accomplish this, Kvant was built with a pressurized pass-through, and included both Kurs and Igla docking systems for the Progress and other vehicles. Kvant was originally designed for the Salyut 7, but after significant delays, was adapted for installation on the Mir base block.

Research modules for the Mir complex were designed after the TKS base module later named the Functional Cargo Block, or FGB (Functionalui Germaticheskii Block). Initially, the research module concepts included the TKS transport craft with

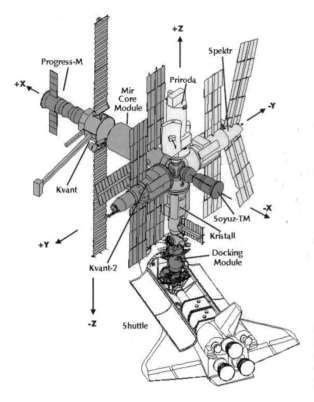

Figure 9.20. Sketch of the Mir module configuration and orientation reference with respect to the docked space shuttle orbiter. Source: Shuttle Mir.

a return capability for equipment, and possibly crew. But more effective research programs could be carried out in dedicated spacecraft that could be built and launched separately, and that had the advantage of being modified according to program needs. Therefore, five separate research modules were developed for attachment to the Mir. The TKS did survive, but as the research module structural prototype and not as a cargo module.

Kvant

Kvant ("quantum") was the first research module added to the Mir core. The small, eleven-ton module was berthed on the aft section of the Mir in April 1987 as an astrophysics research laboratory, and as a pressurized extension of the Mir's aft docking port. To accommodate the Progress and other vehicles, dual automated docking systems were installed on the aft port. The partially automated Igla radar docking system was used for the early Progress cargo vehicles. An improved and fully automated Kurs docking system was also installed to dock the Soyuz T and other vehicles.

Kvant, also known as Kvant-1, was the only Mir laboratory module not derived from the FGB transport spacecraft. The laboratory contained primarily astrophysics instruments, along with life-support equipment and additional gyrodynes for attitude control of the complex.

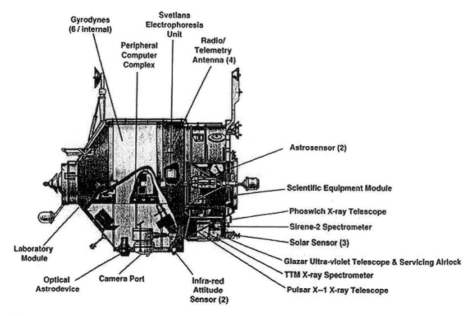

Figure 9.21. Sketch of the pressurized Kvant astrophysics module that included fore and aft docking ports and dual Kurs-Igla docking radar systems. Source: NASA Mir.

- Launch date: April 11, 1987
- Design life: 5 years
- Volume (total): 39.6 m³ (1,400 ft³)
- Mass: 1091 kg (24,000 lb)

Kvant 2

The second research module docked to the Mir complex was Kvant 2, launched on November 26, 1989 with the Proton M. Kvant 2 provided facilities for life support, biological research, Earth observation, and EVA. Additional life support, water recycling, oxygen provisions, and shower and washing facilities were furnished in Kvant 2's three pressurized compartments.

The 19.6-ton module design was based on the FGB cargo vehicle and, like Kvant, contained attitude control Gyrodynes. Unlike Kvant 1, the Kvant 2 module was equipped with dual rotating solar arrays, attitude control thrusters, and an EVA airlock.

Krystall

Krystall was the third research module added to the Mir complex. Launched in June 1990, the laboratory module housed scientific equipment in the larger of two compartments, along with an airlock and androgynous docking unit in the second. The module was primarily designed to investigate materials processing technologies in the

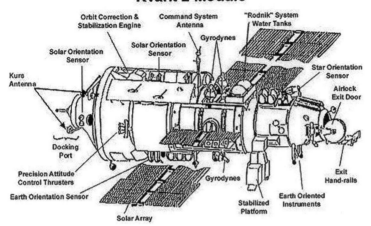

Figure 9.22. Sketch of the Kvant 2 showing layout of the major components including the EVA hatch and platform on far right. Source: NASA Mir.

microgravity environment, but it also supported biological, Earth observation, and astrophysical research.

An unusual pair of retractable solar arrays were attached to the module that were designed to regulate power production by extending/retracting the panels. The outboard docking node was equipped with an androgynous docking mechanism that was originally designed for the Buran shuttle, and later adapted for docking with the U.S. space shuttle orbiters.

Spektr

Spektr ("spectrum") was launched from the Baikonor complex on a Russian Proton M rocket on May 20, 1995, as the fourth research module attached to the Mir complex. Spektr was conceived as a military orbital laboratory before Russia entered into agreement with the United States in the International Space Station program. Russia's participation in the ISS program after the collapse of the Soviet Union benefited the stalled Spektr project because of the funding the U.S. provided for its completion and launch. As a result, a number of U.S. instruments and experiments were flown in the Spektr laboratory in the Shuttle-Mir program.

Scientific research on the Kvant module was focused on Earth observation, but also included material science, biotechnology, life sciences, and space technology studies. Spektr contained a small airlock facility for small experiments that could be deployed externally, and could be used to deploy small satellites.

Priroda

Priroda ("nature") was the fifth and last module added to the Mir complex, and therefore the least utilized. Like Spektr, Priroda was conceived as a military reconnaissance vehicle, but developed and launched with the aid of NASA funding. And like the other Mir complex research modules except for Kvant, the module was based on the Soviet FGB orbital transport structure. Priroda's ultimate objectives were to expand the Earth

Figure 9.23. Cutaway drawing of the Krystall research module with the drogue docking port for the Mir docking node (left) and an androgynous docking port for Buran and space shuttle docking (right). Source: NASA Mir.

remote sensing capability of the Mir complex. Priroda also carried hardware for U.S. and Russian material processing, meteorological and ionosphere research, as well as equipment for French and German experiments.

Docking Module

The last component added to the Mir complex was similar in form and function to the Apollo-Soyuz docking module that connected the Apollo CSM to the Soyuz capsule. Mir's docking module was created with the collaboration of U.S. and Russian engineers for connecting the Mir with the space shuttle. The module also provided the transition hardware to attach the International Space Station's Russian modules to the U.S. modules. A reconfigured version of the Mir docking module was also used to dock the space shuttle to the U.S. modules.

Both the U.S. and Russia would benefit from each other's participation in the International Space Station project because of the reduced cost and the increased research capacity compared to separate stations. But first, NASA and the Russian Space Agency had to prepare for the assembly of the new station by developing hardware for, and practicing rendezvous and docking maneuvers between, the Mir and the space shuttle. Mir proved useful since it was almost identical to the new Russian core module on the ISS. To accommodate the space shuttle, the Mir docking module replaced their Buran docking port already attached to the Krystall.

Russia's Mir docking module design began with their androgynous peripheral docking system called the APAS 89. After fabrication and testing, the unit was delivered and attached to the Krystall on the STS-74 *Atlantis* mission launched in November 1995. This is the same docking adapter used for the space shuttle, but not the same used for the U.S.-ISS module hardware. For that, a modified APAS 89 was

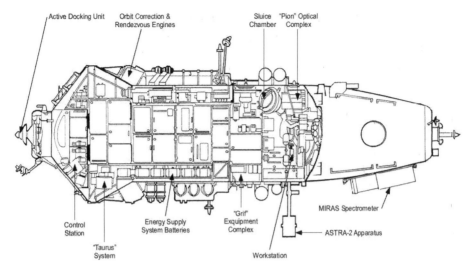

Figure 9.24. Sketch of the Spektr research module with major component layout. Source: NASA SP-4225.

used for the shuttle docking unit, with a U.S.-ISS module docking adapter called the common berthing adapter attached to the opposite end. This new adapter, called the pressurized mating adapter, or PMA, was used for shuttle docking with the U.S.-ISS node, although not with the Russian modules. A total of three PMAs on the U.S.-ISS provide for docking of the space shuttle orbiter with the ISS, and for connecting the U.S. modules at Node 1 Unity with Russia's modules at the Zaria FGB.

Mir Missions

The Soviet Union's Mir space station began as an improved Salyut orbiting laboratory to be used for both military and scientific research, but was considered transi-

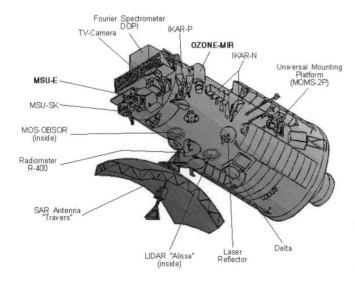

Figure 9.25. Sketch of the instrument layout on the Priroda Earth observation module. Source: NASA SP-4225.

tional. Mir was to be followed by a more ambitious project called Mir II that was halted as the USSR collapsed in 1991. Mir's modular configuration proved to be an ideal solution to building a large research laboratory in space. Research modules and equipment could be added, removed, or replaced in the modular configuration. Crews could also be ferried to and from the station independent of the module additions and separate from the cargo flights. This flexibility allowed planners to schedule Russian and foreign crews for research assignments according to favored scientific and political policies.

Mir even proved useful in transitioning from the Cold War standoff between the Soviets and the Americans to an agreement to build a combined Russian-American space station. And, as both countries began developing hardware and establishing timelines for the construction of the International Space Station, the Mir began receiving U.S. crews and equipment on the space shuttle. These Shuttle-Mir flights created the collaborative environment needed for the astronauts and cosmonauts, the engineers, and the two governmental entities to resolve the myriad of problems and details in the enormous International Space Station project.

The Mir mission record is extensive because of its fifteen-year on-orbit lifetime. For brevity, Mir mission statistics are listed below, along with a brief list of the American astronaut flights on the Mir (S. P. Korolev RSC Energia n.d.).

- Mir core launch: February 19, 1986 (Proton 8K82A)
- Crews: 28 long-duration mission crews (expeditions)
- Occupied: 4,594 days
- On orbit: 5,511 days
- Orbital mass (complex): 124,340 kg

NASA Astronauts Flown on Mir

- Norman Thagard, Mir-18 03/16/95 - 07/04/95
- Shannon Lucid, Mir-21, 22 03/24/96 - 09/19/96
- John Blaha, Mir-22 09/19/96 - 01/15/97
- Jerry Linenger, Mir-22, 23 01/15/97 - 05/17/97
- Michael Foale, Mir-23, 24 05/17/97 - 09/27/97
- David Wolf, Mir-24 09/27/97 - 01/24/98
- Andrew Thomas, Mir-24, 25 01/24/98 - 08/25/98

Mir Flight Records:

- Longest-duration single flight
  - Valery Polyakov—437 days 17 hr 58 min, 1995 (included a study of the effects of microgravity exposure on a cosmonaut during the equivalent time of a round trip to Mars)
- Accumulated Mir flight time
  - Sergey Avdeev—747 days 14 hr 12 min (3 flights)
  - Valery Polyakov—678 days 16 hr 33 min (2 flights)
- Female record holders

- Shannon Lucid (USA)—183 days 23 hr on MIR station (188 days 4 hr, including flight to and from station), 1996
- Foreign citizen flight records
  - Jean-Pierre Haignere (France)—188 days 20 hr 16 min
  - Shannon Lucid (USA)—188 days 4 hr
- EVAs from Mir
  - Total EVAs: 78 (including three egresses into the unpressurized Spektr module)
  - Total EVA duration: 359 hr 12 min
- Spacewalks performed
  - 29 Russian cosmonauts
  - 3 U.S. astronauts
  - 2 French astronauts
  - 1 ESA astronaut (Germany)
- Cosmonaut EVA records
  - Anatoly Soloviev—16 EVAs totaling 77 hr 46 min

Mir's End

In 1994, a cooperative agreement was made between Russia and the United States for building and operating the International Space Station. The agreement laid out specific mission objectives and a timetable that began with the Phase 1 Shuttle-Mir rendezvous and docking flights. Within the agreement was a comprehensive launch, support, and assembly schedule starting with the launch of the first ISS module, the Zarya FGB, on a Russian Proton rocket. America's Unity/Node 1 docking module would then be launched on a space shuttle assembly mission and docked with the Zarya. This was to be followed by the Russian launch of the crew habitation and work module called the Zvezda service module, the redesigned Mir II. These three modules would serve as a home for permanent space station crews, provided there was a docked space shuttle orbiter or a Soyuz TMA ferry to return them to Earth quickly in an emergency. The timetable also spelled out the termination of the Mir station complex to prevent Russia from diverting its resources from the ISS project.

But terminating the Mir space station operations meant that Russia would lose the bulk of its underfunded and shrinking space program. Understandably, the loss of a very successful space program made the Russian leadership reluctant to give up the last vestiges of a once-vigorous space program. Even so, Russia's refusal to deorbit Mir was viewed by many in NASA and in Congress as a breach in their cooperative agreement. Plagued by a series of serious system malfunctions in its later life, Mir was kept in orbit well beyond its useful existence.

Significant delays in the completion and launch of the Zvezda increased the friction already building between Russia, NASA, and the U.S. Congress. Aggravating the Zvezda launch delays was the diversion of Soyuz and Proton launch vehicles intended to support the ISS project to a commercial enterprise called the MirCorp, a privately owned company that provided funds for the continuation of the Mir program.

Progressive failures of the Mir systems and the end of outside funding led to the inevitable decision to deorbit the station in 2000. The planned deorbit was targeted for some time after the station's fifteenth year in orbit, which would be February 20, 2001. A Progress cargo ship was launched in October 2000 and docked to boost Mir's orbit to survive five more months before orbital decay ended its life. The last Progress M was then launched January 24, 2001, to control Mir during its deorbit and reentry for a splashdown in the South Pacific. Preparations were made for the early March deorbit in three stages, although the actual entry time was based on the atmospheric activity and the point at which the Mir reached critical atmospheric density. Several braking burns brought Mir's altitude to a gradual decay rate where a final series of ΔV burns would place the Mir complex in a reliable trajectory back to Earth. On March 22, 2001, Progress M1-5 slowed the station into its final entry ellipse. Breakup of the complex began at about 110 km, with disintegration complete near 90 km. The end came with an estimated 30-35 tons of dense debris surviving the reentry and splashing down in the ocean near the Fiji islands. Although the Mir was lost forever, the complicated and exacting deorbit operations were a tribute to Russia's space station program experience, and will undoubtedly be useful in deorbiting the ISS.

International Space Station

The origin of the International Space Station program reaches back through a long list of orbital station proposals that came from NASA centers and contractors, some of which predated Skylab. Several proposals suggested using the space shuttle as a laboratory by docking it with an orbiting power platform, although most proposals consisted of modular structures that supported commercial operations, satellite launches, or military applications. However, congressional influence directed the primary objectives of NASA's space station efforts toward space exploration and scientific research, as it had in the Apollo and Skylab programs.

While there was no lack of space station proposals in the 1960s and 1970s, there was a lack of funding. An important confluence of opportunity, funding, and a compelling NASA space station proposal appeared in 1984 when President Reagan announced his space exploration initiative. His new initiative attempted to emulate President Kennedy's epic Apollo project by proposing an ambitious space station as well as a bold manned exploration program that were to be in place "… within a decade." Even though the proposed lunar and Mars projects quickly faded, the Space Station Freedom project gained enough support that, within a few years, NASA personnel had completed preliminary design concepts and budget projections. In 1987, the NASA proposal for the dual-keel Space Station Freedom (SSF) went before Congress for funding.

SPACE STATION FREEDOM

NASA's Space Station Freedom proposal was unusual in that it was both practical and extraordinarily ambitious. Practical, because the dual-keel, single truss Space Station Freedom was to be constructed in modules, supported by the new space shuttle. The proposal was overly ambitious because the space station was to serve as an orbiting research laboratory and would have the capability of launching satellites into orbit. It would also provide a platform for assembling and launching cargo and manned missions to the Moon and Mars, and take only seven years to complete and at a cost of only $15 billion. Congressional reaction was swift but skeptical, and clearly favored a reduced-capacity space station launched with fewer space shuttle flights while emphasizing scientific research (Lindroos n.d.).

NASA's original proposal for a full-scale SSF consisted of five construction phases. Phase one would consist of constructing the main truss assembly and connecting the outboard solar arrays to the research and habitation modules. This represented the most basic and the most attractive station configuration because of its simplicity and low cost. These as well as several other attributes of the phase one SSF would survive the budget axe to become the precursor of today's International Space Station structure. SSF's four subsequent development phases added to the capability of the station for launch processing from LEO, from expanding both manned and robotic interplanetary exploration from the orbital platform to developing new launch and booster technologies, but all were too grandiose to gain support outside of NASA.

NASA's Space Station Freedom five-phase development configuration circa 1987 was planned as follows (NASA 1992):

1. Phase One—Assembly Complete: Single-truss baseline space station
2. Phase Two—Enhanced Operations Capability (EOC): Power, experiments, and crew augmentation of baseline

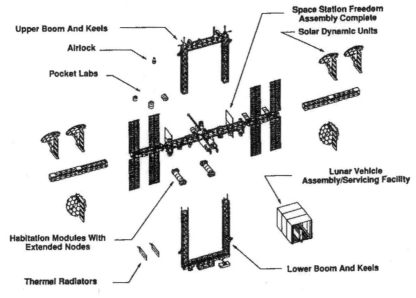

Figure 9.26. Structural and component layout of the fully developed Space Station Freedom. Phase One of the station included assembly of the central truss that connected the outboard solar arrays to the central research modules. The extended design (XOC) included the upper and lower keels, a lunar receiving and servicing facility, and solar dynamic power stations. Source: NASA TM-107981.

3. Phase Three—Lunar Vehicle Capability: Assembly, launch, and recovery of lunar vehicles capability added to EOC
4. Phase Four—Extended Operations Capability (XOC): Expanded crew, power, research capability with continuous lunar operation
5. Phase Five—Mars Vehicle Capability: Mars mission vehicle assembly, checkout, and launch added to EOC

As the station design progressed and the components headed toward fabrication, the SSF budget rose to more realistic levels, further casting the program in doubt. Continued discussion between NASA managers, scientists, and Congress reduced the capabilities of the station overall but provided increased support for research facilities. Power from the photovoltaic arrays was increased by adding additional solar array panels, while the solar dynamic systems were discarded to reduce development time and costs. Somewhat surprisingly, the 1987 proposal was approved by the National Research Council, allowing NASA managers to begin the contract bidding process for the first phase of the project.

SSF's Phase One was considered a far more realistic proposal for the final SSF, but with many changes that favored shorter time-to-flight development and a lower cost, which meant lower on-orbit mass and fewer space shuttle launches for assembly.

In 1989, a new space initiative was proposed by President George H. W. Bush that outlined many of the same goals as President Reagan's initiative in 1984. Those included completion of the Space Station Freedom, returning man to the Moon to stay, and manned flights to Mars. SSF's dual-keel extended operations capabilities were revived once again, and plans for a fourteen- to sixteen-crew station were reconsidered. Budget problems again redefined SSF to the simpler baseline (phase one) configuration. Finally, in 1989, hardware work began on the baseline Space Station Freedom. Several international partners that included the European Space Agency (ESA), Japan, Italy, Brazil, and Canada were actively contributing to the project with hardware and technical assistance. These participants, with the later addition of Russia, formed the core of the international partnership that today makes up the major contributing members of the International Space Station program.

As the first of the SSF hardware design reached the fabrication stage, progress on the program slowed because of a declining budget even though the first launch was only five years away. And as the major hardware designs began to mature, NASA projections for the total costs escalated rapidly, as did the actual on-orbit mass estimates. Increasing the propulsion performance for the space shuttle was considered in order to accommodate the larger payloads. But even with the projected improvements in the SSF project hardware and costs, the Station's future came into serious doubt. Support for the project from outside NASA disappeared because of the uncomfortably high cost, the reduced effectiveness of the Station in supporting research, and the less-than-appealing resource requirements of the project. One of NASA's own reports found that the 500 EVA man-hours quoted for the station assembly could actually be more than 3,000 hours (Lindroos n.d.).

Realizing that much of the support for Space Station Freedom had evaporated and with it the funding, NASA managers began a major redesign of the Station. Their goal was to regain congressional support for the SSF, to build public interest in the project by promoting the project with a simplified station design, and to find other partnerships that could help reduce program costs. In 1991, redesign efforts trimmed the baseline prototype to a smaller crew capacity, along with a reduced research capability, but with increased power. At this point, the SSF became a two-phase project defined by the crew operations—man-tended capability (MTC), and permanently manned capability (PMC). Crews could operate equipment and experiments during the MTC, but only when the space shuttle was docked. In the PMC phase, crews could remain onboard continuously since an emergency crew return vehicle would be docked and available at all times. Then in 1991, a new opportunity for an even larger station emerged from the fall of an old arch-rival.

COLLAPSE OF THE SOVIET UNION

Strict socialist economics mixed with authoritarian rule by a massive bureaucracy led the Soviet Union to its inevitable demise in 1991. The collapse forced the former Soviet countries into independent governance, including Russia, which was the largest state. This was a period of chaos at every level, with not only a collapse of the communist government, but a disintegration of the social and financial structures that supported the entire population that comprised the Soviet Union. Many felt the most serious issues they faced were the results of a ruined economy, and that they were now without any supporting socialistic structure. Of least concern were the organizations, industries, and agencies that did not provide direct aid to the people caught in the disintegrating economy, but were otherwise important in a well-ordered, prosperous nation. In this category were the space agencies and the space industries that ranked far below the level of the public concern, and near the bottom of the list for funding by the new Russian government.

A once-robust space industry was now almost completely unfunded, and for several years what was left was supported only with a minimum of resources. Chaos turned into desperation, and by 1992, NASA managers began to consider Russia's overtures offering their next-generation Mir hardware and the experience of the scientists and engineers as an inexpensive route to expanding the capabilities of Space Station Freedom. It was also an opportunity for the United States to employ at least some of the former Soviet Union's scientists and engineers who might be otherwise drawn into work on nuclear or chemical weapons outside of Russia.

Both NASA and the European Space Agency had shown an interest in the Russian plans for the replacement Mir called Mir-2. Adopting the new Russian space station as part of a larger program offered a lower cost to America's already-planned, long-duration space research. NASA managers' difficulties with the high costs of the SSF project would be solved, at least partly, by joining with Russia's unfunded space station project. Accepting Russia as a major partner in the SSF program was debated in Congress and in NASA through 1993, and by 1994, informal agreements between the two countries began to solidify into the broad partnership that now makes up the International Space Station program. On September 2, 1993, Russian Prime Minister

Figure 9.27. Artist's sketch of an early dual-keel Space Station Freedom design with extended capability. Source: NASA MIX.

Victor Chernomyrdin and US Vice President Al Gore signed an agreement to merge the SSF and the Mir-2 projects into the Space Station Alpha program.

INTERNATIONAL SPACE STATION ALPHA

The International Space Station was a combined effort of the United States, Russia, and more than a dozen international partners that fused Russia's second-generation Mir space station with the U.S. Space Station Freedom. Originally called Space Station Alpha (the first step, as in the first letter of the alphabet), the international orbital space facility was meant to reduce costs and multiply the capabilities of both designs. Before the agreement, Space Station Freedom was close to burial. After the agreement, what may have been a forgotten project grew into the largest space program in history, with more than a dozen participating countries. The joint agreement also saved Russia's Mir-2 from certain death, transforming both condemned projects into Space Station Alpha, and after several modifications in hardware and operations, into the International Space Station.

Although the combined International Space Station project encountered serious disagreements among partners along the way, the ISS has been a successful endeavor to date. Its unusual success came from the near-failures of two smaller station projects. That reality has not been lost to NASA planners, nor to the congressional leadership. To build a space facility of this magnitude required more than any single country could afford, suggesting that broad international partnerships are the most likely solution to future large-scale space programs. But beyond utility, those same international partnerships have also proven to be demanding, and for some program aspects, counterproductive.

ISS Implementation Plan

From the onset, the International Space Station was organized into three assembly and operation phases that were completely unrelated to the SSF phase designation. Phase one consisted of the rendezvous and docking missions, and the hardware development flights of the space shuttle orbiters and the Mir space station. This step was necessary for the U.S. and Russian space agencies to gain experience on the launch and flight systems, in combined space station operations, in rendezvous and docking operations, and in astronaut/cosmonaut exchange and training. Phase two commenced with the launch of the first ISS module, the Zarya FGB, on November 20, 1998. Phase three was coincident with permanent habitation of the station which followed the launch of the Zvezda service module into orbit, then the docking of the Soyuz TMA emergency crew return vehicle.

- Phase One—Shuttle/Mir Flights
  - Shuttle-Mir Program provides flight experience to reduce International Space Station assembly and operations risks
  - Shuttle-Mir Missions
    - STS-60 (Feb. 1994) and STS-63 (Jan. 1995) are the first U.S. space flights to carry Russian cosmonauts
    - Soyuz launches in the spring of 1995 carried the first U.S. astronaut to the Mir for a three-month stay
    - STS-71 (May 1995) first docked with the Mir, bringing up two Russian cosmonauts and returning the U.S. astronaut and two Russian cosmonauts
    - Ten shuttle flights to Mir (1995–1997) provided crew exchange, resupply, and transport of Mir system replacements, including a set of experimental solar arrays
- Phase Two—Human-Tended Capability
  - Space station human-tended capability offered frequent, extended research lab access until permanent operations
  - Use of the human-tended U.S. laboratory (Destiny) begins after fourth U.S. flight
  - Command and control under Mission Control in Houston with backup support from Russian Mission Control in Korolev City, outside Moscow
  - U.S. provides node, laboratory module, central truss modules, along with control moment gyros, and a docking interface for the shuttle
  - Russia provides propulsion, initial power, and docking interface for Russian vehicles
- Phase Three—Permanent Habitation
  - International Space Station begins permanent habitation and operations with Soyuz docked for emergency return
  - Crew size was limited to three except during temporary period while orbiter or second Soyuz transfer vehicles were docked
  - Completed ISS supports a six-person crew

Station at Completion

After completion of the assembly of the ISS, the crew research facilities include four laboratories and two partial habitation modules that support a full-time crew of six. The assembled structure is capable of docking three crew transfer vehicles and two cargo vehicles simultaneously, including two new transfer vehicles built by the ESA and by Japan. The ISS has a combined mass of nearly 455,000 kg (1,000,000 lb) in structure and equipment, with a life expectancy of approximately fifteen years. Combined power generation on the ISS from all modules is roughly 100 kW delivered at two different operating voltages that are shared by power conversion units.

The ISS structure is composed of three basic assemblies; the truss, the solar arrays, and the facility modules. The single-beam truss is divided into nine separate segments, each launched individually on the space shuttle. Located at both ends are the four photovoltaic array modules, two at each end. The four photovoltaic array modules have two opposing panels called channels that rotate in two dimensions to follow the Sun in all station attitudes. Located at the center truss section is the connecting point for the facility modules and the truss frame. The frame is attached to the U.S. laboratory Destiny, which is physically connected to the U.S., Russian, and partner facility modules through Node 1 and Node 2. Facility modules are interconnected with three U.S. six-port nodes and the Russian Zvezda service module.

An exploded view of the major components on the ISS and the international partner contributions is shown in figure 9.28, although there have been several significant modifications since the 1999 drawing was made. Three important changes to the diagram include the X-38 assured crew return vehicle, which has been replaced by the Orion crew exploration vehicle, the U.S. habitation module which has been integrated into the U.S. Laboratory Destiny and Node 3 Tranquility, and the centrifuge accommodation module which has been canceled completely.

ASSEMBLY/CONSTRUCTION

Functional requirements for the first module launched in a multi-module space station require almost all of the supporting systems needed for an orbital space station. Included in the self-supporting module are electrical power, propulsion, attitude control, communications, data management, docking, assembly structures, thermal control, and at least a primitive life-support system. The same basic system requirements applied to the first ISS element launched in 1998 named Zarya ("sunrise"), also known as the Functional Cargo Block, or FGB. Zaria originated from the same TKS module designed for the Salyut and Mir crew and cargo missions, which later served as the prototype for four of the five Mir research modules. Two weeks after the Zaria launch on November 20,1998, the U.S. connecting module Node 1, called Unity, was launched on the space shuttle *Atlantis* and docked with Zaria. After an uncomfortable delay, the second Russian module named Zvezda ("star") used for the crew quarters was launched on July 12, 2000, and berthed at the opposite end of Zaria. This was an important addition since the Zvezda service module provided the

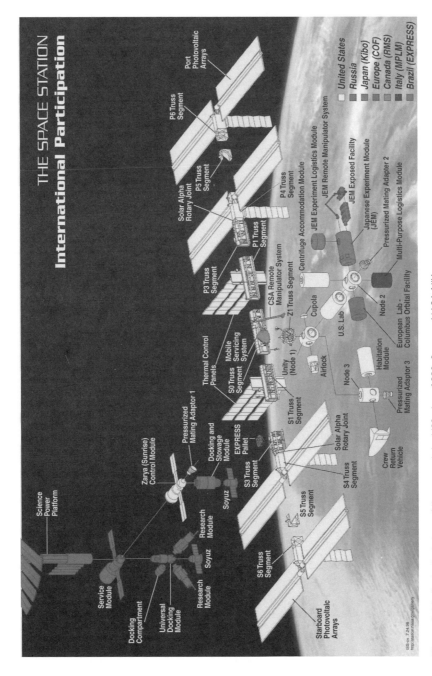

Figure 9.28. Exploded view of the completed ISS circa 1998. Source: NASA NIX.

full-time habitation facilities for a crew plus docking for the Soyuz TMA and Progress M vehicles. Zvezda's launch into orbit was an important milestone since it was the first all-Russian major hardware contribution to the shaky ISS agreement between the United States and Russia—two arch enemies only a decade earlier.

With the crew living quarters now available on the orbiting Zvezda, NASA launched the laboratory module Destiny aboard space shuttle *Atlantis* on February 7, 2001. The Destiny laboratory was docked to the ISS on Node 1, which gave the astronauts and cosmonauts the most advanced equipment and largest orbital research facility since Skylab. In preparation for powering Destiny and Unity, the Z1 truss assembly was launched for docking on Node 1 on October 11, 2000. This was followed by the launch of a combined photovoltaic array and truss segment on November 30, 2000, that was attached to the Z-1 truss. The P6 Photovoltaic Array (PVA) would serve to power Destiny and Unity on its unusual truss mount until it was moved from Z-1 to the outboard port truss assembly on STS-120 in 2007. Z-1's was unique since it was the only truss segment not attached to the nine-element main truss assembly.

Propulsion used for major attitude adjustments and orbit control of the International Space Station was initially provided by Russian modules, beginning with the Zarya, followed by Zvezda after its attachment to the station. On-orbit propulsion is now furnished by either a docked Russian Progress M module or, until 2010, by a docked space shuttle using long, low-thrust burns.

Assembly of the ISS is scheduled to end with the final construction flights in 2010–2011. Following the planned retirement of the space shuttle in 2010–2011, the crew flights and limited cargo missions will use Russian Soyuz and Progress vehicles, with additional cargo delivery by the ESA's ATV and Japan's HTV vehicles. NASA's new Orion crew flight vehicle is scheduled to carry astronauts to the ISS for the first time in 2015.

Modules and Truss Assembly

Habitation, research, and logistics modules furnished by Russian, Italian, Japanese, and ESA contributors make up roughly half of the orbital facilities on the ISS, while the remaining half comes from the United States. For these modules to be compatible, they must have one of five types of berthing/docking ports for attachment to the pressurized station modules. The Russian docking adapters have variations on three different mechanisms. The androgynous peripheral attach system (APAS) is used to connect the Zaria and Unity modules, while their earlier "probe and drogue" units consist of two different configurations and sizes; one for transfer vehicles (Soyuz, Progress), and one for their station modules. The United States has a similar variety used to connect the Unity to Zaria with the pressurized mating adapter (PMA), as well as to dock the orbiter to the PMAs. The other U.S. type, called the common berthing adapter, is used to dock U.S.-compatible modules together. A separate docking mechanism is being designed for the Orion CEV that will accommodate automated docking with the PMA androgynous peripheral adapters designed for docking the orbiters.

The first major ISS assembly operation began with the attachment of the Unity Node 1 to the Zarya module. The attachment employed the versatile PMA adapter for the Zarya-Unity connection and for the docking adapter between the space shuttle and the docking ports on the various nodes. Each of the three PMAs includes thermal conditioning, primarily electric heaters and thermal insulation, and data and power pass-through lines. The end attached to the U.S. modules is a common berthing adapter, while the end attached to Russian hardware, as well as the shuttle docking adapter, is the androgynous docking assembly developed for the Shuttle-Mir flights used in phase one of the ISS program.

For the U.S., ESA, Italian, and Japanese modules, a common device was adopted for the mechanical as well as electrical and thermal connections known as the common berthing adapter. The positive lock and seal features are different from earlier docking and berthing mechanisms found on Apollo and Skylab, and they provide greater strength and rigidity in the mechanical attachment.

Truss-to-Truss Attachments

The U.S. truss segments are locked together using mechanical latches with several alignment devices on both ends of segments to be connected. This reduces EVA requirements, while providing a positive locking mechanism that does not use the traditional nuts and bolts. However, the mechanical truss attachments are accompanied with power, signal, and thermal interconnects that do require EVA activities for complete assembly.

One of the major attachments between the U.S. orbital segments (USOS) and the facility modules is the truss mooring. This module-truss attachment is anchored at the US laboratory Destiny with the S0 center truss segment. A cradle mechanism and brace assembly connects S0 directly to Destiny's external structure. Hence, the entire mass of the assembled truss structure and the entire mass of all habitation, storage, and research modules at completion, and any docked transfer vehicles, are linked at the

Figure 9.29. Pressurized Mating Adapter 2 shown in the Space Station Processing Facility at KSC before attachment to Node 1 being prepared for the STS-88 launch. Source: NASA STS-88.

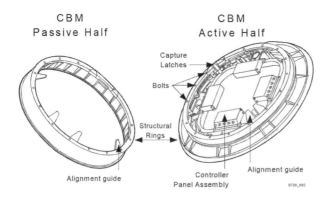

Figure 9.30. Diagram of the U.S. Common Berthing Mechanism used on Unity, Destiny, Quest, and one end of the three PMAs used on the ISS. Source: NASA ISSF.

Destiny-S0 union. Careful analysis of the entire structural assembly was required in the design phase to avoid adverse torques and damaging vibrational modes since any forces throughout the station are translated through that one mechanical connection.

FACILITY MODULES

Zarya FGB

Zarya FGB is a Russian design, constructed by the Khrunichev State Research and Production Space Center in Moscow. But because of Russia's supine economy years after the collapse of the Soviet Union, funding for the module came from the United States in an agreement with NASA management under congressional oversight. Flight designation for this module was 1A/R which meant that both countries were responsible for its development (the A designation is American, R is Russian). Launch of Zaria on November 20, 1998 was on a Russian Proton booster from the Baikonor launch facility in Kazakhstan.

The 19,300 kg (42,600 lb), 12.6 m (39.0 ft) pressurized module is a descendant of the Soviet TKS transport module that originally included a VA return capsule attached to the FGB cargo module. The TKS flew several times in the Salyut mission series, though never with a crew, and became the prototype for Mir's laboratory modules.

- Launch: November 20, 1998 (Proton 8K82K booster)
- Flight designation: 1A/R
- Primary structure: Aluminum alloy
- Length: 12.6 m
- Diameter: 4.1 m
- Mass: 19,323 kg

Node 1: Unity

The first U.S. component attached to the International Space Station was the connecting Node 1 named Unity. The six-port docking unit provides storage, communications,

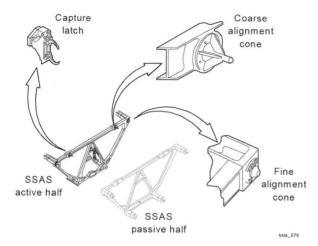

Figure 9.31. Diagram of the typical alignment, latch, and capture bar devices used to connect USOS truss segments. Sketch of the typical segment-to-segment attach system (SSAS) using motorized bolts, coarse and fine alignment features, and a capture latch on the active truss segment. These mate with the adjoining segment using coarse and fine alignment cups, nuts, and a capture bar. Source: NASA ISSF.

pressurized atmosphere, and power connections to the Zarya and the Destiny lab, as well as the Quest Joint Airlock. PMA docking devices at each end of Unity allow physical connection and pressurized pass-through for the crew and cargo. Unity's six docking/berthing ports provide physical attachment for the Z1 truss frame, the U.S. lab Destiny, the joint airlock Quest, the Cupola, Node 3 Tranquility, and the Zarya module.

The two other U.S. nodes attached to the ISS, Node 2 Harmony and Node 3 Tranquility, were built by Italy's Thales Alenia Space and are slightly longer than Node 1.

- Launch: December 4, 1998 (STS-88, *Atlantis*)
- Flight designation: 2A
- Primary structure: Aluminum alloy
- Length: 5.49 m
- Diameter: 4.57 m
- Mass: 11,612 kg

Zvezda Service Module

The Russian Zvezda service module provides more operational functions for the ISS than Zaria, in addition to crew life support and living quarters. From a distance, the Zvezda appears almost identical to the Mir core module launched in 1986. However, many of the systems and components have been redesigned, added, or improved. Zvezda, like its predecessor the Mir core block, can support a full-time crew of three.

Zvezda contains sleeping quarters for the crew, exercise equipment, a toilet and hygiene station, and a galley. The module has a total of fourteen windows consisting of three nine-inch-diameter windows in the forward compartment, a sixteen-inch window in the working compartment, and one in each of the two small crew compartments. It also contains the Elektron system that electrolyzes condensed humidity and wastewater to provide oxygen for breathing, and hydrogen which is dumped overboard.

Oxygen generated in the Elektron water electrolysis system is backed up by solid-fuel oxygen generator canisters that react oxygen compounds with sodium chlorate to

Figure 9.32. Photograph of the Zarya in orbit from the cabin of Atlantis on STS-88. Source: NASA STS-88.

generate pure oxygen at high temperature. Although compact and convenient, these "oxygen candles" proved to be a fire hazard on the Mir complex. CO_2 removal on Zvezda comes from the Vozdukh system, a double amine bed subsystem similar to the Skylab double-bed molecular sieve.

Zvezda became operational and the primary station control module for the ISS on September 11, 2000, when the STS-106 EVA crew attached cables for the handoff of operations from Zaria's systems to Zvezda's.

- Launch: July 12, 2000 (Proton 8K82K booster)
- Flight designation: 1R
- Primary structure: Aluminum alloy
- Length: 13.1 m
- Diameter: 4.15 m
- Mass: 19,050 kg

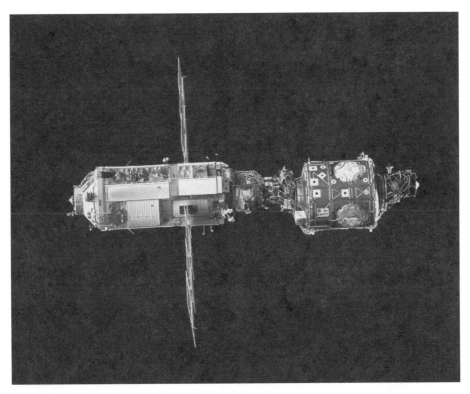

Figure 9.33. STS-88 photograph of Zarya (left) and Unity modules on orbit, connected by Pressurized Mating Adapter 1. PMA 2 on the right of Unity is used for Orbiter docking. Source: NASA STS-88.

U.S. Laboratory Destiny

The second U.S. module launched and docked to the ISS was the primary laboratory facility named Destiny. The research module is berthed permanently on Node 1 and used to house experiment racks and life support equipment for the station crew. Destiny carries a total of twenty-three ISS Standard Payload Racks (ISPRs) to hold equipment, experiments, and life-support components, including a single crew quarters rack—a telephone booth-sized bedroom. Six rack positions are built into each of the port and starboard sides, and in the overhead section, while the floor (deck) region contains five rack slots. A total of thirty-seven ISPR slots are available on the ISS throughout the non-Russian modules.

Destiny's structure is a three-section cylindrical hull that incorporates a waffle-pattern shell between the structural braces and rings. The laboratory includes a 510 mm (20 in) optical-quality glass observation window with a micrometeoroid shutter for protection during its life in orbit. The cylindrical lab is closed with two endcones, each with a hatchway and access window. A debris shield made of Kevlar covers the hull, with a thin aluminum alloy shield placed outside the Kevlar shield for additional protection.

- Launch: February 7, 2001 (STS-98, Atlantis)
- Flight designation: 5A

Figure 9.34. STS-108 photo of the Destiny research laboratory (center) attached to Unity/Node 1 (right) and the Quest/Joint Airlock (top). Pressurized Mating Adapter 2 shown on the left was used for docking the space shuttle Orbiters before Node 2 was berthed. Source: NASA STS-108.

- Primary structure: Aluminum alloy
- Length: 8.53 m
- Diameter: 4.27 m
- Mass: 14,500 kg (32,000 lb)

U.S. Joint Airlock Quest

The Quest joint airlock is used for astronauts/cosmonauts entering or exiting the ISS for EVAs, but also serves as a hyperbaric chamber for decompression emergencies. The structure consists of two chambers, one for equipment such as space suits and EVA gear, the other for the crew undergoing decompression, recompression, and chamber isolation from the station during EVA. The airlock can accommodate astronauts or cosmonauts, but initially only in the U.S. extravehicular mobility unit (EMU) space suit. The Russian Orlon suit is now supported for EVA through the Zvezda service module, the Pirs docking module, and the Quest airlock.

Quest also holds four pressurized gas tanks for U.S. module atmosphere pressurization. Two tanks attached to Quest contain pressurized oxygen and two contain pressurized nitrogen. Pure oxygen prebreathing in preparation for EVA has been replaced with a shorter "campout" in the Quest airlock at reduced pressure. Depressurization time for the airlock chamber is thirty minutes.

- Launch: July 12, 2001 STS-104 (*Atlantis*)
- Flight designation: 7A

Figure 9.35. Quest Joint Airlock being prepared for attachment to Node 1 on the International Space Station during the STS-104 assembly mission. Source: NASA STS-104.

- Length: 5.5 m (18 ft)
- Diameter (max): 4.0 m (13 ft)
- Weight: 6,064 kg (13,368 lb)
- Volume: 34 m³ (1,200 ft³)

Node 2: Harmony

Like Node 1, the second connecting node attached to the ISS named Harmony has six ports for docking and berthing additional modules. Harmony provides attachment ports for the Japanese laboratory Kibo, the European laboratory Columbus, multipurpose logistics modules, the Japanese HTV transfer module, and a PMA for docking the space shuttle orbiter and the Orion. Harmony is constructed in similar fashion to the other USOS modules with an aluminum alloy hull that employs a waffle-pattern to increase strength, and an attached Kevlar debris shield and a thin outer aluminum micrometeoroid shield.

Node 2 was built by Italy's Alenia Spazio, now Thales Alenia Space, in exchange for launching the ESA's Columbus module. In addition to expanding the docking ports on ISS by six, Harmony was designed to help expand the ISS crew capacity to six with the allocation of four ISPR slots for crew quarters racks. After launch on STS-120, Harmony was attached temporarily to Unity's port-side dock, then later relocated permanently to Destiny's forward dock after the departure of *Atlantis*.

- Launch: October 23, 2007 on STS-120 (*Atlantis*)
- Flight designation: 10A
- Length: 7.2 m (23.6 ft)
- Diameter (max): 4.4 m (14.5 ft)
- Weight: 14,288 kg (31,500 lb)
- Volume (pressurized): 5.5 m³ (2,666 ft³)

Node 3: Tranquility

Node 3 is the third and last U.S. connecting node attached to the ISS, and also the largest. Besides its primary purpose of berthing/docking U.S. and partner modules to the ISS, Tranquility furnishes additional life-support including atmosphere revitalization for removing contaminants from the atmosphere and monitoring/controlling the atmosphere constituents in the ISS. A carbon dioxide reduction system employs a Sabatier reactor system which combines carbon dioxide with hydrogen in the high-temperature reactor to generate oxygen for breathing, and methane gas that is dumped overboard. A new wastewater processing unit recycles the valuable resource, in combination with an electrolysis-based oxygen generator system that generates hydrogen and oxygen; the oxygen used for breathing, and the hydrogen routed to the Sabatier CO_2 reduction system. Tranquility also includes two avionics racks, six life-support system racks, and a waste and hygiene compartment (a commode and waste storage unit) installed as a rack unit to supplement the Zvezda toilet and waste facility.

Node 3 is berthed on the Node 1 port-side common berthing adapter, with the six-window observation cupola attached to its nadir (Earth-facing) port before launch. Its remaining ports are available for MPLM cargo modules and docking space shuttle orbiter and Orion crew vehicles. Like Node 2, Tranquility was built by Thales Alenia Space with an aluminum alloy hull, a debris shield, and a thin outer aluminum micrometeoroid shield.

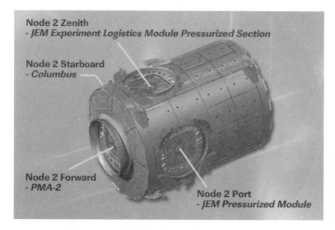

Figure 9.36. Artist's rendering of the Node 2 module showing the six-port configuration. Source: NASA NIX.

- Launch: February 8, 2010, on STS-130 (*Endeavour*)
- Flight designation: 19A
- Length: 6.7 m (22.0 ft)
- Diameter (max): 4.4 m (14.5 ft)
- Weight: 15,000 kg (33,000 lb)

ESA-Columbus

The European Space Agency's Columbus orbital laboratory was patterned after the multipurpose logistics module cargo carrier design and built by the same manufacturer, Thales Alenia Space of Italy. Columbus's waffle-patterned hull is covered with a Kevlar debris shield surrounded by an aluminum micrometeoroid standoff shield. The module was furnished with a common berthing mechanism for attachment to the starboard port of Node 2 Harmony on the STS-122 mission.

Columbus houses ten International Standard Payload Racks that are the same rack units used on the U.S. and Japanese modules. The ISPR units contain experimental and life support equipment and instruments delivered by the Italian MPLM cargo modules that are carried in the space shuttle's payload bay. Eight of the ISPRs are placed in the port and starboard sides with two in the ceiling. The Columbus module also has four attachment points for mounting external payloads for space exposure experiments.

Figure 9.37. Node 3 Tranquility shown in the clean room of the Marshall Space Flight Center before being outfitted for shipping to the Kennedy Space Center for launch preparations. Note the waffle pattern of the outer hull structure. Source: NASA MIX.

- Launch: February 7, 2008 on STS-122 (*Discovery*)
- Flight designation: 1E
- Length: 6.87 m (22.5 ft)
- Diameter: 4.49 m (15 ft)
- Launch mass: 10,300 kg (22,708 lb)

Japanese Experiment Module (JEM) – Kibo

Japan's orbital research module, named Kibo ("hope"), is a multi-module complex developed by the Japan Aerospace Exploration Agency (JAXA). The largest component of the Japanese experiment module is the pressurized module (PM) attached to the port side of Node 2 Harmony. The Kibo PM is also the largest module attached to the ISS—nearly 3 m longer than the Destiny laboratory and about the same diameter. Its aluminum alloy construction is similar to the Destiny, the MPLM, and the Columbus modules but varies in many details. Kibo's PM has a capacity of ten ISPR standard experiment racks, although only four were launched onboard.

The first Kibo-JEM module attached to the ISS was the pressurized section of the experiment logistics module (ELM-PS), launched on STS-123. A pressurized section and unpressurized section permit storage and transfer of experiments and logistics from the ELM-PS to the two exposed facility modules. ELM-PS was first docked to Node 2, then undocked and placed on top of the PM when it was brought up on STS 124 and attached to the same Node 2 port first occupied by the ELM-PS.

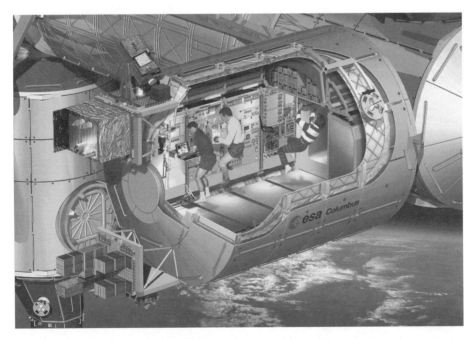

Figure 9.38. Cutaway sketch of the ESA Columbus research module and externally mounted research payloads. Source: NASA NIX.

The third major Kibo component is the Exposed Facility (EF), launched with the attached experiment logistics module–exposed section (ELM-ES) on STS-127 in 2009. The Kibo EF, also known as the "Terrace," is located on the outboard end of the PM module and can be accessed through the PM airlock and hatch assembly. The attached ELM-ES is used to transfer experiments and equipment to/from the EF, then back to Earth in the Orbiter's payload bay.

A key component of Kibo is the JEM remote manipulator system (JEMRMS) which is a grapple-sensor arm similar to the space shuttle remote manipulator system and the ISS remote manipulator system called the space station remote manipulator system, or SSRMS. The JEMRMS was launched attached to the PM on SST-124, and is used to service the exposure facility, as well as for moving equipment to and from the ELM.

- Experiment Logistics Module–Pressurized Section: ELM-PS
  - Launch: March 11, 2008 (STS-123, *Endeavour*)
  - Flight designation: 1J/A
  - Length: 4.2 m (13.8 ft)
  - Diameter: 4.4 m (14.4 ft)
  - Mass: 8,386 kg (18,490 lb)
- Pressurized Module–PM
  - Launch: May 31, 2008 (STS-124, *Atlantis*)
  - Flight designation: 1J
  - Length: 11.2 m (36.7 ft)
  - Diameter: 4.4 m (14.4 ft)
  - Mass: 14,800 kg (33,000 lb)
- Exposed Facility–EF
  - Launch: June 13, 2009 (STS-127, *Endeavour*)
  - Flight designation: 2J/A
  - Length: 5.2 m (17.0 ft)

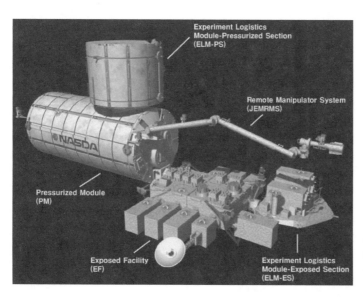

Figure 9.39. Drawing of the Japanese Experiment Module Kibo and the layout of the major module components at completion. Source: NASA NIX.

- Height : 3.8 m (12.5 ft)
- Width: 5.0 m (16.4 ft)
- Mass: 4,100 kg (9,039 lb)

ISS STRUCTURAL ASSEMBLIES

The largest components on the International Space Station are the structural truss and the solar arrays that together span more than a football field in area. Largest of these is the primary truss structure that connects the huge photovoltaic arrays to the habitation and research module complex. ISS's modular truss assembly consists of nine units, launched individually, and centered on the research/habitation module complex. The center section of the truss structure, identified as S0 (starboard-center), is attached to the module complex at the U.S. Destiny laboratory. Outboard of the S0 center truss are four truss modules mounted on the starboard (right) side, and four on the port (left) side. A tenth truss element called Z1 (Z for zenith) was designed as a mount point and is not connected directly to the rest of the truss assembly.

Truss Modules

The station's primary structure, called the integrated truss structure, is comprised of nine modular segments with an assembled length of 108.5 m if the outboard solar panels are included. With a total station mass of 450,000 kg (992,000 lb) and more than 100 m in length, the ISS truss structure calls for a rugged, light-weight, rigid, segmented structure, capable of interconnecting power, thermal, communications, and maintenance functions, while supporting the facility modules. Structural materials also required coating for protection from solar radiation, from micrometeoroid and dust abrasion, and from oxidization from the atomic oxygen encountered in low Earth orbit.

Z1

The first ISS truss module launched was the Z1 unit on October 11, 2000 as part of the STS-92 payload. Z1 that is berthed on the zenith port of Node 1 Unity is the only truss element not physically attached to another truss unit. Z1 served temporarily as a pedestal for the P6, the first photovoltaic array attached to the ISS, which was later moved to the port-side outboard position. The Z1 segment houses a wide variety of systems that includes two plasma contactor units (PCUs) to neutralize the charge buildup encountered in low Earth orbit, two DC-to-DC converter units (DDCUs), four control moment gyros (CMGs), part of the S-band communications system, one of the onboard Ku-band communications systems, power distribution subsystems, thermal control system hardware, and extravehicular robotics hardware.

- Launch: October 11, 2000 on STS-92 (*Discovery*)
- Length: 4.9 m (16.0 ft)
- Width: 4.2 m (13.8 ft)
- Weight: 8,755 kg (19,260 lb)

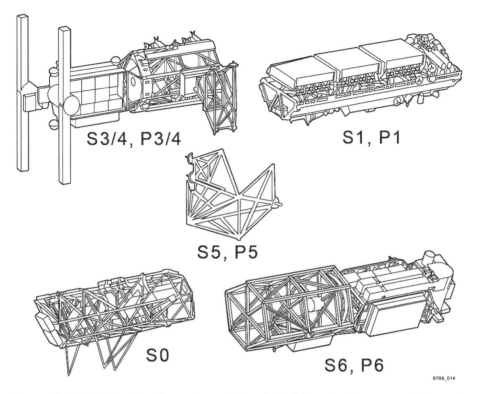

Figure 9.40. Sketch of the four pairs and two single integrated truss assemblies used on the ISS. Source: NASA ISSF.

S0

The centerpiece of the ISS integrated truss is the S0 segment, which could also be called the P0 since either would designate a center segment. S0 has several important roles in the truss structure, the most important being the physical attachment to the module complex. The S0 unit is attached to the Destiny laboratory module which provides a rigid physical connecting point between the entire truss and PVA assemblies, and the entire ISS module complex.

Because of the physical connection between the truss, arrays, and modules, S0 is a connecting point for the PVA power lines into the module complex, and the connecting point for the active thermal cooling loop plumbing that runs to the S1 and P1 radiators. Signal, data, video, and control lines are also routed through the S0 truss from the Destiny laboratory. Mounted on the exterior of the S0 module are mobile transporter rails, the station's four GPS antennas, and two rate gyros.

- Launch: April 8, 2002 on STS-110 (*Atlantis*)
- Length: 13.4 m (44 ft)
- Width: 4.6 m (15 ft)
- Weight: 13,971 kg (30,800 lb)

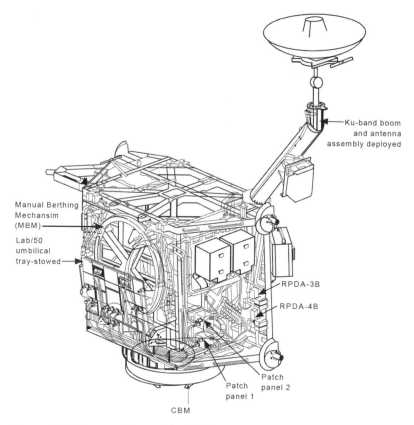

Figure 9.41. Exposed view of the ISS Z1 truss assembly and major components. Source: NASA ISSF.

P1, S1

The P1 and S1 truss elements outboard of the S0 center assembly are nearly identical in structure and function—almost mirror images of each other. S1 and P1 truss modules (also called the starboard and port side thermal radiator trusses) contain the same power, signal and thermal lines as the S0 unit, but each includes a triple set of radiators for the USOS active thermal control system. The P1 truss has a second crew and equipment translation Aid (CETA) cart that can be manually maneuvered along the mobile transporter (MT) rail line.

- Launch:
  - S1—October 7, 2002 (STS-112, *Atlantis*)
  - P1—November 22, 2002 (STS-113, *Endeavour*)
- Length: 13.7 m (45 ft)
- Width: 4.6 m (15 ft)
- Height: 1.8 m (6 ft)
- Weight: 14,124 kg (31,137 lb)

Figure 9.42. Photograph of the S0 truss assembly being extracted from the STS-110 cargo bay for attachment to the Destiny laboratory. Source: NASA STS-110.

P2/S2

ISS's truss structure is based on the original design of the Space Station Freedom, including the photovoltaic arrays, truss modules, and rotary joints. SSF was to have propulsion modules attached to the P2 and S2 truss units for symmetrical thrust re-boost. But with Russia joining the program in 1994, separate propulsion hardware was unnecessary, and the P2 and S2 truss sections were canceled.

P3/P4, S3/S4 PVAs

ISS's four photovoltaic array modules consist of two outboard P6 and S6 units, and the two inboard P3/P4 and S3/S4 units. The P3/P4 and S3/S4 truss modules are combined PVAs (P4, S4) integrated into the P3 and S3 solar alpha rotary joint (SARJ) and truss structure sections. Also included in the hexagonal-shaped truss segment is the unpressurized cargo carrier attach system (UCCAS) for mounting external payloads. The P3 and S3 segments are also the rail ends for the station's mobile transporter.

P4 and S4 are photovoltaic array units called photovoltaic modules (PVMs) that each contain two Solar Array Wings (SAWs). Each solar alpha wing includes two Photovoltaic Blankets (right and left) and an independent power channel. The power channel has three rechargeable secondary battery units, a rotating beta gimbal assembly (BGA) for blanket-sun alignment, and associated power regulation, discharge, recharge, and distribution electronics.

Figure 9.43. P1 truss assembly installation on the ISS during the STS-113 mission. Note the CETA transport cart on the left and the MT rails running the length of the module. Source: NASA STS-113.

Each of the photovoltaic module units contains a separate photovoltaic thermal control system (PVTCS) to remove heat from the battery and regulation/discharge/recharge electronics. The PVTCS incorporates an independent liquid ammonia cooling loop, a radiator, and a pump and flow control assembly. Fully extended, each photovoltaic module has a wingspan of 78 m (240 ft).

- Launch:
  - P3/4—September 9, 2006 (STS-115, *Atlantis*)
  - S3/4—June 8, 2007 (STS-117, *Discovery*)
- Length: 13.8 m (45.3 ft)
- Width: 4.9 m (16.0 ft)
- Height: 4.8 m (15.6 ft)
- Weight: 15,857 kg (34,885 lb)

P5, S5

Truss spacers are placed between the P3/4 module and the P6 truss modules, and between the S3/4 and the S6 truss modules to provide clearance between the inboard and outboard PVMs (P3/4 and S3/4 are inboard, P6 and S6 are outboard). Separation clearance would be insufficient between the P/3/4 and P6, and the S3/4 and S6 module structures. In addition to the basic structure, P5 and S5 carry servicing and transfer hardware for the ammonia cooling loops, and platforms for experiment placement and for storing Orbital Replacement Units (ORUs).

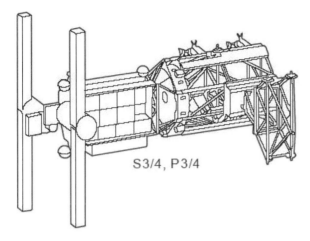

Figure 9.44. Line drawing of the nearly identical S3/4 and P3/4 truss assemblies that include the S4/P4 Photovoltaic Modules (left), each containing two Solar Alpha Wings housed in the canisters, the Solar Alpha Rotary Joint in the center, and the S3/P3 truss structures on the right. Source: NASA ISSF.

- Launch:
  - P5—December 9, 2006 (STS-116, *Discovery*)
  - S5—August 8, 2007 (STS-118, *Endeavour*)
- Length: 3.4 m (11.0 ft)
- Width: 4.5 m (14.8 ft)
- Height: 4.2 m (13.8 ft)
- Weight: 1,864 kg (4,110 lb)

Figure 9.45. P5 truss spacer shown being transferred from the payload bay of Discovery on the STS-115 mission from the space shuttle remote manipulator system (RMS) arm (left) to the ISS RMS arm. An identical truss spacer was later attached to the port-side P4 module on STS-118. Source: NASA 116.

P6, S6 PVA Truss Assemblies

The outboard truss segments on the ISS are the port-side P6 and the starboard S6 photovoltaic modules. To provide power early in the assembly sequence for the Unity and Destiny, an outboard PVM was selected since it was simpler and would not need the SARJ rotational joint. Since a balanced truss assembly would not be finished for several years after the laboratory was attached, the interim Z1 structure was mounted on Node 1 Unity, followed by the launch of the P6 PVM and its attachment on Z1 during the STS-97 mission in 2000. P6 provided early power for Destiny and the rest of the USOS segments until the truss was nearly complete in 2007. Following the retraction of the P6 radiator panel on the STS-118 mission, the STS-120 crew relocated the P6 unit from the Z1 mount to the P5 segment. The STS-120 mission also added the S5 truss unit in preparation for the final S6 PVM addition on STS-119.

Power output from the ISS PVMs is 66 kW each, although total available power for the four PVMs is 84–120 kW.

- Launch:
  - P6—November 30, 2000 (STS-97, *Endeavour*)
  - S6—March 15, 2009 (STS-119, *Endeavour*)
- Flight designation: 4A
- Length: 13.8 m (45.4 ft)
- Width: 5.0 m (16.3 ft)
- Weight: 14,088 kg (31,060 lb)

Mobile Servicing System

A rail and cart transport system is attached to the ISS truss assembly to facilitate maintenance and assembly of the station. The transport rails are built into the truss

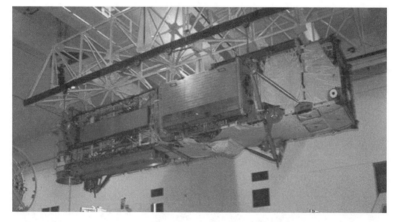

Figure 9.46. S3/4 truss assembly that includes both the S3 truss member and the S4 PVA module, as well as the Solar Alpha Rotary Joint that permits the PVA array complete 360° rotation. This same configuration is used for the port-side P3/4 truss assembly (NASA Station Assembly). Source: NASA Station Assembly.

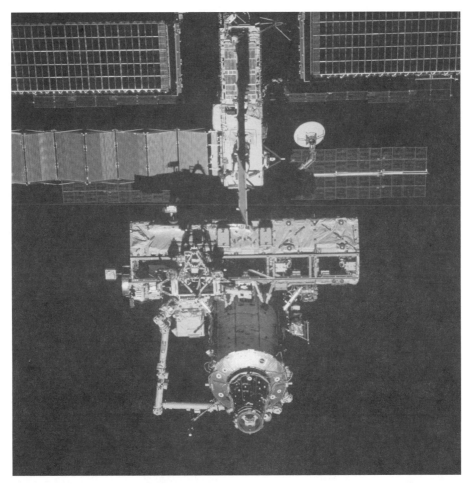

Figure 9.47. Mobile Servicing System unit shown with the Mobile Base System attached to the S0 truss rails and the Canadarm2 attached to the base (Destiny is in foreground). Source: NASA JSC.

assemblies, from the S0 center outward to the ends of the P3 and S3 modules. The combined system, called the Mobile Servicing System (MSS), consists of a powered transport base, or mobile base system that moves astronauts and cargo along the truss structure. Attached to the mobile base system is the space station remote manipulator system, also known as the Canadarm2, which is similar in function to the space shuttle's remote manipulator system, but with far greater dexterity and strength. The robotic arm can not only traverse the truss on the mobile base, but it can also move end-over-end to reach other areas on the station that have power data grapple fixtures (PDGFs). The PDGF furnishes power, data, video, and physical attachment for the MSS when it is not located on the MBS. The mobile transport and servicing system also includes the special purpose dexterous manipulator (SPDM), or Dextre, for remote manipulation and performing tasks previously considered too complex for space robotic systems. Canadarm2 has seven motorized joints and is capable of handling payloads up to 116,000 kg (255,740 lb). Both the SSRMS and the SPDM are the Canadian Space Agency's primary contribution to the International Space Station.

European Robotic Arm

The European robotic arm (ERA) is an additional assembly and maintenance robotic arm that will be attached to the Russian orbital segment of the International Space Station. Design of the ERA is similar to the Canadarm2 in its ability to walk end-over-end between base points located on the Russian modules. Launched on a Russian Proton scheduled for 2011, the ERA will use the future Russian multipurpose laboratory module for an operational base.

Launch and Transfer Vehicles

LAUNCHERS

STS

America's Space Transportation System, the space shuttle, was designed for space station assembly, although not in its initial specifications. The space shuttle orbiter's cargo size and weight limit were established during the planning and budgeting process to accommodate the USAF's need for a winged vehicle with heavy-lift capability for their large surveillance spacecraft. By combining NASA's future space launch needs in a single system with the Air Force's launch and flight vehicle needs, the cost for future launches would be reduced, at least in the arguments presented to the congressional budget meetings. When the USAF ended their participation early in the STS program, NASA was left with a large, costly booster with a large payload capacity. Although it was later cast as a white elephant, the space shuttle's oversized capacity made it ideal for launch and assembly of a modular space station.

Space Station Freedom was introduced in the 1980s as the first of a series of space stations that would provide permanent research and operations in orbit. Throughout its concept development, the SSF modular units were configured for launch in the space shuttle's 15 × 60 foot payload bay. And as the SSF evolved into Space Station Alpha, with the addition of Russia as a central partner and conversion into the International Space Station, the space shuttle remained the primary launch vehicle for the U.S. components. The space shuttle's large cargo capacity also offered both crew and equipment transport to and from the station along with the station modules. This made the STS an ideal, although costly, vehicle for building and supporting the International Space Station.

- Launch Site
  - Kennedy Space Center, FL (Vandenberg AFB launch site built for the Air Force STS launches was later converted into an expendable launch vehicle facility)

Table 9.3. Space Shuttle Specs

First launch	April 12, 1981
Stages	2
Payload weight maximum	25,000 kg (55,000 lb) to 126 mi orbit at 28.3° inclination
Payload size maximum	4.6 m × 18.3 m (15 ft × 60 ft)
Crew Size	7 (10 maximum)
Habitable volume	71.5 m^3 (2,130 ft^3)
Orbiter launch mass (approx.)	105,000 kg (230,000 lb)
Launch inclination	28°–62°

- Landing sites:
  - Kennedy Space Center, Florida (primary)
  - Edwards AFB, California (secondary)
- Alternate and abort landing sites
  - Base Aerienne, Ben Guerir, Morocco
  - Moron AB, Spain
  - White Sands Space Harbor, New Mexico
  - Yundum Airport, Banjul, The Gambia
  - Zaragoza AB, Spain
  - Amberley, Australia
  - Anderson AFB, Guam
  - Amilear Cabral, Cape Verde
  - Hickam AFB, Hawaii
  - Arlanda, Sweden

Proton M (8K82KM)

Russia's heavy-lift launcher named Proton was developed in the 1960s as an alternative ICBM, and as a satellite and interplanetary/lunar booster. Like most other boosters from the Cold War, it proved useful for many space launch tasks, including their space stations. Proton has been used as the only launcher for the orbital station modules, from the Salyut, to the Mir complex, to the Zaria and Zvezda, which are now a part of the ISS. Although still in use, the Proton is expected to be replaced by a variety of boosters under the Angara family name, which are simpler, cheaper boosters that use more efficient RP-1 and LOX first-stage propellants.

Soyuz

Soyuz ("union") is the name given to a number of Soviet and Russian boosters and space flight vehicles, beginning with their first mission plans to send cosmonauts

Table 9.4. Proton Specs

First launch	March 10, 1967
Stages	3 (4th available for interplanetary missions)
Payload weight max	21,000 kg (46,200 lb) to 185 km 51.6° inclination orbit (latitude of Baikonor launch complex)
Payload size (approx.)	7 m × 4 m (21.7 ft × 12.4 ft)
Launch Site	Baikonor Cosmodrome, Kazakhstan

to the Moon. Today, the Soyuz is the name for both the crew transport vehicle, the current model being the TMA, and the three-stage booster used to launch the Soyuz TMA.

Soyuz Launcher

Russia's Soyuz launcher is one variety of their very common two-stage boosters based on the original R-7 ICBM design first launched in 1957. Today's Soyuz launcher has had many improvements and upgrades since the R-7, and includes a third stage to place the Soyuz flight vehicle into low Earth orbit. Launchers related to the Soyuz consist of the same two initial stages that made up the original R-7, but with variations on the third stage boosters.

Zenit 2

The Ukrainian-built Zenit ("zenith") booster used for launching small modules to the ISS has an unusual heritage. Zenit's basic design is a copy of the first stage external boosters used on the huge Soviet-era Energia heavy-lift launcher flown only twice. Zenit's initial design was for a satellite launcher from the 1980s, then adapted for the Energia side-mount boosters and engines, and most recently as a satellite booster launched from a barge facility in the Pacific Ocean with lower costs due to the equatorial latitude. The SeaLaunch partnership includes the

Figure 9.48. Proton launch of the Zvezda ISS module on July 12, 2000. Source: NASA NIX.

Table 9.5. Soyuz Specs

First launch	November 28, 1966
Booster stages	2 originally, 3 for the Soyuz TMA
Payload weight max	7,200 kg (15,800 lb) to 200 km orbit at 51.6° inclination
Launch Site	Baikonor Cosmodrome, Kazakhstan

Boeing Commercial Space Company, the Russian space organization RSC Energia, the Norwegian Company ASA of Oslo, and the Zenit manufacturers SDO Yuzhnoye/PO Yuzhmash of the Ukraine.

TRANSFER VEHICLES

Soyuz TMA Crew Vehicle

Soyuz was originally designed for manned lunar missions as part of the Soviet's race to the Moon, but later it was adopted and modified as a crew ferry vehicle for its Salyut space stations. The three-man vehicle began as a transport system for a crew of two, but over its forty-year life has had improvements in virtually every system on the vehicle. Today's TMA/TMA-1 (Transportnyi Modifitsirovannyi Antropometricheskii) spacecraft has digital flight systems, improved guidance and landing systems, and a number of new accommodations for the flight crews, although the small crew space has been increased only marginally.

The Soyuz is made up of three primary modules: the orbital module, the reentry module, and the service module. A launch escape system similar to those used on the Mercury and Apollo capsules has been a part of the Soyuz vehicle design since its earliest flights.

Orbital Module (BO—Bitovoy Otsek)

The Soyuz spherical orbital module is used for crew operations post-launch and pre-deorbit because of the cramped quarters in the reentry/descent module. Experiments, cargo, hygiene facilities, and eating facilities are housed in the Soyuz orbital module.

Table 9.6. Zenit Specs

First launch	April 13, 1985
Stages	2 or 3 (Zenit 2 is the 2-stage Zenit, Zenit 3 is the 3-stage version)
Payload	13,740 kg (30,290 lb) to LEO at 51.6° inclination
Launch site (ISS support)	Baikonor Cosmodrome

Reentry/Descent Module (SA—Spuskaemiy Apparat)

The reentry/descent module on the Soyuz spacecraft is used for the launch and deorbit-reentry-landing phases of the space flight. This module is separated from the orbital module and service module before the reentry phase and becomes a self-contained crew vehicle similar to the Apollo command module.

Service Module (PAO—Priborno-Agregatniy Otsek)

The service module consists of the aft section of the Soyuz spacecraft which contains the support functions for the crew during the entire flight except reentry and landing. The Soyuz service module, like the Apollo service module, furnishes primary power, orbit-deorbit and attitude propulsion, thermal control, communications, and some life-support functions.

Progress

Russia's Progress was developed as a cargo flight vehicle to support the second-generation Soviet space stations, beginning with the Salyut 6. With modifications, the Progress continues with its legacy as a versatile cargo transport that now supports ISS. The unmanned cargo vehicle is based on the Soyuz crew vehicle using a converted de-

Figure 9.49. Photograph of a Soyuz TMA-6 crew vehicle with the deployed Kurs automated docking system antennas on the orbital module (left). The Descent module is located in the center, the Service Module and attached photovoltaic panels on the right. Source: NASA NIX.

Table 9.7. Soyuz TMA Specs

Crew Size	3
Diameter	2,72 m
Length	7.48 m
Habitable volume	7.2 m³ (BO + SA)
Mass	7,220 kg

scent stage serving as a cargo module. Like the Soyuz, the Progress has an autonomous navigation system that allows automated rendezvous, docking and undocking using the onboard Kurs docking system.

Progress Launcher

Design of the Progress cargo vehicle was derived directly from the Soyuz crew vehicle using the same approximate mass and size. The same booster is also used, the Soyuz (11A511U/2), but named Progress.

Progress M/M1 Cargo Vehicle (7K-TGM)

The Progress M carries up to 2,600 kg of equipment, fuel, and supplies to the station, and because it is expendable, it is also used for removing trash and waste at the end of its mission. The Progress is typically docked for up to six months for unloading, then packed with waste and trash until the arrival of the next ship. The filled Progress is then undocked, deorbited and incinerated during atmospheric reentry.

Table 9.8. Progress Specs

Length	7.23 m
Diameter	2.72 m
Launch weight	7,130 kg
Dry cargo weight	1,500 kg
Liquid cargo weight	1,540 kg

FUTURE TRANSFER VEHICLES

Several new crew and cargo flight vehicles for the International Space Station are under development by the program's major partners. Leading the effort is NASA with its Orion crew exploration vehicle, a replacement for the space shuttle that is scheduled for retirement in 2010–2011. The European Space Agency has developed an equipment transfer vehicle comparable to the Russian Progress M, but with greater payload capacity. ESA's automated transfer vehicle (ATV), which supports both the Russian and the U.S. segments, has an autonomous navigation and automated docking compatible with, but unlike, the Russian Kurs system. A similar vehicle has been developed by the Japan Aerospace Exploration Agency as a cargo vehicle to support the Japanese experiment module complex Kibo, a vehicle that does not have automated docking. A fourth transfer vehicle in early development stages is the Russian Kliper crew transport vehicle planned to replace the forty-plus-year-old Soyuz.

U.S. Crew Exploration Vehicle

The new Orion crew exploration vehicle (CEV) resembles the Apollo command and service module in many ways, but incorporates a larger, modernized crew module that will accommodate four astronauts from Earth-orbit, and possibly 4–6 from lunar orbit. Three versions of the CEV are planned for the ISS support missions. One is the manned vehicle with a dedicated crew module for launch-to-orbit and return. The two others, like Progress, are exclusively cargo modules that can be used for equipment and logistics delivery. The pressurized module can also return equipment to Earth; the unpressurized capsule will be used for waste and equipment disposal on reentry. A larger service module for the unpressurized module would carry added propellant for reboosting the ISS while docked.

NASA's CEV is part of the larger Constellation Program, which includes the Orion flight vehicles and two launchers, the Ares I and the Ares V. Both Ares launchers are derived from space shuttle hardware, including a variation of the solid rocket boosters and technology from the shuttle's external tank. The Orion CEV will be launched atop the Ares I booster, a single SRB that has an added fifth segment to the space shuttle SRB's four motor sections. The second stage consists of lightweight cylindrical liquid oxygen and liquid hydrogen propellant tanks, and a new booster engine derived from the Apollo J-2, called the J-2X. The larger Ares V cargo booster consists of two 5.5-section solid rocket boosters on the first stage with a six-engine second stage similar to the Space Transportation System without the orbiter. A decades-earlier design called the Shuttle-C was nearly the same as the Ares V, but used the space shuttle main engines in a side-thrust configuration. In contrast, the Ares V design includes six bottom-centered RS-68B engines. Although the Ares V does not support the ISS, it is planned for use in conjunction with the CEV for future lunar missions to place cargo and crew on the Moon.

Orion's crew module, originally designed to be reusable with a ten-flight lifetime, is now a single-use capsule with ocean recovery. Orion is wider and longer than the Apollo command module with more than twice the habitable volume of the Apollo CM. Orion has virtually the same shape as the Apollo CM, including the truncated-cone shell and recovery stack at the top. Orion's structure is composed of aluminum-lithium alloys instead of titanium to reduce its weight and cost. An ablative reentry heat shield is used to protect the crew module capsule during reentry, as well as an inflatable, bottom-mounted airbag for a soft touchdown for both land and sea landings.

An abort escape system consisting of solid rockets mounted on top of the crew module is used for separating the capsule from the Service Module and launcher on the pad, or during ascent. Like Apollo's launch escape system, the Orion CEV includes two rocket systems: the larger one to pull the crew module to safety, and a smaller rocket above the larger booster to remove the escape rocket from the CEV after approaching orbit when its no longer needed.

The Orion service module will perform much of the same functions as the Apollo service module, including propulsion, electrical power, and additional communications. However, the Orion SM will be narrower and shorter than the Apollo service module, as well as lighter. The SM structure will also be made of aluminum-lithium alloys, with two attached solar panels instead of Apollo fuel cells which required heavier, more expensive, and more hazardous cryogenic liquid oxygen and liquid hydrogen reactants.

Propulsion on the Orion SM will be different on the three module variants. As the CEV support unit for the ISS, the service module will use the two common hypergolic propellants monomethyl hydrazine and nitrogen tetroxide, which is NASA's second choice since the two are corrosive and highly toxic. On lunar missions, the SM will employ liquid oxygen and liquid hydrogen to improve propulsion efficiency, although more costly than the hypergolic propellants. On Mars missions, the SM will use liquid oxygen and liquid methane because of the availability of methane on Mars. Methane is also a byproduct of the carbon dioxide reduction process in the life-support system's Sabatier reactor where carbon dioxide is combined with hydrogen to produce methane (CH_4) and oxygen. Hydrogen is available from the ISS's two life-support systems' electrolysis units, the Russian Elektron oxygen generator, and the U.S. oxygen generator system.

ESA's Automated Transfer Vehicle

The European Space Agency's automated transfer vehicle is a cargo vehicle designed to support their Columbus research module attached to the International Space Station. ESA's first ATV named Jules Verne was launched March 9, 2008, on an Ariane-5 booster from the Kourou, French Guiana equatorial launch site. Propellants, consumables, equipment, and other supplies were carried in the expendable Jules Verne. Like the Progress, the ESA ATVs are packed with waste, trash, and expendable equipment before being deorbited for incineration during atmospheric reentry.

ESA's ATV carries four solar photovoltaic arrays for flight and docked power, and additional propellants to help in periodic ISS reboost. The cylindrical-shaped 10.3 m long by 4.5 m diameter spacecraft has an automated docking system unlike the Russian Kurs automated docking radar system, but using the same docking ports. The cargo hold in ESA's ATV is pressurized, allowing astronauts and cosmonauts access for unloading and loading its cargo in shirtsleeves. Although Jules Verne serves as a complement to Russia's Progress, it has three times its capacity. Projected launch frequency of the ATV is for fifteen-month intervals.

Conversions of the ATV into other flight vehicles are being considered by ESA and its partners, which include a crew transport vehicle, a mini space station, a payload retrieval system, and a cargo ascent and return vehicle.

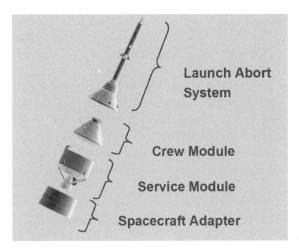

Launch Abort System

Crew Module

Service Module

Spacecraft Adapter

Figure 9.50. Exploded view of the Orion Crew Exploration Vehicle. Source: NASA NIX.

Japan's HTV

Japan's Aerospace Exploration Agency, JAXA, has developed a resupply vehicle for its Kibo Japanese experiment module. The unmanned HTV resupply module is slightly larger than Progress, but without the automated docking system. A manual docking of the HTV is made on Node 2 Harmony using the Canadarm2 after the vehicle completes a close proximity approach.

The HTV, a contraction of the H-II Transfer Vehicle, is launched from the Tanegashima Space Center in Japan using their H-IIB booster. The cargo module consists of a cylindrical structure approximately 10 m long and 4.4 m in diameter. Payload capacity for the HTV is 6,000 kg (13,200 lb), which allows the HTV to deliver a variety of equipment to Kibo, including ISPR experiment racks. Like Progress and the ATV, Japan's expendable HTV is also used for disposal of ISS waste during its burnup on reentry.

Russian Kliper

Russia is also planning a new transfer vehicle to replace the venerable, but small and aging Soyuz. Today's Soyuz TMA is a direct descendant of their transport ferry that has been in service since the maiden flight in 1967. In contrast to the headlight-shaped Soyuz reentry module, the Kliper will include a reusable reentry capsule intermediate between a lifting body and a winged vehicle, somewhat similar to the X-20 Dyna Soar. In addition to the six-crew capacity, the larger Kliper ferry is expected to carry 500 kg of equipment and/or logistics to the ISS. On-orbit lifetime is projected to be six months, the same as the Soyuz TMA, and launched on an upgraded Soyuz 2 booster. First flight for the Kliper is expected in 2011, with the first manned flight expected the year after. Recent proposals for a second vehicle used for propulsion on the Kliper, called the Parom space tug, are less certain.

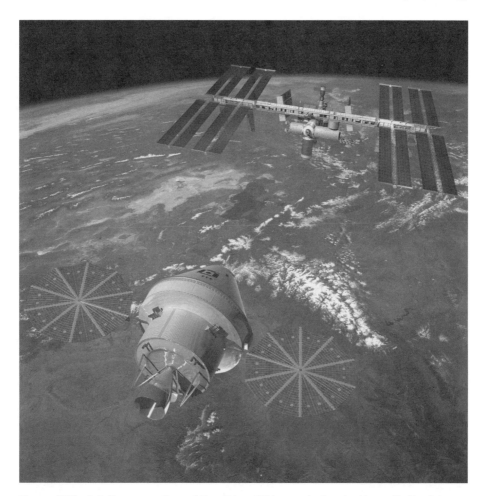

Figure 9.51. Artist's conception of the Orion CEV enroute for docking with the International Space Station. Note the two circular photovoltaic arrays on the Service Module used for primary power on both the Service Module and the Crew Module. Source: NASA NIX.

Table 9.9. ATV Specs

Length	10.3 m (33.8 ft)
Diameter	4.5 m (14.8 ft)
Payload	7,667 kg (16,900 lb)
Launch mass	20,700 kg (45,600 lb)
Mission duration	6 months

Figure 9.52. Jules Verne Automated Transfer Vehicle approaching the ISS on March 31, 2008, for a guidance and proximity test in preparation for final docking. Source: NASA NIX.

References

Baker, David. 1982. *The History of Manned Space Flight.* New York: Crown Publishers.

Belew, Leland F., and Ernst Stuhlinger. 1973. *Skylab: A Guidebook.* NASA EP 107. Huntsville, AL: Marshall Space Flight Center.

Brooks, Courtney G., James M. Grimwood, and Loyd S. Swenson. 1979. *Chariots for Apollo: A History of Manned Lunar Spacecraft.* NASA SP-4205. Washington, DC: NASA Scientific and Technical Information Office.

Compton, W. David, and Charles D. Benson. 1983. *Living and Working in Space: A History of Skylab.* NASA SP-4208. Washington, DC: NASA Scientific and Technical Information Branch.

Disher, John H. 1972 *The Skylab Communications System.* AIAA 4th Communications Satellite Systems Conference, Washington, DC, AIAA 72-543.

Lindroos, Marcus. N.d. "Space Station Freedom," Encyclopedia Astronautica. Retrieved from: http://www.astronautix.com/craft/spaeedom.htm (accessed: 5/30/2009).

Massey, John W. 1960. *Historical Resume of Manned Space Stations.* Army Ballistic Missile Agency report no. DSP-TM-9-60.

NASA. 1971. *Skylab Attitude and Pointing Control System.* TN-D 6068. Huntsville, AL: Marshall Space Flight Center.

NASA. 1974a. *MSFC Skylab Mission Report: Saturn Workshop*. TM X-64814. Huntsville, AL: Marshall Space Flight Center Skylab Program Office.

NASA. 1974b. *MSFC Skylab Orbital Workshop*. Vol. 1. TM X-64813. Huntsville, AL: Marshall Space Flight Center.

NASA. 1992. *Space Station Freedom Media Handbook*. TM-107981. Washington, DC: NASA Office of Space Station.

NASA. 1998. *International Space Station Familiarization*. Washington, DC: NASA Mission Operations Directorate, Space Flight Training Division.

Newkirk, Dennis. 1990. *Almanac of Soviet Space Flight*. Houston, TX: Gulf Publishing Co.

Newkirk, Roland W., Ivan D. Ertel, and Courtney G. Brooks. 1977. *Skylab: A Chronology*. NASA SP-4011. Washington, DC: NASA Scientific and Technical Information Office.

Noordung, Hermann. 1929. *The Problem of Space Travel: The Rocket Motor*. Reprinted as NASA History Series SP-4206, 1995. Washington DC: NASA.

Peebles, Curtis. 1987. *Guardians: Strategic Reconnaissance Satellites*. Novato, CA: Presidio Press.

Portree, David, S. F. 1997. *Mir Hardware Heritage*. NASA RP 1357. Houston, TX: NASA Information Services Division, Johnson Space Center.

Siddiqi, Asif, A. 2000. *Challenge to Apollo: The Soviet Union and the Space Race, 1945-1974*. NASA history series, SP 2000-4408, Washington, DC.

S. P. Korolev RSC Energia. N.d. "Most Long-Duration Manned Space Flights." Retrieved from: http://www.energia.ru/english/energia/mir/mir-long.html

NASA MIX (Marshall Space Flight Center Image Exchange). Retrieved from: http://mix.msfc.nasa.gov/

NASA NIX (NASA Image Exchange). Retrieved from: http://nix.nasa.gov/

USAF image archive. Retrieved from http://www.af.mil/photos/

NASA STS-88 Image Gallery. Retrieved from: http://spaceflight.nasa.gov/gallery/images/shuttle/sts-88/ndxpage1.html (accessed 6/5/2009).

NASA STS-108 Image Gallery. Retrieved from: http://spaceflight.nasa.gov/gallery/images/shuttle/sts-108/ndxpage1.html (accessed 6/5/2009).

NASA KSC Media and Image Archives. Retrieved from http://mediaarchive.ksc.nasa.gov/

NASA STS-110 Image Gallery. Retrieved from: http://spaceflight.nasa.gov/gallery/images/shuttle/sts-110/ndxpage1.html (accessed 6/5/2009).

NASA STS-113 Image Gallery. Retrieved from: http://spaceflight.nasa.gov/gallery/images/shuttle/sts-113/ndxpage1.html (accessed 6/5/2009).

CHAPTER 10

The Space Shuttle

Background

Studies and proposals for a new manned space vehicle originated in NASA and the USAF at about the same time and began converging as the Apollo hardware was being readied for lunar missions. Only a few years later, the project that merged the interests of the USAF, the DoD, and NASA was announced from the Moon by John Young during the Apollo 16 mission. The collaborative and design agreements made in developing the National Space Transportation System, better known as the space shuttle, were important in overcoming the program's many budget hurdles and technology challenges. The same challenges and agreements would have a major influence on the direction of the American space programs for three decades.

NASA engineers and managers had experience in developing large launchers and complex manned vehicles, while the Air Force had little more than the Atlas and Titan ICBMs and the aborted X-20 Dyna Soar project to their credit. Tightening budgets for NASA and the USAF, as well as combined interests in a large-payload manned launcher soon became a marriage proposal between NASA and the Air Force, with the DoD and Congress presiding over the wedding. The proximate justification for the program was that the new manned launch vehicle would offer cost savings, while satisfying all of NASA's needs and Air Force requirements, as well as the needs of space launch customers. To do this, the basic design of the new space transportation system would be comprised of reusable components, have a payload capacity of 60,000 lb, be large enough to carry cargo up to 15 × 60 ft in size, and have a winged structure that could land horizontally like a conventional aircraft. As a result, the new National Space Transportation System (NSTS, later shortened to STS) was ambitious and costly, but would be promoted as the ultimate solution to making space more accessible, and being able to deliver large and small payloads for only $100/lb to low Earth orbit.

Optimism for the new National Space Transportation System spread through NASA and Congress, shortening the approval process. And as the project evolved from concept to proposal, the outcomes of the enormous program were emphasized over technical challenges, aided by NASA's success in the ongoing Apollo program and its predecessors which also hastened approval. Since no Saturn V heavy-lift launchers were available after the completion of Apollo and Skylab, the proposed space shuttle program was even more attractive. But an important element was missing in the rush for approval of the NSTS—the commitment for support funding from the DoD and the USAF (Heppenheimer 1999).

NASA's projection for the flight rate of fifty-five to sixty launches a year for the space shuttle was used to bolster the assertion that the new National Space Transportation System could replace all expendable launchers, including the Atlas, Delta, and Titan. While completely unrealistic, the $100/lb projected space shuttle launch costs for payloads to low Earth orbit was another strong argument for funding the NSTS.

Both the high flight rate and low cost projections for NASA's new manned launcher had a direct influence on the choice of the first stage boosters. Liquid-fuel boosters that could fly back to the launch site for recovery and refurbishment were attractive because of their reusability, but much more costly than solid-fuel rockets. And while the solid-fuel boosters took less time and money to develop, their costs over the entire program were expected to be greater than the reusable liquid-fuel boosters. Ultimately, the shrinking federal budget dictated a minimum in capital development costs for the boosters, which favored the solid rockets. The final decision by the Nixon administration to employ the solid rockets boosters (SRBs) in the space shuttle design appeared to be nearsighted at first, and even more questionable after the *Challenger* accident in 1986. Inherent in the SRB design are combustion vibrations, known as thrust oscillations, which haunt NASA designers even today as the SRB undergoes conversion for use as the first stage of the new Aries 1 launcher.

FINAL DECISION

Between 1971 and 1972, the final specifications were drawn up, and the contracts for the four major components of the space shuttle were awarded. Contracts for the orbiter flight vehicle, the solid rocket boosters, the external tank (ET), and the space shuttle main engines (SSMEs) went to the primary contractors that were responsible for their design and construction, although most of the work was accomplished through dozens of major aerospace subcontractors before integration into the major components.

Orbiters

The primary contract for the Orbiter was given to North American Rockwell Corporation to complete the development of the vehicle based on NASA's initial specifications, followed by the fabrication of four space-qualified orbiters and spares. In addition, Rockwell was to provide a testbed and vehicle prototype initially called *Constitution*. The test article officially known as Orbiter Vehicle 101 (OV-101) was, after effective lobbying by legions of *Star Trek* fans, renamed *Enterprise*. Rockwell's fabrication facility was located in Palmdale, California within towing distance of the Edwards

Air Force Base runways. This was more than coincidence since the completed orbiters would have to be flown to the Kennedy Space Center for launch on the back of a converted B-747 from the Edwards airfield.

Solid Rocket Boosters

Solid Rocket Booster development and fabrication was awarded to Morton-Thiokol, today called ATK Launch Systems Group, with headquarters in Brigham City, Utah. A number of challenges faced the SRB manufacturer since it was by far the largest solid rocket ever built, generating 3.3 million pounds of thrust each. The new SRBs included a number of innovations that included an articulated (vectored) nozzle and construction in four segments which allowed the boosters to be shipped by rail from Utah to the Florida launch site.

External Tank (ET)

The STS's external tank contract was awarded to Martin Marietta (now part of Lockheed Martin) to complete design and fabricate the liquid hydrogen and liquid oxygen tanks that supply the propellants for the space shuttle main engines. Construction of the aluminum-alloy dual tank units that were covered in insulated foam was moved to NASA's Michoud plant in Louisiana to facilitate transportation of the completed ETs to the Florida launch site by barge. The external tank was the largest structure on the STS, and the heaviest when loaded. The ET structure is a light-weight semi-monocoque design with the tanks supplying much of the primary structural strength. Thrust loads from the SRBs and the orbiter's three SSMEs were transferred through the light-weight ET, primarily via the intertank assembly that connected the larger hydrogen tank to the smaller oxygen tank.

Space Shuttle Main Engines

The contract to develop and manufacture the space shuttle main engines was contested on first award, but was eventually given to the Rocketdyne Corporation, the rocket engine manufacturer that has dominated the launch industry since World War II. The STS's high-performance SSMEs established records in a number of performance categories, including engine turbopump power, and the first high-thrust reusable liquid-fuel engine.

USAF Pullout

Unfortunately for the STS program, the Department of Defense's contribution to the STS program was minimal, and mostly in the form of a second launch site needed for the military's polar orbit reconnaissance satellites. The STS launch site built for the USAF at Vandenberg Air Base in California was never used for its intended purpose, primarily because of the problems with the orbiter's weight and the space shuttle's

Figure 10.1. Nighttime photograph of STS-1 on the launch pad at Kennedy Space Center. Columbia is mounted on the white External Tank, which was one of only two that were painted for launch. Source: NASA NIX.

overall performance. Following the *Challenger* tragedy in 1986, a new restriction that no longer allowed the high-performance Centaur liquid fuel booster on the orbiters ended the DoD's and the USAF's feeble participation in the STS program, in spite of imposing design requirements on the STS for a large cargo capacity and winged flight capability. It soon became clear that unreasonably optimistic projections for the flight rate and operational costs of the STS would pose a problem for the project during its operational life. And because the military pulled out of the program after *Challenger*, NASA's budget would be burdened with an enormous manned program with no clearly defined objectives. Combined with these complications were the unexpectedly high vehicle processing costs for each mission which boosted the price for payloads on the space shuttle to more than $20,000/lb—more than 200 times the projected cost in the NASA-USAF program approval request.

Orbiters

Rockwell International began construction of the STS orbiters with the fabrication of two test articles. The first was Shuttle Test Article 099 (STA-099) which later became OV-099 *Challenger*. The second was the OV-101 *Enterprise* used for the aerodynamic flight tests, now on display in the McDonnell Space Hangar at the National Air and Space Museum's Steven Udvar-Hazy Center near the Washington Dulles International Airport in Chantilly, Virginia. Although the *Enterprise* was planned for conversion to a space-qualified vehicle, STA-099 *Challenger* offered a lower-cost conversion and became the second orbiter to fly in space after *Columbia*. Orbiter construction was carried out at Air Force Plant 42 in Palmdale, California, to allow access to the Edwards AFB landing facility for the completed orbiters.

CHALLENGER (OV-099)

The *Challenger* orbiter, named after the British Naval research vessel, was returned to Rockwell's Palmdale plant in December of 1979 for conversion from the test vehicle to a space-qualified orbiter. Major modifications were made to *Challenger* following qualification tests that included wing modification and reinforcement. *Challenger* was rolled out from the Palmdale plant on June 30, 1982, weighing 1,313 kg (2,889 lb) less than *Columbia*. *Challenger*'s first flight was STS-6 on April 4, 1983, followed by eight more missions before its loss on January 28, 1986.

COLUMBIA (OV-102)

Columbia was the first orbiter built to fly in space and was laden with extensive research instrumentation for its early flight missions. *Columbia* was also outfitted with two ejection seats for its first four missions which carried only two crew members. These early missions were more test flights than exploration missions since, for the first time, NASA launched the initial flight of a manned vehicle with a crew onboard. Other *Columbia* firsts included the first scheduled inspection and retrofit program and the first comprehensive inspection called orbiter maintenance down period (OMDP) in October 1994. *Columbia* returned to service in April 1995 after approximately ninety modifications and upgrades.

    Columbia's rollout from the Palmdale plant was on March 8, 1979, followed by a ferry flight to the Kennedy Space Center for completion before its inaugural launch on April 12, 1981. *Columbia* would fly twenty-eight missions before a tragic reentry accident February 1, 2003, on STS-107 which completely destroyed the vehicle. Video of the launch incident that doomed *Columbia* and its crew showed a piece of insulation torn loose from the external tank striking the leading edge of the left wing that went undetected for the remainder of the mission, although suspicions circulated through some of the NASA and contractor offices. The breach in the protective reinforced carbon-carbon panels allowed hot plasma reaching approximately 1,370°C (2,500°F) to enter the forward left wing during reentry. *Columbia* broke up during its hypersonic descent as it passed over the Pacific Coast through Texas.

DISCOVERY (OV-103)

Discovery was the third orbiter to fly in space, which gave the builders the added experience of seven years and two orbiters to make improvements in the structural assembly and systems. In addition to a lighter overall weight, *Discovery*, like *Challenger*, was modified to enable the vehicle to carry the Centaur upper stage on its spacecraft payloads, the highest-energy booster available. Added cooling capacity was also available for RTG-powered interplanetary spacecraft carried aboard. But the additional plumbing for cryogenic loading and purging to accommodate the Centaur was never used on *Discovery* since post-*Challenger* reviews deemed the high-explosive liquid propellants on Centaur too risky to launch on a space shuttle.

Discovery's rollout from the Palmdale plant was on November 16, 1983, with an empty weight 6,870 pounds less than *Columbia*. Following transfer to the Kennedy Space Center (KSC) for completion and preparation, *Discovery* was launched on its maiden flight STS-41D on August 30, 1984. *Discovery* has undergone two OMDP refurbishment and upgrade cycles, the last ending in 2004 with nearly 100 upgrades that included the addition of a fifty-foot inspection boom to check its underside and leading edges for damage.

ATLANTIS (OV-104)

Atlantis had the same contract award date as *Discovery*, but rolled out from the Palmdale plant in only two years since it benefited from the fabrication and assembly innovations learned during the construction of the three earlier orbiters. Assembly time for *Atlantis* was also reduced because of the use of thermal protection blankets on the upper sections of the orbiter that replaced the time-consuming hand fitting of the rigid thermal protection tiles. *Atlantis* was completed on March 6, 1985 with an empty weight 6,974 pounds less than *Columbia*. The inaugural flight of *Atlantis* was STS-51-J on October 3, 1985.

NASA managers had the foresight to have Rockwell build a set of spare structural assemblies including an aft-fuselage, mid-fuselage, forward fuselage, vertical tail and rudder, wings, elevons and a body flap for replacement or repair of a damaged orbiter. By virtue of that request, NASA had a complete set of spares on hand to build the *Endeavour* orbiter as a replacement for *Challenger*.

ENDEAVOUR (OV-105)

Endeavour was built as a replacement for the *Challenger* from spares ordered during the fabrication of *Discovery* and *Atlantis*. Rockwell received the contract award on July 31,

1987, and completed *Endeavour* on April 25, 1991. The inaugural mission of *Endeavour* was STS-49, launched on May 7, 1992.

Because *Endeavour* was the last orbiter built for NASA, its fabrication included modifications and improvements from the previous orbiter builds, and the operational experience that improved performance and its capabilities. Most of these improvements were incorporated into the rest of the space shuttle fleet during subsequent OMDP upgrades. Examples of hardware improvements include the forty-foot drag chute to reduce the orbiter's rollout distance, hardware and electrical connections needed for extended duration orbiter (EDO) modifications for mission durations up to twenty-eight days, updated avionics systems with advanced general purpoose computers, improved inertial measurement units and TACAN systems, enhanced master events controllers and multiplexer/demultiplexers, a solid-state star tracker, improved nose wheel steering, and improved auxiliary power units.

ORBITER QUALIFICATION

Much of the preliminary spacecraft qualification was completed using the *Enterprise* vehicle in the approach and landing tests (ALT). The subsonic airworthiness and model validation test series was scheduled for completion in 1976, beginning with high-speed taxi tests that progressed from mounted flights to landing tests using the first B-747 shuttle carrier aircraft (SCA) to take the orbiter to altitude. All tests were unpowered, hence, were executed with the *Enterprise* mounted on the SCA. Taxi testing was completed in February 1977 reaching a maximum speed of 253 kph (157 mph), followed by unmanned captive tests in February and March of the same year to an altitude of 9,146 m (30,000 ft). Manned captive test flights called captive-active missions took place at the Edwards Air Force Base in June and July, followed by release and landing flights (free flights) between August and October of 1977. A total of five free flights were made for subsonic aerodynamic and operational testing at release altitudes that ranged from 5,793 m to 7,927 m (19,000 ft to 26,000 ft). Three missions were flown with the aerodynamic tailcone, lasting about five and a half minutes, and two without that lasted roughly two minutes from release to landing. Tests and evaluation of the orbiter's systems included terminal area energy management (TAEM) autoland approach capability and the microwave scan beam landing system.

Enterprise's retirement would have to wait until it was flown to Marshall Space Flight Center in Alabama for load and vibration testing, then to Kennedy Space Center in Florida for launch procedures evaluation, then to Vandenberg Air Force Base, California, for similar tests on the new Air Force STS launch pad. *Enterprise* was then refurbished and put on display at air shows in England, France, Italy, Germany, and Canada before being flown to its permanent home at the Smithsonian Air and Space Museum near the Dulles International Airport, Virginia.

SSME Qualification

Qualifying Rocketdyne's SSMEs for use on the STS meant that the hardware had to undergo flight qualification for use on a manned space vehicle. The qualification process

Figure 10.2. Photograph of the enterprise Orbiter during a free flight test with aerodynamic tailcone attached. Source: NASA NIX.

started at Rocketdyne's Santa Susana, California, site, and was then transferred to NASA's Stennis Space Center in Hancock County, Mississippi. SSME engine tests were performed at the Stennis Space Center engine facility beginning in 1975 and continued until 2006. Early testing of the SSMEs was hampered by the high-pressure turbopumps designed to generate extremely high flow rates for both the fuel and oxidizer propellants.

Initial plans to launch the STS in 1979 were delayed, in part, because of the inconsistent SSME engine test results. Successful tests in early 1981 completed the final engine qualification, followed by testing of the three engines slated for *Columbia*'s first launch. The successful launch record of the SSMEs has continued for more than 125 missions with only one SSME shutdown incident.

The Stennis site has been used for testing NASA's large booster engines, including the first two stages of the Saturn V. Several of the Stennis engine test stands are currently in use for NASA's future Ares booster engines, the J-2X and the RS-68.

SRB Qualification

NASA's choice of a solid rocket booster for the first stage of the STS presented many challenges. First, the SRB was the largest solid rocket built for any application. Second, the first SRB flight would also be carrying a crew, which made flight qualification of the booster critical. Third, the SRB was designed with a nozzle joint for the first time, which allowed thrust vectoring of the enormous 3.3 M lb exhaust thrust. Fourth, the SRB was designed to be manufactured in four motor segments that would then require careful, detailed assembly at the Kennedy Space Center. Fifth, the major components on the SRB would have to be designed for reusability, meaning practical recovery methods and high reliability.

SRB motor tests were started at the Morton-Thiokol test range near Brigham City, Utah, in 1977. Included were static firings, as well as cyclic pressure tests to evaluate the steel casing, the segment joints, and the nozzle. Static firings were concluded in 1979 in preparation for the STS-1 mission. SRB drop tests from NASA's B-52 were used to evaluate the SRB structural integrity after ocean impact and completed in 1978. Parachute and recovery testing, and final load and dynamics testing were wrapped up in 1980 when the SRBs were cleared for manned flight.

ET Qualification

While the external tank was the only major element of the STS that was not reusable, it presented a significant design challenge because of its enormous size, its required strength, and its very light structural weight. Martin Marietta's first ET was a test article that served as a prototype for the standard weight tanks that flew on the first five STS flights. Almost immediately, the contractor was asked to design a lighter-weight tank for increased payload capacity, and for the more challenging Vandenberg missions due to of the higher inclination orbits. The first major improvement in the ET design was to simply not paint the ET which saved several thousand pounds. The next series of improvements began with the STS-6 flight using lightweight tanks that incorporated structural modifications, titanium alloys, and improved milling that lowered the tank weight by 5,000 kg (11,000 lb). The super lightweight tank was introduced on the STS-91 mission in 1998 which used lighter aluminum-lithium alloys, reducing the ET weight another 3,175 kg (7,000 lb).

Space Transportation System Primary Elements

There are four primary elements that make up the Space Transportation System flight hardware. All but the External Tank are reusable and include:

1. Orbiter
2. Solid Rocket Boosters (2)
3. External Tank
4. Space Shuttle Main Engines

I. ORBITER

NASA's STS orbiter was designed for vertical launch and orbital flight, but with the capability of aerodynamic flight and horizontal landing, like an aircraft. And though

Table 10.1. Orbiter Specs

Length	37.2 m (122 ft)
Height	17.4 m (57 ft)
Wingspan	23.8 m (78 ft)
Launch weight (approx)	104,545 kg (230,000 lb)
Cargo bay capacity	4.6 m × 18.3 m (15 ft × 60 ft) 27,273 kg (60,000 lb) to 28.5° inclination
Design life	100 flights

built with techniques and materials similar to other spacecraft and aircraft, the orbiter is significantly different. Conventional wings, tail, landing gear, and fuselage allow the orbiter aerodynamic flight in the atmosphere, but at speeds reaching twenty-five times the speed of sound (Mach 25). Aluminum alloys make up most of the orbiter's structure that is covered with insulating tiles and blankets that protect it from the extreme cold of space and the even more extreme heat of reentry.

The orbiter's structure has many similarities to transport-category aircraft, and is even composed of many of the same aluminum alloys. However, higher loads during launch and reentry demand stronger bracing and increased stiffening over the conventional aircraft design to avoid structural fatigue and failures. Added strength in the

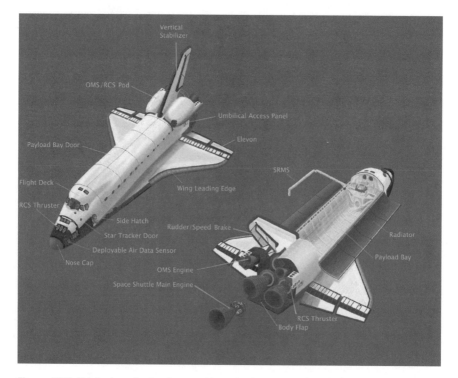

Figure 10.3. Orbiter's main structure and layout of primary assemblies. Source: NASA NIX.

orbiter structure is also needed to prevent heavy payloads from damaging or warping the fuselage, and to prevent damage from the heat and thrust loads of the three main engines. Augmenting the aluminum alloy structure are reinforcements and braces in the aft and mid fuselage sections that include steel alloys and titanium.

Primary Orbiter Structure

The orbiter is comprised of nine major sections or assemblies that house the systems, payloads, and crews. The fuselage structure was constructed using mostly stringer panels, frames, and bulkheads, some of which were cast and machined. Final assembly of the orbiter's large and small elements was completed at Rockwell's Palmdale, California facility.

1. Forward Fuselage The forward fuselage is made up of two separate component shells; the inner crew compartment and the outer top and bottom "clamshell" sections that surround the crew compartment. The inner crew compartment is a pressure vessel built to house the crew and operations areas on the upper deck and in the mid deck, which contains living quarters, a commode, a galley, storage, exercise equipment, an airlock, and sleeping quarters.

Twelve window sets made of triple-pane silica glass are located throughout the crew compartment. Each window set consists of two inner panes framed in the inner crew compartment shell, and an external (pressure) pane fitted to the outer forward fuselage shell. Six window sets are located on the forward flight deck for forward viewing, with two above the pilot seats. The left upper window also contains a pyrotechnic charge release for secondary crew emergency escape on the runway in case the side hatch fails or is blocked. Two window sets are also located on the aft flight deck to view payload operations, and one on each airlock hatch.

2. Wings The orbiter's wing structure is constructed of aluminum alloy skin panels, stringers, ribs, and spars. The wings are attached to the mid fuselage with a torque box, bounded by a leading-edge and a trailing-edge spar. The leading-edge spar is used to mount the reinforced carbon-carbon (RCC) insulation panels. The trailing-edge spar provides a structural mount for the split elevon assemblies used for pitch and roll control during aerodynamic flight. The elevon is a contraction of the combined elevator and aileron surfaces used for aircraft aerodynamic control. Thermal protection on the wing section is one of the most challenging sections on the orbiter since the leading edges encounter the highest reentry temperatures, and the split elevons require protection from the extreme heat of reentry throughout their 65° range of operation

Figure 10.4. Photo of the inner crew compartment of the Orbiter's forward fuselage section in the assembly process. Source: NASA NIX.

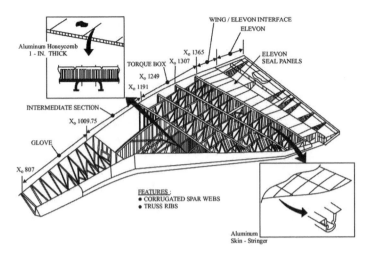

Figure 10.5. Cutaway view of the left Orbiter wing structure. Source: NSTS.

(+40°, -25°). The gap between the rear spar and elevons, as well as the elevon hinges, are protected from high-pressure heat of reentry by aerothermal seals.

3. Mid Fuselage One of the strongest assemblies on the orbiter is also the longest—the 18.3 m (60 ft) long mid fuselage. The mid fuselage structure is designed to carry the mission payloads and serve as the primary load-carrying structure for the orbiter. Construction of the mid fuselage structure is primarily of aluminum alloy main frames and braces, with trusses, longerons, and stringers for rigidity. In addition, steel alloy stiffeners help transfer the major thrust and torque loads throughout the STS without damaging or distorting the payload bay, which could prevent opening or closing the payload bay doors.

Attached to the mid fuselage section of the orbiter are the main wings, the main landing gear, the payload bay doors, and the fore and aft fuselage sections. Mounting brackets and keel assemblies are used to mount the various primary and secondary mission payloads.

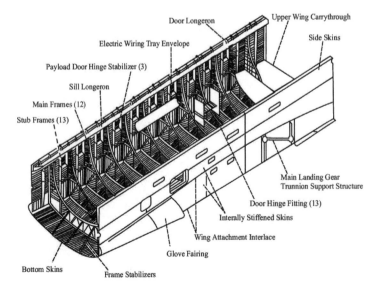

Figure 10.6. Structural elements of the Orbiter's mid fuselage. Source: NSTS.

4. Payload Bay Doors The orbiter's payload bay doors that span the 60-foot length of the mid fuselage and payload bay are composed of lightweight graphite epoxy. Mounted on the inner surface of the payload bay doors are both fixed and extensible radiator panels. A set of two radiator panels is located on the inner surface of both payload bay doors, along with the capability of adding more panels if needed for high heat load payloads. Hinges for the payload bay doors are placed on the longerons located at the top port and starboard rim of the mid fuselage. Electromechanical latches and a motor-driven torque shaft are used to lock and unlock the doors on the ground and in orbit, either automatically or manually.

5. Aft Fuselage The aft fuselage is a load-bearing structure used to mount the orbital maintenance system (OMS) pods, the vertical stabilizer (tail), the three SSMEs, the body flap, and the mid fuselage. The aft fuselage also is used to transfer SSME thrust through the mid fuselage and ET. Aluminum alloy was used for construction of much of the aft fuselage with frame, brace, truss, and skin-stringer panel elements. Titanium was used for the vertical stabilizer support frame and for the interior thrust structure that transfers thrust generated by the three main engines through the Orbiter and ET.

6. Forward Reaction Control System (FRCS) Attitude control of the Orbiter in space is furnished by the forward and aft reaction jets as part of the reaction control system (RCS). Forward RCS jets, in combination with RCS jets on the aft OMS pods, control the attitude and orientation of the orbiter outside of the atmosphere and during most of the orbiter's deorbit, reentry, and descent phases. The FRCS panel includes the RCS primary and vernier (small) thrusters, as well as the propellant tanks and helium pressurization tanks.

7. Vertical Tail The vertical tail is used for the same purpose of stabilizing the orbiter in aerodynamic flight as it is for conventional aircraft. The vertical stabilizer assembly is constructed of aluminum alloys in the form of machined skin panel and stringers, ribs, and spars, and includes a combined rudder and speed brake. The rudder assembly is configured to open, or split, to form a speed brake in addition to its right and left rudder deflection to control yaw forces. The rudder and speed brake combination is also used to augment pitch control during the orbiter's approach for landing phase.

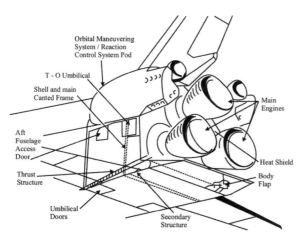

Figure 10.7. Sketch of the Orbiter's aft fuselage main components. Source: NSTS.

8. OMS/RCS System Pods After reaching orbit, the orbiter uses an onboard propulsion system called the orbital maintenance system/reaction control system, or OMS/RCS. Changes in altitude or orbit inclination are made with the OMS engines located in the aft OMS/RCS pods. Smaller RCS thrusters are also located in the aft pods, and used primarily for attitude control and OMS trim. Each of the OMS/RCS pods contains thrusters, propellant tanks, and helium pressurization tanks, as well as associated fill, drain, and feed hardware and plumbing.

9. Body Flap Unique to the orbiter's design is the hydraulically powered body flap used to augment pitch control in aerodynamic flight. The aluminum alloy body flap attached to the bottom aft section of the aft fuselage is also used to protect the SSMEs from the heat of reentry. Aerothermal seals similar to those on the elevons and rudder are used to protect the body flap and hinge mechanism from extreme reentry heat.

Orbiter Systems

The space shuttle orbiter is the most complex spacecraft ever to fly, but with systems that reflect its Apollo-era hardware heritage, updated with modern advances in technology found primarily in the orbiter's avionics. To simplify the maze of orbiter systems, the traditional approach is to group the components, subsystems, and systems into larger functional systems. For the orbiter, these system groups include:

• Orbiter propulsion
• Environmental control and life support system (ECLSS)
• Thermal protection system
• Computer and data handling system
• Orbiter communications
• Guidance, navigation, and control
• Electrical power system

Orbiter Propulsion Propulsion for the orbiter is provided by a variety of onboard thrusters and propellants called the main propulsion system (MPS). The MPS includes the space shuttle main engines, the SSME propellant supply and dump manifolds and components, the orbital maintenance system, and the low-thrust reaction control system used for attitude control.

Propellant feed, and the SSME purge and safe operations after shutdown, include the feed manifold that distributes the fuel and oxidizer coming from the umbilical lines from the ET, to the three SSMEs. Also included are the helium pressurization and purge components that flush the propellants overboard after main engine cutoff (MECO).

The orbital maintenance system is used to change the vehicle orbit, including orbit entry and deorbit. The OMS consists of the two aft-mounted 2,722 kg_f (6,000 lb_f) thrusters, one located in each OMS/RCS pod. Included also are the propellant, propellant tanks, and associated plumbing. Propellants for the OMS engines are nitrogen tetroxide and high-performance hydrazine, which are the same propellants used in the RSC thrusters.

Table 10.2. OMS Specs

Thrust (each)	2,722 kg$_f$ (6,000 lb$_f$)
Fuel	Unsymmetrical dimethyl hydrazine (UDMH)
Oxidizer	Nitrogen tetroxide (NTO)
Delta V (max)	1,000 fps
Thrust vectoring	Electromechanical ± 6° pitch ±7° yaw
Design life	100 missions, 1,000 starts, 15 hr total

The reaction control system is used for orbiter attitude control, for trimming the OMS burns, for small translational thrusts, and for reentry pitch, roll, and yaw control before aerodynamic forces take over. The RCS thruster group consists of forty-four small thrusters, divided into thirty-eight primary thrusters with 395 kg$_f$ (870 lb$_f$) thrust, and six low-thrust vernier thrusters with 11 kg$_f$ (24 lb$_f$) thrust. Fourteen of the thirty-eight primary thrusters are located in the forward RCS panel, and twelve are located in each of the OMS/RCS pods. Two of the vernier thrusters are located in the FRCS, with two more in each OMS/RCS pod. Fuel and oxidizer are the same as that used for the OMS system, but stored in separate tanks. Fill, drain, and feed plumbing allow crossfeeding propellants between RCS and OMS thrusters in the aft OMS/RCS pods, but not between RCS thrusters in the FRCS and the rear pods. Separate helium tanks are used for RCS propellant flow, but pressurization can be shared in the OMS/RCS pods.

Environmental Control and Life Support System The orbiter's ECLSS system provides crew life support functions for missions that range from six to fifteen days in duration, with longer missions supported by the addition of an extended duration orbiter package installed in the payload bay. Because of the vacuum of space and the extreme temperatures that range from -157°C (-250°F) on orbit and 1,371°C (2,500°F) during reentry, the crew compartment is designed as a pressure vessel with an active cooling and heating system. An artificial atmosphere is maintained in the

Table 10.3. RCSS

Thrust (each)	Primary (38)—395 kg$_f$ (870 lb$_f$) Vernier (6)—11 kg$_f$ (24 lb$_f$)
Fuel	Unsymmetrical dimethyl hydrazine
Oxidizer	Nitrogen tetroxide
Operation	1–125 sec continuous, 0.08 sec min pulse
Thrust vectoring	None
Design life	Primary—20,000 starts, 12,800 sec total Vernier—330,000 starts, 125,000 sec total

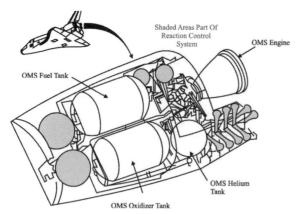

Figure 10.8. Cutaway of the OMS and RCS components in the left OMS pod. Source: NSTS.

crew cabin with the same composition found at sea level on Earth. Carbon dioxide from crew respiration is removed, along with trace gas contaminants, by the atmosphere revitalization system.

Air Revitalization System (ARS)

- Provides oxygen for respiration from cryogenic liquid oxygen storage (LOX)
  - LOX also used to power fuel cells
- Nitrogen supplied by onboard pressurized nitrogen tanks (GN_2)
- Lithium hydroxide (LiOH) canisters are used to remove CO_2
- Long-duration package adds LOX and LH_2 storage and replaces LiOH canisters with dual amine bed CO_2 removal unit
- Trace contaminant removal unit removes CO and other trace contaminants
- Circulation and vent components provide ventilation and thermal control in crew compartment
- Heat exchanger acts as an air conditioner to remove water vapor and cool cabin air
  - Humidity kept at 30%–75%
  - Also used for heat removal from avionics/electronics

Atmosphere Revitalization Pressure Control System

- Supplies crew cabin with a constant pressure of 1 atmosphere (14.7 psi) maintained at approximately 80% N_2, 20% O_2
- Monitors and maintains constant partial pressure of O_2
- N_2 added to maintain 14.7 psi cabin pressure
- Provides overpressure relief venting

Active Thermal Control System (ATCS)

- Four active heat removal systems on the Orbiter
  - Radiator panels on payload bay doors are used to remove heat while in orbit
  - Flash evaporators used above 100,000 ft altitude
  - Ammonia boilers used below 100,000 ft
  - Ground service equipment used when on the launch pad and during processing

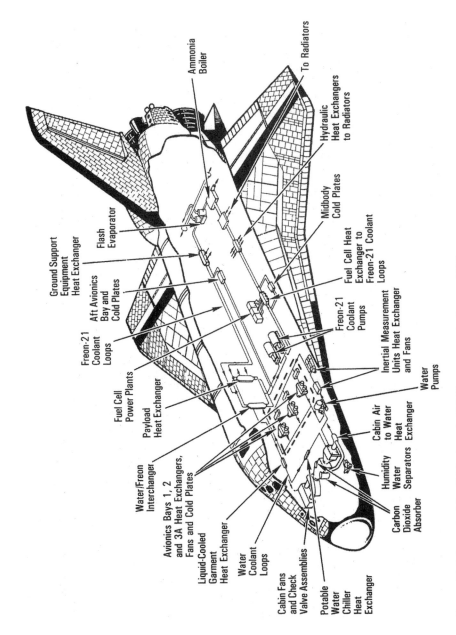

Figure 10.9. Layout of the Orbiter's main ECLSS components. Source: NSTS.

- Two separate internal and external cooling loops
  - Water (inner) coolant loop
    - Collects heat from crew cabin, crew compartment, crew compartment avionics, cabin heat exchanger
    - Transfers heat to ATCS heat exchanger
  - Freon (outer) coolant loop
    - Outer Freon loop pulls heat from the inner loop to circulate through radiator panels, ammonia boiler, flash evaporator, or GSE
    - Payload bay doors must be open to use radiator panels

Water Supply and Wastewater System

- Potable water supplied to the crew by the three onboard fuel cells
- Hydrogen removal and biocide protect drinking and cooking water
- Potable supply stored in tank A, and can also be stored in tanks B, C, D
- Wastewater stored in separate tank, and can be dumped overboard

Waste Collection System (WCS)

- Liquid waste from the Waste Collection System becomes part of the waste water loop
- Solid waste returned to Earth
- Liquid waste generally dumped overboard

Life support functions are also provided for EVA activities, which include the airlock and the space suits. Atmospheric pressure and composition in the airlock varies from vacuum, to 5 psi pure oxygen, to 1 atmosphere $O_2 + N_2$. Suit pressure during EVA and preparation is maintained at approximately 5 psi pure oxygen. CO_2 removal, water, food, thermal cooling, waste removal, and communications are also furnished by the space suit, formally known as the extravehicular vehicular mobility unit (EMU).

Thermal Protection System The orbiter's passive thermal protection system (TPS) covers nearly all of its surface and consists of seven types of insulation. The type of TPS insulation used depends on the highest temperatures encountered at or near a surface, and the aerodynamic load and impact resistance on or near that region. Thermal insulation can also include devices and materials for protecting movable joints, doors, and tile gaps. Heat protection for these components spans a wider range of materials and devices than the tiles and blankets that cover the orbiter's aluminum alloy surface, and includes thermal barriers, aerothermal seals, and gap fillers.

Orbiter passive thermal protection panels, tiles, and blankets include the following:

Reinforced Carbon-Carbon (RCC)

- Protects surfaces reaching reentry temperatures that can exceed 1,260°C (2,300°F)
- Placed on:
  - Leading edges of the wing

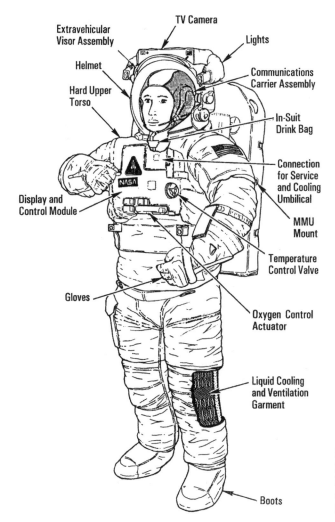

Figure 10.10. Major components of the Extravehicular Mobility Unit used by the space shuttle astronauts. Source: NSTS.

- Nose cap
- Chin panel (beneath nose cap)
- Coated with silicon carbide to prevent oxidation at high altitudes
- Each piece made individually for exact fit

High-Temperature Reusable Surface Insulation (HRSI)

- Black HRSI tiles are placed on orbiter's underside, forward fuselage, OMS/RCS areas, body flap
- Black color used to improve reentry emittance (heat radiation efficiency)
- 649–1,260°C (1,200–2,300°F) operating temperature range
- Approximately 20,000 HRSI tiles on each orbiter
- Stress isolation pad on bottom of this rigid tile reduces damage due to airframe/surface movement

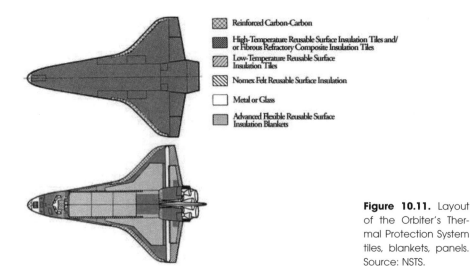

Reinforced Carbon-Carbon

High-Temperature Reusable Surface Insulation Tiles and/or Fibrous Refractory Composite Insulation Tiles

Low-Temperature Reusable Surface Insulation Tiles

Nomex Felt Reusable Surface Insulation

Metal or Glass

Advanced Flexible Reusable Surface Insulation Blankets

Figure 10.11. Layout of the Orbiter's Thermal Protection System tiles, blankets, panels. Source: NSTS.

Toughened Unipiece Fibrous Insulation (TUFI)

- A stronger, more durable tile
- Coating is porous, which helps stop cracks from spreading
- Replacing high tiles in high-abrasion areas
- Insulation properties are inferior to the HRSI and LRSI tiles, which limits their application

Low-Temperature Reusable Surface Insulation (LRSI)

- Originally used on the upper fuselage
- < 649°C (1,200°F) range
- Reduced temperature range means less costly surface treatment
- White color optimum for on-orbit operations
- Most LRSI tiles have been replaced by AFRSI blankets

Advanced Flexible Reusable Surface Insulation (AFRSI)

- Quilted, flexible AFRSI blankets are also known as FIB—Flexible Insulation Blanket
- Fabric/composite blankets
- 649°C (1,200°F range)
- Replaced most of LRSI tiles for efficiency increase, lower cost, greater durability

Fibrous Refractory Composite Insulation (FRCI)

- FRCI blankets used to cover the rounded portions of the orbiter
- Have replaced some of the HRSI 22 lb tiles to provide improved strength, durability, and resistance to coating cracking
- 371–649°C (700–1,200°F) range

Felt Reusable Surface Insulation (FRSI)

- FRSI Nomex insulation blankets are used on the upper regions of the orbiter
- Temperature range <371°C (700°F)
- Same composition as the SIP pad used to bond the HRSI, LRSI, TUFI, and FRCI tiles to the orbiter's skin
- Covers nearly 50% of the orbiter's upper surfaces

Command and Data Handling System Automated and manual flight operations of the orbiter and its systems including communications and payload operations are directed by the onboard digital command and data handling (C&DH) system. Five dedicated IBM AP-101 General Purpose Computers (GPCs) run two different sets of software that command the orbiter, ET, and SRB operations, from launch to the vehicle's return to Earth for its next mission. Design of the C&DH system hails from the post-Apollo era, a hardware heritage that imposed a severe limitation to its cpu speed and memory size. And because of the computers' 128K memory limit, flight and operational software is loaded into the GPS in successive batches by a magnetic tape drive that stores the entire software package in triplicate. Major elements of the space shuttle C&DH system include:

- Five onboard general purpose computers are used for redundancy in the orbiter's flight and operational control during launch, orbit, and landing.
  - Four of the five identical GPC systems have identical/redundant GPC operating software called PASS
  - Usually only one remains powered for on-orbit operations
  - The fifth GPC has different software for launch, flight, reentry, and landing operations, called the backup flight system
- Two magnetic tape mass-media storage units
- Data bus network
- Three SSME interface units
- Four CRT display systems
- Two data bus isolation amplifiers for ground communications
- Two master events controllers
- Master timing unit

Figure 10.12. Photograph of the third-generation IBM General Purpose Computer on the left, with the keypad and CRT monitor center right, and the model it replaced on the far right. Source: NASA NIX.

Orbiter Communications The space shuttle orbiter is the first manned vehicle to have near-continuous communications in Earth orbit, which proved to be one of the major operational weaknesses in the previous manned programs. Crew communications, vehicle and payload telemetry, video data downlink, and command uplink data that were available for fifteen minutes in a ninety-minute orbit, were increased to seventy-five to eighty minutes with NASA's geostationary tracking and data relay satellite system (TDRSS). The TDRSS network consists of two geostationary communications satellites spaced approximately 190° apart, with a third satellite placed across from the vertex at the same 36,220 km (22,500 mi) altitude. The TDRSS ground station is a dual-redundant network located near White Sands, New Mexico. Orbiter communications is also available through the space flight tracking and data network (STDN) ground stations initially installed for the Mercury project and used for the Gemini, Apollo, and Skylab programs. Communications is also possible through the Air Force Space-Ground Link System (SGLS) ground stations.

Three primary frequency bands are used for the orbiter communication, with some overlap in functions. The highest bandwidth communications is also the highest frequency band which is handled only through the TDRSS satellite link. The lowest frequency UHF band is used for audio communications with EVA astronauts and other specialized applications. The workhorse communications link is the S-band communications that transfers telemetry, payload data, commands, and some video data.

Communications Types

- Telemetry
- Command
- Rendezvous and tracking
- Video
- Voice communications
- Documentation

Communications Bands

- S-band
  - Two modulation types
    - FM (frequency modulation)—downlink of vehicle telemetry
    - PM (phase modulation) systems—all other data
  - Communications with ground on two separate uplink and two separate downlink frequencies
    - Designed to allow two shuttles to fly simultaneously
  - Communications links through STDN, TDRSS, SGLS
- Ku-band
  - Ku-band communications assembly is located in payload bay
    - Can only operate while in orbit
  - High data rates available for onboard experiments and instruments, and for video link
  - Includes a Ku-band radar system for rendezvous and a tracking signal for accurate ground tracking
    - Radar cannot be used simultaneously with communications

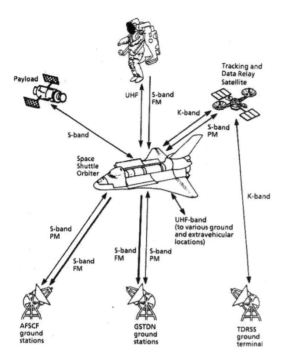

Figure 10.13. Sketch of the Orbiter communications links including the two ground networks (Air Force Satellite Control Facility and the Goddard STDN), as well as the TDRSS satellite network. Source: NASA SODB.

- UHF
  - Used for voice communications, primarily during EVA
  - Data available for biotelemetry
  - UHF voice communications also available through TACAN and other aircraft

Guidance, Navigation, and Control (GN&C) The STS and orbiter flight control systems include guidance and navigation instruments, GN&C software, and guidance equipment. The GN&C computations and operations are controlled by the onboard general purpose computers. Ground-based calculations of the vehicle's position and motion during the STS mission operations and flight segments are integrated into the commands executed by the GN&C software routines for orbit entry, orbit adjustments, rendezvous operations, and deorbit. Because of wide variations in the atmosphere, and communications blackout for part of the reentry, GN&C commands during the reentry and landing are managed by the orbiter's onboard computers.

Guidance Functions

- Thrust phase during ascent to orbit—nozzle steering on SSMEs and SRBs are used to correct trajectory
- Orbit operations—guidance commanded by thrust-vectored OMS engines

Navigation Functions

- Accurate measurement, calculations, and dissemination of vehicle position and velocity (state vector)

- Integration of acceleration measurements made by inertial measuring unit (IMU)
- Integration of star tracker utilities in orbit for accurate attitude alignment
- Integration of crew optical alignment sight (COAS) for accurate orbit measurements and rendezvous
- Integration of onboard GPS for continual input in state vector calculations (primary source for position and velocity data)
- Includes microwave scan beam landing system (MSBLS) precision guidance unit for landing at primary and backup sites

Control

- Gimbaled SSMEs and SRBs during launch and ascent
- OMS and RCS used for orbital flight and reentry
- Aerodynamic surface control during ascent, descent, and landing
- Operation of the flight controls can be either automatic or manual
  - Manual flight operation input controls are:
    - Rotational hand controller
    - Translational hand controller
    - Speed brake/thrust controller
    - Rudder pedals

The orbiter's flight control surfaces are driven by redundant hydraulic systems, each powered by a gas-turbine auxiliary power unit (APU). The rudder, elevons, SSME thrust vector control system, body flap, brakes, nose-wheel steering, and several small actuators for the unbilical doors are directed by the triple-redundant hydraulic systems during the orbiter's mission. Three APUs used to generate hydraulic pressure for three independent hydraulic systems are commanded by several control functions that include the guidance systems, manual braking and steering commands, and sequencing signals sent from the GPCs. APU turbines are powered by hydrazine propellant decomposing into an expanding, hot gas.

Electrical Power System The space shuttle orbiter's electrical system is used as the primary electrical power source for all STS elements, including the Orbiter, the ET, and the dual SRBs. The orbiter's three onboard fuel cells convert hydrogen and oxygen gas into electrical current, water, and heat. The H_2 and O_2 reactants are stored as liquids in cryogenic tanks before being pressure-fed as gas into the fuel cells. Electrical power output for each of the fuel cells is nominally 7 kW under normal loads. Each fuel cell supplies one of three independent, redundant electrical circuits on the orbiter, with one assigned to supply the payloads. Electrical power is supplied to the orbiter by ground service equipment while on the launch pad and during the processing preparations.

The orbiter's electrical power system operates at the standard aerospace bus voltage of 28 Volts dc, with a nominal power output of 7 kW for each fuel cell. Even though the orbiter could function with the power of just one fuel cell, all three fuel cells must be operational for continuing mission operations in orbit. Two of the fuel cells typically support the orbiter's loads, while one supplies electrical power for the payloads.

Each of the orbiter's three independent electrical buses are supplied by one fuel cell, with the capability of being cross-strapped to any other bus for power sharing. In-

Figure 10.14. Photo of a complete Orbiter fuel cell, with the fuel cell stack that provides the power located beneath the white insulated coating shown on the left. The power accessory package on the right of the fuel cell stack manages power output, conditions and regulates reactants, removes water, and controls temperature variables throughout the fuel cell. Source: NASA NIX.

dependent circuitry for each of the fuel cell primary DC buses is distributed throughout the orbiter, with external buses that supply the ET and SRBs.

AC (alternating current) power is generated from each of the three independent electrical buses. The three-phase AC power is used to power onboard motors, specific avionics, and fuel cell operations. Each of the primary buses has interconnect and isolation circuitry to allow independent use of all three power sources (cells) for emergency operations.

The major components of the orbiter's electrical power system are divided into three subsystems.

1. Fuel cell power generators (FCs): The orbiter's fuel cell trio generate electrical power by combining the gas reactants, which produces electricity, potable water, and heat.
2. Electrical power distribution and control (EPDC): DC and AC power is distributed throughout the orbiter, ET, and SRBs by distribution assemblies that include circuit breakers and isolation circuits.
3. Power reactant storage and distribution system (PRSD): Cryogenic reactant storage holds the liquid hydrogen and liquid oxygen that is heated to a gas before combining in the fuel cell stacks. Cryogenic heating, pressure regulation, and flow regulation of the reactant gases are managed by the electrical power accessory package.

II. SOLID ROCKET BOOSTERS

NASA's choice of solid rocket boosters for the STS instead of liquid flyback boosters was made with both development costs and operational costs in mind. Ultimately, the lower development time and costs of the solid boosters were more attractive as NASA's budget came under pressure in the early 1970s. Whether liquid boosters would have offered a lower cost alternative over the program's life is speculative, but the choice of the simple, reusable solid rocket booster has revealed the booster's merits, as well as its weaknesses.

The same attractive simplicity and scalable performance of the three-decades-old SRBs are being adapted for the new Constellation program with less success. The dual booster configuration that served so well as the first stage of the STS is far less well-behaved as a single-stack booster because of the problematic thrust oscillations that are averaged out as dual boosters, and damped by the 4.5 million lb weight of the orbiter, the filled ET, and the loaded SRBs. The dual-booster configuration planned for the

Ares V heavy-lift launcher appears to be less prone to thrust oscillations for the same reasons the STS encountered only modest vibrational loads..

Specifications for the original SRBs design were for each booster to furnish 3.3 million lb of thrust at liftoff for a two-minute burn, and supply 72% of the total initial thrust. Actual performance is close to the original requirements, with typical burn time of 2 minutes, 10 seconds, taking the STS to an altitude of about 45 km (28 mi, or 150,000 ft) before separation.

The high-thrust SRB was designed with the solid propellant filling four separate motor segments, each with a steel alloy casing that makes up the combustion chamber. The articulated nozzle on the bottom section was designed for reuse, with layers of carbon fiber protecting the throat and joint by ablation. Each of the four motor segments propellant charges is cast with a star-shaped hollow core for rapid, controlled, even combustion. The hollow core also allows the ignition flame to propagate down the entire length of the SRB motor at the start sequence. For uniformity, all eight motor segments of the two SRBs used for each STS launch are cast in a single batch.

SRB major components

Motor The SRB motor is comprised of four segments assembled into a single rocket motor pressure chamber (casing) that includes propellant and a nozzle assembly.

- Aft (bottom) segment: includes the aft skirt, solid rocket separation motors, nozzle, and the thrust vector control system including hydraulics. An attach ring that surrounds this casing is used for mounting the SRB to the external tank at the bottom.
- Aft mid segment: installed at the top of the aft segment.
- Forward mid segment: installed above the aft mid segment.
- Forward (top) segment: attached to the forward mid segment. This segment includes the top dome on the combustion chamber and igniter assembly. A top attach

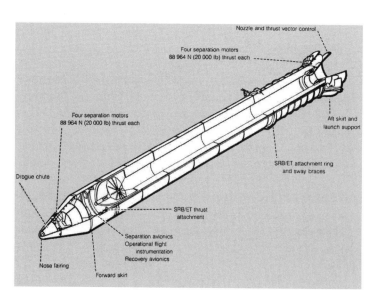

Figure 10.15. Cutaway view of the SRB outlining major components and specifications. Source: NASA SP-407.

Table 10.4. SRB Specs

Weight (each)	589,700 kg (1,300,000 lb)
Empty weight (each)	87,090 kg (192,000 lb$_f$)
Thrust (each)	1,49,800 kg (3,300,000 lb$_f$)
Length	45.5 m (149.2 ft)
Diameter	3.72 m (12.2 ft)
I$_{sp}$	269 sec

ring that surrounds this casing is used for mounting the SRB to the external tank toward the top.

Case The SRB structure consists of four assembled steel alloy casings that hold the propellant, and a propellant-casing insulation that prevents the burning propellant from reaching the steel shell. Each segment casing is attached to at least one other segment with a ring-pin-clevis device that prevents hot gas from escaping between the three segment seams.

Propellant Propellant for the SRB motor segments is cast with an 11-point star-shaped hollow core inside of an insulation layer lining the segment casings. A single batch makes up two complete SRBs for uniform performance. The propellant composition has the consistency of a rubber eraser and consists of:

- Ammonium perchlorate (oxidizer) 69.6%
- Powdered aluminum (fuel) 16%
- Iron oxide (catalyst) 0.4%
- PBAN (polybutadiene acrylic acid acrylonitrile) polymer binder 12%
- Epoxy curing agent 1.96%

Igniter Ignition of the SRB is initiated with the first stage of a three-stage igniter that begins with a NASA standard detonator (NSD). The NSD ignites a larger second stage igniter, that, in turn, ignites the large 1.14 m (45 in) long third-stage igniter that shoots a flame down the hollow core of the entire SRB motor.

Nozzle The SRB nozzle is a reusable, articulated, steel alloy and carbon composite bell-shaped exhaust nozzle used to build combustion chamber pressure and control thrust direction. The nozzle is designed as an integral part of the aft segment, incorporating a nozzle expansion ratio of 7:79. An extension on the aft nozzle is fitted with a shaped charge to separate the nozzle extension from the nozzle joint before splashdown to prevent damage to the expensive joint. Layers of carbon cloth fiber overlay the nozzle throat and joint for protection from the extreme combustion heat by the process of ablation.

Frustum The fifth segment of the SRB called the frustum is attached to the forward (top) motor segment. The cone-shaped frustum contains four solid rocket separation motors, separation pyrotechnics, and the recovery and operational electronics. Included are the communications and sequencer avionics, the triple parachute system, the recovery beacon and lights, the range safety system, and an integrated electronics assembly. The solid rocket separation motors, each 79.0 cm (31.1 in) long and 32.5 cm (12.8 in) in diameter, provide thrust for approximately one second to push the SRBs away from the orbiter and ET after separation, in concert with a set of four identical separation motors located on the aft skirt.

Thrust Vector System Thrust control for the powerful SRB is provided by the articulated nozzle and dual-redundant APU-driven hydraulic systems located in the aft skirt. Under direction of the orbiter's guidance commands, the integrated electronic assembly in the aft skirt drives the dual hydraulic actuators to position the SRB nozzle for trajectory guidance, with a neutral command coming just before SRB separation.

Range Safety Destruct System The SRB's range safety destruct system is a requirement for terminating abnormal flight for any launch vehicle. The pyrotechnic charge consists of a shaped line charge running the length of the SRB motor that is connected to a separate receiver and electronic logic system under the command of the range safety officer. A similar pyrotechnic charge and receiver-initiator system was placed on the external tank during the early STS missions, but has since been removed.

Hold-Down Posts Although not part of the solid rocket booster, each of the SRBs is held to the mobile launcher platform at the aft skirt segment with four bolt-nut pairs. Four large, spring-loaded hold-down posts are positioned below each of the SRB aft skirt structures which pass through the aft skirt hold-down brace. A frangible nut on top of each hold-down post is tightened with enough force to retain the entire weight of the STS on the mobile launcher platform, including the orbiter and filled ET. Each of the four nuts on each SRB has two holes drilled from top to bottom to receive a NASA standard detonator. Following SRB ignition and positive pressure readings from within the nozzle, a command to fire the sixteen NSDs on each of the eight hold-down nuts retaining the two SRBs will split the nuts, releasing the eight hold-down posts, and free the two SRBs.

SRB Operations

SRBs are ignited by signals originating from the master events controller that initiate a three-stage igniter. The third stage shoots a flame down the entire length of the SRB motor segments. Within one second, full thrust is reached in both SRBs. Even though the solid rocket cannot be stopped before the propellant is completely consumed, the thrust output can be modified slightly during the burn. To reduce the SRB's thrust as the STS passes through maximum dynamic pressure (max Q) near Mach 1, the hollow core is designed with a slightly enlarged diameter near the middle of the booster. The diagram in figure 10.16 shows the modest thrust reduction near the middle of the burn, as well as a gradual burnout before SRB separation is initiated.

As the SRBs reach fuel exhaustion, their separation from the external tank is initiated by sensor readings inform within the nozzle reaching a 50 psi threshold. Separation commands coming from the pressure sensors backed up by a timed signal ensure a reliable sequence. Included in the SRB-ET separation is the nut-bolt separation at the top and bottom attachment at the ET, and the ignition of four solid rocket separation motors at the top and bottom of both SRBs. A post-separation apogee of approximately 67 km (220,000 ft) is reached before the SRBs descend for a parachute-slowed splashdown approximately 225 km (140 miles) downrange. Buoyant flotation bags

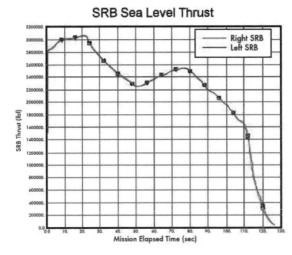

Figure 10.16. SRB thrust data from the STS-107 mission showing the designed thrust reduction near the middle of the burn timed for the vehicle passing through max Q. Source: NASA NIX.

inflate from the frustum before splashdown to prevent the aft section from taking on water and sinking the SRB. Both of the burned-out booster shells are retrieved by ship and towed back to the KSC recovery facility for cleaning. After further processing and disassembly, the SRB steel motor casings are shipped back to Utah by rail for reuse.

III. EXTERNAL TANK

Propellants for the Orbiter's three space shuttle main engines are supplied by the large external tank; the largest component, and the heaviest of the STS components when filled. The unfilled external tank is also the lightest of the main STS components, and the only non-reusable structure on the STS. The ET's extraordinarily light empty weight is an integral part of the tank design that also makes up the ET structure. Separate oxygen and hydrogen thin-walled tank shells provide the external tank's structure, fitted together with a cylindrical intertank assembly. The ET's efficient design is reflected in its structure (empty) weight being only 3% of its loaded weight. Put another way, propellants make up 97% of the filled external tank's weight.

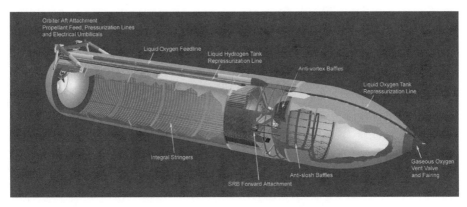

Figure 10.17. Cutaway diagram of the external tank. Source: NASA NIX.

The external tank is constructed from lightweight alloys of aluminum and lithium-aluminum and divided into three units: the H_2 tank, the O_2 tank, and the intertank. Attach points between the SRBs and Orbiter are placed at the top on the intertank, and on brace and truss assemblies at the bottom on the ET. Higher-density liquid oxygen is positioned above the light-weight liquid hydrogen tank for dynamic stability after SRB separation. The SSME thrust line is through the orbiter and outside of the external tank which would allow an undesirable rotation of the ET-orbiter if the lighter fuel tank were atop the heavier oxygen tank.

Because of the ET's semi-monocoque design, propellant lines are routed to the orbiter from the bottom of the hydrogen tank, and from the bottom of the oxygen tank running outside the hydrogen tank. These two propellant lines feed into the orbiter's SSME distribution manifold near the bottom of the ET at the aft attach points. Also included in the propellant plumbing are fuel and oxidizer pressurization lines that run from the SSMEs back to the tanks, and ET tank vents, as well as liquid propellant service ports and lines. Additional components on or in the ET include tank depletion sensors, external thermal insulation, structural attachment points for the orbiter and SRBs, and electronic power and control subsystems. Because the cryogenic propellants are stored at extremely low temperatures, the ET's thermal insulation is also a critical part of the tank design.

Hydrogen Tank

This is the largest component of the ET because liquid hydrogen is the lowest density propellant. Insulation protecting this section from ice buildup is more critical because of the extremely low temperature of the stored liquid hydrogen (-254°C, -425°F). Positive tank pressure is maintained on the pad by the boil off of liquid hydrogen (LH2), and during ascent by gaseous hydrogen pressurization lines and LH2 overflow lines from the SSMEs.

Oxygen Tank

The oxygen tank is placed above the hydrogen tank and connected by the intertank. While dynamical stability is improved with the LOX tank at the top of the ET, the long feed line to the orbiter's umbilical feed at the bottom of the ET introduces pressure, or pogo, oscillations that are suppressed by a dedicated device built into each of the three SSME oxygen feed lines. Positive tank pressure is maintained on the pad by the boiloff of LOX, and during ascent by gaseous oxygen pressurization lines from the SSMEs.

Intertank

The intertank assembly is used to physically connect the hydrogen tank to the oxygen tank, and to house the electronics, control, and power circuitry, and provide much of the structural strength and load transfer capability for the ET. Connecting points for the two SRBs and the orbiter are located on the intertank, which is also responsible for the transfer of significant thrust loads from the SRBs and the SSMEs. The unpressurized intertank is also used to mount the fill and drain ports for the ET.

Table 10.5. ET Specs

Original	
Length	47.0 m (154.2 ft)
Diameter	8.4 m (27 ft)
Weight	Empty: 35,426 kg (78,100 lb) Loaded: 756,600 kg (1,668,000 lb)

Super lightweight tank	
Length	46.9 m (153.8 ft)
Diameter	8.4 m (27.6 ft)
Weight	Empty: 26,540 kg (58,500 lb) Loaded: 762,136 kg (1,680,000 lb)

LOX tank	
Length	16.6 m (54.6 ft)
Diameter	8.4 m (27.6 ft)
Volume (@ 22 psig)	553,355 liters (19,541.66 cubic ft; 146,181 gal)
LOX mass (@ 22 psig)	629,340 kg (1,387,457 lb)
Operation pressure	138-152 kPa (20-22 psia)

Intertank	
Length	6.9 m (22.6 ft)
Diameter	8.4 m (27.6 ft)

LH_2 tank	
Length	29.5 m (97.0 ft)
Diameter	8.4 m (27.6 ft)
Volume (@ 29.3 psig)	1,497,440 liters (52,881.61 cubic ft; 395,582 gal)
LH_2 mass (@ 29.3 psig)	106,261 kg (234,265 lb)
Operation pressure	221-235 kPa (32-34 psia)

IV. SPACE SHUTTLE MAIN ENGINES

The SSME's primary components are similar to the traditional staged liquid propellant engines that include a combustion chamber, turbopumps, injector, and nozzle. Performance of the SSMEs is enhanced by the dual high-pressure turbopumps for both the fuel and oxidizer, and the elaborate engine cooling scheme that incorporates extensive circulation of the liquid hydrogen propellant. Typically, fuel and oxidizer are forced through the injector from feed lines coming directly from the turbopump. For the SSME, added cooling comes from propellants directed through a manifold surrounding the injector head assembly before injection. Liquid hydrogen is heated as it circulates through the nozzle tubing and

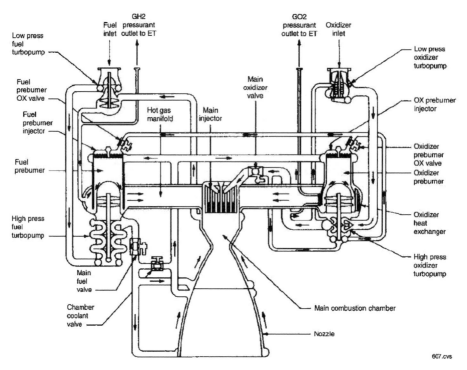

Figure 10.18. Flow diagram of the space shuttle main engines showing the dual turbopump and feed systems for the fuel (left half), and the oxidizer (right half). Source: NSTS.

combustion chamber jacket to cool those areas. The hydrogen and oxygen are then forced through the manifold to the injector assembly by the high-pressure oxidizer and fuel turbopumps. The high-pressure turbopumps themselves are driven by hot combustion gases from small, individual fuel and oxidizer combustion chambers called preburners. These preburners consume a small portion of the oxidizer and fuel gases to drive the turbopumps, with the preburner exhaust routed to the hot gas manifold that feeds into the injector assembly. The injector assembly is also cooled by the part-liquid, part-gas propellant flow. Since engine propellants are used to power the preburner that drives the high-pressure turbopumps, the engine is classified as staged, or dual-stage.

The orbiter's three SSMEs are the second stage of the space shuttle's primary propulsion system, providing 28% of liftoff thrust for the first two minutes, and 100% thrust for the remainder of the eight-and-one-half-minute ascent to orbit. In addition to thrust vectoring, the main engines are also capable of variable thrust ranging from 65% to 109% of the thrust rating for the original Block I engine design. The 65% thrust configuration is programmed into the guidance software to reduce aerodynamic loads on the vehicle when approaching max Q near Mach 1.

The space shuttle main engines were designed specifically for the STS by Rocketdyne, now part of Pratt & Whitney. The design originally included a seven-hour operational life capable of 100 starts with a required breakdown and rebuild after each start because of the extreme mechanical stress and temperatures during engine operation. Liquid hydrogen, the coldest of the cryogenic fuels, is stored at -253°C (-423°F),

Table 10.6. SSME Specs (Block IIA)

Parameter	Condition	Performance Value
Max thrust (109%)	Sea level	189,900 kg$_f$ (418,660 lb$_f$)
	Vacuum	232,670 kg$_f$ (512,950 lb$_f$)
Throttle range		67%–109%
High-pressure turbopump output pressure	Hydrogen	6,230 psia
	Oxygen	7,320 psia
Turbopump power	Hydrogen	51,483 kW (69,040 hp)
	Oxygen	18,754 kW (25,150 hp)
Chamber pressure		3,008 psia
Area ratio		69:1
Weight		3,393 kg (7,480 lb)
Mixture ratio (O$_2$/H$_2$)		6:1
Dimensions	Length	4.27 m (168 in)
	Width	2.29 m (90 in)
I$_{sp}$	Vacuum	452 sec

Note: SSMEs are rated in output performance and matched in sets of three for mission assignments. SSME design life ratings are 100 starts, or 7.5 hours accumulated time (Block II). .

while the temperature in the main engine combustion chamber can reach 3,115°C (6,000°F)—higher than the boiling point of iron.

Payloads and Processing

Processing the complete STS vehicle in preparation for launch is an enormous undertaking that requires about 700,000 man-hours, much of which is focused on ensuring the safe, reliable operation of the largest and most complex manned vehicle ever flown in space. Extensive checks and documentation are also needed because each mission, each crew, each payload, each equipment installation, and each software package is different. To help make the preparations and operations of the flight as safe and reliable as possible, the myriad processing steps for the diverse missions and equipment are standardized as much as practicable.

Initial planning for each mission begins five years or more before the flight with the development of a flight and equipment manifest. Following the approval of the mission and primary payload, the next step in the planning phase is the assignment of an orbiter for the mission. The choice of a specific orbiter may hinge the variations in orbiter performance and equipment such as airlock requirements, vehicle mass, or extended duration capability, or may depend on the requirements of the primary payload. A flight number is then assigned to the mission and orbiter from the sequential STS-XXX numbering convention. Processing or equipment delays can alter the actual launch sequence from the mission number sequence. A recent example is the STS 124, STS-126, STS-119, STS-125, STS-127, STS-128, STS-129 launch sequence that spanned 2008-2009.

Planning and vehicle and payload processing begin to merge a year or more before the launch as the payloads and equipment start arriving at the Kennedy Space Center. The orbiter vehicle arrives for preparations for its next mission at the KSC Shuttle Landing Facility after completing its previous mission, or following its scheduled OMDP maintenance and upgrades. Included in the early planning stages for a mission are the training procedures for the crew.

ORBITER PREPARATION

Processing the orbiter for its next flight takes place in the Orbiter Processing Facility (OPF), with additional steps completed in the Vehicle Assembly Building (VAB), and on the launch pad. Orbiter processing begins when the orbiter is towed into one of the three processing bays available in two buildings at the Kennedy Space Center. A typical period for processing an orbiter in preparation for its assembly on the ET and SRBs is three months from tow-in to the OPF, to roll-over from the OPF to the Vehicle Assembly Buiulding.

Routine maintenance as well as vehicle preparation and servicing are carried out on the orbiter before the processing cycle is complete. Early orbiter processing events include a comprehensive, careful inspection of the fragile insulation tiles and removal of the previous payload equipment and attachments. After reconfiguration of the orbiter's payload bay, the new payload attachments and power, signal, and other services are installed in preparation for the new mission payloads. Smaller and intermediate payloads are then installed in the payload bay, while large payloads are loaded into the orbiter while at the launch pad to avoid excessive loads during the processing cycle.

Processing of the orbiter in the OPF covers the inspection, and, if necessary, the repair or replacement of systems and components down to the level of wiring, nuts, and bolts, although a more detailed inspection and maintenance takes place during the orbiter's OMDP. Fluids, consumables, and propellants are examined, and, if required, topped off or replaced. System checkout for all systems and related components re-

Figure 10.19. Orbiter parked in the OPF following roll-in from the KSC landing strip. Source: NASA NIX.

quire electrical power that is supplied by ground service equipment through the same T-0 umbilical panel on the rear of the orbiter used for launch preparations.

Two of the most time-consuming processing efforts for the orbiter are the inspection and repair or replacement of the thermal protection system tiles, panels, and blankets, and the software installation and checkout. The fragile thermal protection system requires hand-fitting of the tiles, blankets, and panels. About fifty of the HRSI tiles are replaced each cycle, each taking about two weeks. Software is also labor-intensive since each mission differs and each software installation must be tailored for the complete orbiter operations with perfect or near-perfect reliability.

After completion of the processing tasks, the orbiter is placed on a carrier, gear up, and towed into the Vehicle Assembly Building in a brief trip called the rollover.

VEHICLE ASSEMBLY BUILDING

The Vehicle Assembly Building was originally constructed to assemble the Saturn V launcher and crew vehicle during the Apollo program. Assembly steps for the Saturn V were similar to those used for the STS assembly, and include much of the same equipment. The 160.0 m (525 ft) tall VAB covers eight acres, and contains four low bays plus four high bays. Each of the high bays can be used for assembling a complete STS vehicle, though only two are used (bays 1 and 3 on the East side of the VAB). Low bays previously used for assembling the Saturns are now used for storing ETs and other equipment.

Like the Saturn V, the STS is assembled on a mobile launcher platform (MLP) in one of the high bays of the VAB. After assembly, the complete STS and MLP are lifted by the six million pound crawler-transporter for rollout to the launch pad.

Assembly of the STS vehicle begins with the mobile launcher platform base which is refurbished from an earlier launch and placed in one of the two available VAB

Figure 10.20. Rollout of the completed STS from the VAB to the launch pad on the Crawler-Transporter shown beneath the wider Mobile Launcher Platform. Source: NASA NIX.

high-bays by the crawler-transporter. Following the complete checkout of the SRB segments, stacking of both SRBs begins with the attachment of their bottom (aft) sections to the MLP base. A general sequence of assembly is as follows:

SRB stacking: Stacking of the SRBs begins with the bottom section and progresses with subsequent sections stacked on top of the previous ones.

ET mount: After testing and preparations, the ET is lifted into position astride the SRB stack assemblies and mounted with bolts and explosive (frangible) nuts.

Orbiter mate: The orbiter is mated to the external tank after being lifted vertically into the VAB assembly high bay during a process that takes only five to seven days.

Vehicle checkout: The electrical, data, propellant, and vent lines are tested and checked after the orbiter attachment, along with the systems affected by the Orbiter-ET-SRB coupling.

Vehicle rollout: After assembly, integration, and checkout in the VAB, the STS vehicle and MLP used to be taken to either launch pad 39-A or 39-B. Today, the launches are staged from the Launch Complex 39A pad since 39B was deactivated in January 2007 for reconfiguration to accommodate the Ares 1.

LAUNCH PREPARATIONS

After assembly of the STS, the vehicle is moved from the VAB to the launch pad by the crawler-transporter. The vehicle and the MLP are placed on the launch pad using six primary mounts to accommodate the MLP's six steel pedestals 7 m (22 ft) high. Vehicle processing on the launch pad takes about one month to prepare for launch. Final systems checks are made, along with reactant and propellant loading except for the ET propellants which are loaded approximately eight hours before launch.

Payloads that are too large to be safely installed in the orbiter's payload bay while the vehicle is in the OPF are installed on the launch pad. A vertical payload carrier is used to transport the payload to the launch pad, then lift it into the payload changeout room inside the launch pad's rotating service structure. Payload installation and integration are completed during the vehicle's one-month stay on the launch pad.

A final review of the vehicle and launch systems and the launch procedures is run two weeks before the actual launch. This launch countdown demonstration test includes the launch teams as well as the flight crew. Provided that the preparations and paperwork are completed, the STS launch countdown begins three days before the planned launch

Figure 10.21. Endeavour shown being lowered onto the ET-SRB assembly for attachment and integration in preparation for the STS-54 launch. Source: NASA NIX.

with the count beginning at T-43 hours. The total elapsed time is approximately 72 hours for the entire countdown, which includes 29 hours for planned holds.

LAUNCH

The rotating service structure that protects the STS vehicle during launch pad processing is pulled back from the orbiter one day before launch, exposing the completed vehicle for the first time. The launch pad sound suppression system tank is filled with 300,000 gallons of water the day before launch, which is released below the rocket exhaust just before engine start. Following the load operations for the ET propellants, the final systems checks are made, and all personnel are cleared from the launch pad area. At T-5 minutes, the APUs are started and tested, followed by turnover of the STS countdown decisions and flight control to the orbiter's onboard computers at T-29 seconds, although abort commands can be issued by launch processing computers and ground controllers.

Figure 10.22. Liftoff of space shuttle Columbia on the STS-93 mission that placed the Chandra X-ray Observatory in orbit. STS-93 was the first mission commanded by a female pilot, Eileen Collins. Source: NASA NIX.

T-6.6 sec—Orbiter's three SSMEs are started in a 120 ms sequence to reduce thrust loads.

T-0—SRB ignition sequence is initiated, followed by release of the eight hold-down posts, then liftoff of the STS. The ascent sequence of the STS begins with the handoff of vehicle control from the Kennedy Space Center to the Johnson Space Center as the vehicle clears the launch tower.

T+20 sec—An automated maneuver is executed to roll the vehicle over on its back, though it is still rising just 10° from vertical.

T+24 sec—SSMEs are throttled back to 65% thrust as the vehicle passes through max Q, followed by the command to resume 109% thrust ("throttle up") at T+60 sec.

T+2 min, 10 sec—Burnout and separation of SRBs.

T+8 min, 30 sec—Main engine cutoff (MECO) occurs when orbit entry is established which varies with mission. Orbit entry is completed with either 1 or 2 OMS engines burns.

RETURN TO KSC

After deorbit, reentry, and landing at the Kennedy Space Center, the orbiter is towed from the runway to the Orbiter Processing Facility. While on the runway and enroute

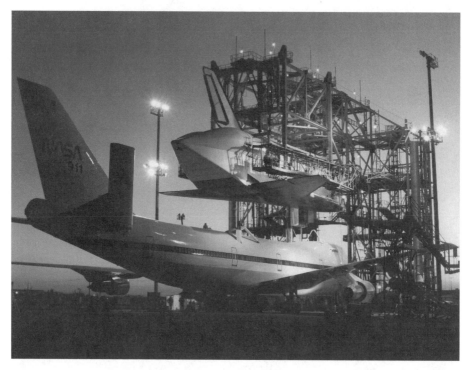

Figure 10.23. Photo of Orbiter Atlantis undergoing a mating to the B-747 Shuttle Carrier Aircraft at the NASA Dryden facility adjoining the Edwards Air Force Base. Source: NASA GRIN.

to the OPF, ground crews begin payload checks, flight crew return preparation, and external inspection and safeing. Approximately four hours after landing, the Orbiter arrives at the OPF and again begins the preparation cycle for its next flight mission.

When adverse weather forces a landing at Edwards Air Force Base, the orbiter goes through preliminary processing before being flown back to the Kennedy Space Center on the B-747 shuttle carrier aircraft. After landing at the KSC landing strip, the orbiter is demated from the SCA and towed into the OPF.

References

Heppenheimer, T. A. 1999. *The Space Shuttle Decision: NASA's Search for a Reusable Space Vehicle.* NASA SP-4221. Washington, DC: NASA Office of Policy and Plans.

NASA NIX (NASA Image Exchange). Retrieved from: http://nix.nasa.gov/

NASA GRIN (Great Images in NASA). Retrieved from: http://grin.hq.nasa.gov/ (accessed 7/2/2009).

CHAPTER 11

Spacecraft Systems

T he space environment represents one of the most hostile operational environments because of the extreme temperatures, the intense radiation levels, the vacuum conditions, and the high-velocity particles and debris encountered. In addition to the extreme environment, the inability to perform maintenance or repairs on spacecraft means that spacecraft structure and system designs are expected to provide the ultimate in durability and reliability. Adding to the adverse environment and the need for absolute reliability are the high costs of development, design, testing, fabrication, and operation that introduce other demands on spacecraft design. As a result, the design of spacecraft and spacecraft systems is highly specialized for specific mission objectives, which makes spacecraft even more costly because of their unique design and single-unit construction. Without economy of scale, spacecraft that are often built in units of two (one to fly, one for ground testing and troubleshooting) are extraordinarily expensive. A crude analogy to building and launching a single spacecraft into orbit would be the construction of a supersonic aircraft to fly a single package across the Atlantic, then scrapping the aircraft after its arrival. The point of the analogy is not in the scrapped vehicle, but in the value of the mission and its justification. While no one would consider building a unique aircraft to deliver a single piece of cargo, the search for the origins of the solar system or the universe, or finding the influence of the Sun's cycle on the Earth's climate can sometimes only be done from space, and with expendable launchers that are typically more expensive than the largest cargo aircraft. It is the extreme environment of space and the enormous cost of space flight that make mission planning, spacecraft design, and space flight operations with near perfect reliability so challenging. done from space—accessible with expendable launchers often more expensive than the largest cargo aircraft.

Project Planning

A typical spacecraft project begins with a conceptual outline of the mission objectives that are further refined toward a definition of the overall project. Research teams and design panels mold the details of the exploration project in several stages of review before the spacecraft design and development phase begins. Following preliminary, intermediate, and final design reviews, details of the spacecraft design and its systems begin progressing to the fabrication and testing phase. Final design of the spacecraft and its systems is often a compromise between advanced technology and the costs of the new technology, and perhaps the time required for its development. Extensive testing as the spacecraft nears completion is used to verify the design expectations and to prepare the spacecraft for integration with the launch vehicle. Final evaluation and tests are carried out before launch, then after deployment, usually in a planned parking orbit.

Spacecraft systems and design concepts will be sketched in this chapter to outline the principles of basic spacecraft systems and their operations. It is by no means a comprehensive review. More complete and detailed discussions on spacecraft systems are available in a variety of textbooks, and in online references too numerous to list.

This chapter is divided into seven basic system groups that represent the makeup of most, but certainly not all, spacecraft.

- Propulsion systems
- Communications
- Guidance, navigation, and control
- Command and data handling systems
- Electrical power systems
- Thermal control
- Structures

Propulsion Systems

Rocket propulsion that is accurately described by Newton's action-reaction principle is the sole means of reaching space and changing orbits or trajectories in space. The action portion of Newton's equation is the thrust generated by the rocket motor; while the reaction is the motion of the rocket opposite to the thrust. The traditional chemical rocket motor produces thrust by the combustion of the fuel and oxidizer either in liquid or solid form. Heat of combustion expands the resulting gas, which forces the exhaust gas out of the nozzle at high velocity. Other types of rocket propulsion, including electric, compressed gas, and nuclear, employ the same action-reaction principle, but a vastly different mechanism to eject exhaust gas, and with much different performance results.

Propulsion principles outlined in chapter 2 introduced the concepts of thrust (force), thrust duration (energy), and specific impulse (propulsion efficiency). These three concepts serve as the foundation for selecting specific propulsion systems for specific space applications. If categorized into the magnitude of thrust required, propulsion systems fall into three basic applications. A more appropriate classification of propulsion capabilities and applications uses ΔV—the change in velocity available from the propulsion system, or the change in velocity necessary to reach the objective orbit or trajectory. Since ΔV can measure both propulsion system performance and the required performance to reach various distances from the point of launch or boost, matching a launch vehicle and payload to the objective orbit or trajectory is simplified. Another measure of performance uses the same ΔV, but as energy which is obtained by squaring the ΔV value. A match between the booster and the target orbit or trajectory can be made in the same way, but using slightly different units. The difference in the two expressions of propulsion system performance is due to convention rather than any underlying physics. In general, propulsion requirements for near-Earth or lunar orbits are listed in ΔV (km/s), while interplanetary propulsion requirements are listed in ΔV squared units known as C3 (energy as km^2/s^2 with mass divided out). Conversion of ΔV into C3, or the reverse, is more than squaring ΔV or taking the square root of C3, since ΔV is a measure of kinetic energy, and C3 is the sum of both kinetic and gravitational potential energy. C3 is also zero at the Earth's sphere of influence, roughly three times the Moon's distance from Earth (see chapter 5 for details).

Commonly used C3 values

Venus—11 km^2/s^2
Mars—17 km^2/s^2
Jupiter—80 km^2/s^2

LAUNCH VEHICLE PERFORMANCE

Matching payload and mission requirements to a launch vehicle and its capabilities can be approximated by using either graphic plots or tabulated values of vehicle performance. But these approximations are very rough estimates since the launch site, launch opportunity, parking orbit, and the position and orbit of the departure and the arrival planets or moons at launch are also important variables in determining the actual booster energy required. Nevertheless, examples of the launch vehicle performance versus target distance are useful for comparing past and present launchers.

Similar data for a wide variety of expendable launch vehicles are listed for various payload weights and shroud sizes, as well as distances and orbit inclinations. The launch vehicle specifications include inclination angles corresponding to the launch latitude (28.3° for Cape Canaveral—Kennedy Space Center), as well as approximate costs in 1994 dollars (NASA 2007).

LAUNCH VEHICLE EXAMPLES

Delta

The Delta launcher had its origins in the same Cold War missiles that were converted into the early expendable launch vehicles including the Atlas, Redstone, Titan, Thor,

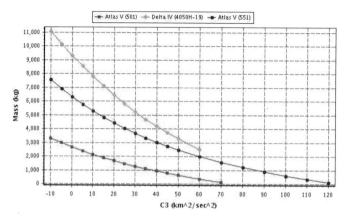

Figure 11.1. Comparison plot of the Atlas V 501 and 551, and the Delta IV Heavy 4050 boosters. The Atlas 501 is a single-core unit with a single-engine Centaur upper stage, while the Atlas 551 is the same core unit and Centaur upper stage, but with five strap-on first-stage boosters. The Delta IV Heavy 4050 model has a single-core second stage with two strap-on first-stage boosters that are identical to the core stage, and a solid rocket third stage. Source: NASA Launch.

and Saturn 1. Delta began simply as a renamed four-stage Thor and quickly became popular as an interplanetary and satellite launcher. Today, Delta is a family of medium- to heavy-lift vehicles that includes the Delta II, the Delta III, and the Delta IV. The Delta IV heavy boosters have little in common with the original missile heritage, which is in stark contrast with Russia's Cold War launchers from the 1950s and 1960s, which have undergone improvements in components and systems, but still serve as military and civil boosters.

Atlas

The Atlas booster also has its origins in the U.S. missile inventory as the first American ICBM, although it was quickly converted into a reconnaissance satellite launcher and a heavy-lift booster that delivered John Glenn to orbit in 1962. The original stage-and-a-half Atlas missile has undergone complete redesign as a series of medium- to heavy-lift boosters that today include the Atlas III, and Atlas V series. In an interesting adaptation of international technology, the new Atlas III and Atlas V, and Atlas V Heavy have replaced the earlier LR-89 and LR-101 engines with a single Russian RD-180 for the core stage. The Atlas V Heavy incorporates three of the RD-180 core stages, with the two outside rockets serving as strap-on boosters. All of the newer Atlas family launchers use the Centaur booster as an upper stage.

UPPER-STAGE AND ORBITAL BOOSTERS

Upper-stage boosters used for launch vehicles, or for boosting payloads to distant orbits, are intermediate in size between launchers and small-thrust attitude control thrusters. These upper-stage and orbital boosters can be simple solid rocket motors

Table 11.1. U.S. Expendable Launch Vehicle Data for Planetary Missions (NASA 2007)

Launch Vehicle Family	Model + booster	Incl. (deg)	Cost (approx.) (FY'94 $M)	Payload weight delivered (kg)					
				LEO	GTO	C3=0	C3=10	C3=50	C3=100
Atlas	I	28.5	77–88	5,820	2,375	1,532	1,185	169	-
	II	28.5	84–88	6,580	2,610	1,999	1,598	434	-
	II-Star 48B	28.5	100–104	4,439	-	1,943	1,650	857	378
	IIA	28.5	90–104	7,280	2,745	2,176	1,752	547	-
	IIA-Star 48B	28.5	106–120	5,139	-	2,157	1,830	945	427
	IIAS	28.5	110–142	8,640	3,379	2,698	2,215	845	-
	IIAS-Star 48B	28.5	126–158	6,499	-	2,625	2,220	1,156	602
	IIIA	28.5	-	-	3,400	-	-	-	-
	IIIB	28.5	-	-	4,500	-	-	-	-
	V (30x, 40x)	28.5			5,100				
	V (5xx)	28.5			8,200				
Conestoga	1229	38	13–14	363	-	-	-	-	-
	1379	38	15–16	771	-	-	-	-	-
	1620	38	18–19	1,179	-	-	-	-	-
	1669	38	15–16	1,361	-	-	-	-	-
	1679	38	21–22	1,497	-	-	-	-	-
	3632	38	21–22	2,141	-	-	-	-	-
	5672	38	27–28	-	-	506	389	136	37

(Continued)

(Continued)

Launch Vehicle

Family	Model + booster	Incl. (deg)	Cost (approx.) (FY'94 $M)	Payload weight delivered (kg)						
				LEO	GTO	C3=0	C3=10	C3=50	C3=100	
Delta	II-7325	28.7	55-60	2,760	-	754	614	270	97	
	II-7920	28.7	49-60	5,045	1,270	692	379	-	-	
	II-7925	28.7	55-60	-	-	1,277	1,041	461	167	
	III	28.7	-	-	8,345	3,800	-	-	-	
	IV (EELV)	28.7	-	-	-	-	-	-	-	
Pegasus	Pegasus	28.0	8-14	455	125	-	-	-	-	
	XL-C Star 24C	23.0	14-14	544	-	98	77	30	-	
	XL-C Star 27	23.0	14-14	544	-	112	90	36	-	
Taurus	Taurus	28.5	21-26	1,420	514	329	263	107	35	
	XL Star 37XFP	28.5	23-26	1,565	595	372	296	120	39	
	XL/S Star 37FM	28.5	23-26	1,980	736	474	378	153	50	

Titan

IIG/Star 37	28.7	33-38	2,655	-	248	192	-	-
IIG/Star 48B	28.7	33-38	2,655	-	610	492	67	-
IIS-SSPS	99.0	44-55	2,445	-	-	-	-	-
IIS-PAM D2	99.0	44-55	2,885	-	-	-	-	-
IIS-4SRM-SSPS	99.0	44-55	3,342	-	-	-	-	-
IIS-4SRM-PAM D2	99.0	44-55	3,665	-	-	-	-	-
IIS-10GEM	28.7	44-55	5,470	-	1537	1116	730	-
III/TOS	28.6	165-245	14,515	11,000	3,610	2,693	3,025	-
IV/Centaur	28.6	435-475	18,144	-	7,477	6,330	3,989	775
IV/SRM/Centaur	28.6	435-475	18,144	-	9,323	7,867	986	1,707
IV/SRM/IUS	28.6	330-435	23,350	2,360	3,988	3,193	-	-

(Continued)

(Continued)

Launch Vehicle

Family	Model + booster	Incl. (deg)	Cost (approx.) (FY'94 $M)	Payload weight delivered (kg)					
				LEO	GTO	C3=0	C3=10	C3=50	C3=100
	40	5.2	71-82	4,900	2,070	-	-	-	-
	42P	5.2	73-85	6,100	2,920	-	-	-	-
	44P	5.2	77-88	6,900	3,380	-	-	-	-
	42L	5.2	100-112	7,400	3,450	-	-	-	-
	44LP	5.2	106-118	8,300	4,170	-	-	-	-
Ariane (ESA)	44L	5.2	130-141	9,600	4,700	2,940	2,341	940	301
	5	28.5	118-130	18,000	6,800	-	-	-	-
Energia (Russia - not in production)	EUS	51.6	120-120	88,000	-	35,680	31,091	17,446	8,006
	EUS/RCS	51.6	120-120	88,000	-	-	-	-	9,735
	II	30.4	157-157	10,500	4,000	2,466	1,950	281	-
	II SPKS	30.4	157-157	10,500	-	2,793	2,338	1,148	472
H (Japan)	IIA	30.4	-	10,000	4,000	-	-	-	-
	IIA LRB	30.4	-	14,000	6,000	-	-	-	-
J (Japan)	1	30.4	-	1,000	-	-	-	-	-

	2C	28.5	24-24	3,500	1,000	–	–	–	–
	2E	28.5	47-47	9,200	3,370	–	–	–	–
	3A	28.5	–	7,000	2,300	–	–	–	–
	3B	28.5	–	13,600	4,850	–	–	–	–
Long March (China)	3	28.5	39-39	5,000	1,330	–	–	–	–
	4B	97.8	–	1,440	–	–	–	–	–
M (Japan)	3SII	31.2	35-37	780	517	–	–	–	–
	V	31.2	45-47	1,950	1,215	–	–	1,917	–
	D-1-e	51.6	59-82	20,000	5,500	5,866	4,845	1,917	–
	D-1-e Star 27	51.6	59-82	20,000	–	5,912	4,884	2,154	624
	D-1-e Star 48B	51.6	59-82	20,000	–	5956	4990	2492	1085
	M-5 (Fregat)	51.6	59-82	–	–	4,838	4,279	2,484	1,061
Proton (Russia)	K	51.6	–	20,900	4,800	–	–	–	–
	M	51.6	–	22,500	5,500	–	–	–	–
Soyuz (Russia)		51.6	18-18	7,000	–	–	–	–	–
Zenit (Ukraine)	2	51.6	77-82	13,740	–	–	–	–	–
	3	51.6	77-82	–	4,300	2,911	2,238	398	–

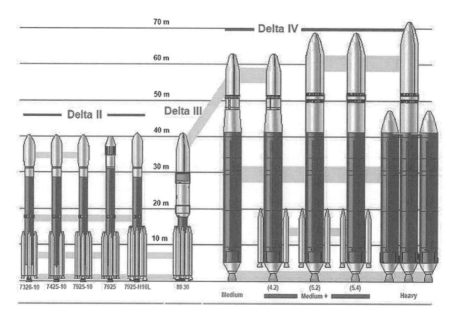

Figure 11.2. Lineup of the current Delta expendable launch vehicle family. Source: NASA NIX.

such as the Star 48, or high-performance liquid bipropellant units like the famed Centaur that was developed for military launchers in the 1960s and is still used today.

Star 48

The solid-fuel Star series of upper-stage boosters is used primarily on the Delta II launcher, and, for a time, on several payloads in the space shuttle orbiter. Star 48, also known as the payload assist module, or PAM, was developed by Morton-Thiokol (now ATK) with two nozzle lengths; the larger used for the Delta II third stage, and the shorter nozzle length for use on earlier space shuttle missions. A Star 48 was also used for the third stage booster on the Atlas V launch of the New Horizons mission to Pluto launched in 2006.

- Diameter: 1.2 m (48 in)
- Length: 2.03 m (80 in) (Delta model B)
- Weight: 2,141.3 kg (4,720.8 lb) (Delta model B)
- Thrust: 68.64 kN (15,430 lb_f) (Delta model B)

Inertial Upper Stage (IUS)

Boeing developed a two-stage solid rocket booster in the early 1980s called the Inertial Upper Stage (IUS). The booster was originally designed for military payloads launched on the Titan III, then Titan IV, and then on a number of space shuttle interplanetary payloads. Following the Challenger accident, new safety precautions prevented the use of liquid-propellant boosters for Orbiter payloads, which made the solid-propellant IUS the default upper-stage for interplanetary missions launched on the space shuttle.

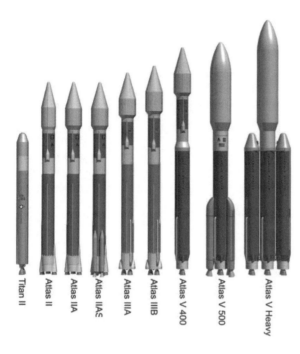

Figure 11.3. Lineup of the newer Atlas commercial space launchers with a Titan II booster on the left for scale comparison. Source: NASA NIX.

A total of twenty-five missions were launched with the Boeing IUS between its first flight in 1982 until its retirement in 2004.

- Diameter: 2.8 m (9.25 ft)
- Length: 5.18 m (17 ft)
- Weight: 14,740 kg (32,500 lb)
- Thrust (first stage): 202.82 kN (45,600 lb$_f$)

Centaur

Centaur originated as a high-performance booster for the U.S. Air Force, but was handed over to NASA to complete the project development for use on interplanetary missions. The Centaur was the first U.S. booster to use liquid hydrogen and liquid oxygen propellants, a high-energy combination that powers most of the upper-stage boosters around the world. Centaur is a lightweight booster with either a single or dual RL-10 engine configuration. The LH2 and LOX tanks also provide the structural strength for the booster unit, like the early Atlas ICBM design. In fact, the Centaur is still used as an upper stage on the new Atlas launchers.

- Model: Centaur-3 DEC (dual-engine configuration)
- Engines: 2 x RL-10A-4
- Diameter: 3.05 m (10.0 ft)
- Length: 11.7 m (38.4 ft)
- Weight: 22,940 kg (50,570 lb)
- Thrust: 198.4 kN (44,602 lb$_f$)

ATTITUDE CONTROL THRUSTERS

Attitude control for spacecraft orientation and instrument pointing is typically furnished by small thrusters since only a small torque is needed to rotate even large vehicles in a zero gravity environment. With only small torques required, a wide variety of small thrusters are available for attitude conrol including low-efficiency compressed gas thrusters, and high-efficiency, low-thrust ion/plasma engines. These small torques can also be provided by internal momentum exchange devices such as momentum wheels, reaction wheels, or control moment gyros. Attitude thrusters range in size from the space shuttle orbiter's 3.88 kN (870 lb$_f$) primary thrusters to Boeing's XIPS xenon ion thrusters that produce 18 millinewtons of thrust for geostationary satellite station keeping and orientation control.

Communications

The basic spacecraft communications system consists of on-board data and intercommunications links, as well as the link with ground communications. The microwave band is used for the space-ground link since the atmosphere is relatively transparent to the 1-10 GHz frequencies. Higher frequencies can and often are used for satellite intercommunications since there is little or no atmospheric interference. Consideration is also given to the system design for sufficient bandwidth to carry the instrument and operations data at the appropriate rates. The spacecraft data links are also designed for reliable communications even during Doppler frequency shift due to the continually changing relative motion between the spacecraft and ground station. Spacecraft communications systems for interplanetary missions must also be designed with enough flexibility to transmit and receive at extreme distances, and often in high-noise environments.

The basic spacecraft communications system is comprised of four functional components:

- Transmitter
- Receiver
- Antenna
- Command system for data handling

For continuous communications, two separate frequencies are used for the spacecraft transmitter and receiver, which correspond to an uplink (to the spacecraft from ground), and a downlink (from the spacecraft to ground). The simultaneous, or duplex, communications use separate frequencies that typically carry commands up to the spacecraft and telemetry from the spacecraft to the ground stations. Larger spacecraft and manned vehicles may have three, four, or even five separate communications bands to carry the various types of data. As an example, the space shuttle orbiters are outfitted with five separate communications channels to transfer voice, video, telemetry, commands, rendezvous, tracking, and documentation data.

Table 11.3. Commonly Used Spacecraft Communications Bands

Band	Frequency
UHF	300 MHz to 3 GHz
L-band	1–2 GHz
S-band	2–4 GHz
C-band	4–8 GHz
X-band	8–12.5 GHz
K, Ka, Ku-band	12.5–40 GHz

Spacecraft communications are typically routed either through orbiting communications satellites such as NASA's Tracking and Data Relay Satellite System (TDRSS) network or through fixed ground stations. Two ground-based space communications networks managed by NASA are the Spaceflight Tracking and Data Network Ground Network (STDN-GN) and the Deep Space Network (DSN). As the name implies, the Deep Space Network is used for communications with interplanetary spacecraft, but it can also be used for spacecraft on lunar missions. NASA's STDN ground network was originally developed for the Mercury manned program to fill some of the large gaps in communications coverage inherent in the first satellite receiver station network built in the late 1950s.

TRACKING AND DATA RELAY SATELLITE SYSTEM

NASA's Tracking and Data Relay Satellite System was designed specifically to provide continuous or nearly continuous communications coverage for manned spacecraft and satellites in low Earth orbit using high-bandwidth, high-frequency Ku-band equipment. Continuous coverage for LEO spacecraft is furnished by two geostationary communications relay satellites on opposite sides of the Earth. A third geostationary satellite orbiting above the ground station at White Sands, New Mexico is used for the ground link. Six more TDRSS satellites are positioned in geostationary orbit for spares and special-use satellites.

Two duplicate ground link stations support the TDRSS satellite communication links with high-frequency Ku-band, S-band, C-band, and UHF frequencies. The TDRSS C-band is used only for the TDRS (Tracking and Data Relay Satellite) operations, and is not available to TDRSS users. Tracking information and duplex communications are available for two spacecraft simultaneously on the TDRSS Ku-band channels, with twenty more satellites supported with lower-bandwidth S-band communications at the same time, but without tracking.

DEEP SPACE NETWORK

NASA's Deep Space Network was developed for continuous or near-continuous coverage of interplanetary missions located anywhere in the solar system. Three communications

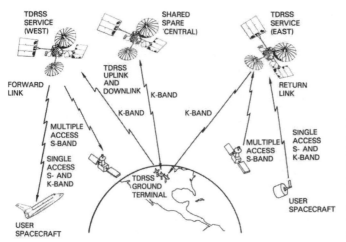

Figure 11.4. Sketch of the TDRSS satellite links between orbiting spacecraft, the primary ground station, and the TDRSS communications and tracking satellites. Source: NASA NIX.

facilities are placed roughly 120° from one another in longitude that provides a communications link to spacecraft virtually anywhere in the sky. In practice, coverage is not available for all positions because the ground facilities are not exactly 120° from one another, nor on the equator. The ground station geometry is further reduced in coverage because of increased attenuation of signals by the atmosphere near the horizon.

NASA's DSN is managed by the Jet Propulsion Lab and consists of three main elements; the three Deep-Space Communications Complexes (DSCC), the Network Operations Control Center (NOCC), and the Network Spacecraft Test and Launch Support Facility. The three DSN complexes are located at:

- Canberra, Australia
- Madrid, Spain
- Goldstone, California

Each of the three DSN communications facilities contains at least four deep space receiver/transmitter platforms, each with a corresponding antenna. These include at a minimum:

- 34 m (111 ft) diameter high-efficiency antenna
- 34 m beam waveguide antenna
- 26 m (85 ft) antenna
- 70 m (230 ft) antenna

NASA's DSN, which is a part of the NASA communications network (NASCOM), is managed by Jet Propulsion Lab and operated by the Goddard Space Flight Center.

SPACEFLIGHT TRACKING AND DATA NETWORK

NASA's Spaceflight Tracking and Data Network is actually a combined space communications network composed of the space segment, or space network (SN), the TDRSS

Figure 11.5. Photograph of the three-beam waveguide antennas and receiver stations located at the Goldstone, California, DSN complex. Source: NASA NIX.

network, and the ground network (GN) consisting of supporting ground stations and operations facilities. The origins of the NASA STDN reach back before NASA existed, even before the first satellite launch, when a network of more than a dozen ground stations was installed to track the first American satellite in support of the International Geophysical Year (1957–1958). The Minitrack satellite network first intercepted and tracked the Sputnik satellite launched in October 1957, then in January 1958, it met its intended purpose by tracking the Explorer 1 satellite. Ironically, the Minitrack tracking network proposal was partly responsible for the selection of the Navy's Vanguard satellite over the Army's Explorer satellite, then called Project Orbiter (Tsiao 2007). A larger space communications network called the Satellite Tracking and Data Acquisition Network (STADAN) replaced Minitrack to accommodate the proliferation of satellites in the early 1960s, including commercial communications satellites.

A parallel space communications network was developed for NASA's new manned space flight program named the Mercury Space Flight Network (MSFN), which provided greater coverage for the critical Mercury capsule operations. The Mercury communications network was later renamed the Manned Space Flight Network in conjunction with the newly developed Manned Space Flight Center in Houston. STADAN and MSFN eventually were brought under the management of NASA's Goddard Space Flight Center in Greenbelt, Maryland. Then in 1975, the two were merged into the Space Flight Tracking and Data Network (STDN) that still exists today (Tsiao 2007). The new STDN space communications network augmented the ground stations with a new geostationary communications and tracking satellite system for continuous coverage for the space shuttle and other high-bandwidth spacecraft. NASA's STDN is now a combined ground station network and TDRSS space network, along with the Network Control Center.

The ground segment of the STDN (the GN) encompasses all of the ground station communications resources, including the Deep Space Network, the Satellite Tracking and Data Network stations, the Wallops Flight Facility, and the Western Aeronautical Test Range (WATR). The Wallops Flight Facility is the lead center for NASA's balloon and sounding rocket programs, and provides management and technical oversight for NASA's ground stations at the White Sands Missile Range, the National Scientific Balloon Facility in Palestine, Texas, the Poker Flat Research Range in Fairbanks, Alaska, and the McMurdo Ground Station in McMurdo, Antarctica. The WATR is the lead organization for NASA's aeronautical flight testing activities, which also provides services for the space shuttle operations, if needed. The WATR's three California ground stations are managed by Ames Research Center's Dryden Flight Research Facility.

COMMUNICATIONS SYSTEM PERFORMANCE PARAMETERS

For reliability, spacecraft communications systems are typically designed with multiple antennas, dual transmitters, dual receivers, and further component and software redundancy that include overlapping functions. Multiple components not only allow for backup in case of failure, but provide operational checks for subsystems and components. Just as important as reliability is the spacecraft communication system's ability to transfer data at the necessary rates under varying conditions. Because signal strength decreases as the square of the separation distance, signal levels must be tailored for the greatest spacecraft distance expected during the mission operations and any anticipated mission extensions. Doppler frequency shifts in the transmitted and received signal must also be designed into the system to accommodate the spacecraft's changing velocity to avoid signal degradation. Noise level models and error rates estimates must be accurate enough to permit science objectives or commercial communications requirements to be met during the entire life of the mission.

Antennas

Antennas play an important role in the spacecraft communications system since they are used to couple the communications signal in space to the communications system electronics. The size and shape of the antennas are their two most identifiable features, and the two most important measures of an antenna's efficiency. Larger antennas can collect or concentrate more signal and are therefore useful to increase the signal level compared to the background noise. One exception to the universal advantage of large antennas for spacecraft is the need for a broad signal path for early deployment of spacecraft. The broad area coverage is useful for the initial deployment of satellites and spacecraft in parking orbit since the uncertainty in the spacecraft position is greatest during deployment. Hence, spacecraft often carry a small, broad-beam, or omnidirectional, antenna to ensure communications in almost any orientation. These small (low gain) communications antenna can also be used as a backup for the primary high-gain antenna. Otherwise, the general rule-of-thumb applies to spacecraft antennas. Larger communications antennas are used on interplanetary spacecraft sent to more distant regions of the solar system.

For deep space missions, the signal loss from the normal $1/r^2$ power decrease is significant, which requires careful antenna design to overcome the loss in signal power with increasing distance. Since power levels for spacecraft communications transmitters are usually on the order of 10–15 Watts, the received signal is extremely weak and must be carefully amplified with as much gain as possible and as little noise as possible. Again, this is improved with a large (high gain) antenna. However, there are physical limitations on the antenna's mass and diameter. The Voyager high-gain antenna, for example, is 3.7 m in diameter—almost 12 feet—which allows it to communicate at a distance of more than 10 billion km.

Antenna diameter is not only going to determine the signal gain, but will also determine how wide the beam is, and as a consequence, how accurately the antenna must be pointed. As antenna diameter increases, gain increases as the diameter squared. The antenna beamwidth decreases in proportion to the diameter increase. In general:

- Large antenna:
  - Higher gain—required for large distances.
  - Smaller beamwidth—more precise pointing required.
- Small antenna:
  - Lower gain—suitable for low Earth orbit.
  - Broader beamwidth—easier pointing.
- Deep-space spacecraft require large, high-gain antennas to improve signal power (both received and transmitted).
- Spacecraft in noisy environments, such as those inside the orbit of Venus or Mercury, may also require high-gain antennas to improve signal power.
- A signal-to-noise ratio greater than 2:1 is a traditional design target for simple systems, but can be less for encoded or repeated (averaged) signals.

Typical spacecraft antennas are parabolic in shape, with the parabola used as a reflector to concentrate the microwave signal on the much smaller antenna element located at the focus. Larger parabolic reflectors concentrate more signal strength, which provides more gain; a relative term used to compare signal strength to the basic dipole antenna element.

- Parabolic antenna gain = $\eta\left(\dfrac{\pi\,d\,f}{c}\right)^2$ where η is efficiency that ranges from 50% to 70%, d is antenna diameter, f is frequency, and c is the speed of light

Calculation example: Gain of a 2.0 m parabolic antenna at 2.35 GHz with 65% efficiency can be calculated using:

$$G = \eta\left(\frac{\pi\,d\,f}{c}\right)^2 = 0.65\left[\frac{\pi \times 2.0\,\text{m} \times 2.35\text{x}10^9\text{Hz}}{2.998\text{x}10^8\,\text{m/s}}\right]^2 = 1{,}577 \text{ in dimensionless units.}$$

Note that the frequency must be expressed in Hz (1 GHz = 10^9Hz).

- Parabolic antenna beamwidth $\Theta° = 63 \, \dfrac{c}{d \, f}$ with the constant 63 used for beam-width in degrees, and c the speed of light in m/s, d is diameter in m, and f is frequency in Hertz.

Calculation example: The beamwidth of a DSN 70 m antenna operating at 3.5 GHz can be calculated using:

$$\Theta° = 63 \, \frac{c}{d \, f} = 63 \, \frac{2.998 \times 10^8 \, \text{m/s}}{70 \, \text{m} \times 3.5 \times 10^{10} \, \text{Hz}} = 0.0077°$$

Transmitter

Signal power levels for spacecraft communications transmitters are typically 5–20 Watts, with higher power levels for larger spacecraft, and much higher power for communications satellites. A 10 W signal traveling from the distance of Jupiter would be on the order of 10^{-24} Watts when received, which is below the threshold of most receivers. To help increase the signal power at the receiver, the transmitter power could be increased, although the practical limit on spacecraft is about 20 Watts. A larger antenna could also be installed on the spacecraft, or an ultra-low-noise receiver design could be used, or a combination. Another technique commonly used to compensate for decreasing signal strength with increasing distance employs variable data transmission rates. With increased distance, the lower-power signal can still be identified, but at a slower transmission rate, which also means a lower bandwidth. What remains constant in the lower received signal strength and the lower data rate is the level of acceptable error in the data, a term known as the bit error rate. Invariably, deep-space spacecraft communications take place at low bandwidths (low data rates).

Signal coding is also used to help discriminate between the signal and the background noise. By encoding the transmitted signal, the combined signal plus background noise at the receiver can be separated, provided the noise power does not overwhelm the signal power and that the noise is random in nature. For nearly all spacecraft, the communications signal coding used is pulse code modulation (PCM). This technique allows recognition of signals at higher noise levels than with other coding/modulation techniques. Without proper coding, reasonable error rates would require that the signal be twice the noise level or more (S/N ≥ 2:1).

Receiver

Because of the extremely weak signals common to spacecraft communications, the receiver must be capable of detecting weak signals while introducing as little noise as possible. To improve weak signal detection, an amplifier is placed at the input of the receiver to boost the signal strength. Unfortunately, it also boosts the noise level. As a consequence, careful noise management and signal-to-noise improvement begins with the selection of the optimum antenna and a sensitive first-stage receiver where noise is most critical. Front-end amplifiers on ground stations can use very low-temperature (cryogenically cooled) components to reduce input noise, while spacecraft receivers are more limited in their sophistication because of weight restrictions. The lowest noise, cooled spacecraft receivers are not found in spacecraft communications systems, but used to detect the extremely weak microwave signals from the cosmic microwave background radiation on the COBE, WMAP, and Planck satellites.

Guidance, Navigation, and Control (GN&C)

Spacecraft guidance, navigation, and control systems include the onboard components and the ground elements used to guide, orient, and change the spacecraft position and orientation according to the mission requirements. Orbital maintenance, orbit transfer, orbit precession, station keeping, pointing, midcourse corrections, tracking, and many other spacecraft operations require complex navigation and control systems. These spacecraft command functions may be controlled autonomously or controlled from ground-based computers and uplinked to the spacecraft. A combination of the two functions is relatively common and includes even the most complicated spacecraft to fly, the space shuttle.

Errors and error corrections are inherent in any space flight operation and are often computed and corrected using ground tracking and onboard sensors. Actual position, velocity, and orientation data are compared to desired mission parameters, but because deep-space missions can be minutes or even hours away in light travel time, ground-based commands are not executed in real time but programmed ahead for critical mission events. Hence, interplanetary spacecraft have sophisticated onboard guidance and navigation systems that can integrate programmed commands from ground for executing planned operations and used for contingencies.

The earliest spacecraft, which were also the simplest, had no control functions for changing attitude or orbit. Stabilization for these early explorers was furnished simply by spinning the spacecraft. Onboard guidance, navigation, and control was first introduced by the Soviets on their Luna 3 mission, which traversed the far side of the Moon, followed by America's Mariner interplanetary spacecraft series. The first manned Mercury and Vostok capsules contained attitude control functions, but could not change orbits except for deorbiting at the end of the mission. Onboard computers soon followed on the Gemini capsules, which could calculate and command orbit changes, but ground tracking and uplinked commands still provided the primary orbit control functions, as they did for the Apollo missions to the Moon.

A few definitions are useful in the following discussion on basic spacecraft GN&C:

- **Guidance**—spacecraft trajectory control during the thrust phase of a launch, or during orbit changes
- **Navigation**—determination of spacecraft position and velocity relative to a specified reference frame, or desired position and velocity
- **Control**—force or torque applied for active or passive control over the spacecraft attitude, orbit, and/or trajectory
- **Stabilization**—establishing a preferred, general, or fixed orientation of a spacecraft using either passive or active control functions, including thruster stabilization, gyroscopic stabilization, spin stabilization, and gravity gradient stabilization

The two primary functions of a GN&C system consist of 1) attitude, trajectory, and orbit determination, and 2) attitude, trajectory, and orbit control.

Attitude, Trajectory, and Orbit Determination

- Determination of a spacecraft's position, motion, and orientation is made by on-board sensors or by ground or satellite tracking.
  - Sensors include linear and rotational accelerometers, three-dimensional laser ring gyros, star trackers, limb or horizon sensors, and magnetic field detectors.
- Calculations of spacecraft attitude, orbit, or trajectory are made with respect to specified reference frame and desired position and velocity.
  - Position and velocity data are compared to a specific or desired position and velocity at specified time to evaluate the need for change.
  - Navigation software computes corrections or changes in spacecraft orientation, position, or velocity, if needed.
  - Commands are formulated and sent to control function components to make appropriate corrections.
- To provide accuracy and versatility for navigation functions, sensors include both relative and absolute position and velocity reference.
  - Relative position and velocity reference measures small changes with sensitive sensors, which provides highest accuracy in position and velocity changes.
    - Examples include accelerometers, laser gyros.
  - Absolute position and velocity reference provides position and velocity reference to a general coordinate system.
    - Examples include GPS, star trackers.

Attitude, Trajectory, and Orbit Control

- Control systems are used to stabilize spacecraft or to alter spacecraft position or attitude.
- Passive stabilization—little or no internal power required. Examples:
  - Gravity gradient force
  - Aerodynamic pressure
  - Solar pressure
  - Planetary magnetic fields
- Active system—onboard power and electromechanical components required for spacecraft position or attitude control. Examples:
  - Thrusters
  - Gyros (momentum wheels, reactions control wheels, control moment gyros)
  - Spin and dual-spin stabilization

EXAMPLES

A practical example of a versatile yet simple GN&C system by today's standards is the Apollo capsule guidance, navigation, and control system. Apollo was the first manned spacecraft to use onboard computers for completely autonomous navigation operations, although there are several qualifications to the innovative Apollo GN&C hardware. First, the earliest onboard computer for controlling the spacecraft in preparation for rendezvous and docking orbit changes and navigation was placed on the Gemini capsule. Second, the commands for refined corrections to the Apollo capsule trajectory and orbit were made from ground tracking data and relayed to the capsule guidance and navigation system to ensure sufficient accuracy and redundancy.

The Apollo capsule carried two types of instruments for navigation input: optical instruments for absolute reference, and inertial instruments for relative and more accurate reference. The two instrument types are sketched in figure 11.6.

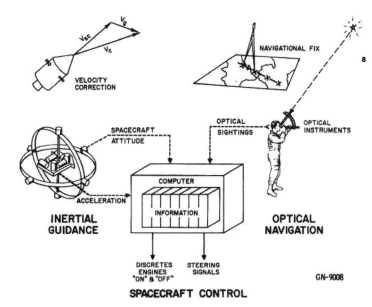

Figure 11.6. Sketch of the Apollo inertial components used for accurate relative motion reference, and optical instruments used for absolute position reference. Source: AGNS.

Command and Data Handling Systems

Digital computers used for onboard spacecraft command processing and data handling typically have a number of specialized processors that perform specific tasks needed for the diverse spacecraft functions. These processors are often, at least for the critical operations, multiple-redundant. The amount of data acquired, transferred, and stored with today's spacecraft command and data handling systems is often so great that dedicated subsystems are used to manage the data flow, data download, and data storage. Large spacecraft, like the space shuttle or the International Space Station, contain dozens of data processors and many of those used for data formatting and data transfer called multiplexers/demultiplexers.

Digital data systems on modern spacecraft can be subdivided into six or seven functional blocks or subsystems that make up the communications, data storage, operations computations, and command operations. One of the most important elements of the data handling process is the management of command instructions for the spacecraft and its systems. Since the commands from ground links pass through the communications system receiver, part of that communications system is dedicated to the command and data handling functions. This unit is commonly called the command processor.

Typical spacecraft command and data handling primary functions include the following:

- Executive control—software and hardware elements used for command and control, resource control, and scheduling
- Command processing—responsible for interpretation/decoding, validation, verification, and distribution of uplink and onboard commands for spacecraft operations
- Computation/calculation—execution of instructions responsible for data manipulation, scaling, calibration, formatting, compression, processing, and analysis
- Data acquisition—sensor and instrument data is typically processed through an A/D (analog-to-digital) converter, then transferred for digital processing and/or storage
- Data storage—bulk digital data storage for payloads, instruments, communications (uplink and downlink data stored temporarily for backup and verification), telemetry, onboard operations, and systems status
- Communications interface
- Payload and mission-specific operations—individual payloads and instruments require specific operations and data handling by the specific hardware and software design of the C&DH system, which must also contain effective fault tolerance and error handling functions

APOLLO COMPUTER

The Apollo computer was first of its kind that introduced a host of innovations that revolutionized the electronics industry. The need for Apollo's unique computer hardware arose from its mission and strict weight and power limitations. The Apollo guidance computer (AGC) had to be much more powerful than previous spacecraft digital computer designs since Apollo missions would involve complex maneuvers at, on, and near the Moon a quarter of a million miles from Earth. Precise navigational accuracy, unsurpassed reliability, and low weight and low power consumption meant that the Apollo computer designers would have to pioneer important advances in the new semiconductor circuitry field. Integrated circuits were used on a large scale for the AGC for the first time, along with new techniques for physically connecting the computer's components and modules.

NASA awarded contracts for development of the Apollo computer in 1961 to the MIT Instrumentation Lab for the AGC design and to the Honeywell Corporation for the production and quality assurance. Although the same computer was used on both of the Apollo flight modules, separate operating systems were developed for the Apollo command module and for the lunar module because of the vastly different operating environments and risk profiles. Logic for the integrated circuits was the resistor-transistor logic manufactured by Fairchild Semiconductor. Memory for the AGC was ferrite core memory that consisted of tiny donut-shaped rings that could be changed from one polarity to the opposite polarity for binary data storage simply by passing current in opposite directions though wires wound around the minute memory core units. Both the rewritable random-access memory (RAM) and the read-only memory (ROM) used the same ferromagnetic cores, but with a different wiring scheme. The AGC's memory contained 2,048 words of 16-bit length for the RAM and 36,864 words for the ROM instructions. Cycle time was 11.72 ms, which was paced by the AGC's 2.048 MHz crystal oscillator (NASA 1988).

The legacy of the Apollo digital computer system was primarily the advancement of the integrated circuitry technology and the miniaturization of electronics. But within the ten years from the first development efforts to the last Apollo lunar flight in 1972, the rapid advances in digital circuitry and computers left the Apollo digital hardware far behind.

Figure 11.7. Apollo guidance computer display keyboard (DSKY) unit. Source: NASA GRIN.

Electrical Power Systems

Reliable electrical power is an absolute necessity for spacecraft operations since all onboard systems are dependent on electrical power for their operations. Because of the importance of the spacecraft power supply, the design goals for these systems cover a fairly long list that is abbreviated here as the basic requirements for spacecraft power systems:

- Lightweight components
- Lowest cost desirable while satisfying the other design and operational objectives
- An output life at least as long as the expected life of the mission and possible extensions
- Provide continuous and reliable electrical power regardless of space environment encountered
- Avoid failure modes that would allow interruption of power to the systems at any point in the mission

Design of the electrical power system begins with the specifications supporting the mission objectives in a process that ultimately leads to the selection of the system components. A simpler power source selection process employs two major variables—power level and mission duration—to identify the appropriate spacecraft power source. As an illustration, table 11.4 outlines the approximate duration and power level used to select a space-qualified power source.

Spacecraft electrical power types include:

- Solar
- Magnetic and electric fields
- Chemical batteries
- Fuel cells
- Nuclear thermal
- Nuclear reactor

Table 11.4. Space Electrical Power Source Duration and Power Level Range

Source	Duration (approximate)	Power Level (approximate)
Batteries (primary)	Minutes to 2 weeks (Gemini was 14 days)	Watts to 1 kW
Fuel cells	Days to months (STS maximum was 17 days)	Watts to 100 kW
Solar arrays	Weeks to 20 years (limited by degradation of array cells)	Watts to 500 kW
RTGs	1 year to 60 years (isotope half-life is 88 years)	20 W to 10 kW
Nuclear reactors	1 year to 50 years (limited by isotope half-life or ability to resupply fuel)	10 kW to 10 GW or more

SOLAR

Solar Photovoltaic (Solar Cell)

Photovoltaic arrays are attractive for spacecraft power systems because of their simplicity, low weight, low cost, and lack of emissions. The few disadvantages for PVAs are their limited operating range in the solar system and the need for rechargeable batteries as a secondary power source for orbital missions, which adds considerable weight to the system.

Solar photovoltaic cells generate electrical current as solar photons mobilize electrons in the conduction bands of semiconductors. This produces a current flow with a voltage equal to the energy required to overcome the band-gap energy for the electron. Typical energy values for semiconductors are 0.5 to 0.7 volts, which represents the work potential of the specific material (volts and work have the equivalent dimensions of energy). Lower-energy photons in the IR band do not have enough energy to generate electron current and only heat the substrate, which reduces power output and can reduce the life of a solar cell. Higher-energy photons in the UV and X-ray bands produce the same electron voltage on the solar cell surface, but deposit additional energy, which can damage the molecular structure of the crystal, reducing its photon-to-electron conversion efficiency. Gallium arsenide (GaAs) semiconductor material offers higher conversion efficiency than silicon germanium (SiGe) and many other silicon-based semiconductor materials. The largest spacecraft photovoltaic array powers the International Space Station using over 260,000 silicon crystalline cells with a thin silica oxide (SiO_2) surface coating with the capability of converting sunlight into electrical power from either side of the array.

Voltage produced by a solar cell is typically 0.6 Vdc, which means that if an electrical power system requires a voltage supply of 100 Vdc, the 0.6 volts/cell output needs a series of 100V/0.6V/cell = 167 cells. Since the current supplied by a single solar photovoltaic cell is on the order of 0.001 amp, the cells must be connected in parallel to combine the electron current to reach the required current; for this example, 1.0 amps. The total number of cells in parallel would be 1.0 amp/0.001 amp/cell = 1,000 cells. The total number of array cells would then be 167 × 1,000 cells to develop 100 volts at 1.0 amp. This amounts to 1A × 100V = 100 Watts of power (1 Watt = 1 Volt × 1 Amp). Typical spacecraft power requirements are on the order of 1,000 W to 2,000 W, while the ISS has a combined output of about 110 kW.

Degradation in PVA power output is caused by particle radiation, thermal cycling from periodic exposure to the Sun, and small micrometeoroid impacts. This power output reduction must be taken into account when estimating PVA size and output requirements for the mission lifetime and possible mission extensions. Assuming a 5% annual degradation in the PVA output from the cumulative effects of space exposure, a ten-year exposure would correspond to a power reduction of 50% of the original (beginning-of-life, or BOL) output. The size of the array, if designed for the end-of-life (EOL) rather than the BOL output in this example, would be twice the size required for the BOL output.

PVA systems on spacecraft that orbit planets, moons, or asteroids also require secondary power supplies because of the interruption in the PVA power for part of the orbit. The size or rating of the battery is matched to the output requirements of the spacecraft, the discharge (eclipse or shadow) period, and the battery efficiency, including its depth of discharge.

Solar Dynamic

Solar energy can also be harnessed to generate electrical power by using mechanical means to convert heat into electricity. Rotary engines that use an expansion and contraction cycle of a gas to rotate a shaft can be harnessed to drive an electrical generator. Three common heat engines have been considered as candidates for spacecraft solar dynamic systems that include:

- Stirling engine—The Stirling engine is the simplest of the heat-to-rotary motion conversion mechanisms. The device is relatively efficient and attractive for moderate-temperature electrical power systems. It is currently being used in terrestrial power applications and is under development for use in space nuclear power as the Sterling radioisotope generator.
- Rankine engine—The Rankine cycle engine is a variation on the Carnot heat engine, with a two-phase liquid-gas expansion and compression cycle, similar to the common steam engine. The expansion cycle absorbs heat, driving a turbine which, in turn, drives an electrical generator. The expanding gas cools and condenses for heat removal with both cycles being adiabatic. Water can be used for the two-phase liquid, but at reactor temperatures, potassium or mercury is more suitable.
- Brayton engine—The Brayton cycle engine is a gas cycle engine that is similar to the turbojet engine in the expansion and compression cycle. The Brayton engine is another variation of the ideal Carnot engine, but with constant pressure heating and cooling. Expanded gas drives a turbine and electrical generator, often with a compressor to pressurize gas for greater efficiency.

MAGNETIC AND ELECTRIC FIELDS

Magnetic and electric fields can be used to generate spacecraft electrical power in two ways. First is the common electromagnetic induction generator where a coil of wire passes through a magnetic field to induce a current, like the common mechanical rotary motion-to-electric power generators on Earth. A spacecraft could produce electrical power in Earth orbit with wire coils oriented perpendicular to its orbital flight path. The drawback to this technique is the force that is required to push the wire circuit through the magnetic field if there is an electrical load in the circuit. The required force removes orbital kinetic energy and gradually reduces the satellite's orbit radius. As a consequence, additional propulsion is needed to re-boost or maintain the satellite's orbit.

The second method uses a conductive tether to take advantage of the Earth's electric and magnetic fields. Charged particles in the upper atmosphere that create a voltage potential can also be used to complete a circuit in a wire deployed vertically from a spacecraft. The dynamics of orbits will put tension on a cable extended above or below a spacecraft, making a conductive tether a potential source of electrical power. A conductive tether experiment was deployed from the space shuttle *Atlantis* on STS-75 that proved the feasibility of generating electric power in space using both the ionosphere and the Earth's magnetic field, although the device failed during deployment of the 20 km tether. Because a return circuit must be completed in the process of generating electrical power, the decreasing density of ions at higher altitudes limits this method of generating electrical power to low Earth orbit altitudes.

CHEMICAL BATTERIES

Two basic types of batteries are available to power spacecraft. Batteries used as a primary source of power are single-use batteries because of their higher energy density compared to rechargeable batteries. Conversely, solar pholtaic systems include rechargeable batteries as a secondary power source to store electrical power for use during orbital eclipse. As with any chemical battery, spacecraft primary and secondary battery power sources have one of the lowest energy densities of any electrical power source.

- Primary power (single-use) batteries
  - Used as a single source of electrical power
  - Often redundant configuration (backup batteries for critical applications)
  - Applications include short-duration missions and single-event action (pyrotechnics, separation events, recovery beacons, etc.)
- Secondary power (rechargeable) batteries
  - Used as electrical power storage for solar photovoltaic eclipse interruptions
  - Contain 60%-90% power density (depth of discharge) compared to single-use or primary batteries
  - Applications are generally for power storage in solar photovoltaic electrical power systems

FUEL CELLS

The fuel cell is a device that generates electrical power from the electrochemical conversion of two reactants into electric current and the byproduct of the reaction. Fuel cells used for manned spacecraft combine hydrogen and oxygen as reactants since the byproduct is pure, potable water. The reaction is the same used for powering the space shuttle main engines, but at a rate much slower and regulated by the exchange of hydrogen ions (protons) and electrons via the hydroxyl radical (OH⁻). Careful regulation of the reaction rate is possible in the hydrogen-oxygen fuel cell by using a catalytic electrolyte to reduce the reaction threshold and allow diffusion of the ions across the electrolytic fluid. Two important variables in the reaction within the fuel cell are the temperature of electrolyte bath and electrodes and the concentration of the electrolyte and ions, which can be diluted with the water byproduct. Water must be continually removed from the fuel cell to avoid quenching the electrochemical reaction and slowing or even stopping the flow of electricity.

Fuel cells were introduced in the American manned program with the Gemini missions to develop and qualify the production prototypes for use on the Apollo lunar missions. Cryogenic LOX and LH2 were stored onboard the spacecraft, then heated to the gas phase before injection into the fuel cell. Because of the fuel cell's many advantages on short-duration manned missions, the technology was again advanced for use on the space shuttle orbiters.

The fuel cell units consist of a series of anode and cathode plates that circulate the hydrogen and oxygen gases in a potassium hydroxide and water electrolyte solution. The hydrogen plates, the anodes, are rich in both hydrogen and water, while the oxygen-side cathode plates are rich in oxygen and hydroxyl ions (OH⁻). Electrons released from the hydrogen form the negative terminal (anode) of the fuel cell. Protons released from the hydrogen combine with the hydroxyl ion diffusing from the oxygen side to form water. At the cathode, electrons entering from the positive circuit side combine with oxygen and water to form the hydroxyl ion, which flows through the potassium-water electrolyte toward the hydrogen-rich anode. The opposing motion of the two ions drives the electrical current in the fuel cell. In operation, the space shuttle orbiter's fuel cells consume the supplied oxygen, while excess hydrogen dissolved in the water byproduct must be removed by a hydrogen extractor.

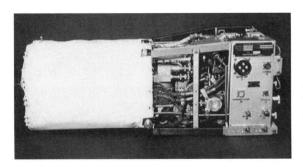

Figure 11.8. Schematic representation of the reactant flow through the hydrogen-oxygen fuel cell is shown, with along with the reactions responsible for generating water, electrical current, and heat. Source: NASA EPS-2102.

NUCLEAR POWER

Nuclear power is typically used for space exploration missions beyond 2 Astronomical Units because of the large surface area needed for a solar photovoltaic array beyond that distance. Although expensive, the compact, durable, long-lived nuclear electrical power sources do not require a secondary electrical source as do most photovoltaic systems. And without the need for a secondary source, space-based nuclear electrical power systems are reduced in total mass, complexity, and launch costs. The relatively simple radioisotope power source can also be used for electrical power on robotic spacecraft operating on the surface of the Moon, Mars, Venus, or even Mercury since the nights can be as long as fourteen days on the Moon, and eighty-eight days on Mercury.

Spacecraft nuclear power is generated by two different methods, although both methods produce heat energy by radioactive decay, which is converted into electrical energy. The radioisotope thermoelectric generator is the simplest and least costly method, but offers the lowest available power. Nuclear reactors that generate the highest power are also the most costly and, by far, the most hazardous.

Radioisotope Power Generator (RPG)

The commercial radioisotope thermoelectric power source has been in use on spacecraft for more than forty years, the first being installed on the Navy Transit navigation satellites launched in the early 1960s. The most common spacecraft RPGs are the radioisotope thermoelectric generators (RTGs) produced for the DoD and NASA spacecraft at power levels ranging from less than 25 W_e (Watts electric) to as much as 400 W_e. Spacecraft requiring more than 400 W_e use a series of RTGs. One of the best examples of RTG-powered spacecraft are the Voyager 1 and Voyager 2 pair that were outfitted with three RTGs, each with a total available power of 470 W_e at launch. Mission lifetime for the Voyager vehicles was originally less than ten years since the first concept objectives were to explore Jupiter and Saturn under the Mariner 11 and 12 designation. Major changes in the Voyager mission objectives routed the Voyager 2 on the "grand tour" past Jupiter, Saturn, Uranus, and Neptune with an expected life of more than twenty years. Because of Voyager 2's success and the RTGs' ^{238}Pu half-life of eighty-eight years, the mission was extended to allow exploration of the outer solar system. The newest extension of the Voyager 2 objectives called the Voyager Interstellar Mission continues today, sending back data on the particles and magnetic fields that undergo changes near the Sun's outer boundaries. Power to the few Voyager instruments that remain turned on should last until 2020, at which time the available power drops below the regulated 30 Vdc bus voltage as the nuclear heat source cools and the solid-state heat-to-electrical power converter performance degrades. An optimistic estimate of 2025 for the Voyager 2 electrical system failure would mean the spacecraft would have explored the solar system and the interstellar medium continuously for a record forty-eight years.

RTG heat energy is created with the spontaneous decay of its ^{238}Pu radioisotope which is then converted into electricity by a cylindrical solid-state thermoelectric converter surrounding the segmented ceramic-radioisotope core. Since the conversion efficiency is proportional to the temperature gradient on the converter, a radiator is placed on the outside of a protective shell surrounding the converter to cool the RTG at the outer surface. Although hazardous, ^{238}Pu is stabilized in a ceramic substrate and protected from accidental breakup. The concentration of the plutonium dioxide is low, and the possibility of generating extreme heat or reaching melting temperatures is nonexistent. Radiation products from ^{238}Pu are gamma radiation (1.09 MeV and 0.04 MeV) and alpha radiation, which are helium nuclei released with an energy of 5.5 MeV.

Several types of radioisotope power generators have been used in the past and are under development for future deep-space missions. The first RPGs were the SNAP (system for nuclear auxiliary power, alternately known as space nuclear auxiliary power) RTGs used for the Navy Transit satellites, and on the Apollo ALSEP experiment packages left on the Moon, as well as on the Viking Mars landers. Small nuclear reactors were also launched in the 1960s with the SNAP designation, but were even numbered (SNAP 10A, for example), while the odd-numbered SNAP power sources were RTGs.

SNAP RTGs were replaced by the multi-hundred Watt RTG (MHW-RTG) with improved radioisotope containment, a more efficient thermoelectric converter, a stronger casing, and higher power ratings. The MHW-RTG was used on four spacecraft, including the Voyager duo.

The MHW-RTG was replaced with the current general purpose heat source RTG (GPHS-RTG) that has resulted in improvements in radioisotope containment, thermoelectric conversion, and significant safety improvements. GPHS-RTGs were flown on the Ulysses, Galileo, Cassini, and New Horizons interplanetary spacecraft.

Two new RTG designs are underway for NASA's future fleet of interplanetary explorers, which include the multi-mission RTG (MMRTG) and the Sterling radioisotope generator (SRG). The MMRTG built by Boeing is the more traditional RTG with a number of improvements in the GPHS-RTG, including higher electrical conversion efficiency, although it still is powered by the GPHS modular radioisotope heat units. The Sterling radioisotope generator is a more advanced, higher-efficiency unit that converts heat into electricity using a Sterling engine. The greater conversion efficiency and the lower temperature heat source of the SRG have many obvious advantages in interplanetary exploration. Yet the MMRTG provides advantages of its own, including higher heat output where the excess heat could be used to counter the ultra-cold temperatures on deep-space missions (NASA JPL 2005). Conversion efficiencies for RTGs are typically 5–8%, which means that an RTG with 400 W_e output at 5% efficiency would generate twenty times that in heat energy, or 8,000 W_t where W_t is thermal energy in Watts.

Nuclear Reactor

Nuclear reactor power for spacecraft is limited to high-power applications because of the high cost of space-qualified reactor units and their inherent hazards of operation. For that reason, reactors have been only flown on military spacecraft, and mostly to power synthetic aperture radar systems launched by the Soviets during the Cold War. The primary advantage of the nuclear reactor power source comes from its high energy

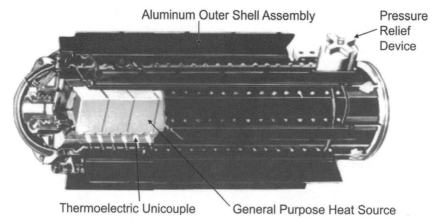

Aluminum Outer Shell Assembly   Pressure
                                       Relief
                                       Device

Thermoelectric Unicouple   General Purpose Heat Source

Figure 11.9. Cutaway diagram of a general purpose heat source (GPHS) radioisotope thermoelectric generator. GPHS is a specific form of the ceramic core segments containing 238 Pu in the form of plutonium dioxide (PuO_2). The pressure relief device at the upper right is used to release the helium gas (alpha particles) generated in the decay of the radioisotope. Source: NASA JPL.

density. Since reactor core energy density is much higher than the core of the radioisotope thermoelectric generator, it can furnish much higher power levels for the same mass which is always a critical concern in spacecraft design. While never flown, reactor power systems have been proposed for powering outposts on the Moon and Mars and for robotic missions to Jupiter and its moons.

The fundamental advantage of the nuclear reactor is in the fissioning radioisotope used as the heat source. These radioisotopes release neutrons as well as other particles and gamma radiation as the nuclei break apart. The neutrons carry no net charge and can therefore induce nearby radioactive nuclei to fission, or break apart, which release more neutrons, setting off a cascade of nuclear fissions. The energy released in each nucleus breaking apart is the energy that binds the nuclear fragments together, which is millions of time greater than electrochemical bonds. This is the reason why nuclear power sources are so much higher in energy per unit mass than chemical energy power sources such as batteries and fuel cells.

Radioisotopes chosen for the radioisotope thermoelectric generators do not release neutrons, and hence decay at a very steady, fixed rate normally measured in units of half-life. Radioisotopes used in nuclear reactors, on the other hand, release neutrons in their breakup, which speeds up the decay rate of nearby nuclei. The decay rate of the radioisotopes that release neutrons, called fissile material, varies with the density of the radioisotope—how closely the nuclei are packed together—and the presence or absence of other material that could slow or absorb the neutrons.

The slowing, or moderation, of the neutrons in a reactor is a necessary and important concept in the reactor operation and design. For spacecraft reactors, the neutron moderation comes simply from the diffusion of neutrons through the radioisotope reactor core. By adding a mechanism around the core to reflect a percentage of the neutrons flowing out, the fission rate can be increased, along with increased energy release. Power reactors on Earth use various moderators, including water, to regulate the rate of stimulated decay and the amount of heat produced. An atomic bomb uses the same fissile material, but instead of moderating released neutrons, an explosive outer shell slams two or more pieces of fissile uranium or plutonium (^{235}U or ^{239}Pu) together to initiate an almost instantaneous release of neutrons and nuclear energy.

Only one American nuclear reactor was flown in space, the SNAP 10A, while the Soviet Union launched thirty-one between 1967 and 1988. The Soviet record of reactor safety in space is similar to its record on the ground, vis-à-vis the Chernobyl reactor meltdown. The USSR's space reactor program was marred by two catastrophic space reactor reentry incidents, and by the careless release of liquid metal reactor coolants expelled at the end of the useful life of their spaceborne reactors. Before these satellites are decommissioned, the reactor is torn from the orbiting spacecraft and boosted into a higher orbit to decay in several thousand years. In the process, liquid sodium and other metal coolants are released to form droplets that can reach 10 cm in size and add significantly to the Earth's space debris hazard.

The American SNAP 10A experimental reactor was launched with a converted Atlas atop an Agena booster on April 3, 1969, from Vandenberg, California into a 800 km orbit. Reactor activation preceded criticality (meaning critical neutron level had been reached) and full power production within twelve hours. Due to an onboard power malfunction, the SNAP 10A reactor was shut down prematurely after forty-three days of continuous operation, then boosted into a 1,300 km polar orbit.

Although never launched, the larger SNAP 100 reactor project was a NASA-DoD initiative to develop powerful space reactors for future exploration missions and lunar and/or Mars outposts. The 100 kWe supply developed 2 MW$_t$ using a lithium-cooled uranium reactor. The SNAP 100 project which began in 1983 was terminated in 1993.

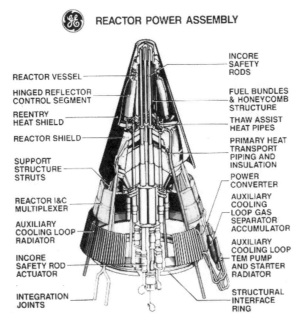

REACTOR POWER ASSEMBLY

INCORE SAFETY RODS

REACTOR VESSEL

HINGED REFLECTOR CONTROL SEGMENT

REENTRY HEAT SHIELD

REACTOR SHIELD

SUPPORT STRUCTURE STRUTS

REACTOR I&C MULTIPLEXER

AUXILIARY COOLING LOOP RADIATOR

INCORE SAFETY ROD ACTUATOR

INTEGRATION JOINTS

FUEL BUNDLES & HONEYCOMB STRUCTURE

THAW ASSIST HEAT PIPES

PRIMARY HEAT TRANSPORT PIPING AND INSULATION

POWER CONVERTER

AUXILIARY COOLING LOOP GAS SEPARATOR ACCUMULATOR

AUXILIARY COOLING LOOP TEM PUMP AND STARTER RADIATOR

STRUCTURAL INTERFACE RING

Figure 11.10. Cutaway of the SNAP 100 nuclear reactor electrical power supply major components. Source: NASA GRIN.

Thermal Control

Because of the extreme temperatures encountered in space, thermal control of space-craft and their components is an absolute necessity, whether in Earth orbit, on a deep-space mission, or in a close-solar orbit. Not all spacecraft are exposed to the most extreme temperatures in space, which range from 3 K to 10^6K ($-454°$F to $10^7°$F), but the temperature extremes encountered by any spacecraft call for both heating and cooling systems to provide a reasonable operating temperature range for the onboard equipment. Spacecraft operating closer to the Sun have a greater heat load, requiring a more robust thermal control system than a similar spacecraft in deep-space. Conversely, deep-space spacecraft require more available heat than inner-solar system missions to avoid the effects of very low temperatures on lubrication, sensitive electronics, fluids, and structural and other materials that can become brittle.

Moderating the temperature extremes on spacecraft is accomplished with passive thermal systems that require no power and active thermal systems that do consume electrical power. Heating cold surfaces or regions on the spacecraft is typically done with either active resistive heaters or passive conductive components or, on deep space missions, with small implanted radioisotope heating units or radioisotope heater units (RHUs). Cooling spacecraft components or surfaces can be achieved with passive radiators; active cooling loops; thermoelectric coolers; or, for very low temperature requirements, cryogenic cooling.

Passive systems

- Attractive since it requires no electric power, but generally limited in heat rejection and/or heat transfer

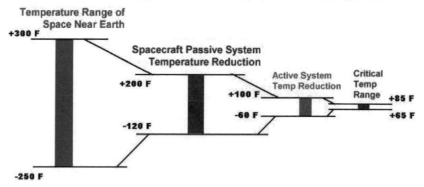

Figure 11.11. An illustration of the extreme temperatures encountered in the near-Earth space environment and the temperature range restrictions on a hypothetical spacecraft. Passive thermal components are intended to moderate the most extreme temperatures to a more limited range. A smaller range of temperatures would require active thermal subsystems to heat cold regions and cool hot components or surfaces. A narrow temperature range requirement for a crew cabin would need even more cooling and heating to moderate the thermal environment.

- Examples are:
  - Passive radiators
  - Coatings
  - Reflective surfaces
  - Heat pipes
  - Insulation
  - Louvers
  - Sun shields

Active systems—Power Required for Operation
(some discrepancies exist in this simple definition)

- Typically used for high temperature differences or high heat loads
- Examples are:
  - Cryogenic cooling
  - Electrical heaters
  - Refrigeration loops
  - Heating/cooling heat transfer loops
  - Shutters

Moderating spacecraft temperature extremes requires more than simply decreasing the temperature difference on components or surfaces, since there are a wide variety of variables in thermal control of spacecraft. The temperature experienced by a spacecraft surface depends on the surface absorption and emission characteristics for the type of heating, which depends on the source of heat energy. Surface heating could form from direct solar electromagnetic radiation, from reflected solar radiation from the Earth's atmosphere, or from the Earth's radiated IR heat. In addition, the Sun's radiation energy peaks in the visible band, with decreasing but still significant energy emitted in the IR, then the UV, and less in the microwave, X-ray, and gamma-ray bands.

- Heat Inputs
  - Solar radiation: 1380 W/m^2 at 1 AU
  - Earth's infrared radiation: Approximately 240 W/m^2
  - Earth's reflected energy: Approximately 30% of direct solar radiation
  - Internal spacecraft heat: Varies
- Heat outputs
  - Radiators: Common passive and active component used to remove heat
  - Evaporation: Used on early manned missions
  - Sublimation: Used for cooling space suits
  - Cryogenic fluid heat absorption: Stored cryogenic liquid is effective, but costly and limited in duration
  - Ablation: Atmospheric entry heat removal using solid-gas phase change

A material will reach a specific temperature based on the intensity of the EM radiation and the material's efficiency in absorbing energy in the various EM bands. Along with the absorption character of a surface, the emission character of the surface determines how efficiently the absorbed energy is given off. Combined, the efficiency of absorption (absorptivity) and the efficiency of emission (emissivity) of a spacecraft surface determine the interplay between energy in and energy out, and the resulting temperature. But the surface heat flow in or out is modified by the thermal conductivity of the surface with respect to surrounding materials, and the increase or decrease of heat energy as the spacecraft orbits and possibly rotates for spin-stability. The complexity of these and other variables makes thermal control design for spacecraft one of the most challenging tasks in the design process. To help with the detailed process, computer models are used for almost every stage of spacecraft thermal system design, from basic concepts to the final fabrication and testing of the thermal control system. Validations of spacecraft thermal models are made with high-temperature and low-temperature vacuum test facilities.

Structures

The spacecraft primary structure has a variety of roles in the spacecraft design, in addition to furnishing the overall structural integrity of the spacecraft. The primary structure, often called the bus, is also used for the attachment of components and secondary structural assemblies, and often used to help manage thermal loads. The spacecraft primary structure is also used to augment electrical conductivity of the spacecraft and reduce charge difference buildup, as well as protect the spacecraft from space debris, and to shield sensitive internal components from radiation. As important as the structural design is, the selection of materials for the various functions critical to the structural design is just as important. Emphasis on cost, strength, and weight for the structure and the materials is complicated by the need to integrate the payload, systems, and subsystems for the mission life, while mating the performance of the launcher to the vehicle for its initial ride to orbit, followed by the spacecraft deployment operations.

For this brief review, simple descriptions of the characteristics and design consideration of spacecraft structures are outlined, followed by examples of some of the more common spacecraft bus shapes.

- Desirable characteristics of the primary structure materials include:
  - High strength
  - Suitable stiffness
  - Low density (minimum weight)
  - High durability
  - Thermal conductivity dependent on heating/cooling requirements
  - High electrical conductivity
  - Low thermal expansion
  - Corrosion resistance
  - High ductility (inhibits cracking)
  - Fabrication ease
  - Low cost

- The structure also supports components and attachments, which needs additional strength and stiffness to counter the effects of:
  - High g load stresses during launch (and reentry, if applicable)
  - Launch forces and vibrations
  - Frequencies generated by the launch vehicle and transmitted to the spacecraft
  - Launch vehicle vibrational and acoustic frequencies, which cannot peak at the same or nearby resonant frequencies of the spacecraft and its components, since resonance between the two could result in damage or destruction of the spacecraft
- Structural design considerations
  - Launch vehicle size, often the greatest constraint on size and mass
  - Payload normally maximized and weight minimized
  - Materials selection
  - Subsystem needs/requirements
  - Space exposure durability
    - High radiation levels—shielding needed for internal/sensitive equipment
    - High thermal gradients are also variable
      - Thermal modeling and evaluation necessary
      - Thermal conductivity can be useful or problematic, or both
      - Thermal expansion is typically minimized
    - Micrometeoroid exposure requires strength and durability
      - Damage risk proportional to flight duration
      - Structure often provides shielding for internal components
    - Charge difference buildup produces electrical field buildup and a possible destructive discharge
  - Magnetic field generation and interactions not good for sensitive equipment
    - EM shielding important
  - Composite structures can be very appealing, but have disadvantages
    - Low weight, strong, high stiffness
    - Low conductivity (possible static charge buildup)
    - High cost
    - Difficult to measure failure modes
  - Rotational dynamics important, especially for spin-stabilized or dual-spun spacecraft
  - Failure modes identified (fatigue, fracturing, etc.)
    - Evaluated for load types and load sources
    - Validation testing required
  - Vibrational modes and resonances evaluated
    - Launch vehicle vibration frequencies are critical
    - Acoustic loads at launch can also be significant
    - Onboard propulsion effects can be significant
    - Reentry loads can be much higher than launch (if survivability is important)
    - Pyrotechnic device shock can be significant
  - Costs
    - Raw and fabricated materials
    - Fabrication/production costs
    - Hazardous materials and/or exotic fabrication techniques with significant costs

VIBRATION AND RESONANCE

Spacecraft and their individual components are subjected to high vibrational loads during launch and, if applicable, entering or reentering an atmosphere. These loads can damage sensitive instruments and electronics but can also damage large structural elements if the resonant frequency of vibration of the structure matches those produced by the launch vehicle or the reentry dynamics. A structure's resonant frequencies are associated with the rigidity, length, and interconnection of the many structural elements and the components. The most critical of these are the longer structures such as the booms, antenna arms, and deployable panels such as PVAs. Tests to identify the resonant frequencies of a spacecraft are typically done on a "shake table" with a frequency range at least as broad as the launch vehicle's vibrational frequencies (and reentry, if planned).

To reduce the resonance response to the high vibrational loads at launch, the resonance frequencies of spacecraft structures, subassemblies, and components can be adjusted to move the critical resonances away from the strongest launch vibration frequencies. This can be done by adding or subtracting mass, with longer or shorter elements, or by stiffening, relaxing, adding, or subtracting structural elements. Since the vibrational load spectrum roughly spans the audio frequencies, from 10 Hz to 10 KHz, the spacecraft must be tested for the range and amplitude expected during the entire ascent. The highest loads are appropriately called max Q, or maximum dynamic pressure (Q), encountered near Mach 1. Both gravity and acceleration can have a profound effect on the spacecraft and component resonances, hence the vibrational loads during launch must be simulated on ground test equipment to provide reasonably accurate analysis.

SPACECRAFT BUS STRUCTURE EXAMPLES

Spacecraft structure design is a reflection of the payload requirements, the mission constraints, launch vehicle restrictions, and weight minimization, as well as a host of other variables. Because of the different payloads, objectives, and systems, spacecraft structures rarely resemble one other, although there are common structural shapes that can be found among the thousands of spacecraft launched in the past half century. Those shapes vary from small, simple shapes such as spheres and boxes, to complex frames providing minimal weight and unique payload integration requirements on larger vehicles. However, these common structural shapes are more accurately described by their load transfer and support elements, which include columns, frames, trusses, plates, and shells utilized in the primary and secondary structural assemblies. A number of examples of the general spacecraft structural shapes are shown in figures 11.12 through 11.21.

SPHERICAL

Figure 11.12. Launch team preparing the spherical-shelled Vanguard satellite for launch in December 1957. Source: NASA NIX.

Figure 11.13. Mockup of the Luna 9 spherically shaped lunar soft lander payload. Source: NSSDC.

RECTANGULAR BOX

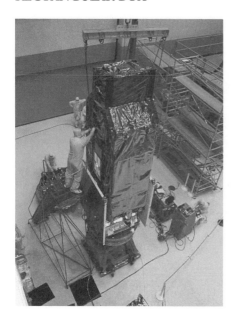

Figure 11.14. Completion of the FUSE (the Far Ultraviolet Spectroscopic Explorer) spacecraft at the Goddard Space Flight Center. Source: NSSDC.

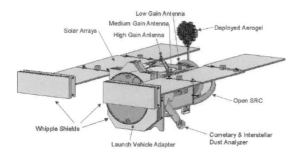

Figure 11.15. Layout sketch of NASA's Stardust sample return spacecraft. Source: NSSDC.

OCTAGONAL BOX

Figure 11.16. Artist's sketch of NASA's Contour comet exploration spacecraft. Source: NSSDC.

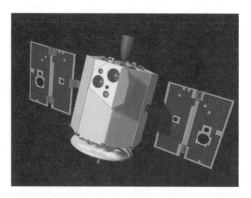

Figure 11.17. Sketch of the Clementine lunar explorer. Source: NSSDC.

CYLINDRICAL

Figure 11.18. Galileo spacecraft being deployed from the payload bay of Orbiter Atlantis in October 1989. Source: NASA NIX.

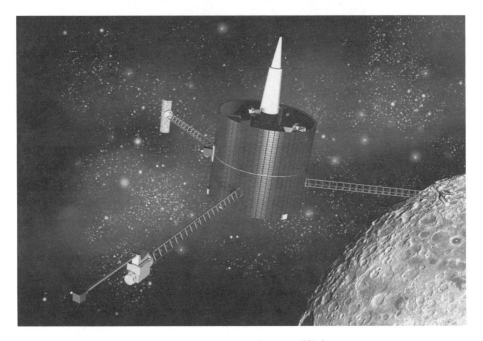

Figure 11.19. Sketch of NASA's Lunar Prospector. Source: NSSDC.

COMPLEX FRAME

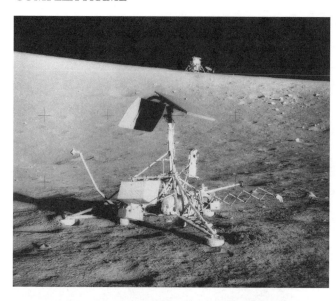

Figure 11.20. Photograph of the Lunar Surveyor 3 spacecraft by the Apollo 12 lander in the background. The three-legged solar powered lander measured approximately 3 m in height. Source: NSSDC.

Figure 11.21. Photograph of a mockup of the Soviet Luna 16 lunar sample return spacecraft. Source: NSSDC.

References

NASA. 1988. *Computers in Spaceflight: The NASA Experience.* NASW-3714. Washington, DC: NASA.

NASA. 2007. "U.S. Expendable Launch Vehicle Data for Planetary Missions." Retrieved from: http://cost.jsc.nasa.gov/ELV_US.html (accessed 7/15/2009).

NASA JPL. 2005. *Expanding Frontiers with Standard Radioisotope Power Systems.* JPL D-28902. Pasadena, CA: NASA Jet Propulsion Laboratory.

Tsiao, Sunny. 2007. *Read You Loud and Clear: The Story of NASA's Spaceflight Tracking and Data Network.* NASA SP-2007-4233. Washington, DC: NASA History Division, Office of External Relations.

NASA NIX (NASA Image Exchange). Retrieved from: http://nix.nasa.gov/

NASA JPL Image Photojournal. Retrieved from: http://photojournal.jpl.nasa.gov

CHAPTER 12

Russian Space History

Background

The Soviet Union's first satellite was a small, hurried payload called Sputnik that was launched on a military booster built by the same designer. The unannounced launch of this small satellite by a nameless engineer might have been a first small step in the exploration of space. In reality, the Sputnik launch was the opening shot in the Space Race that pitted two superpowers in a battle for military supremacy. What shocked the world with the launch of Sputnik were the advances the Soviet Union had made ahead of the United States—the presumed leader in advanced weapons and technology. The belief that the Soviet Union was a nation obsessed with military might but lacking advanced industries and technologies to keep pace with the United States was shaken.

The forces behind the early satellite launches that ushered in the space age were driven by the same military interests responsible for the conflict between the two superpowers. Yet even with massive military budgets and the smothering weight of the USSR bureaucracy, the great successes of the Soviets' early space efforts were, like the United States, the product of a few very influential people. A closer look at the unique history of Russia and the Byzantine political structure that made up the USSR will help in understanding how the powerful political and military authority contributed to the success of the early Soviet space program almost by accident.

EARLY RUSSIA

The earliest name used for the Russian region was Rus, the Viking term for "oarsman." But Russia's Viking settlements in the northwestern Russian regions during the ninth and tenth centuries were only part of the vast expanse of the Russian territories dominated by a shifting mix of cultures and races that today make up the largest and most diverse nation on the planet.

Vikings, or Varangians, captured the largest Slav settlement at Novgorod in 862 and established themselves as the rulers in the northern region. To the south and at about the same time, the Slavic Prince Ka established a stronghold known as the Kievan territory. The capital of the territory, Kiev, was conquered by the Varangians soon after and made the capital of the western region. Kiev has survived over the centuries, and is today the capital of the Ukraine.

The capital of today's Russia is Moscow, founded by Prince DolgorukyIn in the twelfth century. By the fourteenth century it had become the capital of the Russian principality and gained in prominence until Peter the Great moved the capital to St. Petersburg in the early 1700s. Following the Russian Revolution of 1917, Moscow became the capital of the Soviet Union. After the fall of the Soviet Union in 1991, Moscow was again the capital of Russia.

CHRISTIANITY

Russia's influence on the Asian continent is extensive. It is the largest country in the world, extending over eleven time zones, and spanning nearly 1,200 years of history. From its first settlements in the ninth century, the Russian empire gained in size and strength. To solidify his enormous kingdom, Prince Vladimir chose an alliance with the Roman Catholics to establish a form of their religion to help unify the diverse cultures and races. Treaties in the tenth century with the Byzantine rulers, the descendant Roman rulers who then became Christian governors, led to the Christianization of Russia by the eleventh century. Disagreements within the Christian-Roman Church in the next century, however, spurred the formation of an independent Russian Christian Church called the Orthodox Catholic Church of Russia, or the Russian Orthodox Church, that is now the dominant religion throughout much of Russia and the former Soviet states. Scorned by the socialists in the twentieth century and brutalized by Stalin, the religious leaders and their followers and their churches survived. Many of the cathedrals in the Soviet Union were destroyed by Stalin, although the early cathedrals inside the walls of the Kremlin were spared because of their historical prominence. In spite of the heavy hand of the Soviet communists, the Russian Orthodox religion remains a strong cultural influence throughout the Russian region.

Indirectly, Christianity played another important role in the early development of the Russian culture. In the ninth century, a scholar and Christian missionary named Constantine, later known as Saint Cyril, helped develop the Cyrillic alphabet to unify the myriad of languages spread throughout the Russian region. The Cyrillic alphabet used in the Russian language is based, in part, on the Greek alphabet, although there are about a dozen additional letters that represent Slavic sounds not found in the Greek alphabet. Even though the Cyrillic alphabet is found in many of the regional languages (e.g., Belarusian, Bosnian, Bulgarian, Macedonian, Russian, Serbian, and Ukrainian), not all of the alphabet is common to the different languages.

EARLY RUSSIAN RULERS

The first laws of Russia written in the mid-eleventh century were known as the Pravda (not to be confused with the Russian daily chronicle), which were intended to create a civil structure within early Russian society. But in practice, the Pravda ("truth") offered citizens little protection from the abusive, and at times inhumane, rulers. One of the most brutal czars in Russian history was Ivan IV, better known as Ivan the Terrible. His extraordinarily cruel reign (1533–1584) was exemplified by public beheadings in Red Square and the sacrifice of humans and animals that were tossed from the Kremlin wall. Ivan the Terrible's tyranny and cruelty were unparalleled in history, decimating entire cities, obliterating churches and murdering the followers, killing and torturing priests in front of their altars, even disposing of seven of his eight wives. The rule of Ivan and his descendants during the Rurak dynasty ended in the seventeenth century, but the Russian monarchy continued with the Romanovs' rule until the Russian Revolution in the early 1900s.

Wars and conflict were not uncommon throughout the Asian and European continents in the Middle Ages and Russia was no different. Russians battled the Mongols for several hundred years after the establishment of Moscow in the mid-twelfth century. In the late 1200s, the son of Genghis Khan invaded Russia, extending the Mongol Empire throughout much of the Asian continent, including Russia, the Ukraine, Siberia, and much of Europe. For nearly 200 years, the Golden Horde, named for their lavish, gold-threaded tents, ruled Eurasia.

Following the tumultuous rule of the Ivans, Czar Peter I, known also as Peter the Great, began a period of strategic alliances throughout much of Asia and Europe and instituted social reforms and cultural progress that continued from the early 1700s through the early 1800s. One of the progressive rulers of particular note was Catherine the Great, wife of Czar Peter III. She would not only bring social enlightenment and the arts to Russia, but also succeeded in adding to the empire by conquest that included winning several wars against the Turks. Yet in the centuries of czarists' rule of the state, the nobility class retained localized power with control over the regional operations and the common subjects—the serfs. Whether the ruling czar was kind or unkind, or the nobility oppressive or not, the peasantry were little more than slaves. It was this virtual slavery that would later become the soul of the early twentieth century Russian Revolution.

Reasoned governance of the early 1800s gave way to more extreme, oppressive, and indifferent rule by Czar Nicholas II in the last half of the nineteenth century. Although he reinstituted the Dumas ("to think") parliamentary council, he suspended the council during the upheaval of the 1905 Russian Revolution. The czar's oppressive reaction to the public and political demonstrations incited even more rebellion, which led to the newly suspended representatives rallying the political parties against the czar. In effect, Czar Nicholas II turned everyone against his ineffective and oppressive rule by refusing to share his powers in any meaningful way. With the military defeat of the Russian Army in World War I came the inevitable revolution and the tragic end to the Czar and his family, and the collapse of the Russian nobility.

SOCIALIST RULE

A second revolution in Russia in October 1917 was only one of a series of internal wars within Russia that later culminated in the socialist-communist structure known as the Union of Soviet Socialist Republics (USSR). The overthrow of Czar Nicholas II in February 1917 came as strikes and protests in the two largest cities continued to grow until the soldiers who were ordered to put down the uprising joined the revolution. The end of czarist rule was swift, but the new governing body was divided. The Soviet, a revolutionary faction in Petrograd (now St. Petersburg), claimed control over the armed forces and basic rule over Russia. The Provisional Government also arose from the February Revolution claiming similar powers, but creating a weak, parallel-governing body. Another revolutionary force, called the Bolsheviks, was allied with the Soviet under their exiled leader Vladimir Lenin, who advocated seizing land and socializing industry throughout Russia. However, a split between the Soviet, who advocated continuing the war in Europe, and the Provisional Government, who were opposed to the war, delayed the formation of an elected representative body, which was the reincarnated Dumas.

Lenin

A coalition of the three postrevolution governing organizations formed the Congress of Soviets, creating a government that was in almost total disagreement about ownership of power and continuing participation in the European War (World War I). And even though the peasantry and workers, who were largely responsible for the revolution, favored the Bolsheviks, the Soviet Congress forced the popular Lenin back into exile and arrested the prominent leader, Leon Trotsky. Revolutionaries inspired by the Bolsheviks refused to follow the new Congress's dictates. Conflict between the unpopular ruling government and the Bolsheviks led to the second major civil war in 1917, known as the October 24 Revolution (the Russians still followed the Julian calendar in this period, hence the October Revolution was actually 7 November in the Gregorian calendar used worldwide today). The armed insurrection was again inspired by Lenin, but this time led by Trotsky. The dominance of the Bolsheviks in the second Congress in November 1917 was cast in the leadership, which consisted of the three most prominent Bolsheviks—Trotsky, Stalin, and at the top, Lenin.

The first declarations of the new Communist-Bolshevik government transferred all private property to the state and terminated Russia's involvement in World War I. However, Poland, the Ukraine, and Finland had to be given up in the treaty. Peasants and the middle class generally supported complete socialization of the Soviet, but the loss of an important part of the satellite states incited another uprising that set off another revolution. Protests by the new anticommunists, called "Whites," were suppressed by the "Red Terror" campaign ordered by Lenin in a show of newfound power. The war of the Reds and the Whites was as destructive and divisive as the February and October civil wars. Unfortunately for the peasants, Lenin's win introduced a policy of violence and terror against enemies of the Soviet state and established the Red Army as the military arm of the government. The violent clashes in the Red-White war also ushered in the powerful secret police after two assassination attempts on Lenin in 1918. Russia and its Soviet states were now under the ruthless dictatorship of Lenin. But economic revival of the war-torn Russian states was put under a somewhat free-market concept that brought a brief period of relative peace and prosperity to an impoverished, war-torn land.

Lenin's death in 1924 following a series of strokes created an immediate power struggle in the government of the Soviet states, now called the Union of Soviet Socialist Republics, with a similar battle for power within the Communist Party. The Communist political party was considered separate from the Soviet government, although in practice there was no separation since the leadership positions of both the government and the Communist Party were held by the same men. One of the first to rise to power following Lenin's death was Joseph Stalin, first empowered as the Communist Party's general secretary, which was a titular position. Stalin then began a gradual, complete, and ruthless takeover of the leadership of the entire Soviet Union.

Figure 12.1. Lenin's likeness on the Russian 40 kopeck stamp.

Stalin

Joseph Stalin (born Iosif Vissarionovich Dzhugashvili) participated in the 1905 and 1917 Russian revolutionary wars as a Bolshevik, first as a bank robber, which led to his brief incarceration and exile, then by joining the ranks of the Bolshevik-Communist party in 1912. Within a few years, his appointment as general secretary of the Central Committee placed him in a position of modest power. Although the position had little authority, Stalin gained increasing control by pitting political leaders against one another. Ultimately, Stalin removed Trotsky, who was his greatest adversary, and had him expelled from the party and later exiled. By 1928, Stalin was in the top leadership position, dominating political and policy decisions that began moving away from Lenin's socialist economy, which had become diluted by outside market influences. Stalin demanded complete socialization of any and all aspects of Soviet life, including the collectivization of farms throughout the USSR. Rebellious farmers in the Ukraine and elsewhere were simply starved to death when troops removed their entire harvests and embargoed any imports or trade in those regions.

In spite of his dominance of the Communist Party and Soviet government, Stalin did not have absolute and total power until his bloody purges took place between 1936 and 1938. Stalin's drive for complete control began with the purge of members of the party whom he felt were his most dangerous adversaries. This violent period of murders, executions, assassinations, imprisonment, and exile were not only exacted on Stalin's perceived enemies, but also on a number of classes, ethnic groups, communities, and geographic regions unrelated to his power. Because of Stalin's purges that blossomed from his sociopathic paranoia, the Soviet military leadership lost the majority of its officer ranks, while Communist Party members not in Stalin's close circle and intellectuals, scientists, engineers, industrialists, and anyone of prominence, were removed, or exiled, if lucky. Those with a higher public profile that could not be conveniently murdered outright were executed or imprisoned after conviction on trumped-up charges.

Stalin's purges of the Soviet leadership removed not just his adversaries, but also the very people needed to save the Soviet Union from the invasion by the Germans in World War II. Two of those who were lucky enough to avoid execution were brought out of prison camps at the end of the war to rebuild the Red Army armaments. Near war's end, Stalin released the man who would ultimately be responsible for Russia's successful space program, Sergi Korolev, and his mentor, Andrei Tupolev, who would build military and civil aircraft fleets for decades to come.

Figure 12.2.
Stalin's likeness on the Russian 40 kopeck stamp.

POST–WORLD WAR II

Stalin's ruthless behavior is thought to have originated in childhood from the abuse by his father, then aggravated in his school years by his peers and reinforced by the "Red Terror" policy instituted by Lenin after the October Revolution. Stalin, who is reported to have killed more Soviets during his rule than the 20 million killed in World War II by the Germans, created an aggregate culture that was ruled by fear and distrust, whether within the Soviet government or within Soviet society. Secrecy and distrust dominated the Soviet Union's international affairs as well. Stalin's perverse paranoia helped take the Soviet Union directly from World War II into the Cold War.

To rebuild the Soviet Union after the devastation of World War II, and to compete with the Western Allies, Stalin began a broad industrialization program, as well as a massive weapons buildup. By building an atomic bomb that matched America's, Stalin also had the capability of holding the Western Allies at bay on the European continent. But Stalin also wanted to develop a missile that could protect the Soviet Union by launching an atomic bomb to virtually any target in the world—the intercontinental ballistic missile. After capturing part of Germany and a number of their V-2 rockets, Stalin unknowingly set a course for rocket development that would later lead to the first satellite in orbit, and the beginning of the space race.

Soviet Organizations

The Soviet Union Communist Party was the dominant force in all spheres of Soviet life in spite of being classified as a political structure. The Communist Party was officially outside the USSR government, but in practice, the Communist Party leaders controlled all functions of the Soviet Union, from the highest levels to the bottom-level factories, communes, and institutions.

Below are some definitions of Soviet political structures.

Politburo—A contraction of political bureau. The Soviet Politburo was the executive organization of the Communist Party and represented the leadership of the Soviet Union.

Supreme Soviet—The highest legislative body in the Soviet government was composed of two chambers, one representing the general population and one representing ethnic populations that were elected by each of the fifteen Republics. The Supreme Soviet was, in essence, a rubber-stamp parliament of the Soviet leadership.

Presidium—The Presidium of the Supreme Soviet was a government body elected by the Supreme Soviet to carry out its legislative tasks. The organization consisted of the president, a secretary, fifteen delegates (one from each of the republics), and twenty members.

Council of Soviet Ministers—The Council of Soviet Ministers had the responsibility of governing the day-to-day tasks of executing the policy and directives of the Politburo and the Central Committee. Its membership was established by the Central Committee.

COUNCIL OF CHIEF DESIGNERS

Far down in the Soviet bureaucracy were functional organizations that included the Council of Chief Designers initiated by the new chief missile designer, Sergei Korolev. The informal committee of members of various divisions responsible for missile design was organized for planning missile production after World War II. The Council discussed problems and reached agreements on design, development, and production issues to streamline their first missile production efforts. The process and the council lasted through the space race under the direction of Korolev during his career at NII-88 and OKB-1, although it did vary in membership along the way. The informal group became more structured when its effectiveness was recognized and became a model for other applications in Soviet industries. After the death of Korolev, the Council came under the direction of Valentin Glushko.

Rocket Design and Development

GIRD The Group for the Study of Reactive Motion (Gruppa Isutcheniya Reaktivono Dvisheniya, or GIRD) was co-founded in 1924 by Fredrik Tsander to develop the first advanced Soviet rockets. Korolev replaced Tsander after he became ill in May 1932. GIRD members were the first to launch a Soviet liquid-fuel rocket called the GIRD-9 in August 1933, which used LOX and gelled petroleum fuel propellants. The second launch was the GIRD-10 launched in November of 1933 and the first true liquid-propellant rocket launched in the USSR.

GDL The Gas Dynamics Laboratory (GDL) began as a rocket-powered weapons lab in Leningrad in 1925 by Nikolai Tikhomirov, who was joined by Georgi Langemak, Valentin Glushko, and others in the late 1920s. GDL attempted its own liquid-propellant rocket design, but was unsuccessful before its absorption in October 1933 into the Leningrad branch of the RNII Jet Propulsion Research Institute, where the development of the ORM family of the liquid-propellant engines continued.

RNII In October 1933, the Soviet military decided to merge the GIRD organization in Moscow with the GDL propulsion laboratory in Leningrad to create the Jet Propulsion Research Institute (Russian acronym RNII). The merger joined Korolev, Glushko, and others in an influential design enterprise that was later decimated by Stalin's purges. Although Korolev was installed as deputy chief of the institute, personality conflicts placed Georgi Langemak in the position, for which he was later tried and executed during the Stalin purges. Korolev survived the demotion with his life, although he was tortured and imprisoned for six years in the same purge.

NII-88 The Scientific-Research Institute No. 88 (NII-88) was established by Stalin's decree in 1946 to convert captured V-2s into Russian missiles with the assistance of captured German engineers. NII-88 later formed the core of the Soviet space program design organizations, such as NII-88 Department Number 3, a division that became OKB-1 later in the year, headed by Sergei Korolev.

OKB-1 Special Design Bureau Number 1 (OKB-1) was established first as Department 3 of RII-88 in 1946, which was the rocket and missile design group headed by Sergei Korolev. OKB-1 remained in existence until Korolev's death in 1966 when it was turned over to Korolev's deputy, Vasily Mishin, and renamed TsKBEM. The OKB-1 bureau would again be reorganized in a merger with other design bureaus and renamed NPO Energia in 1974. NPO Energia became the S.P. Korolev Rocket and Space Corporation (RKK) Energia in 1994, with Valentin Glushko at the helm.

OKB-52 Vladimir Chelomei was assigned the position of chief of the new design bureau OKB-52 in 1955, a bureau created to design missiles for the Soviet Navy. After some adroit political maneuvering by Chelomei, the bureau expanded into space launchers and spacecraft projects until the fall of Khrushchev, whose son was hired by Chelomei. OKB-52 was absorbed into NPO Energia in 1974, along with OKB-1 (TsKBEM), becoming part of the RKK (Raketno-kosmicheskaya korporatsiya) Energia that exists today.

OKB-456 Glushko was placed in the leadership position of the rocket propulsion bureau OKB-456 at the same time that the other missile and propulsion bureaus were established by Stalin in 1946, following the capture of the German V-2 materials. Glushko remained as the head of the OKB-456 propulsion design bureau until 1974 when major reorganization of the missile, space, and rocket teams placed him in charge of the combined bureaus called NPO Energia.

NPO Energia Following the failure of the N-1 heavy-lift lunar rocket and the Soviet manned lunar landing program headed by Korolev and his successor, the rocket and space design teams under Mishin, Chelomei, and several others were combined into the new NPO Energia (Energia Scientific-Production Association), headed by Glushko. NPO Energia was again reorganized and renamed RKK Energia in 1994.

RKK Energia The Energia Rocket and Space Corporation, formally named the S. P. Korolev Rocket and Space Corporation Energia, is the successor to Korolev's rocket and missile design bureau OKB-1 and Chelomei's OKB-52 bureau.

Soviet Space Pioneers

Some of the greatest gains in rocket technology came from Germany's V-2 ballistic missile program, and many of those innovations originated from Robert Goddard's designs. At the close of the World War II in 1945, both the United States and the Soviet Union began reproducing captured German V-2 rockets and parts to bolster their respective weapons technology. Under the direction of Joseph Stalin and the Soviet Politburo, the Russian scientists and engineers who designed or constructed rockets before the war were released from prison to develop the simple V-2 design into a larger missile, and ultimately the first ICBM. Along with the captured German weapons, scientists, and documents, the Americans and Soviets had their own theoretical work that supported the early missile development. Robert Goddard's rocket research and theory had been taken surreptitiously by the German rocket designers, which helped accelerate the V-2 design. Soviet scientists had also developed rocket and space flight theory, including Konstantin Tsiolkovsky and Yuri Kondratyuk. While Russians had the advantage of access to the research and writings from the West, including Robert Goddard and Herman Oberth, the West was closed off from any of the theory and work that took place in the USSR.

KONSTANTIN TSIOLKOVSKY (1857–1935)

The foundation of much of Russia's rocket and space flight theory was developed well before the first liquid-fuel rockets were launched by a math teacher and tutor named Konstantin Tsiolkovsky. Although Tsiolkovsky had no formal education and was almost completely unknown outside of the Soviet Union, his contributions to the Russian space program were influential in molding some of the early missile, orbital satellite, and manned spacecraft design concepts.

Konstantin Tsiolkovsky was born to a middle-class family on September 17, 1857 in the village of Izhevskoe, Russia. Konstantin was fortunate to have an educated mother who inspired his lifelong learning since he was but one of seventeen brothers and sisters. After moving to Ryazan, Konstantin contracted scarlet fever at the age of nine or ten, resulting in the permanent and almost complete loss of his hearing, which ended his early schooling.

At the age of sixteen, Tsiolkovsky traveled to Moscow to enter the Moscow Technical College, but failed to pass the entrance exams. Instead of returning home, he decided to further his self-education in math, physical sciences, engineering, and philosophy at a number of libraries throughout Moscow. His focus turned to aeronautics and mechanics as he developed an increasing interest in space flight.

Figure 12.3. Early photograph of Konstantin Tsiolkovsky in his library. Source: NASA GRIN.

After three years of study, Tsiolkovsky returned home to help support his family as a tutor while continuing his self-education. As soon as he received a teaching certificate, Konstantin began tutoring and teaching math in the district of Kaluga. His greatest interests, however, were in designing aircraft and lighter-than-air ships. Tsiolkovsky was later promoted, then moved to the city of Kaluga where he would spend the rest of his life pursuing his passion for lighter-than-air vehicles, aircraft, rocket propulsion, and spacecraft. Tsiolkovsky's home in Kaluga, where he and his wife raised their three daughters and four sons, and where he completed much of his practical and theoretical work in aeronautics, rocketry, and space flight, is now a museum.

Tsiolkovsky began working on space flight concepts well before publishing one of his important articles on the reaction principle of rockets in 1903 as "The Exploration of Cosmic Space by Means of Reaction Motors." Tsiolkovsky's analyses were some of the first works to detail the application of Newton's action-reaction principle to rocket propulsion and space travel. Some of his most insightful works were on the use of liquid hydrogen and liquid oxygen as rocket propellants published two decades before the launch of Robert Goddard's first liquid-fuel rocket in 1926 and more than a half century before the launch of the first LOX-LH2 rockets. It is important to note that his LOX-LH2 propellant combination is today the most efficient propellant mix available and is used on most booster upper stages, including the Saturn V, the space shuttle, and the Ares rockets.

Tsiolkovsky's workshop in his Kaluga home is on display, along with many of the laboratory tools and equipment used in his aeronautical designs. These included a press for corrugating thin sheet tin and a wind tunnel mockup also on display in the Tsiolkovsky Museum in Kaluga—the Konstantin E. Tsiolkovsky State Museum of the History of Cosmonautics. Konstantin's first space flight experiment in his laboratory was not a rocket, but a poultry centrifuge that was one of the first research projects in an area today considered as one of the most important in long-duration space flight. Other interesting design concepts Tsiolkovsky introduced were several methods to reduce high acceleration/deceleration forces encountered during launch to orbit and during atmospheric reentry. His hydro-bath that could have reduced the high acceleration/deceleration loads on the body was briefly considered for the capsule design in early Russian manned vehicles, but never used (Harford 1977).

Tsiolkovsky's difficulties with the Russian czar and the Soviet regime were not unlike other Russian scientists and intellectuals in that period. And in somewhat the same tradition, he was persecuted for his scientific work and later recognized by the Soviet government as an important figure in Soviet science. Revolutionary party distrust of his writings and anti-Soviet accusations by others led to his arrest and brief incarceration in Moscow. Following publication of Goddard and Oberth on rocketry and space flight theory, the Soviets recognized his important contributions with a number of awards, some of which included direct grants that allowed him to work on his research and publications instead of supporting himself with his teaching salary. Tsiolkovsky was also recognized by the Soviet scientific community as a gifted scientist in a field that he helped define, an honor that earned him membership in the Soviet Academy of Sciences in 1919.

Although the Soviet leadership embraced Tsiolkovsky's theoretical work in space flight, there were obvious conflicts between the Soviet doctrine and his published philosophy. Much of his later works dealt with human exploration of the solar system and its colonization, as well as the human adaptation to a spiritual universe. These were concepts that were not just missing from the party dogma, but put many Russians in prison or labor camps, or worse.

Tsiolkovsky did live long enough to see the progress in Soviet rocketry begin to catch up with his theoretical work. The first Russian liquid-fuel rocket was launched two years before Konstantin's death in 1935, which allowed several of those pioneer missile and rocket engineers, including Valentin Glushko, to meet with him to discuss their work. In his last years, Tsiolkovsky's recognition for his scientific efforts came from those more important to him than the Soviet leadership—the rocket designers. For Glushko, this was an important affiliation that he knew would later advance his influence within the Soviet government (Harford 1997).

Tsiolkovsky pursued more than purely scientific interests in space flight. Later in his life, Tsiolkovsky wrote "The Cosmic Philosophy," an outline of the development for human habitation of the solar system. The ambitious plan had sixteen steps in the exploration efforts that would lead to "Universal Happiness" and a new direction for human evolution. The basic progression in exploration published in 1926 as the sixteen steps were:

1. Creation of rocket airplanes with wings
2. Progressively increasing the speed and altitude of these airplanes
3. Production of real rockets without wings
4. Ability to land on the surface of the sea
5. Reaching escape velocity and the first flight into Earth orbit
6. Lengthening rocket flight times in space
7. Experimental development of plants to make an artificial atmosphere in spaceships
8. Using pressurized space suits for (EVA) activity outside of spaceships
9. Making orbiting greenhouses for plants
10. Constructing large orbital habitats around the Earth
11. Using solar radiation to grow food, to heat space quarters, and for transport throughout the solar system
12. Colonization of the asteroid belt
13. Colonization of the entire solar system and beyond
14. Achievement of individual and social perfection
15. Overcrowding of the Solar System and the colonization of the surrounding stars
16. The Sun begins to die and the people remaining in the Solar System's population go to other suns (Rynin 1931)

Tsiolkovsky's many contributions to early flight included the development of concept and theory in areas of aeronautics and space flight. A partial list of his efforts and concepts includes:

- Design of lighter-than-air vehicles
- Applied Newton's action-reaction principle to develop important rocket equation for reaching orbit
- Orbital space flight concepts
- Use of atomic energy propulsion and solar propulsion
- Gyroscopic stabilization of rockets
- Extravehicular activity
- First liquid-fuel rocket propulsion systems
- Identified liquid hydrogen and liquid oxygen as ideal liquid propellants

- Flotation device for accommodating humans on high-g launches
- Spacecraft pressurized systems
- Space suits for the space vacuum environment
- Staged rocket concepts to reach orbit
- Space station development and operations
- Life support systems
- Artificial plant growth

SERGEI KOROLEV (1906–1966)

Considered the father of the Russian space program, Sergei Korolev led the Soviet Union to its many firsts in the space race, including the first spacecraft to orbit Earth, the first man and the first woman in orbit, the first spacecraft to fly by, impact, soft-land on, and orbit the Moon, and many more. His efforts were monumental, but he remained obscure even to the Russian people until his death in 1966. His importance was recognized outside of the USSR only after the fall of the Soviet Union in 1991 when the records of their space program began to surface. But those records also paint a dark picture of a communist dictatorship under the murderous leader Joseph Stalin. Even though Stalin ruled for more than two decades with violent repression, many Russians remained loyal to his leadership. Sergei Korolev was an example of a loyal, respected Soviet citizen who spent more than six years in Siberian prisons simply because he was a successful manager and engineer during the Stalin purges. Although his health was seriously affected (his jaw was broken and he lost his teeth during his incarceration), his interests and dedication to rocketry and space flight and his loyalty to the Soviet government never wavered.

Childhood

Sergei Korolev was born to parents that were joined by marital contract and therefore had little hope of remaining together. His birthplace in Zhitomit, a small city in the Ukraine near Kiev, was little different from other parts of Western Russia at the turn of the century. Young Korolev watched the civil strife that swept the region from his window—not only the Czarist revolution and the Bolshevik struggle, but the Communist and Soviet war for national control and the effects of World War I. Endless conflict during his childhood, along with the departure of his father when he was three, were instrumental in molding his independent character. His independence was reinforced by his mother who left the family to attend school in Kiev and was with him only during the weekends. Korolev was raised by his grandparents as a child, which added to his solitude and likely contributed to his disciplined academic study habits both in and out of school. Like Tsiolkovsky, Korolev leaned toward diverse subjects in his studies. At the age of six, Korolev's new stepfather, Grigory Balanin, provided the much-needed parental nurturing, as well as inspiration for his future education. Grigory's influence generated even greater interest in physical science and engineering, although young Korolev also enjoyed eclectic as well as classical works (Harford 1997).

Sergei was intrigued with the idea of flying at the age of sixteen, or perhaps even earlier, when a seaplane detachment in the port of Odessa, Ukraine, became his extra-curricular focus. His earned his first flights by trading his time and skills in helping to repair several of the bi-wing float planes. His love of flight led him to an aeronautical engineering career that began with an early glider design while in school in Odessa. This was the same year that he met a girl who would become his wife. But young Korolev's distractions—his passion for flight and his love of Xenia Vincentini—brought pressure from his stepfather to spend more time with his formal education, which ultimately divided the two (Harford 1997).

Korolev's strengthened principles derived from an allegiance with a new and grow-ing socialist government led him to apply for admission to the Zhukovsky Academy, an institution involved with the design of military aircraft. After being rejected because he was not a military pilot, Korolev decided to enter the Kiev Polytechnic Institute.

Education

As a student in Kiev, Korolev redesigned and built four gliders that he used to set a number of gliding records, then transferred to Moscow's Bauman High Techni-cal School (MTVU) to pursue his aviation studies. It was there that he met Andrei Tupolev who would become one of the greatest aircraft designers in history. Tupolev was responsible for Korolev's academic inspiration and at least partly responsible for Korolev's release from prison after the Stalin purges. Following his engineering assign-ment in an aviation bureau, Korolev's SK-3 Red Star glider was accepted for produc-tion by Sergey Ilyushin, another famous designer of a multitude of Soviet military and civil aircraft, some of which are still flying today.

Korolev's career in rocket design began after his graduation from MTVU when he joined the Central Aero and Hydrodynamics Institute (TsAGI) and met Fredrik Tsander. Both were members of the Moscow Group for the Study of Reactive Pro-pulsion (MosGIRD) amateur rocket group: Tsander, the founder, and Korolev, the lead engineer. This group was the first to launch a liquid-fuel rocket in 1933, though Tsander did not live to see the launch.

In 1933, the Gas Dynamics Lab and the Group for the Study of Reactive Propulsion were merged by military decree into the RNII. In the next three years, advances in rockets, gliders, and engine design led to a number of new vehicles that included Russia's first liquid-fuel rocket. But the missile and powered glider successes proved treacherous. RNII designers were somewhat isolated from politics even though they were members of the Communist Party. They were not, how-ever, immune from the brutal paranoia of Stalin and his nearly indiscriminant orders for the imprisonment and execution of Russia's military leaders, scientists, engineers, and intellectuals. In 1936, a number of members of RNII attended a conference held by the Soviet Academy of Sciences. As a consequence, nearly ev-eryone in the RNII group was arrested and tried for collaborating with anti-Soviet organizations in Germany. Even the Soviet spies responsible for passing intelli-gence to Russian contacts about German rocket development were imprisoned or executed (Harford 1997).

Imprisonment

Glushko and other managers in RNII who were arrested before Korolev had testi-fied at several tribunals and were given lighter sentences. Following a mock trial by a troika, a three-member panel notorious for their predetermined judgments, Korolev and other engineers were sent to prison; Korolev for a minimum of eight years. An apparent conflict with another engineer about the focus of the RNII being on liquid-fuel winged rockets versus solid fuel rockets spread suspicion of Korolev's loyalty to the Soviet Union, which led to his arrest. Patron-leader of the RNII, Tukhachevsky, as well as others who admitted to false crimes were executed. Glushko and Korolev denied involvement in crimes and were simply imprisoned (Korolev had been replaced as chief designer two years earlier by Georgi Langemak who was executed since he was in the top position). Even Tupolev was arrested and imprisoned, although he was soon returned to a prison factory to continue building military aircraft.

Imprisonment during Stalin's purges generally meant exile to the most inhumane gulags (GULAG is the Russian acronym for Main Directorate for Corrective Labor Camps). After a brutal beating and forced confession, Korolev was sentenced to prison for ten years and the loss of his property. His imprisonment included serving in several of the worst gulags in Russia, which included the Kolyma gold mine in eastern Siberia (Harford 1997).

Conditions in the Soviet gulags were appalling, with little food, infrequent, if any, medical care, often unheated housing even in Siberia in the winter, and ten to fifteen hours of daily labor. It was reported that 10% to 20% of the prison population died each year, adding to the twenty to thirty millions of Russians that were starved, executed, ethnically purged, or forced into a delayed death by Stalin.

Korolev survived imprisonment and forced labor from 1938 to 1943, but suffered from the physical brutality. In 1939, he was resentenced to eight years after a brief reinvestiga-tion, then allowed to leave the Kolyma camp after a number of others, including Andrei Tupolev, intervened to get him transferred to an industrial penal factory (collectively called sharashkas) located at Tupolev's aircraft facility. But life and living conditions in the sharashkas was little more than a prison. Inmates worked on specific projects during the day but were confined to prison barracks without outside contact, including mail and visitation.

Figure 12.4. Pho-tograph of Korolev taken during his im-prisonment. Source: NASA GRIN.

World War II and the V-2

At the close of World War II, Korolev and others were released from the prison fac-tories and sent to Germany to secure V-2 missiles and German engineers in order to advance Russia's long-range weapon plans. Stalin ordered the creation of several design bureaus to handle the development of future missile projects, beginning with the rep-lication of the V-2 using Soviet-built parts. The rocket design bureau named OKB-1

(Opytnoye Konstruktorskoye Byuro) was headed by Korolev. The new rocket engine design bureau named OKB-456 was headed by Glushko.

Korolev remained head of the OKB-1 throughout his career as chief designer of many of the Soviet military missile projects, as well as a number of important space launch vehicles and most of the early exploration spacecraft programs. But because of the Soviet bureaucracy, most of his efforts as chief designer were spent securing funding and approval for his programs rather than applying his management and engineering skills toward space projects. The Politburo's conservative attempt to ensure that important projects did not encounter any significant setbacks was to duplicate the design bureaus and devise several layers of responsibility. One of the primary influences on Korolev's missile and rocket design efforts was the military leadership's desire to modernize weapons for the Cold War's arms race with the United States. As chief designer of the new rocket and missile bureau, Korolev had to battle the Politburo and the military authorities while developing advanced missiles because of a parallel pursuit of his true interests in space exploration. Even more taxing to Korolev's efforts were three design bureau chiefs who worked continually to compete for missile, rocket, and spacecraft assignments.

In 1953, Stalin approved Korolev's request to develop an intercontinental ballistic missile capable of launching a nuclear warhead to Europe, Asia, and North America. Korolev's assignment was to integrate all of the ICBM design efforts, including the engine development efforts headed by his nemesis, Valentin Glushko. In spite of disagreements with Glushko over the fuel and engine design, the R-7 ICBM was launched successfully in 1957 carrying a dummy warhead over 8,000 km. The successful August ICBM launch was followed by the October 4, 1957 launch of the first Earth satellite, Sputnik 1. Under the orders of Stalin's successor, Nikita Khrushchev, Korolev placed a dog in orbit just one month after Sputnik 1, and an enormous space laboratory in orbit just three months after that. America responded to Korolev's feats with a Vanguard launch attempt in December 1957 that ended in a spectacular failure, followed by a successful launch of the Explorer 1 into orbit on January 31, 1958. Even so, the Soviets proved their superior weapons and space technology to the world under the guidance of Sergei Korolev. The unknown rocket, missile, and spacecraft designer was responsible for the long list of space firsts that followed. Because of his successes, Korolev was allowed latitude no one else thought possible. His manned capsules launched the first man and woman into space, and orbited the first multi-crew mission in the race with America to the Moon. Korolev alone maneuvered projects and funding through the tangled Soviet bureaucracy with remarkable success, while battling other competing designers for many of his successful space projects. A handful of engineers were considered capable of designing the vehicles that Korolev managed, but no one else had the understanding, the strength, and the skills to push programs through the Soviet political and military bureaucracy .

Korolev's Death

Korolev's career was brought to an early end during routine surgery to remove polyps in his colon on January 5, 1966. A team headed by one of the USSR's best surgeons discovered several more serious medical problems during the surgery that included a large intestinal tumor. Complications extended the operation for nearly eight hours. Internal bleeding, insufficient anesthesia because of Korolev's broken jaw, and heart failure complicated the surgery. Even though Korolev's surgical procedures were a success, Korolev suffered cardiac arrest and never recovered (Harford 1997).

News of Korolev's death surprised everyone in the Soviet Politburo and shocked the missile and space design teams. An immediate assessment of the space programs in progress was made, including Russia's largest and most troubled space program, the manned lunar landing project. The consequences of Korolev's death were clear, but the response was chaotic. Even though Russia's best rocket and spacecraft designers were still in place, no one had the respect, force, diplomacy, and concentrated will of Korolev. The combined effects of a monolithic Soviet bureaucracy, uncommitted leadership, continual undermining of the program by Korolev's adversaries, poor industry performance and standards, opposition by key military figures, a lack of program funding, and a myriad other problems crushed the manned lunar project that even Korolev had trouble keeping alive. And without him, the program was doomed. Fierce program opposition and embarrassment over the program failure kept details of the project from reaching the outside world until the collapse of the Soviet Union in 1991. The failed Soviet manned lunar program would also end with the destruction of nearly all of the records and hardware related to the project.

Korolev's status as chief designer and architect of the Soviet's successful space program was unknown outside the inner Soviet government and the design teams. His death, however, afforded him recognition as one of the great heroes of the Soviet Union. Not only did he have one of the very few private homes in Moscow (private ownership of property was not allowed in the socialist state and is only now being gradually accepted), but his status allowed him burial in the Kremlin Wall, a distinction limited to Russia's greatest heroes and leaders of the last century. Nikita Khrushchev is one of the intriguing omissions from the Kremlin Wall mausoleum in central Moscow.

Korolev's funeral procession included a number of friends and allies, including Yuri Gagarin, Gherman Titov, Valentina Tereshkova, Vladimir Komarov, Pavel Belyaev, Alexi Leonov, Mstislav Keldysh of the Soviet Academy of Sciences, Leonid Sedov, and many others. A large number of party, military, and government officials were also in attendance (Harford 1997).

FREDRIK TSANDER (1887–1933)

Fredrik Tsander was one of the original pioneers of rocketry in Russia. He was responsible for the design of the first Soviet liquid-fuel rocket and organizing the innovative GIRD rocket club in 1931. Tsander, who was born in Riga, Latvia, was educated as an engineer and particularly drawn to the ideas of Konstantin Tsiolkovsky. Tsander's interests resembled Tsiolkovsky's in several ways, including theoretical publications on orbital and interplanetary space flight. But Tsander was more ambitious in his desire to reach space and, like Robert Goddard, extended his theories to practical rocket design. In addition to creating the GIRD rocket group, Tsander also founded the Society for Studies of Interplanetary Travel.

Figure 12.5. Photograph of the GIRD members in the early 1930s. Korolev is seated at center front, and Fredrik Tsander at right front. Source: Sidiqqi.

Tsander's GIRD-9 rocket design was the first liquid-fuel rocket launched in the USSR, but he did not live to see the launch. Tsander died in 1933 of typhoid fever at the age of forty-six with little recognition for his pioneering work. His influence did not end with his death, however. Tsander's efforts and interests continued through his inspiration of Korolev, Glushko, and other members of the early GIRD and GDL rocket teams.

VALENTIN GLUSHKO (1908–1989)

Valentin Glushko began his engineering career with the design of the early liquid-fuel engines at the Gas Dynamics Lab in the early 1930s. In the next fifty years, he would become one of the three most influential rocket designers in the Soviet Union's space race with the United States. Glushko, like Korolev, grew up and attended school in Odessa, Ukraine, both with technical trade skills. He then attended Leningrad (St. Petersburg) State University, then joined the Gas Dynamics Lab to work on liquid-fuel rocket engines. Glushko was also a member of the GIRD located in Leningrad, while Korolev became a member of the Moscow-GIRD at about the same time.

While working at GDL, Glushko was involved with the production of the first liquid-fuel rocket engine, the ORM-1 (Opytnyi Reaktivnyi Motor No. 1) that completed its first test firing in 1931. GDL's engineers continued to make advances in liquid fuel rocket motors with governmental recognition, but without funding. These GDL engines not only would propel the first Russian liquid-fuel rockets, they also powered the first winged rockets.

Glushko was transferred to the RNII in 1933 to produce missiles, winged rockets, and rocket-powered gliders, along with the GDL, Korolev, and the GIRD. But in 1938, Stalin's purges trapped Glushko and others into testifying against other high-level designers in the RNII organization, including Korolev. Glushko's testimony was likely insignificant in the sham trials intended to remove any perceived threat to Stalin's absolute control. Even so, Glushko's sentence of imprisonment was later converted to incarceration at an aircraft design bureau, while Korolev was sentenced to 10 years of imprisonment and shipped to work camps in the Siberian gulags until his transfer to an aviation industrial camp six years later.

Following Stalin's decree that established the missile and weapon development bureaus in 1946, the GDL design group was reorganized into the OKB-456 design bureau, headed by Glushko. He would remain the chief designer in OKB-456 until Korolev's successor, Mishin, was removed as head of TsKBM. Glushko took over as director of a larger design group that absorbed the TsKBM and Chelomei's OKB-52, and went on to develop a variety of spacecraft, heavy-lift boosters, and missiles, including the Energia. The collective spacecraft and rocket design organization named NPO Energia was led by Glushko from 1974 until his death in 1989.

VLADIMIR CHELOMEI (1914–1984)

Vladimir Chelomei was an aviation engineer who headed an aircraft design bureau after World War II, then began promoting cruise missile concepts similar to Korolev's winged rockets. His early success came to an end when he was removed as chief designer of the bureau because of a successful competing cruise missile proposal. Artem Mikoyan, the chief designer of Soviet MIG fighter jets, had employed the son of Stalin's security chief and won. In the process, Stalin then placed Chelomei's bureau under Mikoyan.

Soon after the death of Stalin in 1953, a special design bureau for cruise missiles was decreed, then expanded in 1955 to develop submarine-launched cruise missiles. This became the Experimental Design Bureau number 52, or OKB-52. Vladimir Chelomei, who had been instrumental in promoting the submarine cruise missiles, advanced within the bureau and gradually ascended to chief designer. His bureau would eventually join with Glushko to produce heavy-lift vehicles that competed with Korolev's R-7 and N-1 designs. One of those was the successful UR-500 Proton heavy-lift launcher originally built as a military launcher for large surveillance satellites and manned observation platforms. The Cold War and race to the Moon served to redirect Chelomei's Proton to support Russia's manned lunar missions, and later to launch the highly successful modular space stations. Chelomei's bureau was also responsible for the Almaz military space station design that was a competing design to Mishin's Salyut space station series.

VASILI MISHIN (1917–2001)

Vasili Mishin entered his career in missile development as Korolev's deputy in OKB-1, assisting in the design and fabrication of the R-7 ICBM used for the launch of the Sputnik satellites. Mishin aided in the design and development of Korolev's launchers and spacecraft, including the ill-fated N-1 heavy-lift booster developed specifically for their manned lunar missions. Upon Korolev's death in 1966, Mishin assumed command of the OKB-1 bureau, and all of the problems that came with the manned lunar project. Because of the rush to put Soviet cosmonauts on the Moon ahead of the United States, the N-1 was never tested before launch. The N-1 engines were also first-generation designs because Glushko refused to work on the project's propulsion system. The Korolev-Glushko conflict proved fatal to the manned lunar program, as did the inability of anyone else, including Mishin, to successfully orchestrate major space programs within the Soviet bureaucracy.

Four failures of the N-1 booster led to the termination of the project and the entire manned lunar program as conceived by the OKB-1 bureau. At the same time, Mishin was relieved as director of the TsKBM bureau (formerly OKB-1). The TsKBM (Tsentralnoye Konstruktorskoye Byuro Eksperimentalnogo Mashinostroyeniya) design bureau was renamed NPO Energia and taken over by Glushko who served as its director until his death in 1989.

DIMITRIY USTINOV (1908–1984)

One of the most important figures in the era of the Soviet space race was not a designer, but a minister of military and space programs who served as Secretary of the Central Committee for Space and Defense from 1965 to 1976. Ustinov's control of the missile, rocket, and space programs promoted the interests of space exploration, as well as military defense, although not in equal portions. From directing development of the first R-1 and R-2 missiles, to forging the development of the Zenit surveillance satellite while Korolev was building the Vostok manned capsule, Ustinov succeeded in strengthening the USSR's position in space while building military might using the same missiles. Perhaps Ustinov's greatest achievement was that his successes came during the treacherous reign of Joseph Stalin. Ustinov's service as Minister of Defense in 1976 ended his participation in the Soviet civil space efforts which allowed him to reach his lifelong ambition of joining the Politburo.

MIKHAIL TIKHONRAVOV (1900–1974)

Mikhail Tikhonravov was one of the Soviet Union's most productive rocket and spacecraft engineers. His work on rocketry began in the GIRD organization where he designed the liquid fuel engines for the first Soviet liquid-fuel rockets, the GIRD-9 and GIRD-X. Tikhonravov was also a division leader in the RNII rocket and missile group. As part of Korolev's OKB-1, Tikhonravov was responsible for the development of the Sputnik 3 orbital laboratory and numerous other early Soviet satellite projects. The first Sputnik satellite and interplanetary probes were developed in his division of the OKB-1. Korolev also used Tikhonravov's expertise in spacecraft design on the manned Vostok program.

Soviet Missile Development

At war's end, Stalin recognized the utility of the V-2 missiles and ordered the capture of the German V-2 scientists and their rockets, a decision hastened by intelligence on how much hardware and personnel were already in American hands. His directives were passed through the ministries to bring the missiles, personnel, and materials back to Russia so that copies of the V-2 could be made entirely from Russian parts. Stalin's clever approach to missile development took Soviet design efforts on a much faster track than if they had simply used the captured V-2s for testing. Experience with large military projects also taught the Soviet bureaucrats that multiple design and production facilities (and directors) kept bottlenecks to a minimum, but at the expense of efficiency and innovation.

Responsibilities for the V-2 conversion project were separated into three major design bureaus with several other groups participating in other industrial and military programs. The most prominent of the design bureaus were the OKB-1 missile design group headed by Sergei Korolev and the OKB-456 headed by Glushko who had the responsibility for designing propulsion systems, including reproduction of the V-2 engine.

R SERIES

The first Soviet long-range missile, called the R-1, was an exact copy of the V-2 with the same appearance and performance but constructed from Russian parts. Flight accuracy, reliability, and durability were improved in the years following with assistance from the captured German rocket scientists. The R-series (R for raketa, "rocket") of Russian rockets soon increased in thrust and range that made them versatile enough to serve as both a weapons platform and a suborbital space research vehicle. An important addition to the early launch rockets was the addition of a second stage that extended the reach into the upper atmosphere. A launch site for the R-series boosters was prepared at Kapustin Yar that was deep in the Russian interior, with R-1 and R-2 production nearby at Podlipki.

Major improvements in the R-1 included a higher-thrust engine developed by Glushko's bureau and identified as the RD-101. The new engine produced 32 tons thrust compared to the R-1's 25 tons and incorporated an integrated fuel tank into the upper skin structure for the new R-2. Maximum payload for the R-2 missile was increased from 1 ton to 1.5 tons, with a range increase from 270 km to 600 km. Flight testing of the R-2 missile began in October 1959 (RussianSpaceWeb n.d.).

From the R-2, the Soviet design bureaus began planning, designing, and building more versatile and long-range missiles that included the R-3, R-5, R-9, R-10 and R-11 series, with Korolev managing much of the work from the R-3 onward. But Soviet missile production was not without internal rivalry. In 1960, Mikhail Yangel's R-16 was completed with a number of improvements over Korolev's R-7 including the use of hypergolic propellants more easily stored than cryogenic liquid oxygen. The R-16 also offered a more compact design and the ability to launch from an underground silo which made them much less susceptible to counterattack. The first test of the R-16 on October 24, 1960 came after numerous delays but ended in the greatest launch disaster in history. During countdown, a rupture in the fuel and oxidizer tanks spilled fuming nitric acid and UDMH that are both extremely toxic and hypergolic. The propellants ignited spontaneously, exploding into an inferno that killed 165 workers, technicians, and managers, including Marshal Mitrofan Nedelin. "The Nedelin disaster" was named for the head of the Soviet strategic forces, who had brought a chair to the launch and whose hubris ended the lives of

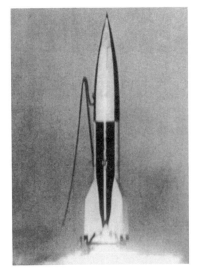

Figure 12.6. Launch of an R-1 missile from Kapustin Yar in the Russian interior. Source: NSSDC.

Table 12.1. Comparison of R-1 and V-2 Performance

	R-1	*V-2/A-4*
Length	14.3 m	14 m
Diameter (max)	1.65 m	1.65 m
Weight (fueled)	13,400 kg	12,800 kg
Fuel	Alcohol/water mix	Alcohol/water mix 3,710 kg
Oxidizer	Liquid oxygen	Liquid oxygen 4,900 kg
Thrust (liftoff)	26,700 kg	25,000 kg
Max speed	5,400 km/hr	5,400 km/hr
Apex altitude	90 km	90 km
Burn time	65 sec	65 sec
First launch	1948 Oct. 18	
Launch sites	Kapustin Yar	Peenemunde, Nordhausen
Flight range	270 km	320 km nominal
Impact velocity	Mach 3	Mach 3
Warhead mass	1,000 kg	1,000 kg (738 kg explosives)
Engine	RD-100	
Burn time	65 sec	65 sec

many of the project's technical staff. The R-16 did complete its development and was deployed as an ICBM beginning in the mid-1960s, along with Korolev's R-9.

R-1A, B, V, AND D RESEARCH ROCKETS

The R-1's success as a long-range weapon stirred interest in the Soviet Academy of Sciences geophysicists because of its ability to reach above the atmosphere. To accommodate the research flights, a separable payload container was outfitted with scientific instruments in place of the 1.1 ton warhead. Separation mechanisms and recovery parachutes added weight to the launcher which reduced the payload to about 800 kg. A more vertical trajectory was programmed which reduced the range but increased the peak altitude to about 100 km. The first of the six R-1As launches took place on April 21, 1949.

Figure 12.7. R-1D research rocket in preparation for launch from the Kapustin Yar missile site. Source: NSSDC.

Animal experiments that paralleled the U.S. Bumper (V-2 + Aerobee second stage) research flights began in July 1951. The pressurized container was capable of atmospheric reentry, separation, and recovery. The first success of the R-1B animal flight was in July 1951, with the canine crew named Dezik and Tsygan. An added parachute for recovery of the R-1B rocket body was also flown under the designation R-1V.

The R-1D rocket launched a more elaborate experimental package than the R-1B, with a pair of dogs ejected from the capsule separately. Both dogs were protected by individual space suits that included a life support system and a protective cradle for a parachute landing. Three launches of the R-1D between June and August 1951 were successful.

R-7 ICBM (8K71)

By 1948, Soviet missile development programs included the first long-range surface-to-air missile; the long-range and later submarine-launched R-11; and the first nuclear warhead missile, the R-5. All had evolved directly from the basic V-2/R-1 design. But none were capable of reaching other continents or able to reach Earth orbit. Korolev knew that the same performance was needed to reach other continents as it would to put a satellite in orbit and requested permission to build the first intercontinental ballistic missile. In 1952, Korolev's request to build the first nuclear ICBM was approved by decree, although little was known of its utility for space research outside of Korolev's and Glushko's design bureaus. Approval of the R-7 ICBM meant many things to Korolev, including a satellite and an interplanetary launcher when he finished

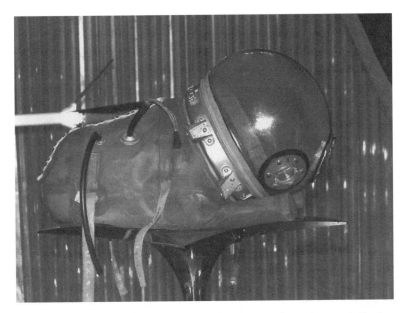

Figure 12.8. Photograph of an early canine space suit used on early Russian suborbital flights. Source: Memorial Museum of Cosmonautics, author.

building the ICBM that could also put a cosmonaut into orbit. Korolev's impressive new R-7 was a completely new missile design, and the first Russian heavy-lift rocket not derived from Germany's V-2 design.

Following the government decree of February 13, 1953, the design bureaus under Korolev's OKB-1 began a preliminary design of the 170-ton, two-stage long-range missile. The R-7 specifications called for a payload capacity capable of launching a 3,000 kg nuclear warhead 8,000 km, making it not only the first intercontinental ballistic missile, but also the most threatening weapon in the early Cold War era.

Korolev designed the R-7 Semiorka ("Little 7") ICBM as a two-stage booster with all engines starting at the same time. The simultaneous start made the R-7, like the Atlas, a stage-and-a-half launcher. His design had four external booster segments for the second stage called Stage B. The first stage, called the central core or Stage A, separated from the four first-stage boosters at burnout.

Both stages of the R-7 employed a four-engine cluster designed by Glushko's propulsion group, although Korolev's original specifications called for single high-thrust engines rated at 65 tons for each of the five units. Instead, Glushko combined four 25-ton thruster units into quad-cluster for the four boosters and the core stage. The RD-107 and RD-108 quad engine units used a single turbopump to feed fuel and oxidizer to all four combustion chambers, which simplified the design, but reduced the reliability of the engines. The four-engine design also needed additional steering thrusters to achieve the needed warhead impact accuracy during the thrust phase, and for steering the missile after booster cutoff. In the completed design, both the RD-107 and RD-108 units had vernier guidance thrusters; four for the core RD-108 engine, and two on each of the RD-107 booster engines. This configuration meant that all thirty engines were started for liftoff, but staggered slightly so the ignition and startup was not simultaneous for all thirty engines. Engine start and operational reliability of the thirty engines was a headache for Korolev and further strained the relations between him and Glushko.

Glushko's quad cluster design provided 100 tons of thrust from the four main engines, all fed by the same turbopump unit. Regenerative cooling of the exhaust nozzle was accomplished by circulating kerosene through an external network surrounding the nozzles before injection into the combustion chamber. The choice of liquid oxygen (LOX) and kerosene was made by Korolev because of its high specific impulse. Korolev's choice of propellants for the ICBM design was one of the bitter disputes between Korolev and Glushko that grew more acrimonious over time. Korolev wanted higher specific impulse, while Glushko believed that propellants should be hypergolic for greater missile-launch flexibility. This disagreement festered until the death of Korolev in 1966, at which time Korolev's cryogenic-fed boosters were replaced by Glushko's new missile designs that used hypergolic propellant. Though the dispute between the designers ended in 1966, Korolev's cryogenic propellants would reemerge in some of Glushko's later advanced booster designs.

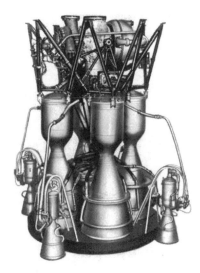

Figure 12.9. Sketch of the RD-108 quad-chamber engine used on the R-7 core booster that produced 100 metric tons of thrust. A single fuel and oxidizer turbopump was used to feed the four primary and four secondary (guidance) thrust chambers which simplified the engine operations. Steering on the RD-108 was accomplished by articulation of the four small outboard thrusters (2 on the RD-107 first-stage booster engines) that are commanded by the missile's guidance control system. Source: NSSDC.

R-7 TESTING

Three years of development and testing of the R-7 after its formal approval led up to the first launch attempt of the ICBM on May 15, 1957. The flight proceeded normally until one of the strap-on boosters separated early, creating an instability that destroyed the vehicle. A second test launch on June 11, 1957 had similar results, but for a different reason. A third test launch a month later on July 12 again failed thirty-three seconds after liftoff. A critical fourth launch on August 21, 1957 was made with Nikita Khrushchev in attendance. Fortunately for Korolev and his future plans, the launch was a success. Although the R-7 performed as designed, a warhead separation malfunction ended with a warhead-booster collision. The Soviet media announced their successful launch of the world's first ICBM. With that announcement U.S. military planners and the Eisenhower administration recognized they were years behind the Russians with their Atlas ICBM program. The specter of this "missile gap" would drive the nuclear arms race between the United States and the USSR to new heights in the 1960s.

Korolev continued tests of the R-7 and used the second verification launch on September 7 to qualify the booster for his Sputnik satellite launch. Additional test

Table 12.2. R-7 Initial Specs

Design (overall)	OKB-1 (Korolev)
Engine design	OKB-456 (Glushko)
Vehicle length	28 m (91.8 ft) for the core
Weight	280 metric tons (fueled) 27 tons empty
Fuel	Kerosene
Oxidizer	Liquid oxygen (LOX)
1st stage	4 strap-on (external) boosters, each with 4-engine cluster RD-107 (8D74) + 2 steering engines
2nd stage	Core (center) module—4-engine cluster RD-108 (8D75) + 4 steering engines
Range	8,500–8,800 km (5,280–5,466 mi)

flights of the R-7 to certify the booster for its primary role as an ICBM were made until its acceptance and deployment were approved in 1960. Yet even after the test launches on the R-7 proved the vehicle a reliable, accurate ICBM, the missile was never deployed in significant numbers. Fueling was time-consuming, and a loaded LOX tank could be maintained for only twenty-four hours, after which it would have to be drained and inspected and, if necessary, parts would be refurbished or replaced. Preparation for another launch required another cycle, which took another thirty-six hours. The R-7's three-day launch cycle tripled the number of R-7 ICBMs that had to be in place for strategic deployment. In addition, the R-7 required above-ground launch facilities which made it vulnerable to attack, or even sabotage. These and other shortcomings led to the R-7's removal from the Soviet ICBM inventory. In 1967, the world's first ICBM was replaced by missiles that used non-cryogenic propellants that could be stored in silos. However, the R-7 continued its career as a space launch

Figure 12.10. Photograph of R-7 launch preparations at the Baikonor launch complex. Source: NASA GRIN.

vehicle to become the most prolific launcher on record under several different designations. Today's R-7 derivatives are used to launch the Soyuz crew ferry vehicles and the Progress cargo carrier to the ISS, as well as the Molinya satellites and a host of other missions. At the end of 2005, there were a total of more than 1650 R-7-derived booster launches.

Russia's First Satellites

The first Earth satellite launches were targeted for the International Geophysical Year, which started in January 1957. IGY committee meetings used to plan the coming global research projects were held in 1953 where both the United States and the USSR declared their intentions to launch an artificial satellite in support of IGY. Concerns over using a military launcher for a research satellite hobbled the American efforts but not the Soviet's since all of their launchers were military missile designs. By 1954, Korolev started developing a satellite for the IGY launch on his newly approved R-7 called Object-D. The satellite was a massive 3,000 kg orbital laboratory that would be capable of detailed studies of the Earth-space environment. But the Soviet bureaucracy that could slow even simple tasks delayed the fabrication and testing of Object-D past the target date set for a mid-1957 launch. When Korolev realized that the satellite would not be ready in time, he organized the development of a small, simple satellite that would transmit telemetry of the spacecraft and its environment. That small spacecraft that served as the stand-in for Object-D was Sputnik 1.

SPUTNIK 1

Launch of the Sputnik ("fellow traveler") was not preannounced, either within Russia or to the outside. Even so, clues could be found in a variety of publications on the impending launch, and even on the payload. In the weeks before Sputnik's launch, several press releases from Moscow mentioned an impending orbital satellite launch. Outlandish claims from endless Soviet propaganda were so common in that era that claims of an Earth-orbiting satellite launch by the USSR were easily dismissed. Other announcements of the satellite launch were far more credible, including Soviet press conference notes from June 1957, which listed the altitude and velocity of the Sputnik launch. A relatively accurate description of the Soviet satellite was relayed by the *New York Times* detailing the 100-pound, eighteen-inch sphere that would have an orbital period of ninety minutes. A letter to the U.S. representative to the National Committee for IGY mentioned Russian-sounding rocket and animal launches, as well as a satellite launch soon after the start of the IGY in July 1957. An article on the satellite's reflective brightness was written by the chairman of the Astronomical Council of the Soviet Academy of Sciences. The paper encouraged participation by the astronomical community in reporting visual observations of the first artificial satellite (Dickson 2001). Another announcement was made by a Russian scientist at an atmospheric conference in Boulder, Colorado, on August 27 that listed the transmitter frequency of the satellite as 24 MHz. Then, in mid-September, a Radio Moscow transmission claimed that Russia was ready to launch the satellite, although the date was not included in the announcement (Dickson 2001).

A Defense Department alert on an impending satellite launch on Russia's new R-7 Sapwood ICBM was presented to Congress and the Secretary of Defense with little response, other than disbelief. In addition, CIA surveillance photos from U-2 flights showed the scale of the rocket being prepared for launch at the Tyuratam (Baikonor) site, confirming Russia's ability to launch a satellite. Although the Eisenhower Administration was briefed on the launch preparation, no change was made in the planning for a Vanguard satellite launch, and Eisenhower had no interest in beating the Russians into space.

Even more surprising, Russian emissaries invited the United States to place instruments on their upcoming launches. Several memos from mid-1957 show little interest within the Defense Department for cooperative research for several reasons. Although the scientists from both Russia and the United States had mutual interests in sharing research and rocket payloads during IGY, the administration and the Defense Department were not convinced of the importance of being the first in space and discounted the scientific and strategic importance of Earth satellites and orbital launchers (Dickson 2001).

Sputnik Launch

Launch of Sputnik 1 by the Soviet Union was targeted for Tsiolkovsky's 100th birthday on September 17, 1957, but was delayed several times before its launch on October 4th, well within the time frame of the International Geophysical Year that spanned eighteen months instead of twelve because of the peak in solar cycle activity. Little fanfare was given the first satellite launch, even though there was interest in scientists from both sides to be first. Reaction from the Soviet press was muted until the launch ignited an overwhelming response worldwide. More the worldwide reaction than the actual launch triggered, the rapid series of events in both the United States and the USSR that marked the beginning of the space race. Because of the tactical advantage that space-based weapons represented, the U.S. military leadership and the Eisenhower administration accelerated their disjointed space satellite programs.

USSR's Sputnik satellite represented more than a threat to U.S. national defense. It exposed a glaring gap between the two nations in advanced weapons technology, as well as space exploration technology. The Eisenhower administration had a subdued response to the launch in an effort to avoid alarming the public. After the second Soviet launch of the larger and more advanced Sputnik 2, several major decisions were made by the Administration, including the transformation of the National Advisory Committee for Aeronautics into the National Aeronautics and Space Administration under the National Aeronautics and Space Act of 1958.

The Russians took full advantage of their lead in space launch vehicles and space flight technology, guided by their chief designer Sergei Korolev. Engineers and scientists in his OKB-1 division began planning space exploration programs for interplanetary research, as well as Earth exploration, lunar exploration, solar exploration, and manned orbital flights. Korolev had already developed the booster to launch these spacecraft, and now had the go-ahead from Soviet Premier Nikita Khrushchev to begin more ambitious projects because of their tremendous propaganda value.

Sputnik Satellite

The 0.6-meter-diameter Sputnik 1 was constructed with two aluminum alloy spherical shells and four externally mounted antennas. Sputnik's outer shell was polished to provide a reflective surface that could be spotted more easily. The inner sphere was pressurized with nitrogen gas with internal and external temperature sensors that doubled as a micrometeoroid detector. A significant internal temperature drop would have meant pressure loss had occurred from a micrometeoroid penetration of the shell. Two RF transmitters were used for telemetry communications, one at 20 MHz and one at 40 MHz, with alternating transmissions 0.3 seconds in duration. Sputnik's data transmissions contained useful telemetry, and gave it the famous pulsating beep that served as proof of the satellite's existence in orbit.

The Sputnik 1 spacecraft was placed in a 200 × 900 km, ninety-minute orbit that decayed rapidly because of the significant atmospheric drag at its perigee and because of its small mass. The onboard batteries operated for twenty-one days before being depleted. While transmitting, the spacecraft provided data on the density of the upper layers of the atmosphere and the propagation of radio signals through the ionosphere. Sputnik 1 reentered the atmosphere and burned up on January 4, 1958 after three months in orbit.

SPUTNIK 2

Even though the launch of the first Sputnik was no surprise to the Eisenhower administration or to the intelligence community, opportunism took hold of the political landscape. Nikita Khrushchev was surprised by the worldwide acclaim of the Soviet technology superiority after Sputnik 1's launch and recognized its value almost immediately. Khrushchev had little difficulty taking advantage of the Sputnik launch

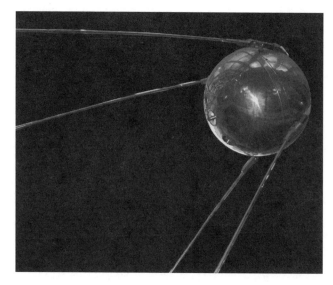

Figure 12.11. Photograph of a Sputnik 1 satellite mockup. Source: NASA NIX.

Table 12.3. Sputnik 1 Specs

Launch date	October 4, 1957
Launch vehicle	Sputnik—8K-71 (R-7) 2-stage booster
Mission duration	3 weeks (limited by battery life)
Deorbit	January 4, 1958
Mass	83.6 Kg (184 lb)
Diameter	0.58 m (22.8 in)
Structure	Dual-shell aluminum alloy sphere approximately 2 mm in thickness
Power	3 silver-zinc dry batteries
Communications	• 1 W transmitter • 4 transmitter antennas • 20.005 and 40.002 MHz signal (approximate wavelength 15 m and 7.5 m)
Orbit	• Semimajor Axis: 6,960 km (4,323 mi) • Eccentricity: 0.05201 • Inclination: 65.1° • Orbital Period: 96.2 min • Apogee: 7,310 km (4,540 mi) • Perigee: 6,586 km (4,090 mi)
Instrumentation	Nitrogen pressurization + internal and shell thermometers for space environment measurements including micrometeoroid detection

because of world's fascination with the artificial satellites, and because of the irrational remarks made by several prominent scientists and politicians including Edwin Teller, father of the hydrogen bomb, who commented that America had "lost a battle more important and greater than Pearl Harbor."

President Eisenhower felt that outward concern about the Sputnik launch by the administration would only provoke negative public reaction to his statements. Responding that America's space efforts were on course and that the Russian satellite launch he called "one small ball in the air" would not alter the direction of America's civil space program proved to be a feeble retort after the USSR launched a much more sophisticated satellite with an animal onboard only one month later (Dickson 2001). Continued Soviet political opportunism and an inherently weak space policy forced the President to act more decisively. His reluctance to acknowledge the Soviet's lead was not indifference, but a desire to launch a research satellite after a Soviet launch in order to create an "open sky" policy that would then allow any satellite into orbit without protest of military intentions. Eisenhower was also more concerned about developing a reconnaissance satellite to monitor new and dangerous atomic weapons buildup in the Soviet Union. To avoid the perception that America's first satellite was a military satellite launch, Eisenhower decided two years earlier to turn down von Braun's Redstone proposal for a satellite launch in 1956 in favor of the Vanguard satellite launch during IGY. His decision to pursue both projects was noble, but proved to be a national embarrassment when the televised Vanguard satellite launch attempt exploded on the launch pad.

Origin

Sputnik 2 arose from an informal edict from Khrushchev directing Korolev to produce another spectacular satellite launch on the November anniversary of the Russian Revolution just one month after Sputnik 1. As short as the time frame was, developing a vehicle that could support an animal in orbit was made easier by the success of several biological experiments flown earlier on suborbital missions. Many of the systems used on Sputnik 2 were already developed for those suborbital flights, albeit for much shorter duration.

Hurried design and fabrication teams worked around-the-clock on the new spacecraft without the quality checks Korolev considered necessary for safety and success. The carrier and life support chamber was designed to fit within the Sputnik launcher's conical payload shroud, along with the backup Sputnik 1 satellite that would provide communications and telemetry. The launch vehicle was also designated Sputnik since the Soviet launchers took on the name of the payload.

Selection of the canine passenger was based on temperament and on training skills. Laika ("The Barker"), a 6 kg female part-Samoyed terrier originally named Kudryavka ("Little Curly"), was picked for the flight. The carrier and chamber fitting were scaled around her physical dimensions. Room was made for her to move and possibly stand erect. Biotelemetry for Laika was programmed to transmit important physiological responses including heart rate and temperature. In addition, a slow-scan camera was placed in the chamber to observe her during the mission.

Sputnik 2's structure was placed in the same-sized shroud as Sputnik 1 that consisted of a four-meter-long cone-shaped capsule with a two-meter base diameter. Capsule systems included a sealed life-support pressurized cabin supplied with pure oxygen, CO_2 removal, water and food supplies (gelatin fed through tubes placed in Laika's stomach), engineering and biological telemetry and communications, and a battery power supply for approximately one week's operation. Instruments included biological sensors, thermal control, UV and X-ray spectrophotometers for measuring solar radiation and cosmic rays, and a magnetometer.

Because there was no reentry capability for spacecraft at the time (warhead reentry was proven, but at much higher temperatures, pressures, and g-forces than animals or humans could survive), Laika was provided with life support only for a one-week mission. The certain death of Laika spurred protests from the public, which was almost unheard of in the post-Stalin era. After the oxygen supply was depleted, she would have been fed poison, or died of oxygen deprivation. However, during the actual orbital entry, the propulsion module did not separate as planned and some of the thermal insulation was torn loose. Temperatures in the cabin reached 40°C (104°F) within several orbits, and Laika died from hyperthermia within several hours. News of her untimely death was never released, and details of the flight became available only after the fall of the Soviet Union thirty years later.

Figure 12.12. Photograph of the Sputnik-2 mockup located in the Tsiolkovsky Museum, Kaluga, Russia. Laika's capsule is located in the polished chamber near the bottom. The backup Sputnik 1 used for telemetry communications is shown positioned near the top of the shroud (Sputnik 1's antennas are pointed downward). Source: Energia Museum, author.

After launch into a 212 × 1,660 km elliptical orbit, Sputnik 2 decayed in altitude over a period of six months. It reentered Earth's atmosphere on April 14, 1958, after 162 days in orbit.

SPUTNIK 3

Sputnik 3, originally named Object D, was the third and most impressive of the early satellites launched into Earth orbit by the Soviets. The large, 1.3-ton satellite was originally scheduled to be the first satellite launched for the International Geophysical Year in 1957. But because of fabrication delays, Korolev substituted the much simpler Sputnik 1. The Sputnik 3 orbital geophysical laboratory was meant to surprise the world with its sophistication and size, but was upstaged by Sputnik 1, Sputnik 2, and Explorer 1. Nevertheless, Sputnik 3 was another compelling exhibition of the lead that the Soviets held over the United States in space technology at the beginning of the space race.

Table 12.4. Sputnik 2 Specs

Launch	November 3, 1957
Launch vehicle	8K71 Sputnik (R-7 with third stage booster)
Crew	Laika, a 6 kg part-Samoyed terrier
Mass	508 kg (1,120 lb)
Length	4 m (13.1 ft)
Diameter (max)	2 m (6.6 ft)
Power	Battery (lifetime was six days)
Communications	Tral-D biotelemetry system
Instrumentation	• Two spectrophotometers for measuring ultraviolet and X-ray solar emissions • Geiger tube for measuring charged particles and cosmic rays • Magnetometer for measuring Earth's magnetic field • Biotelemetry instruments included a heart rate and respiration rate monitor, and blood pressure sensor
Orbit	• Semimajor axis: 7,315 km (4,543 mi) • Eccentricity: 0.098921 • Inclination: 65.33° • Orbital Period: 103.7 min • Apogee: 8,031 km (4,988 mi) • Perigee: 6,583 km (4,089 mi)
Deorbit	April 14, 1958 (162 days)

Launch of the 3.6 m ×1.7 m 1,327 kg cone-shaped vehicle on May 11, 1958, was only six months after the Sputnik 1 launch. The spacecraft was Russia's contribution to the 1957-1958 IGY with an unparalleled orbital research laboratory containing a dozen instruments for studying the Earth-space environment, including the Earth's ionospheric pressure and composition, radiation belt particle densities, cosmic rays/heavy nuclei, magnetic fields, electric fields, micrometeoroids. Temperature sensors were also placed inside the spacecraft and on its surface. A failed tape recorder prevented complete data from being sent back on the Earth's van Allen radiation belt, although the outer reaches of the belt were measured by virtue of its highly elliptical orbit. Sputnik 3 reentered Earth's atmosphere two years after launch on April 6, 1960.

Sputnik-named spacecraft were not limited to the early Soviet orbital exploration satellites. Sputnik 5, for example, was a validation flight for the Vostok manned capsule in 1960 that carried two dogs, Belka ("Squirrel") and Strelka ("Little Arrow"), into orbit and, perhaps more important, to a successful reentry and landing. Strelka lived to produce several litters of pups, one of which was presented to President Kennedy by Nikita Khrushchev for his daughter, Caroline. Pushinka ("Fluffy") herself produced a litter of pups that the president at times called the "Pupniks."

Figure 12.13. Photograph of a mockup of the Sputnik 3 orbital lab at the Energia Museum, Russia. Source: author.

Table 12.5. Sputnik 3 Specs

Launch	May 11, 1958
Launch vehicle	8A91 Sputnik (modified R-7 with third stage booster)
Mass	1,327 kg (2,926 lb)
Length	3.57 m (11.7 ft)
Diameter	1.73 m (5.67 ft)
Power	Dry-cell battery
Communications	• Tral-D biotelemetry • Multi-channel telemetry • 20.005 MHz (7.5 m wavelength)
Orbit	• Semimajor axis: 7,419 km (4,608 mi) • Eccentricity: 0.110932 • Inclination: 65.2° • Orbital Period: 105.9 min • Apogee: 8,235 km (5,115 mi) • Perigee: 6,588 km (4,092 mi)
Deorbit	April 6, 1960

Manned Flight

Korolev's passion for space exploration was fanned by his growing success in building launchers, interplanetary probes, and satellites in the early space race. Both he and Khrushchev were also interested in orbiting the first cosmonaut before the American Mercury manned flights scheduled for 1961. Korolev had the advantage of the R-7 booster which was capable of placing more than 1,000 kg into low Earth orbit, and as much as 6,000 kg into orbit with a third-stage booster. On the other hand, the military was pushing for an orbital surveillance satellite that could be designed to use the same capsule structure as the spherical manned capsule Korolev requested approval for earlier. Another threat to Korolev's manned project effort was a competing design for a winged suborbital manned bomber that could also be used as a ballistic manned reentry capsule. In the end, Korolev won approval for the spherical Vostok ("east") manned capsule, but had to build the military reconnaissance satellite called Zenit ("zenith") that was approved at the same time as the Vostok.

Vostok's basic design was tailored to the most critical flight operations constraints that included mass, aerodynamic loads, and reentry heating. Simple ballistic reentry capsules provided little or no lift during reentry, which would translate into entry deceleration of more than 10 g's from Earth orbit and 20 g's from the Moon. To solve the reentry load problem, Korolev's engineers used the same solution developed for the Mercury, Gemini, and Apollo capsules. By displacing the center of mass from the center of aerodynamic pressure on a symmetrical capsule, a net lifting force would be produced in the direction of the offset. That solution generated a modest lift-to-drag (L/D) ratio of 0.2 to 0.3, which lengthened the reentry period and reduced the maximum reentry temperature, and limited Vostok's reentry deceleration to approximately 8 g's.

Vostok's mass limitation came from the launch vehicle payload capacity to low-Earth orbit. Calculations for the R-7 with an added third stage booster showed a maximum capsule mass on the order of 4,500–5,500 kg. From these estimates came the flight duration of the capsule and the type of landing system. Preliminary calculations showed that a soft landing by parachute could not be made with the Vostok. To overcome this, an ejection seat was integrated in the capsule design for the cosmonaut to exit the capsule for a separate parachute descent to touchdown. At approximately 7,000 m (23,000 ft), the egress hatch would be blown off, followed by ejection of the pilot and seat, followed by separation of the pilot and seat for a parachute landing. A separate parachute was deployed from the capsule to slow its descent for recovery even though the capsule was not reused.

Extreme conditions encountered during reentry made the accuracy and reliability of the manned capsule guidance system crucial. Calculations of ballistic entry and landing indicated that the greatest danger to the crew and capsule was from incorrect orientation during the firing of the deorbit retrorockets. To ensure the system's reliability, the attitude control unit designed for the capsule consisted of two redundant systems: an automated solar orientation system and a manual-visual orientation system. The manual orientation system which included a Vizor periscope system for alignment observations was the sole flight control for the pilot, although other commands such as the deorbit burn could be made manually. All other operations were

Figure 12.14. Yuri Gagarin's Vostok capsule on display at the Energia Museum located in the Moscow suburbs, along with a replica of a suited cosmonaut showing the scale of the vehicle. The large, circular window on the left is the egress hatch opening for pilot ejection. Source: author.

commanded from ground or were automated. Either of the orientation controls could operate the two onboard redundant cold nitrogen gas thruster systems. Deorbit was provided by a retrorocket pack that was fired in the direction of orbital motion, which made capsule reversal the first order of business after reaching orbit for emergency reentry. An ablation layer surrounded the spherical capsule to protect it from reentry heating, although the thickest layers and highest temperatures were on the capsule bottom (Hall and Shayler 2001).

An instrument unit was attached to the bottom of the crew capsule which carried additional life support, power, and propulsion. A deorbit thruster was mounted on the bottom-center of the conical instrument module. Ten minutes after the deorbit burn, the instrument section was jettisoned for reentry. Vostok's cold gas thrusters continued to operate during the reentry and descent to prevent abnormal attitudes, oscillations, and spin.

A pressurized flight suit was provided for the cosmonaut's survival in the event of a capsule depressurization, and for capsule ejection. During the initial design, consideration was given to a liquid bath to absorb the high-g loads during launch and landing, as suggested in several of Tsiolkovsky papers, but it was never adopted (Hall and Shayler 2001).

TESTING

Korolev's Vostok went through extensive testing on the capsule structure, mechanisms, and systems, in preparation for its first flight in 1960. Hatch seals, the ejection seat, life-support systems, thermal system components and operations, recovery operations and training, abort procedures, and many more details were tested in an effort to qualify the capsule for the first manned flight in 1961. Seven test capsules were built and flown for the preliminary testing of the Vostok, as well as the Zenit reconnaissance satellite. Complete versions of the Vostok were then launched for qualification flight tests with canine crews. The 3KA capsule test flights, designated Korabl Sputnik, were launched in order to collect additional data on the effects of space environment on biological specimens, especially solar radiation, and to further test spacecraft systems (Hall and Shayler 2001).

The first of these qualification flights was made with canine research passengers Belka and Strelka. The Vostok capsule's one-day orbital flight was outfitted with individual pressurized space suits for the dogs. The Korabl Sputnik 2 mission provided the first evidence of space adaptation syndrome (space sickness) in animals, later found to also be common to over 50% of humans exposed to zero-g. Deorbit, reentry, and landing were successful, and the landing proved to be very accurate—within 3.3 miles of the planned site. Ejection and parachute landing of the animals and other biological experiments were also successful.

VOSTOK MANNED FLIGHTS

Vostok 1 (Gagarin)

The first manned Vostok flight was made on April 12, 1961, with Air Force Lieutenant Yuri Gagarin at the controls. Gagarin's one-orbit flight went as planned, but was punctuated by several noteworthy events. Although training for the mission was underway for the top three candidates for months, Gagarin was announced as pilot-cosmonaut for the mission only one day before the flight. Air Force Lieutenant Gagarin was also promoted to major while in flight. His *Vostok 1* spacecraft lifted off from the Baikonor launch site on a modified R-7 three-stage booster for a single-orbit, ninety-eight-minute flight that landed near the small village of Smelovaka in Western Russia 108 minutes after launch. This Soviet space flight first was another blow to the U.S. space program and became a national celebration for the Soviet Union. Although he flew in space only once, Yuri Gagarin became one of the most heralded figures in Russian history.

Figure 12.15. Photograph of Gagarin and Korolev seated outside Korolev's Moscow home. Source: NASA GRIN.

Yuri Gagarin was killed in an aircraft accident on March 27, 1968, during a training flight with test pilot Vladimir Seregin. Although no conclusive evidence has been made available, most accounts of the accident cite wake turbulence from a passing aircraft in instrument conditions as the cause of the Mig-15 crash.

Because of his historic space flight, Gagarin received even more recognition after his death. Honors bestowed on Gagarin included a prominent far-side lunar crater named after him, his birth town of Czhatak renamed after him, the Cosmonaut Training Center at Star City named after him, and the Memorial Museum of Cosmonautics in Moscow named after him. He is also buried in the Kremlin wall, along with the most celebrated Russian heroes. Gagarin remains an international icon of space exploration, with a Yuri's Night celebration held annually on April 12 to commemorate his historic flight, underscored by the launch of the first space shuttle on April 12, 1981.

Vostok 2 (Titov)

The second Vostok flight was made on August 6, 1961, with Gherman Titov in orbit for twenty-four hours with a mission total of sixteen orbits. The *Vostok 2* flight was a much more demanding flight than the first because of the physiological and psychological stress of microgravity exposure for twenty-four hours, and because of almost total isolation, and for the first time, induced space motion sickness. This was the first time that rapid head motion in microgravity was recognized as the cause of disorientation and space adaptation syndrome. The next Vostok flight was more than a year later, a delay most likely due to the concern over Titov's space sickness.

Vostok 3 (Nikolayev), and Vostok 4 (Popovich)

Soon after the first two Vostok flights, Korolev's engineers increased the capabilities and the operational duration of the capsule to explore the effects of extended space flight. Although the Vostok 3 and 4 flights were not well publicized, they carried far more instrumentation and experiments onboard and provided more life support for the occupants than the previous flights. *Vostok 3* was launched on August 11, 1964, one day before *Vostok 4*, making it the first time that multiple manned spacecraft were in orbit at the same time. Nikolayev's *Vostok 3* completed a four-day mission, while Popovich's *Vostok 4* remained in orbit for three days. Both of the capsules deorbited and landed within seven minutes of each other on August 15.

Although the two Vostok spacecraft were close enough to identify each other visually and the cosmonauts could communicate with each other by radio, neither spacecraft was capable of altering its orbit. But without the capability to change orbits, other than for deorbit, the term rendezvous that is often used for the mission press releases is incorrect. The first true Soviet spacecraft rendezvous and docking was the Soyuz 4 and 5 dual mission launched in January 1969.

Vostok 5 (Bykovsky), and Vostok 6 (Tereshkova)

The next Vostok flight missions 5 and 6 were the final single-capsule missions, which provided an impressive accumulation of Soviet space flight experience and logged several space flight firsts. Like Vostok 3 and 4, the Vostok 5 and 6 were joint missions, but this time launched two days apart; *Vostok 5* on June 14, 1964, and *Vostok 6* on June 16. Both missions ended on June 19 with *Vostok 5* logging five days, and *Vostok 6* in orbit for three days. Even though more Vostok missions were planned, all were canceled in order to concentrate on Korolev's Soyuz lunar manned capsule.

Figure 12.16. Valentina Teresh-kova shown with her rescue gear after her Vostok 6 land-ing. Source: NSSDC.

The last Vostok mission carried the first woman into space, Valentina Tereshkova, who made several space flight records during her twenty-four-hour mission in *Vostok 6*. Valentina was not just the first female in space, but she logged more time on her flight than all of the astronauts in the Mercury program combined.

Following the second flight of the Vostok, an effort was made to increase the crew size of the manned capsules to beat America's first multi-crew Gemini launch that was scheduled for 1964. Khrushchev was also interested in placing the first woman in space for yet another space record. Working on both projects was not just a challenge to Korolev's schedule, but was further complicated by delays in the Vostok development with the military leadership's demand that a Vostok conversion be made for a reconnaissance/surveillance satellite. But within months, the design teams began work again on converting the Vostok to a three-man capsule. Korolev was also planning for much more difficult manned lunar flights to beat the Americans to the Moon with the development of the Soyuz crew vehicle, and held a second selection of lunar mission cosmonauts in 1962.

VOSKHOD

Khrushchev developed a craving for scoring as many space flight firsts as possible, partly as policy, but mostly as propaganda. While NASA engineers were making progress in the Mercury, Gemini, and Apollo programs, Korolev was working to complete a design to replace the Vostok with a larger manned capsule for future manned lunar missions. The Soyuz capsule flight tests were scheduled to begin in 1964, with lunar mission prototype testing projected for 1966. NASA's announcement of the impending flight of the first multi-crew Gemini spurred Khrushchev to order Korolev to launch a multiple crew flight to beat the Americans. Since Gemini was scheduled for launch in late 1964 or early 1965, Korolev's target date for the launch was set for the 1964 anniversary of the October Revolution. Khrushchev also specified that three cosmonauts were to fly, not just two (Harford 1997).

Korolev was given the task to launch a multi-crew mission in one year, which excluded using the untested Soyuz capsule and left the Vostok as the only space-qualified

Table 12.6. Vostok Flights (RussianSpaceWeb n.d.)

Vehicle	Designation	Pilot	Date	Duration	Notes
Korabl-Sputnik 1KP	Sputnik IV		May 15, 1960	5 years	4,340 kg prototype test vehicle included 2,500 reentry capsule with life-support systems for test and evaluation. Telemetry included video of dummy passenger. Error in orbit entry resulted in low decay rate and an unpredictable reentry.
Korabl-Sputnik 2	1K	Belka and Strelka	Aug 19, 1960		Recovered the next day. First successful satellite recovery. Passengers were Belka and Strelka, forty mice and smaller life.
Korabl-Sputnik 3	1K Sputnik 6	Pcheka and Mushka	Dec 1, 1960		Burned up on reentry due to steep entry angle. Pcheka and Mushka perished.
Korabl-Sputnik 4	3KA (#1)	Chernushka the dog, and Ivan Ivanovich the dummy	Mar 9, 1961		Qualification flight, with successful recovery of Chernushka and Ivan.
Korabl-Sputnik 5	3KA (#2)	Zvezdochka the dog and dummy	Mar 25, 1961	One orbit	Qualification flight, with successful recovery of Zvezdochka and dummy.
Vostok	3KA (#3)	Yuri Gagarin	Apr 12, 1961	1h 48m	First man in space, single orbit
Vostok 2	3KA (#4)	Gherman Titov	Aug 6-7, 1961	25h 18m	
Vostok 3	3KA (#5)	Andrian Nikolayev	Aug 11-15, 1962	72h 22m	

(Continued)

(Continued)

Vehicle	Designation	Pilot	Date	Duration	Notes
Vostok 4	3KA (#6)	Pavel Popovich	Aug 12-15, 1962	48:23:02	
Vostok 5	3KA (#7)	Valeri Bykovsky	Jun 14-19, 1963	119h 02m	
Vostok 6	3KA (#8)	Valentina Tereshkova	Jun 16-19, 1963	70h 48m	First female in space
Vostok 7		Vladimir Komarov	Apr 1, 1964		Canceled
Vostok 8		Pavel Belayev	Jun 1, 1964		Canceled
Vostok 9		Boris Volynov	Aug 1, 1964		Canceled

Table 12.7. Vostok Specs

Vehicle ID	3KVA
Launch vehicle	8K72K (3-stage R-7)
Crew size	One
Flight duration (max)	10 days
Length	4.4 m (14.4')
Diameter (max)	2.4 m (7.9')
Total mass	4,730 kg (10,430 lb)
Deorbit engine propellants	Nitrous oxide and amine
Electrical power	Batteries—24.0 kW-h total
Cabin atmosphere	Oxygen + nitrogen at 1 atmosphere
Landing system	Ballistic reentry with ablative heat shield, parachutes for descent, crew ejected at 7–10 km altitude
Reaction control system	Cold gas
Propellant	Nitrogen

vehicle. For the designers, the vehicle choice was fixed and its limitations serious. The first constraint for the Vostok conversion was the R-7 payload capacity. The second was the cabin size. The third was the equipment and consumable mass needed to orbit three cosmonauts. The first problem was solved by placing stringent limitations on the third, and the third was solved by limiting the vehicle's flight duration and safety equipment.

To accommodate three cosmonauts in the Vostok cabin designed for one, the ejection seat was removed and replaced with three seats. For more weight reduction, three small cosmonauts were selected for the mission. Even more weight was saved by the most questionable safety-related decision in any manned program: eliminating pressurized space suits for the flight crew. Adding to the program's curious character was the project engineer who opposed the unsafe conversion, but volunteered to fly on the three-man mission if selected. Konstantin Feoktistov was a capable engineer and one of Korolev's favored designers at the time, but he could not qualify as a cosmonaut because of his imperfect health. Nevertheless, Korolev, who also directed cosmonaut selection and training, placed Feoktistov on the crew roster if he would support the program (Vladimirov 1973).

The first Voshkod crew was not given space suits or ejection seats, therefore a soft landing system had to be designed for the spherical capsule and tested for the flight. Drop tests for qualification showed that using a larger parachute decreased descent rates, but increased deployment "g-shock." The decision was made to use slightly smaller parachutes along with retrorockets placed on the capsule bottom that fired just before landing to cushion the touchdown. The retrorocket design was successful enough to be adopted for the Soyuz vehicle, but was never qualified before the first flight of the multi-crew manned capsule. Korolev's new crew vehicle was renamed

Voskhod ("sunrise"), and designated 3KV for the three-crew version and 3KD for the two-man crew. But the capsule had limited utility, and even though it was scheduled for a number of later flights, it was used for only one more manned mission before being terminated.

The significant risks to the crew on the first Voskhod flight were recognized but ignored by the Soviet leadership, even though there was a much safer approach using two cosmonauts instead of three (Hall and Shayler 2001). Fortunately, the launch and flight of *Voskhod 1* was successful, although limited to a twenty-four-hour flight because of the restrictions on life-support provisions. Final weight savings for the mission were attained by putting the cosmonauts on a low-fat, low-starch diet. By the October 1964 launch, the *Voskhod 1* vehicle and payload were reduced to the 5,340 kg limit (Vladimirov 1973).

Voskhod 1

Voskhod 1's three crew members consisted of commander Vladimir Komarov, engineer Konstantin Feoktistov, and doctor Boris Yegorov. Launch of *Voshkod 1* and her crew from the Baikonor facility was on 11 October 1964, with only a few lightweight research experiments onboard because of the severe weight restrictions. In spite of the risks and limitations, the entire twenty-four-hour, seventeen-orbit mission was a success. More by luck than by design, the launch, orbital flight, and landing were made without incident, with the retrorockets providing the final cushion just before touchdown. After the landing, commander Komarov jettisoned the parachutes to avoid being dragged by the prevailing winds.

Russia's first Voskhod flight—the riskiest of all manned flights on record—was heralded by the Soviet press as an important mission that introduced soft-landing capsule technology and new lightweight instrumentation. The most noteworthy development in the lucky and modestly successful *Vostok 1* flight was the dismissal of Nikita Khrushchev by the Politburo following the mission. Premiere Khrushchev, who was responsible for muscling the project through to its completion, was replaced by Leonid Brezhnev. A parade for the cosmonauts in Moscow's Red Square was delayed for three days until Khrushchev's successors, Brezhnev and Kosygin, were officially in place (Vladimirov 1973).

Voskhod 2

A second flight of the Voskhod was conceived as a multi-crew mission to make the first-ever space walk and log yet another Soviet space first. The mission was to include two cosmonauts on a flight of one to two days; one assigned to capsule operations while the other executed the space walk. Capsule design of *Voskhod 2* was less of a challenge than the first Voskhod, but still proved to be demanding. For simplicity, and for continuous avionics cooling, the cabin would not be depressurized and then repressurized during the EVA. Instead, an inflatable pressure chamber (an airlock called the Volga) was attached on the exterior of the ingress/egress capsule hatch to

Figure 12.17. Voskhod 1 crew from left to right are Vladimir Komarov, Boris Yegorov, and Konstantin Feoktistov shown making their way to the launch pad in lightweight clothing instead of the traditional bulky flight suits. Source: NASA GRIN.

isolate the cabin from the vacuum of space. An umbilical cord connected to the same lines as the airlock provided life support and communications, as well as a physical connection between the floating cosmonaut and the vehicle. The inflatable airlock was designed to be pressurized before capsule egress, then depressurized for the EVA. It was repressurized after the EVA for ingress back into the capsule, then depressurized for the last time before separation from the capsule by pyrotechnic release, followed by reentry and burnup.

The crew selected for the *Voskhod 2* EVA mission were Air Force Lt. Colonel Alexei Leonov as pilot and EVA cosmonaut, and Colonel Pavel Belyayev as commander. Preparations were not extensive and training was shortened by the timetable set by NASA's planned launch of the Gemini IV EVA mission in late March 1965. Probably the greatest risk to the mission was introduced simply by mission planners neglecting to test Leonov's space suit in a vacuum environment, an omission that became a near-fatal problem for Leonov on his EVA.

Voskhod 2 was launched on March 18, 1965, as Cosmos-57 to hide its identity, just five days before the Gemini IV flight. The launch was punctuated by warning alarms in the flight control center, provoking the nonsmoking Korolev to light up a cigarette (Harford 1997). Orbit entry was successful, as were the preparation and cabin exit by Leonov. The planned EVA went as scheduled, but the ten-minute spacewalk was followed by a exhaustive twelve-minute attempt to reenter the capsule, requiring Leonov to reduce his suit pressure several times in order to fit into the air lock chamber. Leonov's heart rate climbed dramatically, leaving controllers even more worried about the crew's survival (Hall and Shayler 2001).

Deorbit and reentry were as problem-riddled as the space walk, beginning with the failure of the hatch to completely close. The pressure loss triggered a signal to increase gas flow, but flooded the cabin with pure oxygen and created an extreme fire hazard. An even greater problem arose when the ground control initiation of the automated deorbit system failed. The crew was ordered to make a manual deorbit, which had not been executed on any Vostok or Voskhod flights before. The deorbit command came for the following orbit, which displaced the landing site by approximately 2,500 km. Reentry was complicated by the failure of the instrument module to completely separate from the capsule, inducing gyrations in the capsule during the reentry phase until the cables burned through and the module separated.

A delay in the deorbit firing moved the touchdown point northward because of the Earth's rotation during the ninety minute orbit period. Landing overshoot was several thousand kilometers, putting the capsule and crew in a forest in the Ural mountains. The landing was further complicated by the parachutes being snagged in the upper limbs of the trees. Fortunately, the landing sensors, which were wires trailing below the capsule to initiate the retrorockets and sever the parachute lines, did not contact the tree branches. Had that happened, the parachutes would have been severed following the incorrectly sensed touchdown, allowing the capsule to free-fall from tens of meters above ground. But the mission woes did not end there. The capsule landing in the forest prevented the rescue crew from reaching the stranded flight crew until the next day. A cold night in a small capsule was followed by another night in the forest since the rescue helicopter could not retrieve the capsule in the dense cover, although it did drop food and cold-weather clothing and supplies. A rescue team arrived at the landing site on skis to lead the Voskhod crew out of the forest to the waiting rescue helicopter (Hall and Shayler 2001).

Plans for two more manned Voskhod flights were put on hold and eventually canceled after the death of Korolev in January 1966. A single flight with the canine crew Veterok ("Breeze") and Ugolek ("Blackie") was made with the Cosmos 110 designation on February 22, 1966. Mission duration for the flight was twenty-two days, which was far longer than the manned missions, leading to the discovery of bone loss in the animals due to the lack of gravity. The profound calcium level changes and other related illnesses after only twenty-two days in orbit was the first indication that humans would suffer the same physiological changes, which was verified on the Gemini and Apollo flights. Extensive research on the related microgravity exposure problems continues, but as yet, only minor remediation has been found.

FIRST COSMONAUTS

Informal plans for manned space flight emerged in discussion and on sketch pads well before the space race, inspired by a century of science fiction. But it was the space race

Table 12.8. Voskhod Flights

Voskhod vehicle	Pilot/crew	Launch / landing dates	Duration	Notes
Cosmos-47 3KV		Oct 6, 1964	1 day	First use of 11A57 launcher. Successful qualification flight, but the capsule was dragged by the wind after landing.
Voskhod 1 3KV	Vladimir Komarov, Konstantin Feosktistov, Dr. Boris Yegorov	Oct 12, 1964 / Oct 13, 1964	1 day	First multiple-crew space flight
Cosmos 57		Feb 22, 1965		Errant signal from a tracking station initiated an incorrect command for deorbit firing. The capsule began to tumble rapidly, stranding it in orbit. A self-destruct signal from ground was used to destroy the spacecraft.
Voskhod 2 3KD	Pavel Belayev, Aleksi Leonov	Mar 18, 1965 / Mar 19, 1965	25hr 55min	First EVA in space
Voskhod 3	Georgi Shonin, Boris Volynov	Jun 1, 1966		Canceled
Voskhod 4	Georgi Beregovoi, Georgiy Katys	Sep 1, 1966		Canceled
Voskhod 5	Valentina Ponomaryova, Irina Solovyova	Dec 1966		Planned all-female, ten-day space flight; Canceled
Voskhod 6	Yevgeni Khrunov	Jan 1967		Canceled

that gave birth to the first manned space flight projects as the two adversaries worked on their long-range missile projects. Wernher von Braun had worked for three decades on manned flight concepts before launching the first astronauts on his converted Redstone missile. And Sergei Korolev had worked toward building the first manned vehicles beginning with the R-1, although his ambitions had to weave through Stalin's torturous control. Sputnik's launch and the beginning of the space race gave Korolev and von Braun the license to start planning how to put the first men in orbit. Just two years after Sputnik 1, both designers were starting the integration of the crew capsules with the launch vehicles. In only two more years, the first manned flights were entered into the space history records.

Unlike von Braun, Korolev's responsibility for his space projects was a broad charge that included cosmonaut selection and training, as well as design and development of the launch and flight vehicles. As work began on the Vostok project, Korolev initiated the first cosmonaut selection in October 1959 from a pool of young military pilots with only modest flight experience. This was a contrast to a minimum of 1,500 hours high-performance jet experience required for NASA's first astronauts. A second requirement for the cosmonaut candidate's selection was an education in the field of aeronautics, engineering, or navigation. In early 1960, Korolev made the final selection of the twenty cosmonaut pilots. Of those selected, twelve were launched into space, while the other eight were dismissed, injured, or killed in accidents.

Air Force Group 1

Selection date: March 7, 1960

- Lt. Ivan Anikeyev (27)
- Maj. Pavel Belyayev (34)
- Lt. Valentin Bondarenko (23)
- Lt. Valeri Bykovsky (25)
- Lt Valentin Filatiyev (30)
- Lt. Yuri Gagarin (25)
- Lt. Viktor Gorbatko (25)
- Capt. Anatoli Kartashov (27)
- Lt. Yevgeni V. Khrunov (26)
- Capt. Vladimir Komarov (32)
- Lt. Aleksei Leonov (25)
- Lt. Grigori Nelyubov (25)
- Lt Andriyan Nikolayev (30)
- Capt. Pavel Popovich (29)
- Lt. Mars Rafikov (26)
- Lt. Georgi Shonin (24)
- Lt. German Titov (24)
- Lt. Valentin Varlamov (25)
- Lt. Boris Volynov (25)
- Lt. Dmitri Zaikin (27)

Figure 12.18. A 1960 photograph of the first cosmonaut selection group, along with Korolev (center front) and his wife and child, as well as several training directors. Seated in front from left to right are Pavel Popovich, Viktor Gorbatko, Yevgeniy Khrunov, Yuri Gagarin, sergi Korolev, his wife Nina Koroleva and daughter Natasha, Cosmonaut Training Center Director Yevgeniy Karpov, parachute trainer Nikolay Nikitin, and physician Yevgeniy Fedorov. Standing in the second row from left to right are Aleksey Leonov, Andrian Nikolayev, Mars Rafikov, Dmitriy Zaykin, Boris Volynov, German Titov, Grigoriy Nelyubov, Valeriy Bykovskiy, and Georgiy Shonin. Standing in the back row are Valentin Filatyev, Ivan Anikeyev, and Pavel Belyayeu. Not shown are Vladimir Komarov, Anatoliy Kartashov, and Valentin Varlamov who were dropped from training because of injuries, and Valentin Bondarenko who died in a training accident. Source: NASA GRIN.

Yuri Gagarin

Yuri Gagarin (1934–1968) was one of about 100 pilots selected from nearly 3,000 prospective candidates with an aeronautics, engineering, or navigation background. The young Air Force Lieutenant was then picked from the final twenty cosmonaut candidates by Korolev and one of the six to train on the Vostok simulator in preparation for the first manned space flight (Hall and Shayler 2001). The interview and selection process was reportedly the beginning of a close friendship between Gagarin and Korolev.

A training center for the Soviet cosmonauts was established outside of Moscow at a site named Zvezdny Gorodok (Star City). A one-year training program was set up for the Vostok pilots that included academic subjects as well as training in launch vehicles, capsule systems, flight operations, ground control operations, and survival and recovery operations. Like the NASA astronauts, extensive flight training was an important component of the preparations for flight in space.

Selection of the pilot-cosmonaut for the first-ever space flight was announced internally on April 10, 1961 when Yuri Gagarin was picked as the top candidate of the final three. The Soviet press announced the selection of Gagarin to the outside world the following day, just one day before the launch. Gagarin's launch on the Vostok three-stage booster in the Vostok 1 capsule on April 12, 1961, lasted 108 minutes—the total time for one orbit, plus launch, and reentry. Launch and orbital operations were uneventful, and the deorbit and reentry were made by the automated commands as planned. Minor control inputs were made by Gagarin for capsule deorbit alignment. *Vostok 1*'s instrument module failed to separate completely which induced gyrations in the reentry phase until the cables burned through, the same problem that later plagued Leonov's EVA mission. Gagarin ejected for a successful parachute landing in farmland, but the details were not released for years because of the requirement that the pilot remain with the craft until its landing to satisfy FAI (Fédération Aéronautique International) criteria for establishing world records in aviation and space flight.

Although Gagarin would not fly in space again, his historic flight would elevate him to the highest status in the Soviet Union. In 1963, Gagarin was appointed deputy director of the Cosmonaut Training Center, and in 1966 he served as backup crewmember for the Soyuz 1 flight that ended in the tragic death of fellow cosmonaut Vladimir Komarov. Gagarin was killed in a Mig-15 flight training accident along with his instructor on March 27, 1967. Even after the accident reports and investigations were completed, controversy swirled over the cause of the accident which continues today.

Valentina Tereshkova

Cosmonaut selection entered a new direction in 1961 when plans for manned lunar flights and for female space flights took shape. The first female cosmonaut selection was completed in March 1962, while the lunar cosmonaut selection was delayed until 1963. The female cosmonaut selection produced five candidates even though there was only one Vostok flight proposed at the time. Training was focused on preparation and training similar to their male counterparts, though the selection criteria were distinctly different since there were no female pilots in the Soviet military. Due to the ejection-sled and parachute landing required on Vostok missions, the candidates came from aviation and parachute clubs, and all were required to be parachutists. Candidacy also required academic or academy degrees (Hall and Shayler 2001). Training for the female cosmonauts included copilot experience on the Mig-15, parachute training, weightless simulation in aircraft parabolic flights, and centrifuge training.

The five female cosmonauts selected in 1962 and their ages at selection were:

- Tatyana Kuznetsova (20, youngest female cosmonaut/astronaut ever to be selected)
- Valentina Ponomaryova (28)
- Irina Solovyova (24)
- Valentina Tereshkova (24)
- Zhanna Yorkina (22)

Final selection for the female Vostok flight was made in April 1963 for a launch in June of the same year. Of the five women who began training, two were disqualified for medical reasons. Valentina Tereshkova was chosen with the highest "social and moral" rating, as well as higher academic and training scores. Because of the uncertainties and lack of flight experience for the female cosmonauts, the decision was made to fly the final two Vostok vehicles (5 and 6) at the same time (Hall and Shayler 2001).

Vostok 5 was launched with Valeri Bykovsky on June 14, 1963, for a five-day flight which remains a single-pilot space flight record. Valentina was launched on June 16 with the call sign Chaika ("seagull") for a two- to three-day mission. The two Vostok capsules would come as close as 5 km in their separate orbits, but no rendezvous was possible since the Vostok capsules lacked orbital thrusters.

Discomfort was reported on these flights due to the chafing from the flight suits and the lack of many basic hygiene provisions (no toothbrush or toothpaste, for example). Experiments and observations were carried out by both cosmonauts for the flights, including practice at manual deorbit orientation and photography (Hall and Shayler 2001).

Both cosmonauts deorbited on June 19, with Bykovsky landing two orbits later than Tereshkova. Deorbit and reentry for Valentina's three-day (forty-eight-orbit) flight was uneventful, including the parachute landing. Her recovery was timely, although local residents surrounded her in a spontaneous celebration of the new Soviet heroine. After the flight and debriefing, it was suggested that Valantina did not perform as well as her male counterparts and that she became sick on her flight. A number of sources also mention that Korolev was unhappy with her performance, but few include that her own mission partner failed to perform a number of assignments on his Vostok 5 flight and was also briefly incapacitated while space sick. Suggestions that the female cosmonauts were somehow inferior to male cosmonauts have all the appearances of gender bias rather than factual support in the space flight record.

Valentina successfully demonstrated the capability for females to train for and perform in space, yet it would be nineteen years before another Russian female would fly in space. That was not surprising since Russia's female cosmonauts were not considered equals to their male counterparts, and did not receive flight assignments beyond their use for propaganda. Nonetheless, a female-only mission on the Voskhod was considered for another in the long list of Soviet space firsts. Valentina Ponomaryova and Irina Solovyova were scheduled for the dual flight in 1966 before it was canceled due to the Voshkod 2 flight problems and the pending Soyuz test flights. A crew of three females was scheduled for a Soyuz mission for International Women's Day in 1985, but was canceled due to problems with the Salyut 7 space station.

Valentina Tereshkova was married to cosmonaut Andrian Nikolayev on November 3, 1963, in a high-profile marriage. What started as rumor about their romantic involvement had little substance, but the couple was ensnared in a grand wedding ceremony that Nikita Khrushchev was reportedly responsible for. A year later on June 8, 1964, Valentina gave birth to their daughter Alyona. Medical teams closely monitored the child since she was the first to be born from parents that had both flown in space. Yelena went on to become a medical doctor. Valentina and Andrian divorced in 1980.

Valentina would become very active in the Communist Party, and received the Order of Lenin and Hero of the Soviet Union awards. She later served as the president of the Soviet Women's Committee and became a member of the Supreme Soviet, the USSR's national parliament, and a member of the Presidium.

Gherman Titov

Gherman Titov (1935–2000), who was selected in the first Air Force cosmonaut group, was the second cosmonaut to fly in space. Like Gagarin, Titov flew only once in space. His one-day mission on Vostok 2 was launched on August 6, 1961. At age twenty-five, Titov was the youngest person to fly in space, and also the first to experience space adaptation syndrome (space sickness). Titov went on to spend the rest of his career in the Soviet and Russian space programs, then, after his retirement, served in the State Duma until his death in 2000.

Alexi Leonov

Alexi Leonov was selected in the first group of twenty cosmonauts and was one of the most important figures in the USSR's early space program. Born in Listvyanka in the Altay region of Siberia, Leonov is probably best known for his spacewalk in 1965 when he became the first human to perform an EVA outside of his *Voskhod 3* capsule.

Leonov graduated from Chuguyev Higher Air Force School in 1957 before he joined the Soviet Air Force as a fighter pilot, then became a parachute instructor for the Air Force. Leonov went on to obtain a graduate degree in 1968, and was promoted to the rank of Major General in the Soviet Air Force after his flight on the Apollo-Soyuz project in 1975.

Leonov's initial space flight assignment was aimed at adding another first in the growing list of Soviet space records—the spacewalk. Because of the importance and the uncertainty of the EVA, Leonov and his suit were instrumented to study the effects of a human exposed to space before, during, and after the spacewalk. In planning Leonov's mission, Korolev and the Soviet Politburo knew of the impending Gemini EVA flight, which was a clear advantage since the Americans had little, if any, information on the Soviet launches.

Leonov's *Voskhod 2* capsule had an attached airlock to maintain cabin pressure during the EVA that was collapsible to allow placement of the entire vehicle within the launch vehicle's aerodynamic shroud. Though the mission was plagued by a series of life-threatening problems, Leonov's space walk was a success that earned him the Hero of the Soviet Union award and several assignments on later high-risk flights. Leonov was also awarded the Order of Lenin and the Order of the Red Star.

Leonov's second mission assignment was the commander of the first manned circumlunar mission. The unsuccessful program was ultimately canceled, which contributed to the delay in his next assignment as commander of the Soyuz 11 expedition to the newly launched Salyut 1 space station in 1971. That mission was canceled just days before launch when one of Korolev's crew members was suspected of having tuberculosis and the backup crew was sent instead. The Soyuz 11 expedition to the Salyut was successful, but all three crew members died tragically due to a cabin pressurization relief valve malfunction during reentry. Leonov was reassigned to the Soyuz 12a which was canceled on January 8, 1971, following the death of the Soyuz 11 crew. He was reassigned again to the Soyuz 12b which was also canceled, this time due to the failure of Cosmos 557 to reach stable orbit. Launch failure of the third DOS space station just three days before Skylab was unannounced and obscured by the nebulous Cosmos designation.

Leonov would wait another four years for his second space flight on *Soyuz 19*, the Apollo-Soyuz Test Project (ASTP), which was called the Soyuz-Apollo Project in Russia. *Soyuz 19* was launched July 15, 1975, on a six-day mission with Leonov as commander. *Soyuz 19* and crew landed safely on July 21, less than 10 km from its target touchdown point. The next docking between American and Russian spacecraft would not take place until 1996 when the Mir and space shuttle began docking, with operations in preparation for the International Space Station construction project.

After his last space flight on the ASTP, Leonov served as chief cosmonaut from 1975 to 1982, followed by the position of deputy director of the Yuri Gagarin Cosmonaut Training Center. Leonov then served as chief of the International Detachment, training guest cosmonauts for space travel after his retirement in 1991. As of 2007, Leonov was an investment banker and president of the U.S. investment corporation Alpha Capital in Moscow.

Leonov is well known for his EVA and near-fatal encounters on the *Voskhod 2* flight, as well as his artistic skills, but less so for his other brushes with death. He survived drowning when his car skidded and plunged into a deep ice-covered lake where he pulled his wife and driver to safety. He also survived an assassination attempt on Premier Brezhnev when their car was ambushed, killing the driver.

Soviet Space Exploration

Space exploration did not begin with a specific project, but as a vision, and not with one man or woman, but with many. For the Soviet Union, those visionaries are considered to be Konstantin Tsiolkovsky; designer of the first liquid-fuel rocket, Mikhail Tikhonravov; and Sergei Korolev, chief designer that launched the first satellite. In reality, it was these individuals and many more. Tsiolkovsky laid the foundation for some of the important space flight theory in the USSR and served as an inspiration as much as he did as a scientist. Tsander, who founded the first rocket society in Russia, also helped translate theory into the first liquid-fuel rockets. Tikhonravov was inspired, skilled, educated, and capable of building GIRD's first rocket, and responsible for much of the design of the Sputniks and the Vostok, but he is one of the least known of the space race leaders. Korolev, who managed the successful ICBM project, the first orbital satellites, and the first manned flights, was probably one of the greatest space exploration visionaries, comparable to Wernher von Braun. But Korolev could not have built the R-7, or the Sputnik 1, or the Vostok by himself. It was his talent for management that was responsible for creating the hardware and opportunities for Russia's space program. In sum, the people most important to the success of Russia's early space exploration were those that made others succeed in the effort. Were that the only category of success, Korolev would reside at the top because of his ability to manage his design teams, to organize and direct the early Council of Designers, and to influence a number of ministers in the top-heavy Soviet leadership.

Korolev's success in building the first ICBM and launching the first satellite was a prelude to his interests reaching far beyond Earth orbit. Even before the launch of Sputnik, Korolev began planning deep-space exploration missions as well as manned missions to the Moon and Mars. The success of his early Sputnik satellites also opened the door for the development of sophisticated instruments, equipment, and technologies that were needed for space research. However, competing interests in military missiles and weapons forced Korolev to ensure that his research efforts succeeded, not for science's sake, but for propaganda purposes. Not all of his early space projects were fruitful until the surprising success of his Sputnik satellites, which secured significant support from the Soviet leadership. And as grave as the Cold War was, the early space exploration programs trumped many of the missile and surveillance programs, since military space operations required the fundamental research that could only be supplied by exploration spacecraft. Generous funding and rapid program development were the hallmarks of the early space age for both the Soviet Union and the United States because both countries' space programs were also tied directly to military missile, spacecraft, and weapons projects.

LUNAR RESEARCH

Space exploration began first with modest suborbital flights that reached above the Earth's atmosphere, then on to the first Earth-orbiting satellites in 1957 and 1958. Almost immediately, space exploration efforts were underway to reach the Moon for two reasons. First, the Moon was the closest celestial body and the next logical step in space exploration. Second, the Moon was an unwritten goal for the two leaders of the U.S. and Soviet space programs' early manned mission plans. Before the launch of Mercury and Vostok, a new dimension opened up in the space race—the race to the Moon.

Korolev began the lunar exploration program with a series of surface impact missions named Luna ("Moon"), often referred to as "Lunik" because of the popularization of the "nik" suffix following the launch of the Sputniks (muttnik, flopnik, beatnik, etc.), and obviously not the choice of the Soviets. Access to the Moon for the Soviets was simplified by the R-7 two-stage booster that could reach the Moon just by adding a modest third stage with an advanced guidance system. Later missions with larger, more complex probes used a heavier booster with a more accurate guidance system and a delivery bus. Later and larger Luna missions used the larger Proton booster in order to lift the heavy sample return and the rover spacecraft, beginning with Luna 15—the Soviet's first soil-sample return attempt.

The Luna mission series originated in Korolev's OKB-1 bureau under his direction, but as the Soyuz and manned lunar projects absorbed more and more of his time, he turned the Luna probe development over to the Lavochkin design bureau, beginning with the first lunar lander, Luna 9. These probes were more than research missions to further science. They were, like the U.S. Ranger, Orbiter, and Surveyor missions, exploration flights in preparation for the manned lunar program.

Korolev's Luna mission series started off with three failures before the successful launch of the Luna 1 spacecraft on January 2, 1959. This was the first spacecraft to leave the Earth's gravity, and although its mission objective was to impact the Moon, it missed its target by 6,000 km. As Luna 1 left the Earth's atmosphere, a command was

Table 12.9. Soviet Lunar Missions (compiled from NSSDC and Wade)

Designation/ design	Spacecraft type	Launch date	Launch vehicle	Mission objectives	Mission results
Unnamed	E-1	September 23, 1958	Modified R-7	Lunar impact	Launch failure
Unnamed	E-1	October 12, 1958	Modified R-7	Lunar impact	Launch failure
Unnamed	E-1	December 4, 1958	Modified R-7	Lunar impact	Launch failure
Luna 1/ OKB-1	E-1 361 kg	January 2, 1959	8K72 Modified R-7	Lunar impact	Almost successful lunar impact mission, but missed the Moon. Data returned on the Earth's radiation belt and solar radiation; discovery that the Moon had no magnetic field; solar wind from the Sun extended to the Moon. Spacecraft released sodium gas in upper atmosphere for magnetic field and particle measurements.
Unnamed	E-1A	June 18, 1959	8K72 Modified R-7	Lunar impact	Launch failure
Luna 2/ OKB-1	E-1A 390 kg	September 12, 1959	8K72 Modified R-7	Lunar impact	Successful impact on the lunar surface west of Sea of Serenity. Released sodium gas in upper atmosphere for magnetic field and particle measurements. More extensive solar wind ion measurements; found no significant magnetic field on the Moon.
Luna 3/ OKB-1	E-2A 279 kg	October 4, 1959	8K72 Modified R-7	Lunar flyby	Photographic camera and scan system returned first images of the far side of the Moon (29 total). Discovered fewer and smaller mare covering the Moon's far side than the near side.
Unnamed	E-3	April 15, 1960	8K72 Modified R-7	Lunar flyby	Launch failure

Missions outlined in gray were complete or substantial failures.

(Continued)

(Continued)

Designation/ design	Spacecraft type	Launch date	Launch vehicle	Mission objectives	Mission results
Unnamed	E-3	April 16, 1960	8K72 Modified R-7	Lunar flyby	Launch failure
Unnamed	E-6	January 4, 1963	8K78L Modified R-7	Lunar lander	Stranded in low Earth orbit
Unnamed	E-6	February 3, 1963	8K78L Modified R-7	Lunar lander	Launch failure
Luna 4/ OKB-1	E-6 second generation lunar probe 1,442 kg	April 2, 1963	8K78L Molniya (4-Stage R-7)	Lunar lander	First attempt at a soft landing on the Moon—missed by nearly 9,000 km and entered Earth orbit. Spacecraft 5 times mass of first generation. Soft landing failed
Luna 1964A	E-6	March 21, 1964	8K78M Molniya	Lunar lander	Launch failure
Luna 1964B	E-6	April 20, 1964	8K78M Molniya	Lunar lander	Launch failure
Zond 1964A	E-6	June 4, 1964	8K78M Molniya	Lunar lander	Launch failure
Cosmos 60	E-6	March 12, 1965	8K78L Molniya	Lunar lander	Stranded in low Earth orbit Cosmos designation typically assigned to Earth-orbit spacecraft, whether a planned orbit or stranded orbit.
Unnamed	E-6	April 10, 1965	8K78L Molniya	Lunar lander	Launch failure
Luna 5/ OKB-1	E-6 1,474 kg	May 9, 1965	8K78M Molniya	Lunar lander	Soft landing failed—descent rockets failed; crash landed in the Sea of Clouds

Name	Variant	Mass	Date	Rocket	Mission	Description
Luna 6/ OKB-1	E-6	1,440 kg	June 8, 1965	8K78 Molniya	Lunar lander	Error in mid-course correction, missed the Moon by 159,600 km. Soft landing failed
Luna 7/ OKB-1	E-6	1,504 kg	October 4, 1965	8K78M Molniya	Lunar lander	Soft landing failed - impacted on the surface in the Sea of Storms
Luna 8/ OKB-1	E-6	1,504 kg	December 3, 1965	8K78 Molniya	Lunar lander	Soft landing failed—descent rocket firing failed; crash landing in the Sea of Storms
Luna 9/ GSMZ Lavochkin	E-6	1,550 kg	January 31, 1966	8K78M Molniya	Lunar lander	First successful soft landing on the Moon; touched down in the Ocean of Storms February 3. The 99 kg payload was separated before touchdown, then bounced to a landing and stabilized with the opening of four petal doors. Television camera and extended rotating mirror provided a panoramic view of nearby rocks and the horizon several km from the spacecraft.
Cosmos 111	E-6S		March 1, 1966	8K78M Molniya	Lunar orbiter	Stranded in low Earth orbit
Luna 10/ GSMZ Lavochkin	E-6S	1,582 kg	March 31, 1966	8K78M Molniya	Lunar orbiter	First spacecraft to orbit the Moon. Measured weakness of Moon's magnetic field, micrometeoroid density, lunar radiation belt. Found lunar rocks comparable to terrestrial basalt rock. Gravitational studies found the first evidence of mass concentrations (mascons). Mission ended from battery depletion on May 30.

(Continued)

(Continued)

Designation/ design	Spacecraft type	Launch date	Launch vehicle	Mission objectives	Mission results
Luna 11/ GSMZ Lavochkin	E-6LF 1,640 kg	August 24, 1966	8K78M Molniya	Lunar orbiter	Similar spacecraft to Luna 10 but instrumentation included gamma-ray and X-ray detectors to determine the lunar surface composition, and high-energy particle detectors. Radio transmission Doppler data used to measure lunar gravitational anomalies. Mission termination due to battery depletion on October 1.
Luna 12/ GSMZ Lavochkin	E-6LF 1,620 kg	October 22, 1966	8K78M Molniya	Lunar orbiter	Lunar orbiter returned images of the lunar surface from orbit for the first time. High-resolution television system contained 1,100 scan lines with a maximum resolution of 15–20 m. Transmissions ceased on January 19.
Luna 13/ GSMZ Lavochkin	E-6M 1,620 kg	December 21, 1966	8K78M Molniya	Lunar lander	Soft lander similar to Luna 9 but included more sophisticated equipment including stereo panoramic cameras, mechanical soil penetrometer, backscatter densitometer, dynamograph, radiometers, radiation densitometer for measuring cosmic-ray reflectivity of the lunar surface. Transmissions from Oceanus Procellarum ended on December 28, 1966
Cosmos 146	7K-L1P (stripped Soyuz)	March 10, 1967	Proton 8K82K (UR-500) + Block D 4th stage	L1 test	Successful launch of the L1 (a stripped Soyuz used to test the vehicle that was later planned for manned flight) into heliocentric orbit. Proton's Block D fourth stage was the fifth stage of the ill-fated N-1 booster.

Name	Spacecraft	Launch vehicle	Date	Mission	Result
Cosmos 154	7K-L1P	Proton 8K82K + Block D 4th stage	April 8, 1967	L1 test	Stranded in Earth orbit
Cosmos 159	E-6LS	8K78 Molniya	May 17, 1967	Lunar orbiter	Successful radio communications relay satellite for later lunar missions
Unnamed	7K-L1	8K82K + Block D	September 28, 1967	Circumlunar L1	Launch failure
Launch of first Saturn V on Apollo 4			November 9, 1967		
Unnamed	7K-L1	8K82K + Block D	November 22, 1967	Circumlunar L1	Launch failure
Unnamed	E-6LS	8K78M Molniya	February 7, 1968	Lunar orbiter	Launch failure
Zond 4	7K-L1	Proton 8K82K + Block D	March 2, 1968	Circumlunar L1	Entered heliocentric orbit (launched away from Moon to avoid gravitational influence)
Luna 14/ GSMZ Lavochkin	E-6LS 1,700 kg	8K78M Molniya	April 7, 1968	Lunar orbiter	Similar to Luna 12 in its instrumentation. Gravitational studies of the Earth-Moon system and the Moon's field and lunar motion were included, along with measurements of solar-charged particles and cosmic rays. This was the final flight of the second-generation E-6 Luna series
Unnamed (Zond 1968A)	7K-L1	Proton 8K82K + Block D	April 23, 1968	Circumlunar L1	Launch failure

(Continued)

(*Continued*)

Designation/ design	Spacecraft type	Launch date	Launch vehicle	Mission objectives	Mission results
Zond 5	7K-L1 5,375 kg	September 15, 1968	Proton 8K82K + Block D	Circumlunar L1	First successful circumlunar mission with reentry and recovery. Biological payload included turtles, wine flies, meal worms, plants, seeds, bacteria, and more.
Launch of first manned Apollo mission, Apollo 7		October 11, 1968			
Zond 6	7K-L1	November 10, 1968	Proton 8K82K + Block D	Circumlunar L1	Partially successful circumlunar mission intended to qualify cosmonaut circumlunar mission before Apollo 8. Included biological samples. Reentry was abnormal, and cabin decompression killed samples. Parachute deployed early
Launch of Apollo 8 manned circumlunar mission		December 23, 1968			
Unnamed	7K-L1	January 20, 1969	Proton 8K82K + Block D	Circumlunar L1	Launch failure
Unnamed	E-8	February 19, 1969	Proton 8K82K + Block D	Lunar rover	First Lunokhod lunar rover launch Launch failure
Unnamed	7K-L1S	February 21, 1969	N1/L3	Circumlunar L3	First N1 manned lunar mission booster launch Launch failure
		March 3, 1969			Launch of Apollo 9 all-up hardware test in Earth orbit
		May 18, 1969			**Apollo 10 launch of manned lunar mission at the Moon and vehicle test (landing mission executed except for actual landing)**
Unnamed	E-8-5	June 14, 1969	Proton 8K82K + Block D	Lunar sample return	First sample return mission Launch failure

Name	Designation	Date	Booster	Mission	Notes
Unnamed	7K-L1S	July 3, 1969	N1/SL	Circumlunar	Second N1 launch failure
Luna 15/ GSMZ Lavochkin	E-8-5 5,700 kg	July 13, 1969	Proton 8K82K + Block D 4th stage booster	Lunar sample return	First of the new E-8 series lunar spacecraft. The Luna 15 soil sample return flight was intended to beat Apollo 11 back with samples but crashed during descent. Vehicle included a descent stage for landing and ascent stage to return soil sample capsule back to Earth. First instance of the USSR releasing mission flight plans beforehand—in this case, to avoid collision with Apollo 11. Impacted in Mare Crisium July 21, one day after Apollo 11 landing

Apollo 11 launched on first successful manned lunar mission

July 16, 1969

Name	Designation	Date	Booster	Mission	Notes
Zond 7	7K-L1	August 8, 1969	Proton 8K82K + Block D	Circumlunar	Successful circumlunar flight of the Zond spacecraft with a skip-reentry descent and landing
Cosmos 300	E-8-5	September 23, 1969	Proton 8K82K + Block D	Lunar sample return	Stranded in Earth orbit
Cosmos 305	E-8-5	October 22, 1969	Proton 8K82K + Block D	Lunar sample return	Stranded in Earth orbit

Apollo 12 manned lunar lander mission launched

October 14, 1969

Name	Designation	Date	Booster	Mission	Notes
Unnamed	E-8-5	February 6, 1970	Proton 8K82K + Block D	Lunar sample return	Launch failure

(Continued)

(Continued)

Designation/ design	Spacecraft type		Launch date	Launch vehicle	Mission objectives	Mission results
Luna 16/ GSMZ Lavochkin	E-8-5	5,727 kg	September 12, 1970	Proton 8K82K + Block D	Lunar sample return	Nearly identical to Luna 15. First successful Soviet lunar sample return mission, bringing back approximately 100 g (4 oz) of soil on September 24. First soft landing in darkness. Landed in Mare Foecunditatis (the Sea of Fertility) after 2 days in lunar orbit for landing preparations. Remained on the surface for 26 hours.
Zond 8	7K-L1	5,375 kg	October 20, 1970	Proton 8K82K + Block D	Circumlunar	Last of the circumlunar Zond missions. Launch, flight and Earth reentry successful.
Luna 17 GSMZ Lavochkin	E-8	5,700 kg	November 10, 1970	Proton 8K82K + Block D	Lunar rover	First lunar rover to land on the Moon's surface. Solar-powered, 8-wheeled Lunokhod rover studied the lunar surface in the Sea of Rains for 11 lunar days (11 months), although mission was planned for 3. Lunokhod rover traveled a total of 10.5 km, returned more than 20,000 images and 206 panoramas; performed 25 soil analyses with X-ray fluorescence spectrometer and used its penetrometer at 500 different locations.
			January 31, 1971			**Apollo 14 manned lunar lander mission launched**
Unnamed/ OKB-1	7K-LOK		June 27, 1971	N 1	Circumlunar	Third failure of the N 1 launcher
			July 26, 1971			**Apollo 15 manned lunar lander mission launched**
Luna 18	E-8-5		September 2, 1971	Proton 8K82K + Block D	Lunar sample return	Lunar sample return mission failure—crashed on descent.

Luna 19	E-8LS	September 28, 1971	Proton 8K82K + Block D	Lunar orbiter	Successful lunar exploration orbiter, mapping gravity anomalies (masscons) and radiation energy and distribution.
Luna 20	E-8-5	February 14, 1972	Proton 8K82K + Block D	Lunar sample return	Second successful lunar soil sample return mission. Soil and drilled rock samples of approximately 30 g (1 oz) were returned on 25 February.
April 16, 1972					**Apollo 16 manned lunar lander mission launched**
Unnamed	7K-LOK	November 23, 1972	N 1	Circumlunar	Fourth and final N1 launch failure
December 7, 1972					**Apollo 17 launched, the last manned lunar lander mission**
Luna 21	E-8	January 8, 1973	Proton 8K82K + Block D		Second successful Lunokhod rover mission, lasting approximately 4 lunar days and covering 37 km. A polonium-210 isotopic heat source was used to keep the rover warm during the lunar nights.
Luna 22	E-8LS	May 29, 1974	Proton 8K82K + Block D		Successful lunar orbiter studying lunar surface composition and solar wind.
Luna 23	E-8-5M	October 28, 1974	Proton 8K82K + Block D	Lunar sample return	Sample return mission failure from excessive descent rate.
Unnamed	E-8-5M	October 16, 1975	Proton 8K82K + Block D	Lunar sample return	Launch failure
Luna 24	E-8-5M	August 9, 1976	Proton 8K82K + Block D	Lunar sample return	Third successful sample return mission. Returned 170 g sample from Sea of Crises on August 22.

sent to release a cloud of sodium gas which marked the probe's trajectory. The colorful path clearly showed the interaction between the ionized sodium and the Earth's magnetic field, along with the fingerprint of the solar wind.

UNMANNED LUNAR EXPLORATION

Soviet Lunar Exploration Vehicles

Soviet lunar exploration hardware was more diverse in its mission objectives and makeup than the U.S. counterparts. Korolev's spherically shaped Luna 1 and Luna 2 impact probes were soon surpassed in sophistication and in size. By the end of the Soviet lunar program, spacecraft had become robotic marvels, although none of the vehicles would carry its cosmonauts beyond Earth orbit. Some examples of the vehicles are outlined in figures 12.19–12.22.

MANNED LUNAR PROGRAM

America's Apollo manned lunar program was more than a threat to Russian leadership in space technology. Because of the tremendous advances made developing the sophisticated Apollo launch and flight hardware and its advanced vehicle navigation

Figure 12.19. Mockup of the Luna 2 spacecraft that was the first to impact on the Moon. Source: Energia Museum, author.

capabilities, the program also presented a threat to the Soviet military's weapons planners. Earlier, Sergei Korolev and the other designers succeeded at combining military long-range missiles and space launch efforts, creating a complementary relationship that opened many Soviet funding resources to the space exploration programs, including the military weapons budget. The combined efforts also advanced military weapons development at a faster pace and lower cost than could be done through the normally convoluted Soviet procurement process. However, the success of the sometimes contentious working relationship between space exploration planners and the military weapons directors was dependent on the success of the space conquests. Korolev logged many space firsts, and because of his successes, he gained the support of the Soviet premier, and just as important, begrudging support from influential military weapons planners. After placing cosmonauts in Earth orbit, Korolev's next step was to send them to the Moon.

In 1961, the same year as the first manned flight and President Kennedy's announcement of the Apollo program, the Soviets had approved lunar exploration projects and several related heavy-lift launch vehicles. But because of a bloated

Figure 12.20. Mockup of the first lunar lander, Luna 9, with the flight bus comprising the lower section and the lander located under the "cap" attached at top. Source: Tsiolkovsky Museum, author.

bureaucracy, the Soviet's response to America's Apollo program was not approved as a directive for nearly three years. In addition, the Soviet government chose several spacecraft and rocket design branches to develop the manned lunar flight hardware in parallel. Disorganized policy and duplicative planning in the Soviet centralized government resulted in competing designs for lunar launch and flight hardware that also divided the already limited budget. Their robust space exploration program launched in 1957 became a victim of the Soviet government's bureaucratic struggle for supremacy in, and the absolute control of, their manned lunar program.

A Divided Program

Between 1961 and 1964, Korolev and Chelomei proposed several variations on launchers and flight vehicles that could take cosmonauts to the Moon and back. Korolev's N-1

Figure 12.21. Photograph of a mockup of the Luna sample return vehicle (Luna 15, 16, 20, 24). The largest section at the bottom is the descent module with tripod landing struts. The ascent module is located in the center, with the reentry capsule at the top. Source: Tsiolkovsky Museum, author.

booster, like the R-7, was a kerosene and liquid oxygen rocket, but much larger and with four stages. Korolev's Soyuz was under development to serve as both a replacement for the Voskhod, and as the manned lunar mission vehicle. Chelomei offered a much different version of the lunar hardware that embraced a different role. Chelomei proposed a multistage booster using Glushko's new engine designs incorporating volatile hydrazine and nitrogen tetroxide propellants.

Chelomei's initial booster design was attractive for several rea-

Figure 12.22. Mockup of the Soviet Lunokhod lunar rover launched on the Luna 17 and 21 missions. Note the stereoscopic imaging lenses on the left. Source: NSSDC.

sons, but also too ambitious. The large booster concept was whittled away by the Soviet leadership until a more modest launcher emerged, the UR-500, which later became the

Proton. The booster was designed to carry a smaller vehicle to the Moon and back, but the payload was too small to allow a landing. After evaluating the proposals, Khrushchev divided his approval of the manned lunar program into two phases: a more easily attainable circumlunar manned mission assigned to Chelomei's OKB-456 bureau, and a more ambitious manned lunar landing project to beat Apollo to the Moon assigned to Korolev's OKB-1. In the final design configuration, Chelomei's project hardware was designated UR-500/L1, and Korolev's N-1/L3. And as if doomed from the beginning, no name was ever assigned to the Soviet manned lunar program.

Perhaps the greatest weaknesses in the Soviet lunar efforts were the lack of a comprehensive program plan and the failure to establish a chief design bureau to undertake the difficult and complex endeavor. Disorder in the program's organization slowed progress until the death of Korolev in 1966. Korolev's passing completely derailed the program, although that did not bring the program to an end. His Soyuz spacecraft was adopted in different forms for both manned lunar projects, and his enormous booster continued to completion several years later.

More than a dozen launches of the lunar flight hardware were made to qualify the equipment for manned missions, but without success. The USSR's ambitious and costly manned lunar program never carried a cosmonaut beyond Earth orbit. There were a number of reasons that the Soviet manned lunar program failed, not the least of which was that no one but Korolev had the mastery of program management and the influence on policy to accomplish the enormous task within the Soviet bureaucratic jungle.

Lunar Mission Hardware

Korolev's N-1 (N for Nositel or "carrier") heavy-lift rocket design was not originally selected for the lunar launch vehicle. It was originally funded as a booster to put a Zvezda military surveillance space station into orbit. When the N-1 approval did come, Korolev had no one to design the engine because of a bitter and irreconcilable conflict with Glushko the expert Soviet rocket engine designer. With only a short time available to develop the launcher, Korolev had little choice but to use an aircraft engine designer named Nikolai Kuznetsov of OKB-276. Kuznetsov had turbojet engine experience, but limited experience in rocket engine design. Following the agreements, Kuznetsov went to work developing a modest engine that was configured slightly different for each stage to compensate for the different atmospheric pressures at altitude. His basic engine design for the N-1 project was called the NK-15.

Korolev passed away before completion of the N-1, leaving the design and fabrication teams to complete the work and begin testing. Because there was little time before the scheduled test flights, the decision was made to test the engines but not the three assembled stages. Each of the three stages would require a new test station, a process that would be both expensive and time-consuming. With the first test flight looming, the final N-1 design was frozen with a total of thirty NK-15 engines on the first stage (Block A), eight NK-15V engines on the second stage (Block B), and four NK-19 engines on the third (Block V). The fourth stage was a booster attached to the Soyuz and the lunar lander.

The inaugural N-1 test launch skirted the preliminary testing, which, along with other shortcuts, would ultimately lead to four successive failures of the booster and the termination of the N-1/L3 program. It is noteworthy that the problems with the first stage could have been remedied given enough time and resources since the engine design, along with later improvements, is being used today. But that is not to say that the N-1 would have been successful in getting the mission payloads to the Moon and back.

Success of the N-1 rested on the engine's performance and reliability, and as the N-1 project progressed, Kuznetsov's NK-15 engine underwent improvements, resulting in the NK-33. This engine design was the first dual-stage engine, and the highest-performance kerosene-LOX rocket engine built. Kuznetsov's NK-33 was one of the very few successes in the Soviets' manned lunar program. Today, the rocket engine is being marketed worldwide by the GenCorp Aerojet in California.

Chelomei's UR-200, UR-500, and UR-700 boosters were based on engine designs provided by Glushko in their collusion to take over Korolev's dominance in military and space launchers, and compete in space flight vehicle design. Mikhail Yangel was also a competitor against Korolev, but primarily for intermediate- and long-range missiles. Yangel was placed in charge of a second and independent missile production facility in the Ukraine to guarantee a missile design and production facility even if Russia came under attack. But his efforts were not as successful as those of Chelomei and Korolev, although his designs proved useful for early missile deployments that included the R-12, R-14, R-16, and R-36. Yangel did eventually participate in the manned lunar program when his OKB-586 bureau developed the propulsion system for the LK lunar lander module.

Chelomei began broadening his proposals from a 1960 decree that opened competitive development and production of missiles and space flight hardware. Both his UR-200, which did not survive and his UR-500, which began as an ICBM prototype used non-cryogenic propellants consisting of hydrazine and nitrogen tetroxide. Approval of the UR-500 ICBM came in 1962 by another decree that approved development of a wide variety of space and missile launch designs, and spacecraft and hypersonic vehicles.

Khrushchev's overthrow in October 1964 dramatically changed the direction of Chelomei's OKB-52 design bureau. The UR-200 booster was canceled and the spacecraft design branch headed by Lavochkin was separated. The robotic lunar and interplanetary spacecraft design programs were awarded to Korolev's bureau. Nearly all of Chelomei's branches were severed from his OKB-52 bureau, although the UR-500 was spared, becoming the launcher for the new circumlunar program. What was one of Chelomei's greatest assets, Khrushchev's son working for him as an engineer, was now a detriment to his survival as the bureau chief (Siddiqi 2000).

The UR-500 was a unique design, with large, staged combustion engines developed by Glushko for the project. The two-stage ICBM consisted of a central core second stage, surrounded by detachable boosters with tanks that carried only fuel instead of carrying both fuel and oxidizer. This heavy-lift launcher proved reliable as an ICBM, but even more practical as a space launcher with the addition of a third stage. The UR-500 is still being flown for space launches today as the Proton rocket. A fourth stage is available in various sizes, which furnishes the Proton with a payload capacity comparable to the Arianne 5, Delta IV, and Atlas V.

Figure 12.23. Launch of the Proton M to deliver the Zvezda Service Module payload to the International Space Station, July 12, 2000. Source: NASA NIX.

LUNAR FLIGHT VEHICLES

Korolev proposed a set of manned lunar vehicles in 1962 to compete with the Apollo hardware, beginning with a four-module flight vehicle complex. The first series was named Sever ("south"), then Soyuz ("union"), after which they were designated L1, and ultimately L3. In its first incarnation, the four-module vehicle was made up of a two-man Soyuz crew flight module and three propulsion modules. A Vostok would be launched with the crew to dock with the lunar complex in orbit. After the crew transferred to the Soyuz, the Vostok would have been deorbited to burn up in the atmosphere, and the lunar complex boosted to the Moon with the trans-lunar injection engine located in the last propulsion module.

The following year, Korolev proposed a more robust lunar complex in his paper "Assembly of Space Vehicles in Earth Satellite Orbit." The design resembled the earlier L1 design, but without the Vostok and with larger booster sections fueled in Earth orbit. Along with the flight vehicle proposal, Korolev was completing a follow-up N-1 design for the Politburo's approval.

Competing designs for both lunar flight vehicles and launch vehicles by Glushko and Chelomei forced Korolev to simplify the Soyuz complex. His new design would be

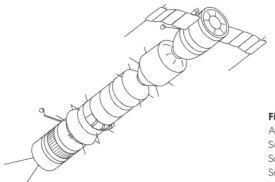

Figure 12.24. Sketch of the Soyuz A+B+C circumlunar vehicle with the Soyuz-A crew vehicle at right, the Soyuz-B booster in the center, and the Soyuz-C tanker-booster at left. Source: Portree.

comprised of a three-vehicle unit resembling the earlier L1 but made up of a manned spacecraft, the Soyuz-A, and the trans-lunar booster and docking stage that was designated Soyuz-B. At the aft was a three-tanker booster called Soyuz-C. This multi-Soyuz vehicle was designed with enough versatility to allow flights to and from a space station planned by Korolev called the OS-1. Korolev's OS-1 would later be scaled-down and transformed into the Almaz-Salyut space station design (Lindroos n.d.).

L1

The rapid pace of the U.S. Apollo program in 1964 forced the Soviet Central Committee to make a firm and final decision on the flight and launch vehicle designs for their manned lunar program. Korolev's proposed lunar lander project included the Soyuz crew vehicle, three booster modules for trans-lunar and trans-Earth injection, for lunar orbit entry, and for a lunar descent and ascent module. Because the N-1 had less payload capacity than the Saturn V, Korolev's crew size was limited to two. And of those two sent to the Moon, only one cosmonaut could reach the Moon's surface.

With assistance from several others allied against Korolev, Chelomei offered a competing design for a circumlunar module. Chelomei's proposed lunar hardware included a third stage of the new UR-700/500 with a service and instrument module, much like the Soyuz, and his own design for the habitation and flight module. Approval for Chelomei's circumlunar project came from the Central Committee in August 1964, and was thereafter known as UR-500/L-1 (Hall and Shayler 2001). Undeterred, Korolev and his engineers proceeded with the development of the more challenging L3 lunar lander complex using the N-1 launcher in a combination designated N-1/L3.

L3 Complex

Korolev's N-1/L3 would include two booster stages added to the basic three-stage N-1. The Block G booster was the fourth stage of the N-1 designed to place the L3 in a departure trajectory to the Moon, or trans-lunar injection (TLI). Block D, which could be considered the fifth stage of the N-1, would be used for the lunar orbit entry (braking). Individual propulsion systems would also be added to the Soyuz-based LOK (Lunniy Orbitalniy, Korabl, or lunar orbiter), and the single-cosmonaut lunar lander called the LK (Lunniy Kabina, or lunar cabin).

The N-1/L3 program was directed by an organization roughly equivalent to NASA called the Ministry of General Machine Building. The target date for the first Soviet manned lunar landing was set for 1968, preceded by a new cosmonaut class selection in 1965 to fly the UR-500/L1 circumlunar and N-1/L3 lunar lander missions.

L3 Complex Modules

- LOK manned lunar orbiter with Block I booster
- LK manned lunar lander with Block E booster
- Block D booster L3 braking for lunar orbit entry
- Block G booster for trans-lunar boost from Earth orbit

LOK Korolev's Soyuz-A manned lunar transport module graduated from a proposal in the early 1960s to working vehicle designs that included the Soyuz, the L1, and the LOK. The Soyuz crew vehicle (7K-OK) is now used for space station crew flights, while the L1 and LOK were derived for manned lunar mission flight modules. The L1 (7K-L1) was much simpler and considerably smaller than the LOK (7K-LOK), since it had no orbital/living module and more modest circumlunar missions. Common to all three crew vehicles was the descent module used for reentry and landing that differed primarily in systems and reentry thermal protection. The most advanced of the three, the LOK, was also the most complex. Power for the LOK was generated by fuel cells that used liquid oxygen and liquid hydrogen reactants which was also the source of potable water for the crew. Additional LOX was supplied for crew breathing oxygen Power for the Soyuz 7K-OK and the L1 7K-L1 was furnished by solar arrays with secondary rechargeable batteries.

The LOK also differed from the other Soyuz vehicles in its unique docking mechanism. The LOK had a mechanical docking unit to mate with the LK docking unit for lunar orbit rendezvous maneuvers, but no transfer tunnel for the crew. The cosmonaut could reach the LK from the LOK by donning his space suit and performing a spacewalk to and from the LK. The simpler L1 had no docking mechanism, while the Soyuz 7K-OK had a docking mechanism and pressurized tunnel section for crews transferring to and from the module (Wade n.d.).

The Russian LOK was equivalent in function to the Apollo Command and Service Module with an enlarged instrument and propulsion module for its extended mission and trans-Earth boost from lunar orbit. The dual-thruster Block I propulsion unit used unsymmetrical dimetlhyl hydrazine and nitrogen tetroxide propellants with small thrusters for mid-course corrections.

LK The Soviet LK lunar lander module was similar in function to the Apollo lunar module, but differed in many aspects. LK was designed to carry only one cosmonaut to the Moon from lunar orbit. Unlike the Apollo LM, the LK had no docking tunnel, which saved weight but required a hazardous spacewalk to get from the orbiting LOK into the LK. A second spacewalk was required to exit the LK and enter the LOK for the return trip to Earth. The mass of the LK was also much less than the LM, and for a reason. Their N-1 had 30% less payload capacity to low Earth orbit compared to the Saturn V, which reduced both the payload and the crew capacity.

The LK did not have a separate descent and ascent stage like the Apollo lunar module, but instead used a separate deorbit rocket to slow the vehicle before the booster was discarded. The LK then used the three onboard Block E engines for touchdown and for liftoff to rendezvous with the LOK after completion of the surface mission. In an effort to limit risks on the lunar surface, an unmanned L3 flight was planned to precede the first manned flight to ensure a safe return of the cosmonaut from the lunar surface. If the cosmonaut's LK was disabled in any way, the backup LK could be used to return the stranded cosmonaut to the LOK (Wade n.d.).

Weight limitations also reduced the cosmonaut's mission duration to three days. A cabin pressure of 0.74 atmospheres of oxygen and nitrogen reduced the risk of fire, but increased the weight of the pressure vessel compared to 0.3 atmospheres of pure

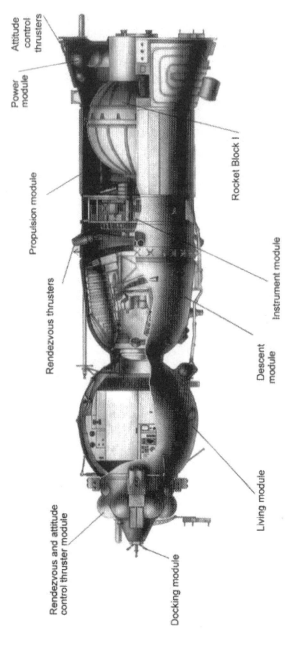

Figure 12.25. Cutaway sketch of the LOK lunar orbiter with the habitation module at left, the deorbit/reentry module at center, and the service/support module and trans-Earth booster on the right. The L3 was functionally equivalent to the Apollo Command and Service Module. Source: NSSDC.

Figure 12.26. Photograph of the LOK descent module partially disassembled (bottom), and an exposed view of the lunar orbital module (top). The attitude control section was attached to the face plate at the top of the orbital module. These are approximately the same scale as the Soyuz space station crew vehicle, but without the forward docking module. Source: Moscow Aviation Institute, author.

oxygen environment on the Apollo LM. Multiple cameras inside and outside the cabin allowed both manual and ground control landings. The 5.5 m tall vehicle had an approximate orbital mass of 5,500 kg, compared with the 6.4 m LM, which had an orbital mass of 14,600 kg.

Initial testing of the LK was done separately from the L3 complex using modified R-7 Soyuz boosters. Three flights of the LK (designated T2K) test vehicle were conducted successfully in 1970 and 1971 as Cosmos 379, 398, and 474 (Wade n.d.).

TEST PREPARATIONS—1965

L1

The Soviets intended to get their cosmonauts to the Moon first on a swing-by circumlunar mission before the Apollo 8 flight using Chelomei's L1 project hardware and the UR-500 (Proton) booster. Preparations for the qualification flights for the manned missions included a test launch of the UR-500 Proton on July 16, 1965, that orbited a 12.2-ton scientific satellite, far heavier than any previous Soviet payload. However, Chelomei's inexperienced capsule design teams encountered difficulties and fell behind in the development schedule. With America's first lunar flights looming and the recent

Figure 12.27. Photograph of the LK lunar lander hardware on display at the Moscow Aviation Institute. Source: author.

dismissal of Khrushchev, Chelomei's bureau and project were placed under the direction of Korolev.

L3

Korolev took advantage of Chelomei's project delays by proposing that his L3 upper stage and the Soyuz manned vehicle be approved and given priority. His proposal was accepted, and even more important, Korolev was placed in charge of the UR-500/L1 program. But with Korolev's death in January of 1966 came the inevitable and swift unraveling of the manned lunar program since none of the design chiefs nor Kremlin leaders knew what was required to complete either project or how to succeed in manipulating the Soviet bureaucratic structure to complete the program. Although there was cooperation between many of the design bureaus, there was no clear definition of the final flight hardware because of the bureaucratic indecision.

FIRST TEST LAUNCHES—1966

Preparation for the manned lunar missions involved many of the Luna robotic spacecraft sent to explore the Moon and its environment. In 1966, two very important exploration missions were completed: the Luna 9 lander and the Luna 10 orbiter.

Radiation, thermal, and gravitational data were gathered for the coming unmanned L1 circumlunar missions. Luna 11 and 12 orbiter missions were also launched successfully in 1966, providing Russian scientists with close-up images of the landing sites planned for the manned LK landers scheduled for 1969.

Late in 1966, the Soviet leadership issued directives that allowed Mishin—Korolev's successor—to go ahead with the N-1/L3 project. But with only two years remaining before the target launch dates, the design and testing schedule became unrealistic. Chelomei countered with a proposal to develop a heavy-lift UR-700 launch vehicle with a capacity of 150 tons to low Earth orbit, some 50% more than the N-1 could lift, allowing two or more cosmonauts to land on the Moon. The Keldysh Commission turned down Chelomei's request, but approved the UR-500 booster for the L1 (Sadiqqi 2000).

On November 28, 1966, Korolev's Soyuz capsule, which was the cornerstone of the L3 complex, was launched successfully into orbit. The manned Soyuz test flight was not so fortunate.

L1 Tests

Progress on the competing L1 circumlunar and L3 lunar lander programs in 1967 was promising for the two respective design bureaus, and for the party and government officials. Their expectation for a circumlunar launch in 1968 before Apollo 8 was buoyed by a successful launch of a UR-500/L1 vehicle into a highly elliptical orbit, although it was a trajectory that took it away from the Moon to avoid the complications of lunar gravity. Especially important was the trans-lunar Block D propulsion engine on the L1 that fired twice as planned. The flight, designated Cosmos 146 (Zond 4), was launched on March 10, 1968, with flight hardware that included a launch escape system and communications test equipment. However, the descent module was destroyed on reentry after a fatal guidance error. A second test of the UR-500/L1 in April was unsuccessful because of a second-stage booster failure. Only two weeks later, the flight test of the first manned Soyuz mission ended in the death of cosmonaut Vladimir Komarov. Just as the Apollo 1 fire delayed America's lunar program, the death of Komarov stalled the Soviet manned lunar program until questions could be answered about the multiple failures on the Soyuz and the launch failures on the first UR-500/L1 flights.

Following critical evaluation of Komarov's accident, and after making the necessary corrections, the first successful circumlunar launch was made on September 15, 1968, with the UR-500/L1 hardware designated Zond 5. The L1 capsule passed behind the Moon and came within 2,000 km of the surface before returning to Earth. The biological specimens were recovered despite a reentry system failure that required manual ballistic reentry that subjected the payload to 20 g's and with a splashdown in the Indian Ocean short of the target.

Qualification of the UR-500/L1 circumlunar mission for a crew flight was scheduled for December 1968 and required near-perfect flight operations. Zond 6 was launched November 10, 1968 on what was expected to be the L1 qualification flight. Zond 6's lunar encounter was a success, returning useful data and photos of the far side of the Moon. On the return leg, the main cabin lost pressure which would have been fatal for the crew. Reentry was a successful skip entry, but early parachute deployment and its partial failure destroyed much of the capsule and payload on impact.

Another launch of the UR-500/L1 on a circumlunar mission in January 1969 failed soon after launch, although the abort system landed the L1 capsule safely for recovery and future use. Without qualification of the vehicle and launcher to carry a crew, and because Apollo 8 successfully took astronauts Anders, Borman, and Lovell to the Moon and back in December 1968, there was no hope of even matching the Apollo circumlunar flight within the following year. As a result, the manned UR-500/L1 missions scheduled for early 1969 were canceled.

The final UR-500/L1 flights on Zond 7 and Zond 8 were launched August 8 and October 20, 1969. While both missions returned to Earth, the Zond 7 mission was the first completely successful Soviet lunar return flight including a skip entry and landing. Zond 8 encountered a guidance failure on the return leg and suffered a ballistic reentry that subjected the payload to a force of more than 20 g's, which would likely have been fatal for a human crew.

L3 Testss

By 1969, the preparations for a manned flight to the Moon shifted from the L1 (7K-L1) circumlunar missions to the lunar landing missions that incorporated the N-1/L3 hardware. Partial success of the L1 circumlunar vehicle tests allowed project planners to use the L1 and booster module without the same level of testing as the LK lander module and the larger Block G booster that made up the rest of the L3.

Testing the L3 complex was not possible on the UR-500 since the L3 mass exceeded the payload capacity of the booster. And to conserve resources, the first launch of the N-1 was prepared with an L1 capsule instead of an LOK, along with a dummy lunar lander in place of the LK designated L3S.

N-1

The Soviet N-1 launcher evolved from a heavy-lift military launcher concept from the early 1960s to an even larger booster for their manned lunar program. Proposals for much larger launchers based on the N-1's early design were never approved, although variations on the heavy-lift booster threaded through Korolev's OKB-1 until the demise of the entire manned lunar program in 1974. Korolev's early design of the N-1 first stage consisted of twenty-six NK-15 engines placed in a circle and fed by two spherical tanks that contained the LOX and kerosene propellants. After his death, the N-1 design was modified by Mishin's OKB-1 engineers to include thirty NK-33 engines (improved NK-15 units) on the first stage, and in two circles. The combination of two rings of 24 + 6 engines produced a total thrust of 4,500 metric tons at liftoff. Four roll control engines augmented the stability of the first stage, which had a burn time of approximately 110 sec.

All too obvious in the design of the N-1 first-stage was the potential for just one engine failure to induce a thrust imbalance which could overwhelm the steering engines. To correct for this, Korolev and Mishin designed an engine controller called the KORD that would shut down an engine diametrically opposed to any engine that failed to retain thrust symmetry. However, the rush to complete the N-1/L3 project pushed OKB-1 managers to omit an all-up test of the entire first and second stage units. Even though the NK-33 engines did go through operational tests, the test stand for and testing of the entire first stage were never completed.

The second stage of the N-1 (Block B) had eight NK-43 engines, producing a total thrust of 1,400 metric tons with a burn time of 130 seconds. Fuel and oxidizer were also kerosene and liquid oxygen. Third stage (Block V) included four NK-39 engines producing 164 metric tons of thrust using the same LOX and kerosene propellants. The L3 complex included the fourth (Block G) and fifth stage (Block D) propulsion units for trans-lunar boost and lunar orbit injection respectively, which also used kerosene and liquid oxygen propellants (Wade n.d.).

Figure 12.28. Launch of the N-1/L3S at Baikonor. Source: NASA GRIN.

N-1 Launches

Inaugural launch of the newly completed N-1/L3S (vehicle 3L) was on February 21, 1969, using an L1 payload instead of the L3 complex. Two of the first-stage Block A engines were shut down erroneously by the KORD just seconds into the ascent. Sixty-six seconds into the flight, an oxidizer line to one of the NK-33 engines ruptured, triggering a fire that resulted in all engines being shut down by the KORD. Although the N-1 was completely destroyed, the emergency escape system extracted the L1 payload to a safe landing by parachute several kilometers from the pad (Wade n.d.).

The second N-1/L3S launch attempt was made on July 3, 1969 (vehicle 5L). Nine seconds into the flight one of the NK-33 engines exploded after ingesting a metal object, shutting down or destroying other engines and subsequently destroying the launch pad as it fell back to the ground. The pad destruction was the first solid evidence of the Soviet's manned lunar program based on the scale of destruction viewed in the U.S. reconnaissance satellite images. Other details of the Soviet manned lunar program were unknown to the outside because of the Soviet leadership's obsessive secrecy.

A number of design changes were made to the N-1 before the next launch two years later. NK-33 main engine improvements, propellant line filters, and improvements in the KORD were made, but the delay put the test flight well after the first Apollo astronauts landed on the Moon.

A more ambitious N-1/L3 program was proposed by Mishin that involved a larger lunar complex launched with two N-1 vehicles, along with a promise of improved N-1 reliability and greater thrust performance. The N-1/L3M vehicle again carried mockup payloads for testing, but again without success. The June 27, 1971 launch experienced severe roll torque after liftoff. As the vehicle lost stability, the KORD shut down all first-stage engines approximately fifty-one seconds into the flight. The L3 payload was attached to an escape rocket system, but that too was a dummy and the entire vehicle was destroyed.

As the U.S. was nearing completion of the Apollo program, the Soviet leadership felt that a manned lunar base was still promising, and within reach with the new N-1 design. Therefore, a fourth test was scheduled and launched on November 23, 1972. Ninety seconds into the flight, the central six engines were shut down as programmed, but the resulting propellant pressure surge ruptured several of the feed lines. The rupture caused a widespread fire in the first stage that destroyed the vehicle within twenty seconds. An actual lunar orbiter and lunar lander prototype were saved by the launch escape system, but the N-1 never flew again.

By 1974, interest in the L3 project disappeared, which resulted in a complete shakeup of the spacecraft and launch vehicle design bureaus. Mishin was relieved from his post as chief designer of OKB-1 and the N-1/L3 program, which were transferred to Glushko. The two remaining test flights of the N-1 scheduled for 1974 were canceled by Glushko, and the equipment and vehicles were destroyed with a vengeance. However, Glushko remained interested in a lunar flight program and proposed a Vulkan heavy-lift launcher for manned lunar and Mars missions. Glushko's Vulkan was a precursor of the Energia booster built later by NPO Energia—the descendant of OKB-1. Glushko's control over the OKB-1, OKB-52, and a number of other design bureaus allowed him to turn the major Soviet space programs away from Korolev's emphasis and toward his own. Swift approval came from above for Glushko's plans to end the N1/L3 project and for the development of the much-larger Energia booster and the Buran space shuttle. But those projects had an even shorter lifetime. Energia flew only twice before the fall of the Soviet Union in 1991, and the Buran only once. Because of their costs, both of Glushko's grandest successes contributed more to the downfall of the Soviet Union than to Russia's space exploration legacy.

MARS EXPLORATION

Even though the greatest gains in Soviet propaganda were made with their early manned programs, the greatest gains in science came from their robotic missions. The advantages of robotic space exploration were known by many in the Soviet leadership, ignored by the military command, acknowledged by the Council of Designers, and promoted by the Soviet Academy of Sciences. Like the Americans, the Soviets quickly expanded their space exploration missions after reaching the Moon. Missions to Mars began in 1960 for the Soviets, although unsuccessfully, and in 1964 for the United States with the launch of Mariner 3. Missions to Venus began in 1961 for the Soviets and in 1962 for the United States with the first successful interplanetary mission, Mariner 2.

The Soviet's Mars exploration record was disappointing but the missions provided invaluable data on the Martian and interplanetary environment, and on the strengths and weaknesses of their exploration projects. As with the American exploration efforts, the Soviet Union's lunar, Mars, and Venus exploration programs ran in parallel. A brief summary of the Soviet Mars projects is listed below.

Mars exploration missions were generally subordinate to the lunar missions because of the emphasis placed on the Soviet manned lunar program during the race to the Moon. But three years before the Soviet Union collapsed, their most ambitious project and their largest robotic spacecraft ever built were launched to Mars. Russia's Phobos 1 and Phobos 2 spacecraft sent to explore Mars were launched just five days apart, although both ended in failure. A similar project based on the Phobos spacecraft design survived

Table 12.10. Soviet Mars Mission Summary (NSSDC n.d.; Wade n.d.)

Spacecraft	Launch	Mission	Results
Mars 1960A Marsnik 1 (flyby)	November 10, 1960, on Molniya 8K78 (R-7 derivative)	Mars flyby, imaging, observation; interplanetary observations	Launch failure
		The spacecraft was a cylindrical body about 2 meters high with two solar panel wings, and carried a 10 kg science payload consisting of a magnetometer, cosmic ray counter, plasma-ion trap, a radiometer, a micrometeorite detector, and a spectroreflectometer.	
Mars 1960B Marsnik 2 (flyby)	October 10, 1960—Molniya 8K78	Mars flyby, imaging, observation; interplanetary observations	Launch failure
		Spacecraft structure and systems were the same as Mars 1960A.	
Sputnik 22 (flyby)	October 24, 1962—Molniya 8K78	Mars flyby, imaging, observation; interplanetary observations	Earth-orbit breakup
Sputnik 24 (flyby)	November 4, 1962—Molniya 8K78	Mars flyby, imaging, observation; interplanetary observations	Mars transfer orbit explosion (failure)
		Failure of booster to Mars, Earth orbit decayed and re-entered Earth's atmosphere 19 January 1963	
Mars 1 (flyby)	November 1, 1962—Molniya 8K78	Mars flyby, imaging, observation; interplanetary observations. Instrumentation was to provide data on cosmic radiation, micrometeoroid impacts and Mars's magnetic field, radiation environment, atmospheric structure, and possible organic compounds.	Interplanetary data transmitted until communications lost before reaching Mars

(Continued)

Spacecraft	Launch	Mission	Results
Zond 2 (flyby)	November 30, 1964—Molniya 8K78	Mars flyby, imaging, observation; interplanetary observations	Interplanetary data transmitted until communications lost in early May 1965 before reaching Mars due to a power failure. Mars flyby was on August 6, 1965, at a distance of 1,500 km.
Mars 1969A (orbiter)	March 27, 1969—Proton 8K82K	Mars orbital observations, interplanetary observations	Launch failure 439 sec after liftoff
Mars 1969B (orbiter)	April 2, 1969—Proton 8K82K	Mars orbital observations, interplanetary observations	Launch failure 41 sec after liftoff
Mars 2 (orbiter + lander)	May 19, 1971—Proton 8K82K	Mars orbital observations, Mars lander surface and atmosphere observations; interplanetary observations	Orbital mission successful; lander crashed although it returned images and atmospheric data before impact
Mars 3 (orbiter + lander)	May 28, 1971—Proton 8K82K	Mars orbital observations, Mars lander surface and atmosphere observations; interplanetary observations	Orbital mission successful; lander soft-landed but stopped transmitting data 20 sec after touchdown
Mars 4 (orbiter)	July 21, 1973—Proton 8K82K	Mars orbiter; interplanetary observations Retrorockets never fired to slow the craft into Mars orbit. Mars 4 flew by the planet at a range of 2,200 km	Mars orbit entry failed, entered heliocentric orbit
Mars 5 (orbiter)	July 25, 1973—Proton 8K82K	Mars orbiter; interplanetary observations	Mars orbit entry and data transmissions successful but ended after 9 days
Mars 6 (lander)	August 5, 1973—Proton 8K82K	Mars carrier flyby and Mars lander; interplanetary observations	Mars bus successful flyby, but lander ended transmission at/near touchdown

Mission	Launch	Description	Outcome
Mars 7 (lander)	August 9, 1973—Proton 8K82K/11S824	Mars carrier flyby and Mars lander; interplanetary observations. Spacecraft and mission were the same as Mars 6.	Bus reached and passed Mars but the lander separated early, missing planet. Both lander and bus continued on into heliocentric orbits
Phobos 1 (orbiter, lander)	July 7, 1988—Proton 8K82K	Mars orbiter, Phobos landers, Mars and interplanetary exploration	En route power failure resulting from software command error, which deactivated the attitude thrusters.
Phobos 2 (orbiter, lander)	July 12, 1988—Proton 8K82K	Mars orbiter, Phobos landers Same spacecraft and similar mission as Phobos 1.	Systems and communications loss in orbit because of onboard computer malfunction
Mars 96 Orbiter (orbiter, lander, penetrator)	November 16, 1996—Proton 8K82K	Mars orbiter, two Mars landers, two surface penetrators	Launch failure

Missions outlined in gray were complete or substantial failures.

the fall of the USSR, but encountered myriad delays until its launch in 1996. External funding saved the Russian Mars 96 project, but not its mission. A failed booster while in Earth orbit placed Mars 96 in a highly eccentric orbit, with a fatal reentry the next day. Reported splashdown in the Pacific may have actually been an impact of the nuclear-powered spacecraft on land in South America (Oberg 1999). Today, Russia continues to plan interplanetary exploration projects that include Mars missions, but they suffer from the same lack of funding that delayed or canceled many projects in the past.

Mars Spacecraft

MARS 3

Figure 12.29. Mockup of the Mars-3 orbiter-lander spacecraft located at the Tsilokovsky State Museum of the History of Cosmonautics. Source: author.

MARS 3 LANDER

Figure 12.30. Photo of the Mars 3 lander mockup at the Memorial Museum of Cosmonautics in Moscow. Note the 4-petal arrangement on the spherical capsule that closely resembles the early Soviet lunar landers. Source: author.

PHOBOS

Figure 12.31. Mockup of the Phobos lander and orbiter spacecraft that contained a lander for the Mars moon Phobos and an orbital laboratory. The four large tanks at the bottom and the torrodial tank near the center stored propellants for orbital operations and attitude control. Source: NSSDC.

VENUS EXPLORATION

The Soviet Venus exploration program named Venera ("Venus") ran in parallel with their Luna and Mars exploration programs, which began in 1961 and lasted until 1984. The first Venus flyby and the first Venus lander were made by the Soviets, as well as the first images and only from the surface of Venus. Measurements of the density, pressure, temperature, and composition profiles of the atmosphere made with the Venera probes and landers showed an extremely hostile environment with searing temperatures of almost 470°C (878°F) at the surface, a surface pressure of about 90 atmospheres and sulfuric acid rain. Images returned for the Venera 9, 10, 13, and 14 landers showed basaltic lava beds strewn with smaller, sharp rocks littering the surface, an indication that Venus's surface was relatively new and suffered little erosion. Venera missions ended with the flight of a pair of successful synthetic aperture radar orbiters that mapped a large part of the surface of Venus, which is permanently obscured by dense cloud cover.

BURAN

As America's space shuttle began taking shape in the mid-1970s, the Soviet leadership ordered the development of a space shuttle project of their own. Their intention was to keep pace with the American space shuttle and its military potential with their program named the Reusable Space System (Russian acronym MKS). But Soviet concern went beyond duplicating the American shuttle out of the unfounded fear that the space shuttle had a sophisticated manned weapons and antisatellite capacity that could overwhelm their defenses. Protests and international proposals to restrict the Soviet shuttle's military activities surfaced between 1976 and 1984 but tapered off when the Buran flight testing was underway (NASA 1988).

The Soviet space shuttle project is commonly and incorrectly referred to as the Buran program since Buran ("snowstorm") was the name of the first space-qualified orbiter, like the *Columbia* orbiter in America's space shuttle fleet. Authorization for the MKS project and several others was made by decree from the Soviet Central Com-

Table 12.11. Soviet Venera Missions (NSSDC n.d.; Wade n.d.; RussianSpaceWeb n.d.)

Spacecraft	Launch	Mission	Results
Sputnik 7 (IVA)	February 4, 1961 Molniya 8K78 (modified R-7)	Venus flyby mission	Launch failure
Venera 1 (IVA) (flyby)	February 12, 1961 Molniya 8K78 (modified R-7)	Venus flyby	First spacecraft to successfully fly by Venus (passed within 100,000 km of Venus and entered a heliocentric orbit)
Unnamed (2MV-1)	August 25, 1962	Venus lander	Fourth stage failure
Unnamed (2MV-1)	September 1, 1962	Venus lander	Stranded in LEO
Unnamed (2MV-2)	September 12, 1962	Venus flyby	Fourth stage failure
Unnamed (3MV-1A)	February 19, 1964	Venus flyby	Launch failure
Cosmos 27 (3MV-1)	March 27, 1964 Molniya 8K78M	Venus lander	Departure booster failure stranded spacecraft in LEO
Zond 1 (3MV-1)	April 2, 1964 Molniya 8K78M	Venus flyby	Passed Venus at 100,000 km then entered a heliocentric orbit. Communications from spacecraft ended on May 14, 1964. No data returned from Venus.

Missions outlined in gray were complete or substantial failures.

Spacecraft	Launch	Mission	Results
Venera 2 (3MV-4)	November 12, 1965 Molniya 8K78M	Flyby mission carried a TV system as well as interplanetary experiments.	The spacecraft system had ceased to operate before the planet was reached and returned no data. Passed Venus at a distance of 24,000 km then entered heliocentric orbit.
Venera 3 (3MV-3)	November 16, 1965 Molniya 8K78M	Venus lander	Venera 3 communications systems failed before planetary data could be returned. Spacecraft impacted Venus on March 1, 1966, making it the first spacecraft to impact on the surface of another planet.
Cosmos 96 (3MV-4)	November 23, 1965	Venus lander	Spacecraft stranded in LEO
Venera 4 (V-67)	June 12, 1967 Molniya 8K78M	Orbiter and atmospheric probe, interplanetary measurements. Tyazhelly Sputnik bus used to carry the capsule to Venus.	Data were returned by the probe after entering the Venusian atmosphere until it reached an altitude of 25 km.
Cosmos 167 (V-67)	June 17, 1967 Molniya 8K78M	Same mission as Venera 4	Booster failure stranded spacecraft in low Earth orbit for approximately 8 days before reentry.
Venera 5 (V-69)	January 5, 1969 Molniya 8K78M	Same mission as Venera 4. Probe was similar to Venera 4 but with a more robust design.	Data from the Venusian atmosphere were returned for 53 min.
Venera 6 (V-69)	10 January, 1969 Molniya 8K78M	Same spacecraft and mission as Venera 5	Data returned for 51 minutes during atmospheric entry.

(Continued)

(Continued)

Spacecraft	Launch	Mission	Results
Venera 7 (V-70)	17 August, 1970 Molniya 8K78M	Atmospheric probe/lander and orbiter, similar to Venera 5 and 6.	The capsule was the first man-made object to return data after landing on another planet. Data returned for 35 min and another 23 min of very weak signals were received after the spacecraft landed on Venus.
Cosmos 359 (V-70)	August 22, 1970 Molniya 8K78M	Orbiter and atmospheric probe/lander, similar to Venera 5, 6, and 7.	Probable probe failure
Venera 8 (V-72)	March 27, 1972 Molniya 8K78M	Atmospheric probe/lander and orbiter, similar to Venera 5, 6, and 7.	Aerobraking used to reduce descent speed, then parachutes from about 60 km to the surface. Refrigeration system was used to cool the interior components. Wind speeds of less than 1 km/s were measured below 10 km. Continued to send back data from the surface for 50 minutes after landing.
Cosmos 482 (V-72)	March 31, 1972 Molniya 8K78M	Atmospheric probe/lander	Transfer booster failure
Venera 9 (4V-1)	June 8, 1975 Proton 8K82K	Orbiter + lander, camera payload for surface images. Heat dissipation and deceleration from protective hemispheric shells, three parachutes, a disk-shaped drag brake, and a compressible, metal, doughnut-shaped, landing cushion.	First spacecraft to send images from the surface of Venus. Data showed cloud layers containing HCl, HF, Br, and I. Surface pressure measured at 90 atmospheres with a surface temperature of 485°C. TV images showed no apparent dust in the air, and a variety of 30–40 cm rocks which were not eroded.

Spacecraft	Launch Date / Vehicle	Description
Venera 10 (4V-1)	June 14, 1975 Proton 8K82K	Same mission and spacecraft as Venera 9.	Data confirmed pressure and temp readings from Venera 9. TV images returned during descent showed large pancake-like rocks with lava or other weathered rocks in between. Surface wind speed measured at 3.5 m/s consistent with dense atmosphere and no wind erosion on rocks.
Venera 11 (4V-1)	September 9, 1978 Proton 8K82K	Similar mission and spacecraft as Venera 9 & 10.	Data returned during descent showed evidence of lightning and thunder, a high Ar^{36}/Ar^{40} ratio, and the discovery of carbon monoxide at low altitudes. Orbiter relayed transmissions from the lander for 95 minutes after touchdown. Imaging system failed to return surface data.
Venera 12 (4V-1)	September 14, 1978 Proton 8K82K	Similar mission and orbiter and lander spacecraft as Venera 9, 10, and 11.	Instruments on board included a gas chromatograph to measure the composition of the Venus atmosphere, instruments to study scattered solar radiation and soil composition, and a device named Groza designed to measure atmospheric electrical discharges. Images were not returned to Earth.
Venera 13	October 30, 1981 Proton 8K82K	Similar mission and orbiter and lander spacecraft as Venera 9, 10, 11, and 12.	The Venera 13 lander survived for 127 minutes (design life was 32 minutes) in an environment with a temperature of 457°C and a pressure of 84 atm. The lander utilized a camera system, an X-ray fluorescence spectrometer, a screw drill and surface sampler, a dynamic penetrometer, and a seismometer to conduct investigations on the surface .
Venera 14	November 4, 1981 Proton 8K82K	Identical spacecraft and mission as Venera 13.	The Venera 14 lander survived for 57 minutes in an ambient temperature of 465°C and a pressure of 94 atm.

(Continued)

Spacecraft	Launch	Mission	Results
Venera 15	June 2, 1983 Proton 8K82K	Venera 15 and Venera 16 were dual spacecraft designed to use 8 cm band side-looking synthetic aperture radar to map the surface of Venus.	Used polar orbit with 4° orbit plane separation for imaging Northern hemisphere during eight months of mapping.
Venera 16	June 7, 1983 Proton 8K82K	Venera 16 was the second of the duplicate Venera 15–16 pair used for imaging Venus's surface.	Same as Venera 15
Vega 1 (5KV)	December 15, 1984 Proton 8K82K	Venus lander and flyby. Comet Halley flyby. International instrument suite onboard.	Lander and balloon descent probe deployed at Venus. Lander similar to Venera 9–14. Flyby of Comet Halley returned comet core images and composition data after passing Venus. Balloon was carried 12,000 km over 3 days before losing communication.
Vega 2 (5KV)	December 20, 1984 Proton 8K82K	Same as Vega 1	

mittee in February 1976 in response to Glushko's proposal. Within the same order was the approval for development of the Energia booster, the Buran shuttle (as the Reusable Space System), the Mir space station, and the Luch satellite communications system similar to the U.S. TDRSS satellite system. The same decree also ordered the termination of the troubled N-1 launch vehicle and manned lunar project, although it was not officially canceled until Glushko gained control of the original OKB-1.

Glushko's ambitions spread far and wide after gaining control over virtually all of the Soviet space programs. Glushko started the planning and development stages for the shuttle and launcher within the newly merged engineering bureau NPO Energia before the MKS project was officially approved. His inherited UR-500 Proton launcher that was a scaled-down version of the larger UR-700 concept Chelomei first envisioned and called the Vulkan was replaced with a heavy-lift booster using much larger engines than had been built before. Glushko's new Vulkan heavy-lift booster that was related to Chelomei's only in name was designed with first-stage engines that used Korolev's cryogenic liquid oxygen and kerosene mix. These were the same propellants that Korolev demanded but Glushko refused to incorporate in several earlier designs. The second-stage engine design on the new booster was also the first of a kind for the Soviets, which used liquid oxygen and liquid hydrogen propellants to improve thrust efficiency.

Approval of Glushko's MKS space shuttle and the Vulkan booster in 1976 came four years after the United States announced their Space Transportation System, which put development time at a premium. Glushko had already begun the design of the Vulkan booster by the time it was approved, but even with the head start it would take more than ten years to prepare the first Vulkan for launch.

Glushko hoped to reestablish a dominant Soviet presence in space with his vision of large manned orbital stations, manned missions to the Moon and Mars, and large interplanetary laboratories. To this end, Glushko began developing a versatile heavy-lift booster based on the Vulkan concept called the Energia, which was one configuration of the original Vulkan. These booster variants ranged from a simple, four-engine core module, the Energia M, to the same four-engine core with eight strap-on liquid fuel boosters—the Vulkan. Intermediate in scale was the Energia comprised of the core and four strap-on boosters that was tailored specifically to support the Buran program.

Project Development

Although Glushko's NPO Energia group managed the entire MKS project including the orbiter, launcher, and support systems, the design of the orbiter was turned over to a new design bureau spun off solely for the project. The new bureau, designated NPO Molinya, was headed by a new director, Gleb Lozino-Lozinsky, a Mig aircraft designer. The new design bureau put in place to develop the 100-ton Buran shuttle was a contentious decision since it passed over experienced spacecraft designers who had worked on hypersonic vehicles. Chelomei and his teams were not only left out of the MKS project, but stripped of space-related program opportunities and, within several years, relegated to building smaller missiles (Siddiqi 2000). Nonetheless, Glushko was successful in orchestrating the project's more than 1,200 subcontractors and 100 government ministries to build and fly the first Soviet space shuttle.

Orbiter tests began with scale model wind tunnel tests of the orbiter and continued with larger models that progressed to suborbital flights in 1983. Scaled lifting body designs were also launched for orbital reentry tests of the heat shield tiles and to validate the hypersonic aerodynamics for the orbiter. The first full-scale test vehicle to fly was the Aero-Buran that began flight tests three years after the first U.S. space shuttle was launched in 1981. Aero-Buran, known also as the "analog," was powered by four turbojet engines for its horizontal takeoff thrust and to reach altitude for low-cost automated landing tests and aerodynamic design validation.

Preliminary design of the Buran orbiter focused on a winged space plane for the same reasons the United States had adopted the design for the space shuttle orbiters, which was the 1,000 to 2,000 km cross range capability—the landing distance deviation available for a specific reentry profile. This prevented reentry data from being used to predict a landing site for classified missions. It also offered a one-orbit operational capability for a quick weapons delivery mission. Russian engineers used nearly all of the space shuttle orbiters' aerodynamic design features, but incorporated a number of major non-aerodynamic changes that took advantage of Russia's late entry into the shuttle race, with some improvements coming from their differing and occasionally superior technology.

One of the major differences in the Buran orbiter design and the Space Shuttle Orbiter was the onboard propulsion systems. While the U.S. orbiters housed three SSME engines fed by an external propellant tank, the Buran was designed to be launched without ascent engines attached, making it lighter and increasing its payload. Ascent thrust was delivered entirely by the two-stage Energia booster. Another significant difference was the four ejection seats planned for a nominal crew of four on the Russian orbiters, while the space shuttle orbiter had none except in the first four manned test flights.

Fabrication of the first Russian orbiters began in 1980 just as the United States was about to launch its first STS flight. A total of five of the Soviet shuttle orbiters were scheduled for construction, but because of delays and the mounting debt and weakening economy in the Soviet Union during the late 1980s, the program was reduced in scope and then canceled in 1993. The one orbiter that did fly, OK-1.01 Buran, was launched on November 15, 1988, in a completely automated, two-orbit flight, without fuel cells or flight instruments to bypass development delays. Landing was also automated and went without complications, even in the 61 km/h (38 mi/h) crosswind. However, the enormous expense of the project including the Energia launcher totaled more than 14 billion rubles which contributed to the program's collapse before its second flight. Buran OK-1.01 was mothballed at the Baikonor fabrication facility, although it did fly again in 1989 to the Paris Air Show atop the largest aircraft to fly, the Antonov 225. After a worldwide tour, Buran OK-1.01 was returned to a museum at the Baikonor Cosmodrome, mounted on an Energia launch vehicle mockup, and placed on display. In 2002, the Baikonor assembly building that housed Buran OK-1.01 collapsed from the weight of a heavy snowfall, crushing and completely destroying the vehicle and the Energia launcher mockup.

Of the four other Buran vehicles, only the second, OK-1.02 called Pitchka ("little bird"), was completed, or nearly so. This vehicle contained the life support systems

Table 12.12. Buran Specs

Maximum weight	105,000 kg (231,490 lb)
Payload weight	30,000 kg (66,140 lb)
Landing weight	82–87,000 kg (180,800 - 191,800 lb)
Crew	2–10 (4 nominal)
Flight duration	7–30 days
Orbit inclination range	50.7°–110°
Orbit altitude range	250–1,000 km (155–620 mi)
Landing speed (82 ton landing weight)	312–360 km/hr (194–224 mi/hr)
Maximum reentry cross range	1,700 km (1,060 mi)
Length	36.4 m (119 ft)
Body width	5.5 m (18.04 ft)
Height	16.4 m (53.8 ft)
Wing span	23.9 m (78.4 ft)

and avionics that were to carry the first cosmonauts to orbit on a launch scheduled for 1993. Pitchka and the following three were designed as transportation vehicles to the Mir-2 station. Newer series vehicles included OK-2.01 (named Baikal, meaning typhoon), OK-2.02, and OK-2.03. These orbiters, which were in early stages of fabrication, were either completely dismantled, or scattered throughout Russia in various stages of cannibalization and neglect.

ENERGIA

Glushko's Energia was the largest Soviet booster ever built and comparable with the Saturn V in LEO and trans-lunar payload. The Energia design was also Glushko's first major success in developing a single-chamber high-thrust engine, which had plagued him since the R-7 engine assignment. Combustion instabilities were a challenge for all engine designers, and more so for the Russian teams because of the difficulty in scale for engines beyond twenty tons of thrust, and because of their lack of the computer technologies available to the American engine designers. Before Glushko's breakthrough, his high-thrust engine designs consisted of four smaller combustion chambers and nozzles tied together with a single turbopump and packaged as a single propulsion unit. This four-chamber engine scheme succeeded in delivering the 100-ton thrust required for the R-7 core and booster units but were considered inferior to single high-thrust engines because of the increased system complexity.

Glushko's high-thrust RD-0120 developed for the Energia used individual turbopump and thrust assemblies, although not all of the Energia engines were designed

Figure 12.32. Comparison sketch of the major features of the Russian Buran-Energia on the right and the U.S. Space Transportation System on the left. Source: NASA NIX.

Table 12.13. Buran Orbiter and Space Shuttle Orbiter Comparison

Characteristics	Buran	Space Shuttle
Operations	Crew, ground control, or full automation	Crew or ground control only
Boosters	Four liquid strap-ons for first stage; single core second stage (all liquid propellants)	Two solid boosters for first stage and three orbiter liquid propellant engines (SSME)
Payload to orbit	30,000 kg (66,000 lb)	25,000 kg (55,000 lb)
Payload return	20,000 kg (44,000 lb)	15,000 kg (33,000 lb)
Reentry L/D ratio	6.5	5.5
Orbital altitude (max)	1,000 km (621 mi)	450 km (280 mi)
Length	36.4 m (119 ft)	37.3 m (122 ft)
Wingspan	23.9 m (78.4 ft)	23.8 m (78.0 ft)
Wing Sweep back	45°	45°
OMS/RCS propellant load (max)	14,500 kg (31,970 lb)	14,100 kg (31,090 lb)
Payload bay dimensions	4.65 × 18.55 m (15.3 ft × 60.8 ft)	4.57 m × 18.29 m (15 ft × 60 ft)

as single-chamber thrusters. The first-stage (strap-on) boosters incorporated a quad-thruster RD-170 design fed with LOX and kerosene propellants for each of the four engine units. Maximum thrust for the RD-170 was 7,784,400 kN (1.75 M lb$_f$) which was 15% more than the Saturn V's F-1 single-chamber engine. Although the engines were of Russian design, the boosters were built by a Ukrainian design group that also received approval in the 1976 decree to build a replacement ICBM called the Zenit. Today, Zenit is the name for a family of space launchers available from the Ukrainian manufacturer Yuzhnoye Design Bureau, formerly OKB-586, and part of the Boeing-ILS Sea Launch program.

The Energia's second-stage core booster consisted of four single-chamber RD-0120 engines, each producing 1,961.000 kN (440,850 lb$_f$) of thrust using LH2 and LOX propellants. This was the same core module envisioned for the other Energia family boosters, with the exception of the Energia-M, which was a single RD-0120 engine core with two outboard Zenit boosters. Glushko's earlier Vulkan concept consisted of the same core as the Energia with eight booster sections attached instead of four, supplying a 230-ton capacity into low-Earth orbit.

Energia Launches

The Soviet heavy-lift Energia was successfully launched twice, and with live payloads instead of test articles. Energia's first launch was the Polyus ("pole") military satellite on May 15, 1987. Polyus was the ultimate in diabolical spacecraft created for the Cold War. Outfitted as a reconnaissance satellite, Polyus contained an on-board cannon to be used as an anti-satellite defense system, and a space mine launcher. It was also designed with the capability to be docked with the Buran shuttle. Fortunately, the Polyus launch to orbit was unsuccessful because of a failure in its onboard guidance system that was unrelated to the Energia booster.

Polyus was based on the FGB orbital module design with an added operational section, but placed backwards on the Energia that required a 180° turn after separation from the Energia for orbit insertion. A full 360° rotation after booster separation doomed the spacecraft by subtracting orbital ΔV instead of adding it.

The second and final launch of Energia was made with the Buran payload from the Baikonor Cosmodrome on November 15, 1988. The flight was unmanned and contained

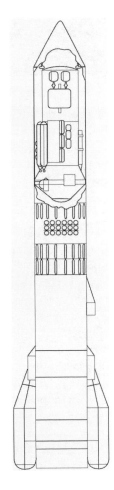

Figure 12.33. Cutaway drawing of the Soviet Polyus satellite that consisted of a FGB orbital and reentry module located at the top of the sketch, and an operational weapons/defense section located at the bottom. Not shown are deployable solar panels attached near the top. Farings attached at the bottom were target dispensers for tracking the satellite in orbit. Source: Portree.

Figure 12.34. Launch preparations for the November 15, 1988, launch of the Buran space shuttle. Source: Molinya.

no life-support system, CRT display, or display instrumentationonboard, and it had only minimal software installed. The vehicle was launched into orbit roughly 250 km in altitude with an inclination of 51.6°, and orbited the Earth twice before firing its thrusters for reentry. The Buran 1.01 mission was reportedly limited to two orbits due to computer memory limitations.

Russia's future

Russia's current space projects are supported by an aging stable of boosters from the Soviet Union era. Efforts to replace the two major booster designs with advanced technology, lower-cost space launch vehicles continue. Three families of boosters are being considered for replacement, the oldest being the Soyuz/Molinya/Progress derivatives of the R-7. Although the basic design is more than fifty years old and out-of-date, the boosters are also reliable and inexpensive. Because of their modest performance, the replacement booster is targeted for a low-cost design that can use proven technology that is already built in to the aging R-7 fleet.

The second oldest Russian launcher in service is the larger Proton, which serves as their heavy-lift booster. A redesign of the Proton would aim at improving its performance, as well as replacing the Zenit boosters which are purchased from the Ukraine—the third booster family. Further stretching their limited funds, the Russians expect to enlarge the Plesetsk Cosmodrome to reduce its dependence on the current primary launch site at the Baikonor Cosmodrome in Kazakhstan.

Table 12.14. Energia Specs

Number of stages	2
Length	58.8 m (193')
Diameter (max)	20 m (66')
Weight (fueled)	2,400,000 kg (5,290,000 lb)
Launch site	Baikonor
Payload from Baikonor	95–100,000 kg (209,000–220,000 lb) to 51° LEO 18,000 kg (39,700 lb) to geostationary transfer orbit (GTO) 32,000 kg (70,500 lb) to trans-lunar injection 28,000 kg (61,700 lb) to Venus/Mars
Stage 1	4 strap-on boosters
Fuel	Kerosene (RP-1)
Oxidizer	Liquid oxygen
Stage 1 burn time	156 seconds
First stage propulsion	1 four-chamber RD-170 engine 7,784,400 kN (1.75 M lb_f) thrust
Stage 2	Central (core stage)
Fuel	Liquid hydrogen
Oxidizer	Liquid oxygen
Stage 2 burn time	470 seconds
Second stage propulsion	4 one-chamber RD-0120 (11D122) engines 1,961.000 kN (440,850 lb_f) thrust each

Proposals for the new heavy-lift vehicle came from several design and manufacturing groups within Russia that include RKK Energia's proposal for a scaled-down Energia M rocket. Although the Energia M was passed over, the RD-0120 and RD-170 engine designs were adopted with modifications for use in the successful Khrunichev Angara rocket proposal. The RD-170 may be used for the Angara booster's first stage, although the RD-0120 quad-chamber would be split into a single-chamber engine designated RD-0124. Interestingly, a two-chamber RD-0120 was already built for use on the American Atlas III by NPO Energiomash and could have been adopted in the RKK Energia design to limit development costs.

The decision was made by the Russian Ministries of Defense in 1994 to award the new booster contract to the Moscow-based Khrunichev State Space Scientific Production Center. The Khrunichev rocket called Angara, like the Proton and the Energia, consists of a core stage surrounded by first-stage boosters that vary in number based on the payload requirements. Angara's proposed variable configuration has a payload capacity ranging from 2,000 kg to 24,500 kg to LEO, with the largest being a direct replacement of the aging Proton. Projections for the first Angara launches were 2000–2005, although more recent estimates place the first Angara launch in the 2010–2012 era. The Angara may also be scheduled for future launches from the Baikonor Cosmodrome under recent agreements between Russia and Kazakhstan.

A replacement for the Soyuz crew vehicle that supports the Russian cosmonauts on the ISS is also under development, with a projected inaugural flight in 2012. The Russian Kliper is a reusable crew-flight vehicle shaped like a lifting body, somewhat like the canceled American X-38 crew-emergency-return vehicle that has been replaced with Orion CEV. Kliper will be launched vertically on an Angara or upgraded Soyuz booster, with an aircraft-like horizontal landing. Maximum crew size for the proposed vehicle is six, with the potential for 500 kg of cargo payload to the ISS.

References

Dickson, Paul. 2001. *Sputnik: The Shock of the Century.* New York: Walker & Co.

Hall, Rex, and David J. Shayler. 2001. *The Rocket Men: Vostok and Voskhod, the First Soviet Manned Spaceflights.* New York: Springer-Praxis.

Harford, James. 1997. *Korolev: How One Man Masterminded the Soviet Drive to Beat America to the Moon.* New York: John Wiley & Sons.

Lindroos, Marcus, ed. N.d. "The Soviet Manned Lunar Program." Retrieved from: http://www.fas.org/spp/eprint/lindroos_moon1.htm (accessed 7/7/2009).

NASA. 1988. *Soviet Reusable Space Systems Program: Implications for Space Operations in the 1990s.* Intelligence Assessment. Retrieved from: http://fas.org/irp/cia/product/sovsts88.pdf

NSSDC (NASA Space Science Data Center). N.d. Retrieved from: http://nssdc.gsfc.nasa.gov/planetary/

Oberg, James. 1999. "The Probe That Fell to Earth." *New Scientist*, March 6.

Portree, David S. 1995. Mir Hardware Heritage. NASA RP-1357. Houston, TX: Johnson Space Center, Information Services Division, March.

RussianSpaceWeb. N.d. News and History of Astronautics in the Former USSR. Retrieved from: http://www.russianspaceweb.com/ (accessed 7/3/2009).

Rynin, N. A. K. E. 1931. "Tsiolkovsky: Life, Writings, and Rockets." *Interplanetary Flight and Communication* 3, no. 7, Leningrad Academy of Sciences of the U.S.S.R (translated and published in NASA TT F-646).

Siddiqi, Asif. 2000. *Challenge to Apollo: The Soviet Union and the Space Race, 1945-1874.* NASA SP-2000-4408. Washington, DC: NASA History Division.

Vladimirov, Leonid. 1973. *The Russian Space Bluff: The Inside Story of the Soviet Drive to the Moon.* Translated by David Floyd. New York: Dial Press.

Wade, Mark. N.d. "Encyclopedia Astronautica." Retrieved from: http://www.astronautix.com/

NASA GRIN (Great Images in NASA). Retrieved from: http://grin.hq.nasa.gov/

NASA NIX (NASA Image Exchange). Retrieved from: http://nix.nasa.gov/

Index

About the Author

Lance Erickson is an educator, astronomer, and pilot who has been teaching undergraduate and graduate classes in space exploration for more than twenty years. He is currently professor of applied aviation sciences at Embry-Riddle Aeronautical University in Daytona Beach, Florida. His work, which includes more than a dozen manuals, texts, and electronic publications on the subject of space flight, also includes a Web-based course on the space shuttle developed for NASA's Kennedy Space Center.